PRINCIPLES OF
CLINICAL CHEMISTRY

BOOKS OF RELATED INTEREST

Eggert: Electronics and Instrumentation for the Clinical Laboratory, 1983, 0471-86275-4

Campbell: Medical Mycology Handbook, 1980, 0471-04728-7

Hayhoe: Color Atlas of Hematological Cytology, Second Edition, 1982, 0471-86868-X

PRINCIPLES OF
CLINICAL CHEMISTRY

KENNETH E. BLICK, Ph.D., A.B.C.C.

Associate Professor of Pathology
College of Medicine
Adjunct Professor, Medical Technology
College of Health
University of Oklahoma Health Sciences Center
Oklahoma City, Oklahoma

SUZANNE M. LILES, M.T.(A.S.C.P.)S.C.

Educational Coordinator
Medical Technology Training Program
Deaconess Hospital
Evansville, Indiana

A Wiley Medical Publication

John Wiley & Sons

New York / Chichester / Brisbane / Toronto / Singapore

Library of Congress Cataloging in Publication Data

Blick, Kenneth E.
  Principles of clinical chemistry.

  (A Wiley medical publication)
  Includes index.
  1. Chemistry, Clinical.   I. Title.   II. Liles,
Suzanne M. (Suzanne Martin)   III. Series.   [DNLM:
1. Chemistry, Clinical.   QY 90 B648p]
RB40.B58   1985   616.07′56   85-3175
ISBN 0-471-88502-9

Printed in the United States of America

10 9 8 7 6 5 4 3 2 1

To my wife, Veida,
and my children, David, Sharon, and Brian
                                    K.E.B.
To my sons, Conrad and Richard and
my mother, Mary
                                    S.M.L.

PREFACE

This book provides a concise presentation of the important topics in clinical chemistry. It is our impression that the other major books in the field serve primarily as reference works that do not lend themselves easily for use as textbooks. Therefore, this book is written in a practical format that will help students and instructors retain and organize their material.

Principles of Clinical Chemistry is written as a textbook for medical technology students in clinical chemistry courses and as a self-study guide for those students taking the registry examination. This book can be used also as a review for experienced laboratory professionals who are preparing for specialty certification or board examinations.

Principles of Clinical Chemistry consists of 14 chapters, and each includes chapter outlines, learning objectives, glossaries of terms, post-tests, and answers. The chapter outlines provide structured learning formats for students and lecture guides for instructors. The glossaries define key words and medical terms for each chapter. The post-tests can be used to measure comprehension of the material. Out-of-date methodologies and detailed procedures are not included. To minimize lengthy descriptions of subject matter, we have added solved problems (or examples), figures, and case studies whenever possible. We have also attempted to maintain a similar style and organization within each chapter, a feature that should add coherence between the chapters.

Chapters 1 and 2 present a review of basic material in chemistry, statistics, and quality control. The more experienced laboratory worker may wish to proceed immediately to the post-tests of these chapters or begin with Chapter 3. Chapter 3 describes the management aspects of a computerized laboratory. Computer programming and general concepts of computer functions are included. Chapter 4 discusses laboratory instrumentation and immunoassay procedures. Chapters 5 through 14 address all other areas of clinical chemistry including carbohydrates; proteins; lipids; inorganic ions; acid–base and electrolytes; enzymes; liver, pancreas, intestinal and gastric functions; endocrinology; NPN substances and body fluids; and therapeutic drug monitoring.

The authors are grateful to students and colleagues whose comments and suggestions have been most valuable. Special gratitude is due to Thomas A. Webb, M.D., Mary Jo Commens, M.T.(A.S.C.P.)S.C., and H. Veida Blick, M.T.(A.S.C.P.) for their support and guidance in manuscript preparation.

KENNETH E. BLICK
SUZANNE M. LILES

Oklahoma City, Oklahoma
Evansville, Indiana
April 1985

CONTENTS

PRINCIPLES OF
CLINICAL CHEMISTRY

CHAPTER **1**

REVIEW OF BASIC CHEMISTRY AND LABORATORY MATHEMATICS

2. Unit Conversion Problems
3. Calculations Involving Dilutions
   a. Preparation of Standard Solutions
   b. Dilute Solutions from Solid Material

XVI. MOLARITY, NORMALITY, EQUIVALENT WEIGHT, AND MOLECULAR WEIGHT
   A. Chemical Reactions
   B. Molecular Weigh (MW)
   C. Avagadro's Number
   D. Equivalent Weight (EW)
      1. Milliequivalent Weight
   E. Molarity and Normality
      1. Dilution Problems Based on Molarity and Normality
         a. Preparation of Standard Solutions
      2. Preparing a Molar or Normal Solution from a Solid
   F. Molality
   G. Problems Based on Chemical Reactions
      1. Oxidation–Reduction Reactions
      2. Preparation of Standard Solutions from Hydrated Salts

XVII. OTHER LABORATORY CALCULATIONS
   A. Temperature Conversion
   B. Clearance Calculations
   C. Logarithms
      1. Logarithms and pH

OBJECTIVES

On completion of this chapter, you will be able to:

1. Define, illustrate, or explain the following terms:
   a. molecule
   b. oxidation/reduction
   c. oxidation numbers
   d. metal/nonmetal
   e. acid/base
   f. molarity
   g. molality
   h. equivalent weight
   i. normality
   j. buffer
   k. equilibrium constant
   l. half-life
   m. logarithm

2. Discuss the various oxidation states for metals and nonmetals along with the formation of ionic and molecular compounds.
3. Describe the nature of nuclear changes and associated radioactivity of unstable isotopes.
4. Describe the various functional groups of organic compounds.
5. List the rules regarding significant figures, rounding off, and expressing numbers in scientific notation.
6. Describe the metric system of units and the Standard International Units and convert measurements from one system of units to another.
7. Perform calculations involving the preparations of standard solutions.
8. Define, in terms of units, molarity, normality, molecular weight, and equivalent weight, and solve problems based on these concepts.

GLOSSARY OF TERMS

Density. The mass per unit volume of a substance.

Element. Basic unit of matter that cannot be broken down into simpler substances by chemical means.

Equilibrium. A chemical reaction that never reaches completion because the products of the reaction combine to form the reactants. A point of equilibrium is reached when the rate of the "forward" reaction becomes equal to the rate of the "reverse" reaction.

Indicator. A substance, added to a chemical reaction, that undergoes a color change to indicate an equivalence point between two reactants (e.g., an acid and base).

Isotopes. Elements that have the same atomic number but different atomic weights.

Radioactive decay. The spontaneous breakdown of unstable nuclei resulting in the loss of high-energy particles and/or energy in the form of electromagnetic radiation.

Scientific notation. Allows all numbers to be expressed as a number between 1 and 10 with an exponential term (base 10) to account for the position of the decimal point.

Strong electrolytes. Chemical compounds that are completely converted to ions when mixed with water. Weak electrolytes are only partially "ionized" when mixed with water.

In order to solve problems in clinical chemistry, a brief review of basic chemistry may be needed.

PERIODIC TABLE OF ELEMENTS

Listed in Figure 1.1 are the major elements of the Periodic Table including major groups of elements: IA, alkali metals; IIA alkaline metals; IIIB to VIIIB, transi-

Bold symbols - nonmetals

Row	IA	IIA	IIIB	IVB	VB	VIB	VIIB	VIIIB			IB	IIB	IIIA	IVA	VA	VIA	VIIA	0
1	1 H +1 −1																	2 **He** 0
2	3 Li +1	4 Be +2											5 **B** +3	6 **C** +4 0 +2 −2 −4	7 **N** +5 −1 +4 −3 +3	8 **O** −2 −1	9 **F** −1	10 **Ne** 0
3	11 Na +1	12 Mg +2											13 Al +3	14 **Si** +4 +2 −4	15 **P** +5 −1 +3 −2 +1 −3	16 **S** +6 +2 +4 −2	17 **Cl** +7 +1 +5 −1 +3	18 **Ar** 0
4	19 K +1	20 Ca +2	21 Sc +3	22 Ti +4 +3 +2	23 V +5 +3 +4 +2	24 Cr +6 +2 +3	25 Mn +7 +3 +4 +2	26 Fe +3 +2	27 Co +3 +2	28 Ni +2	29 Cu +2 +1	30 Zn +2	31 Ga +3	32 Ge +4 +2 −4	33 **As** +5 +3 −3	34 **Se** +6 −2 +4	35 **Br** +5 −1 +1	36 **Kr** 0
5	37 Rb +1	38 Sr +2					43 Tc			46 Pd	47 Ag +1	48 Cd		50 Sn +4 +2	51 Sb +5 −3 +3	52 **Te** +6 +4	53 **I** +7 +1 +5 −1	54 **Xe** 0
6	55 Cs +1	56 Ba +2								78 Pt +4 +2	79 Au +3 +1	80 Hg +2 +1		82 Pb +4 +2	83 Bi +3 −3			86 **Rn** 0

Figure 1.1. Periodic table of selected elements.

Table 1.1. Elements and Atomic Weights

	Symbol	Atomic Number	Atomic Weight
Actinium	Ac	89	(227)[a]
Aluminum	Al	13	26.9815
Americium	Am	95	(243)[a]
Antimony	Sb	51	121.75
Argon	Ar	18	39.948
Arsenic	As	33	74.9216
Astatine	At	85	(210)[a]
Barium	Ba	56	137.34
Berkelium	Bk	97	(247)[a]
Beryllium	Be	4	9.01218
Bismuth	Bi	83	208.9806
Boron	B	5	10.81
Bromine	Br	35	79.904
Cadmium	Cd	48	112.40
Calcium	Ca	20	40.08
Californium	Cf	98	(251)[a]
Carbon	C	6	12.011
Cerium	Ce	58	140.12
Cesium	Cs	55	132.9055
Chlorine	Cl	17	35.453
Chromium	Cr	24	51.996
Cobalt	Co	27	58.9332
Copper	Cu	29	63.546
Curium	Cm	96	(247)[a]
Dysprosium	Dy	66	062.50
Einsteinium	Es	99	(254)[a]
Erbium	Er	68	167.26
Europium	Eu	63	151.96
Fermium	Fm	100	(253)[a]
Fluorine	F	9	18.9984
Francium	Fr	87	(223)[a]
Gadolinium	Gd	64	157.25
Gallium	Ga	31	69.72
Germanium	Ge	32	72.59
Gold	Au	79	196.9665
Hafnium	Hf	72	178.49
Helium	He	2	4.00260
Holmium	Ho	67	164.9303
Hydrogen	H	1	1.0080
Indium	In	49	114.82
Iodine	I	53	126.9045

Table 1.1   (*Continued*)

	Symbol	Atomic Number	Atomic Weight
Iridium	Ir	77	192.22
Iron	Fe	26	55.847
Krypton	Kr	36	83.80
Lanthanum	La	57	138.9055
Lawrencium	Lr	103	(257)[a]
Lead	Pb	82	207.2
Lithium	Li	3	6.941
Lutetium	Lu	71	174.97
Magnesium	Mg	12	24.305
Manganese	Mn	25	54.9380
Mendelevium	Md	101	(256)[a]
Mercury	Hg	80	200.59
Molybdenum	Mo	42	95.94
Neodymium	Nd	60	144.24
Neon	Ne	10	20.179
Neptunium	Np	93	237.0482[b]
Nickel	Ni	28	58.71
Niobium	Nb	41	92.9064
Nitrogen	N	7	14.0067
Nobelium	No	102	(254)[a]
Osmium	Os	76	190.2
Oxygen	O	8	15.9994
Palladium	Pd	46	106.4
Phosphorus	P	15	30.9738
Platinum	Pt	78	195.09
Plutonium	Pu	94	(242)[a]
Polonium	Po	84	(210)[a]
Potassium	K	19	39.102
Praseodymium	Pr	59	140.9077
Promethium	Pm	61	(147)[a]
Protactinium	Pa	91	231.0359[b]
Radium	Ra	88	226.0254[b]
Radon	Rn	86	(222)[a]
Rhenium	Re	75	186.2
Rhodium	Rh	45	102.9055
Rubidium	Rb	37	85.4678
Ruthenium	Ru	44	101.07
Samarium	Sm	62	150.4
Scandium	Sc	21	44.9559
Selenium	Se	34	78.96
Silicon	Si	14	28.086

Table 1.1 (*Continued*)

	Symbol	*Atomic Number*	*Atomic Weight*
Silver	Ag	47	107.868
Sodium	Na	11	22.9898
Strontium	Sr	38	87.62
Sulfur	S	16	32.06
Tantalum	Ta	73	180.9479
Technetium	Tc	43	98.9062[b]
Tellurium	Te	52	127.60
Terbium	Tb	65	158.9254
Thallium	Tl	81	204.37
Thorium	Th	90	232.0381[b]
Thulium	Tm	69	168.9342
Tin	Sn	50	118.69
Titanium	Ti	22	47.90
Tungsten	W	74	183.85
Uranium	U	92	238.029
Vanadium	V	23	50.9414
Xenon	Xe	54	131.30
Ytterbium	Yb	70	173.04
Yttrium	Y	39	88.9059
Zinc	Zn	30	65.37
Zirconium	Zr	40	91.22

[a] Mass number of most stable or best known isotope.
[b] Mass of most commonly available, long-lived isotope.

tion metals; IB to IIB metals; IIA to VIIA, nonmetals and metals; and 0, the noble (or inert) gases. The atomic number for each element [i.e., the number of protons in the nucleus (or the number of electrons in the atom)] is also shown in Figure 1.1. Listed in Table 1.1 are the atomic weights and chemical symbols of each element.

ELEMENTS AND ATOMS

An element is the basic unit of matter, which cannot be broken down into simpler substances by chemical means. An atom is the smallest component of an element that retains the properties of the element.

COMPONENTS OF ATOMS

As shown in Figure 1.1, each element has an atomic number (e.g., H-1, He-2, Li-3), etc.). The atomic number is the number of protons in the nucleus of the atom. Protons have mass (equal to one atomic mass unit, AMU) and a +1 charge. Neutrons, also located in the nucleus of the atom, have essentially the same mass as protons but no charge. Electrons, which have a mass of only a small fraction of the mass of a proton (1/1837), have a −1 charge. The atomic number also equals the number of electrons, hence the net charge of a neutral atom is zero.

Structure of Atom

Protons and neutrons form the central core (or nucleus) of the atom; electrons travel in energy-related "orbitals" around this core.

Isotopes

The number of protons in the nucleus is fixed for a particular element; however, the number of neutrons in the nucleus may vary. For example, two isotopes of carbon can have an atomic weight of 12 or 13:

$$^{13}_{6}C \qquad ^{12}_{6}C$$

Hence, isotopes have the same atomic number but different atomic weights owing to different numbers of neutrons in the nucleus. By subtracting the atomic number from the atomic weight (see Table 1.1), the number of neutrons can be determined. The carbon 13 nucleus has six protons and seven neutrons; carbon 12 has six protons and six neutrons.

Ions

Electrons can be lost or gained by elements. If electrons are lost, the "extra" protons in the nucleus cause the atom to become a positively charged ion, or "cation." If an atom gains electrons in a chemical reaction, the "extra" electrons cause the atom to become a negatively charged ion, called an "anion."

CHANGES IN THE ATOM

NUCLEAR CHANGES (RADIOACTIVITY)

Stability of the nucleus of the atom is usually determined by the number of protons (P^+) and neutrons (N) making up the nucleus. In stable nuclei, the number of neutrons is usually equal to or greater than the number of protons.

Radioactive Decay

Unstable nuclei are associated with radioactive decay where a rearrangement of the nucleus takes place to form a more stable nucleus. This process often results in the loss of high-energy atomic particles (e.g., neutons, protons, electrons) and energy in the form of electromagnetic radiation (e.g., X-rays from changes in the electronic shell; gamma rays from nucleus rearrangement).

Neutron-rich nuclei usually decay by giving off beta particles where neutrons disintegrate into a proton (P^+), with concomitant production of a beta particle [an electron (e^-)] and a massless, chargeless neutrino. Proton-rich nuclei spontaneously disintegrate by (1) electron capture (where a proton on the nucleus "captures" an orbital electron and converts to a neuron) or (2) positron decay [where a positron (e^+) converts to a neutron].

Half-Life

Half-life of a radioisotope is the time required for one-half of the atoms present to decay.

Activity

The radioactivity of a radioisotope is expressed in curies (Ci) where 1 millicurie (mCi) equals 3.7×10^7 dps (where dps is "disintegrations per second"). Gamma radiation and X-rays can be measured by a gamma counter as a number of counts relative to the dps (or activity) of the isotope present. The ratio of counts per second (cps obtained by a gamma counter) to dps (absolute activity) is the efficiency of the gamma counter (usually about 80%):

$$\frac{\text{dps}}{\text{cps}} \times 100 = \text{counting efficiency (\%)}$$

ISOTOPES USED IN CLINICAL CHEMISTRY

In clinical chemistry, ^{14}C, ^{3}H, ^{125}I, ^{57}Co, and ^{131}I are used to label or tag compounds to be measured (or counted). The ^{125}I and ^{57}Co isotopes are used for the measurement of blood levels of hormones, enzymes, viral antigens and antibodies, drugs, and vitamins. Iodine 125 has a half-life of 60.2 days and a specific activity of 17.3 mCi/mg, and emits gamma radiation by electron capture at energy levels of 28 and 35 keV. As shown above, specific activity of a substance is the radioactivity of the substance per unit mass.

CHEMICAL REACTIONS

Chemical reactions primarily involve the loss, gain, or sharing of electrons (e^-) between atoms with no change whatsoever in the composition of the nucleus.

Deciding how many electrons an atom will lose, gain, or share in a "chemical reaction" between two atoms can be readily understood by examining Figure 1.1, where we have listed the various oxidation states each element can exhibit in chemical compounds.

OXIDATION STATES

Positive Oxidation States or Oxidation Numbers

The most positive oxidation state an element can have equals the number of electrons it must lose in order to have the same number of electrons as the closest inert gas (Group 0). For example, A1 (13 electrons) loses three electrons in chemi-cal rections in order to have the same number of electrons as Ne (10 electrons). The highest positive oxidation on A1 then is +3. Groups IB, IIB, and IIIA-VA metals (see Figure 1.1) lose enough electrons to be electronically similar to the inert metals of group VIIIB.

Negative Oxidation States

The most negative oxidation number (or state) equals the number of electrons an atom must gain in order to have the same number of electrons as the closest noble gas (Group 0 elements). Arsenic (33 electrons) gains three electrons to have the same number of electrons as Kr (36 electrons). Hence, the most negative oxida-tion state of As is −3.

Chemical reactions involve the loss and gain of electrons between atoms with no changes in the nuclei of reacting atoms. Chemical reactions involving elements can (1) form ionic compounds held together by ionic charges or (2) form molecular compounds held together by chemical bonds.

REACTIONS BETWEEN METALS AND NONMETALS

Ionic Compounds

There are two major subdivisions of elements, metals and nonmetals. Metals generally lose electrons in chemical reactions and hence are *oxidized* (this makes them reducing agents). Nonmetals gain electrons and hence are *reduced* (this makes them good oxidizing agents). Elements in Group 0 (or the inert gases) do not generally lose or gain electrons; therefore, we can reason that the number of electrons present on these atoms must be optimum for stability. It seems logical, then, that other elements react with each other to lose or gain enough electrons to give the desired stable electronic configuration of the noble gases. Accordingly, when sodium and fluorine react, sodium (Na) *loses* one electron to be electroni-cally similar to Ne; and fluorine (F) *gains* one electron to be electronically similar to Ne as well.

$$
\begin{array}{l}
\text{Na} \quad \rightarrow \text{Na}^+ + \text{le}^- \\
\underline{\text{F} + \text{le}^- \rightarrow \text{F}^-} \\
\text{Na} + \text{F} \rightarrow \text{Na}^+\text{F}^-
\end{array}
$$

The resulting ionic compound (or salt) is held together in a crystalline structure by the opposite charges of the sodium cation (Na^+) and the fluoride anion (F^-). Note that the oxidation numbers correspond to the ion charges for sodium (+1) and fluorine (−1) as shown in Figure 1.1. Hence, the oxidation numbers listed in Figure 1.1 are based on the number of electrons lost, gained, or shared in chemical reactions.

Generally, salts such as NaF are compounds that form when a metal reacts with a nonmetal as described above and are invariably crystalline solids that have unique physical properties of salts, notably hardness and high melting tempera-tures. In naming a salt, the negative ion (or the nonmetal) gets an ''ide'' suffix (or fluorine becomes fluoride). Also, crystalline salts like NaF are most often soluble in polar solvents like water with the dissolved NaF called the *solute*. The water serves as the *solvent*.

EXAMPLE 1.1.   Develop formulas for three ionic compounds using groups IA and IIA metals reacting with group VA nonmetals.

Answer:

$$Ca \rightarrow Ca^{2+} + 2e^-$$
$$N + 3e^- \rightarrow N^{3-}$$

Charges (or oxidation numbers) must balance:

$$(Ca^{2+})_3(N^{3-})_2 \rightarrow Ca_3N_2 \text{ (calcium nitride)}$$

Others:

$$Mg_3P_2 \text{ (magnesium phosphide)}$$
$$Na_3As \text{ (sodium arsenide)}$$

Molecular Compounds

In reactions of two highly electronegative (electrophilic, "electron loving") non-metallic elements, neither nonmetal element will entirely lose electrons in chemical reactions. Indeed, two nonmetals react to gain additional electrons to be electronically similar to the inert gases by *sharing* electrons.

Valence Electrons
Only the outermost electronic shells participate in chemical reactions. Figure 1.1 indicates the total number of valence electrons in element groups with column designations (e.g., IA, one valence electron; IIIB, three valence electrons; and IVA, four valence electrons; etc.)

Octet Rule
We have observed that all of the inert gases have a stable octet (eight electrons) in their valence shell; hence, nonmetallic elements strive for an octet of electrons in chemical reactions with other nonmetals.

Formation of Molecules

Fluorine is in group VIIA with seven valence electrons, one short of the octet. If another atom reacted with the fluorine atom, for example, chlorine, we would get a compound, ClF, with fluorine gaining one electron to be electronically similar to Ne and Cl being electronically similar to Ar. The additional electron is achieved on both atoms by sharing a pair of electrons. This sharing of a pair of electrons represents the formation of a "single bond" between the atoms (see Figure 1.2).

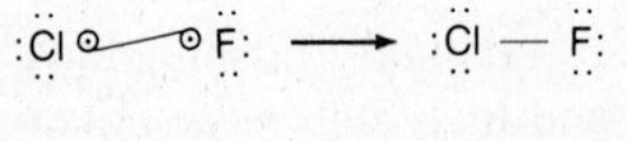

$$:\!Cl \circ\!\!-\!\!\circ F: \longrightarrow :\!Cl - F:$$

Figure 1.2. Reaction between elemental fluorine and chlorine. Valence electrons are indicated by (··) and (—) denotes a chemical bond.

Chemical Bond
The resulting compound, ClF, is called chlorine fluoride; the latter compound is a discrete molecular unit called a *molecule* with the compound being held together by the shared pair of electrons between the atoms. The shared pair of electrons is also called a "chemical bond" between the atoms.

Electronegativity
Since nonmetals have an affinity for electrons, we say nonmetals are *electronegative*. The bonding electrons, or shared electrons, spend more time on the more electronegative atom, hence, the "ide" suffix is given to fluorine, changing the name to fluoride. Some values of electronegativity are as follows: $F = 4.0$; $O = 3.5$; $N = 3.0$; $Cl = 3.0$; $Br = 2.8$; $Si = 2.5$; and $H = 2.1$. As expected, metals have low electronegativities, with values ranging from 0.7 for Fr (francium) to 2.4 for Au (gold).

EXAMPLE 1.2.   Describe a chemical reaction between oxygen (Group VIA) and fluorine (Group VIA).

   Answer:

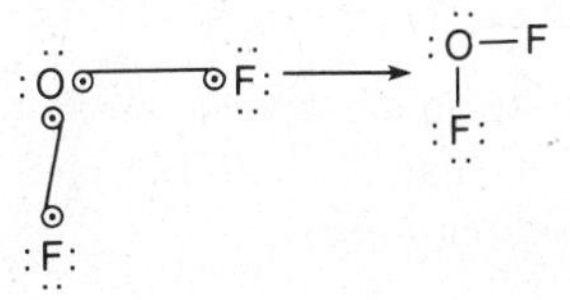

Figure 1.3

In the figure ($\cdot\cdot$) denotes valence electron pairs and (—) depicts a chemical bond. The chemical reaction is

$$O + 2F \rightarrow OF_2 \text{ (oxygen difluoride)}$$

Polar Molecules
Whenever there is an electronegativity difference between two bonded, nonmetallic elements, the bond is described as polar. A polar molecule has a center of positive charge and a center of negative charge, which do not coincide.

Multiple Bonds
Double and triple bonds are seen where more than one pair of electrons are shared between atoms (see Figure 1.4).

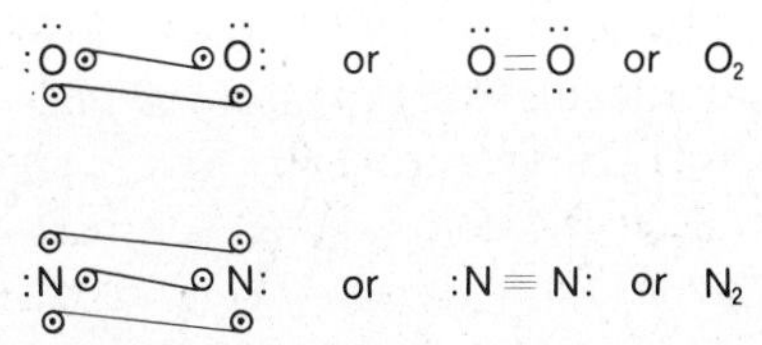

Figure 1.4.   Double and triple bonding between atoms.

Nonpolar Molecules

These two previous compounds, O_2 and N_2, are nonpolar because electrons are shared equally between two atoms that have the same electronegativity. This is important, because polar molecules attract one another and form close molecular aggregates (e.g., liquids and solids), especially at high pressures and low temperatures. Nonpolar molecules have little intermolecular attraction, hence, they tend to form gases.

Variable Oxidation Numbers

Depending on the compound, nonmetallic elements can have an oxidation number equal to the total number of valence electrons. For example, in phosphorus pentachloride, P has a +5 oxidation number. The halogen group, for example, chlorine and iodine, can have oxidation states of +7, or equal to the total number of valence electrons. However, these high oxidation numbers are only achievable when chlorine is reacted with highly electronegative elements like oxygen or fluorine. For example, in CIF_7, the Cl has a +7 oxidation number. In general for nonmetals, the highest possible oxidation number is equal to the total number of valence electrons. The lowest possible oxidation number is equal to the negative of the number of electrons required to achieve the number of electrons of the closest inert gas (see Figure 1.1).

Also, elements, in various compounds, can have oxidation numbers ranging from the highest to lowest oxidation number (e.g., CIF_7, CIF_5, CIF_3, CIF, HCl with oxidation numbers on Cl of +7, +5, +3, +1, −1 respectively).

Some representative oxide ions are shown in Table 1.2.

Electron Dot Picture or Lewis Structures

Drawing Lewis electron structures for molecules and ions is helpful in understanding structure and bonding in compounds formed from nonmetallic elements. Lewis structures for SO_3 and SO_3^{2-} are shown in Figure 1.5.

Some rules to remember in drawing Lewis structures: (1) give each nonmetallic element (except B and H) an octet of electrons; (2) third row and lower elements usually have high oxidation numbers, hence, they are usually the "central" atom in the structure; (3) the sum of the valence electrons of the elements making up the compound minus the total number of electrons in the Lewis structure will give the charge (if any) on the structure; (4) forming a double bond reduces the number of electrons required for the octet rule by 2; and (5) hydrogen (a nonmetal) will only share two electrons in Lewis structures.

Oxidation Numbers for Atoms in Lewis Structures

Since there are no apparent charges on atoms in molecules or polyatomic ions, oxidation states are less apparent in Lewis structures. However, (1) by giving all shared electrons in bond(s) to the most electronegative element and (2) subtract-

Table 1.2. Oxide Anions of Nonmetals and Selected Metals

Group IIIA		*Group VIIA*	
BO_3^{3-}	Borate	ClO_4^-	Perchlorate
		ClO_3^-	Chlorate
Group IVA		ClO_2^-	Chlorite
CO_3^{2-}	Carbonate	ClO^-	Hypochlorite
SiO_4^{4-}	Silicate	BrO_3^-	Bromate
		BrO_2^-	Bromite
Group VA		BrO^-	Hypobromite
NO_3^-	Nitrate	IO_4^-	Periodate
NO_2^-	Nitrite	IO_3^-	Iodate
PO_4^{3-}	Phosphate	IO^-	Hypoiodite
PO_3^{3-}	Phosphite		
AsO_4^{3-}	Arsenate	*Group VIB*	
AsO_3^{3-}	Arsenite	CrO_4^{2-}	Chromate
		$Cr_2O_7^{2-}$	Dichromate
Group VIA			
SO_4^{2-}	Sulfate	*Group VIIB*	
SO_3^{2-}	Sulfite	MnO_4^-	Permanganate
$S_2O_3^{2-}$	Thiosulfate		
		Organic Anions	
		CHO_2^-	Formate
		$C_2H_3O_2^-$	Acetate
		$C_2O_4^{2-}$	Oxalate

ing the original number of valence electrons (see Figure 1.1), the oxidation number can be determined. Hence, oxidation numbers on molecular compounds are determined by assuming the molecule is "temporarily ionic." For example, nitrogen, a VA nonmetal, has five valence electrons as a neutral atom. However, in the NF_3, two electrons remain on the nitrogen, hence, nitrogen has $+3$ oxidation number with the oxidation number on the fluorine being -1 (see Figure 1.6).

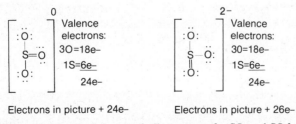

Figure 1.5. Lewis "electron-dot" structures for SO_3 and SO_3^{2-}.

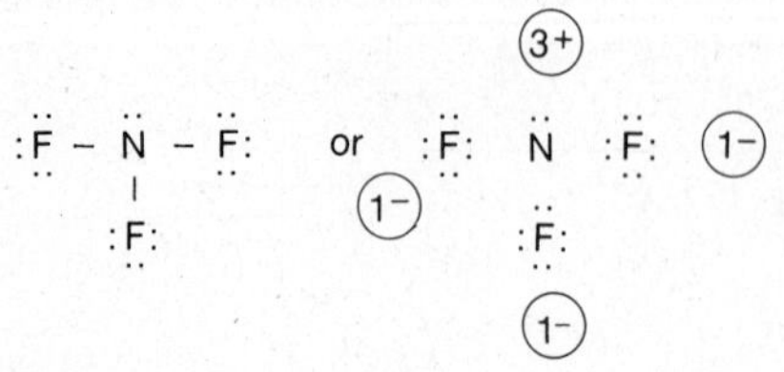

Figure 1.6. Oxidation numbers of individual atoms assuming shared electrons belong to the most electronegative atom.

ACIDS

Generally, oxides of nonmetallic elements are acid anhydrides which in turn form acids when combined with water:

$$2S + 3O_2 \rightarrow 2SO_3 + 2H_2O \rightarrow \underset{\text{sulfuric acid}}{2H_2SO_4}$$

NOMENCLATURE

Generally, the unique element in any acid is the nonmetallic "central" atom, since most acids contain H and O. We name acids based on this unique nonmetal and use an "ic" suffix if the nonmetal is in its highest possible oxidation number (e.g., sulfur*ic* acid, nitr*ic* acid, chlor*ic* acid, selen*ic* acid, phosphor*ic* acid, carbon*ic* acid, etc.)

WATER SOLUTIONS OF ACIDS

Any compound that combines with water to give H_3O^+ (sometimes indicated as H^+) is, by definition, an acid:

$$H_2SO_4 + H_2O \rightarrow H_3O^+ + HSO_4^-$$

In Lewis structures of acids, the hydrogens are usually bonded to the oxygens; the oxygens are bonded to the central nonmetal. Acidic hydrogens are removed (or hydrated) in water solutions of strong acids. In the presence of compounds called bases (to be discussed later), more than one hydrogen can be removed from acids containing more than one "acidic" hydrogen. Since all strong acids completely dissociate in water to give one H^+ and an anion (e.g., HSO_4^-), equal molar amounts of strong acids all have the same "acid strength" in water. Water, then, is referred to as a "leveling" solvent for acids.

FORMULAS OF ACIDS

The formula of an acid can be constructed by (1) writing the central atom with the highest possible oxidation number (N^{5+}); (2) adding enough O^{2-} ions to form a negative ion [$(NO_1)^{3+}$, $(NO_2)^+$, $(NO_3)^-$]; and (3) adding enough H^+ to cancel the negative charge formed (HNO_3, nitric acid).

"Ous" ACIDS

An "ous" acid suffix is given to acids with lower oxidation numbers on the central nonmetal, usually two less than the highest possible oxidation number. For example, sulfur*ous* acid has the formula H_2SO_3 with a 4+ oxidation number on the central sulfur atom.

ACIDS OF CHLORINE

For acids with multiple oxidation states (e.g., Cl = 7, 5, 3, 1), we need the prefixes "hyper" and "hypo" (in addition to the two suffixes "ic" and "ous") to name acids. For example, "hyper" and "ic" (+7 oxidation state) hyperchloric acid ($HClO_4$), chloric ($HClO_3$), chlorous ($HClO_2$), and "hypo" and "ous" (+1 oxidation state) for hypochlorous ($HClO$).

BASES

Oxides of metals are base anhydrides, which combine with water to form bases:

$$Na + O_2 \rightarrow Na_2O + H_2O \rightarrow \underset{\text{sodium hydroxide}}{2NaOH}$$

The NaOH is an ionic compound (like NaCl); hence, in water, NaOH exists as separate ions, Na^+ and OH^-. The OH^- ion is called a hydroxide ion. A base, then, is a compound that, when dissolved in water, yields hydroxide ions (OH^-).

ACID–BASE REACTIONS

NEUTRALIZATION

Hydroxide ions (OH^-) react with acid protons (H^+) to form water, a process called neutralization:

$$\underset{\text{acid}}{H_2SO_4} + \underset{\text{base}}{NaOH} \rightarrow \underset{\text{salt}}{NaHSO_4} + \underset{\text{water}}{H_2O}$$

Increasing the amount of base (twofold) removes the second hydrogen on the acid:

$$H_2SO_4 + 2NaOH \rightarrow Na_2SO_4 + 2H_2O$$

Lewis structures of the reactants and products are shown in Figure 1.7.

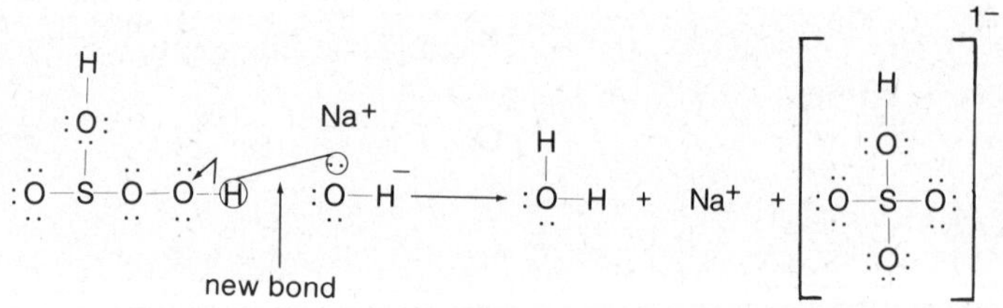

Figure 1.7. Lewis structure of an acid–base reaction.

BASIC SALTS

Upon neutralization, an ''ic'' acid forms salt with an ''ate'' suffix while an ''ous'' acid forms a salt with an ''ite'' suffix:

$$H_2SO_4 + 2NaOH \rightarrow 2H_2O + Na_2SO_4$$

sulfuric   sodium
acid   sulfate

$$HNO_2 + NaOH \rightarrow H_2O + NaNO_2$$

nitrous   sodium
acid   nitrite

OXIDATION–REDUCTION REACTIONS

Chemical reactions can involve loss and gain of electrons where the central non-metallic element in a polyatomic grouping (e.g., H_2SO_4) undergoes a change in oxidation number. The oxidation number may be reduced by the gain of electrons or increased by the loss of electrons. Some acids also serve as oxidizing agents in chemical reactions:

$$H_2SO_4 + 2H^+ + 2e^- \rightarrow H_2SO_3 + H_2O$$
$$\downarrow$$
$$SO_2 + H_2O$$

Note that the oxidation number on the S changes from 6+ to 4+ in the ''half reaction'' above owing to the ''gain'' of electrons. The H_2SO_4 is reduced, hence H_2SO_4 is an oxidizing agent.

Copper dissolves in nitric acid (forms Cu^{2+}) in an oxidation–reduction reaction with reduction of HNO_3 to NO:

$$3Cu + 2HNO_3 + 3H^+ \rightarrow 2NO + 4H_2O + 3Cu^{2+}$$

Hence HNO_3 is an ''oxidizing acid.''

CHEMICAL EQUATIONS

Just as the word ''equation'' in mathematics signifies equality on both sides of the equal sign, a chemical ''equation'' implies equality (in terms of mass and charge) between reactants on the left of the reaction arrow and products on the right:

$$\text{reactants} \rightarrow \text{products}$$

Hence, the number of atoms and charges on both sides of the "chemical equation" must balance.

EXAMPLE 1.3.   Balance the oxidation–reduction equation

$$Cu + HNO_3 \rightarrow Cu^{2+} + NO$$

Answer: Recall that balancing a chemical equation first requires balancing each element present first by mass (or number of atoms), then by charge, or number of electrons. Balancing an oxidation–reduction reaction can be accomplished by the following:

1. Write down the oxidation and reduction half reaction separately:

$$HNO_3 \rightarrow NO$$
$$Cu \rightarrow Cu^{2+}$$

2. Balance the central nonmetal (here the nitrogen is balanced).
3. Balance the excess oxygens by adding water molecules:

$$HNO_3 \rightarrow NO + 2H_2O$$
$$Cu \rightarrow Cu^{2+}$$

4. Balance excess hydrogens with H^+ ions

$$3H^+ + HNO_3 \rightarrow NO + 2H_2O$$
$$Cu \rightarrow Cu^{2+}$$

5. Balance excess charge with electrons:

$$3e^- + 3H^+ + HNO_3 \rightarrow NO + 2H_2O$$
$$Cu \rightarrow Cu^{2+} + 2e^-$$

6. Multiply both half reactions to make the number of electrons lost equal to the number gained:

$$2 \times (3e^- + 3H^+ + HNO_3 \rightarrow NO + 2H_2O)$$
$$3 \times (Cu \rightarrow Cu^{2+} + 2e^-)$$

7. Add the two half reactions and subtract any common reactants and products (usually H_2O and electrons):

$$6e^{-1} + 6H^+ + 2NHO_3 + 3Cu \rightarrow 2NO + 4H_2O + 3Cu^{2+} + 6e^{-1}$$

This procedure works for oxidation–reduction reactions in acid solutions. Calculations based on chemical equations must be performed on "balanced" equations.

VOLATILE REACTION PRODUCTS

Metals in a reduced elemental state are insoluble in water. However, many metals react explosively with acids (owing to hydrogen gas formation). Alkali metals (Group IA) react explosively with water because of hydrogen gas formation. Also, small molecular molecules like NO, SO_2, CO_2, which are nonionic when formed in a chemical reaction, will often escape the solution as gaseous products. Some product gases that appear over reaction solutions are quite explosive [e.g., hydrogen (H_2), ammonia (NH_3), etc.]; others are poisonous [e.g., hydrogen cyanide (HCN), carbon monoxide (CO)]. Hence, extreme care should be exercised in mixing various substances.

STRONG AND WEAK ELECTROLYTES

STRONG ELECTROLYTES

Strong electrolytes are completely converted to ions when mixed with water. Most metal hydroxides (bases), strong acids (H_2SO_4, HNO_3, H_3PO_4, etc.), and salts (NaCl, KNO_3) all completely dissociate to form ions when combined with water. Weak electrolytes are only partially converted to ions in water. Nitrogen bases like ammonia (NH_3) and organic acids (acetic acid) are examples of weak electrolytes. The terms "weak base" and "weak acid" are commonly used in clinical chemistry.

WEAK ELECTROLYTES

Amine Bases

Ammonia (NH_3) and organic amines are weak bases because the nitrogen atom has a "lone pair" of electrons that bind H^+ ions of water:

$$NH_3 + H_2O \rightleftharpoons NH_4^+ + OH^-$$

Other ions can serve as bases in water because of free lone pairs of electrons on oxygen binds the H^+ ion of water:

$$CO_3^{2-} + H_2O \rightleftharpoons HCO_3^- + OH^-$$

$$CH_3COO^- + H_2O \rightleftharpoons CH_3COOH + OH^-$$

$$HCO_3^- + H_2O \rightleftharpoons H_2CO_3 + OH^-$$

Note that conjugate anions of weak acids (e.g., acetate from acetic acid or HCO_3^- from H_2CO_3) are relatively strong bases.

WEAK ACIDS (HA)

A weak acid (HA) only partially dissociates in water; hence, we have an equilibrium existing between the dissociated and undissociated acid:

$$HA \rightleftharpoons H^+ + A^-$$

The equilibrium constant Ka is

$$Ka = \frac{[H^+][A^-]}{[HA]}$$

Solving for $[H^+]$ or hydrogen ion concentration

$$[H^+] = \frac{Ka\,[HA]}{[A^-]}$$

Taking the negative log of both sides of the equation,

$$-\log[H^+] = -\log Ka + \left(-\log \frac{[HA]}{[A^-]}\right)$$

Since $pH = -\log[H^+]$ and $pKa = -\log Ka$,

$$pH = pKa + \left(-\log \frac{[HA]}{[A^-]}\right)$$

$$pH = pKa + \log \frac{[A^-]}{[HA]}$$

Hence the pH of weak acid solutions is largely determined by the ratio of the concentration of the acid salt $[A^-]$ of the weak acid and the undissociated acid $[HA]$. If the $[A^-]$ equals $[HA]$, the pH equals the pKa since log 1 is zero.

BUFFER SOLUTIONS

A buffer solution is a combination of a weak acid and its conjugate salt or weak base and its conjugate salt. Because a buffer solution contains both an acid and its conjugate base, addition of small amounts of strong acids or bases to a buffer solution does not greatly affect the pH of the solution. A buffer solution essentially neutralizes the added acid or base. The salt/acid ratio must alter significantly before the pH changes.

EXAMPLE 1.4.   Prepare a 2 M phosphate buffer with a salt/acid ratio of 4.

   Answer: The sum of the concentrations of the acid and conjugate base (or salt) equals 2 M.

$$\text{salt} + \text{acid} = 2\ M$$

$$\text{salt} = 2\ M - \text{acid}$$

The salt/acid ratio is 4. Hence,

$$\frac{\text{salt}}{\text{acid}} = 4$$

Substituting

$$\frac{2\ M - \text{acid}}{\text{acid}} = 4$$

and solving yields

$$2\ M - \text{acid} = 4(\text{acid})$$

$$5(\text{acid}) = 2\ M$$

$$\text{acid} = \tfrac{2}{5}\ M = 0.4\ \text{mol/L}$$

$$\text{salt} = 2\ M - \tfrac{2}{5}\ M = \tfrac{8}{5}\ M = 1.6\ \text{mol/L}$$

A 2 M phosphate buffer solution will result if 0.4 mol of acid ($H_2PO_4^-$) are mixed with 1.6 mol of salt (HPO_4^{2-}) in 1 L of solution.

Since over 15,000 mEq of volatile acid (CO_2) along with 100 mEq of nonvolatile acid are produced each day in a typical person as a result of metabolism of carbohydrates and fats, the plasma and cell buffers must properly neutralize this acid to maintain the pH between 7.35 and 7.45. Major plasma buffers include the bicarbonate and phosphate buffers; cellular buffering systems rely on hemoglobin and other proteins for buffering the acids of metabolism.

ORGANIC COMPOUNDS

In our chemistry review to this point, we have discussed molecules and ions that are generally classified as "inorganic." Organic molecules and ions are composed of carbon, hydrogen, and oxygen, and generally have more atoms with more complex structures than inorganic molecules and ions. The term organic in clinical chemistry generally applies to compounds of carbon, particularly those found in living beings.

FUNCTIONAL GROUPS

In organic chemistry, molecules are often classified with respect to their functional group or their particular function in biological systems. The molecular (or

carboxylic
acids

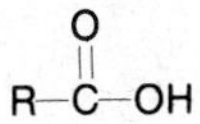

$$R-\overset{\overset{\displaystyle O}{\|}}{C}-OH$$

amine bases $R-NH_2$

alcohols $R-OH$

ketone

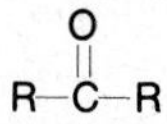

$$R-\overset{\overset{\displaystyle O}{\|}}{C}-R$$

aldehydes

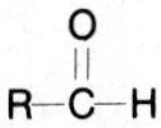

$$R-\overset{\overset{\displaystyle O}{\|}}{C}-H$$

esters

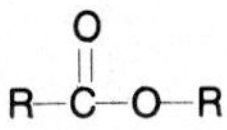

$$R-\overset{\overset{\displaystyle O}{\|}}{C}-O-R$$

amino acids

$$R-\overset{\overset{\displaystyle H}{|}}{\underset{\underset{\displaystyle NH_2}{|}}{C}}-\overset{\overset{\displaystyle O}{\|}}{C}-OH$$

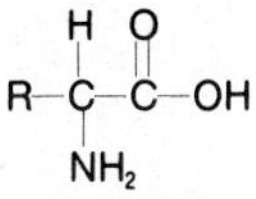

Figure 1.8.   Functional groups of various organic compounds where "R-" represents carbon–hydrogen chains or rings.

ionic) structure of organic compounds may be long chains of carbon atoms as seen in fatty acids or single or multiple rings of carbon atoms as seen in carbohydrates and steroids. Other nonmetallic elements (e.g., N, S, O) may replace a carbon atom in these rings or chains, hence an enormous number and variety of compounds are possible. Selected functional groups of organic compounds are shown in Figure 1.8 with "R" denoting the carbon chain or ring.

Classes of organic compounds of significant interest in clinical chemistry include carbohydrates (Chapter V), proteins (Chapter VI), lipids (Chapter VII), enzymes (Chapter X), hormones (Chapter XII), and drugs (Chapter XIV).

CHEMISTRY CALCULATIONS

"NUMBER" AND "UNIT" COMPONENTS OF MEASUREMENTS

Clinical chemistry is primarily concerned with the measurement of analytes (substances) in human body fluids and the relationship of the results of these measurements to various states of disease or health. All measurements consist of two important components: (1) the actual number associated with measurement (e.g., 8.9 for a serum calcium) and (2) the "units" on that number [e.g., milligrams per deciliter (mg/dL)]. Accordingly, a serum calcium level includes both a "number" and "unit" components (i.e., 8.9 mg/dL). The measurement is of no scientific significance without both components.

NUMBER COMPONENT

Significant Figures

The number portion of an analytical measurement should be written in a manner that reflects the reliability of the measurement. For example, a measurement of 10.5 g has three significant figures with the last digit indicating that the measurement is correct to the nearest tenth of a gram. The actual weight lies between 10.45 and 10.55 g. A more sensitive balance may give a weight of 10.50 g with implied limits of 10.495–10.505. Hence the zero in the one-hundreths place is significant in that it reflects the sensitivity of the measurement. If the number is less than one, zeros following the decimal point and preceding the number are not significant; zeros following the number are significant.

EXAMPLE 1.5.   How many significant figures are there in (a) 0.018, (b) 0.180, and (c) 1.018?

   Answer:
(a)   0.018 contains two significant figures.
(b)   0.180 contains three significant figures.
(c)   1.018 contains four significant figures.

Mathematics of Significant Figures

When adding, subtracting, dividing, and multiplying, certain rules in retaining significant figures apply. To subtract or add a number, round all numbers involved to match the number that has the least number of significant figures to the right of the decimal point, then subtract or add the numbers in the group.

EXAMPLE 1.6.   Add the following three numbers and report the results with the correct number of significant figures:

$$0.0212 + 29.64 + 1.056931$$

   Answer: Prior to addition, round all numbers to the nearest one-hundreth, then add:

$$0.02 + 29.64 + 1.06 = 30.72$$

To multiply or divide numbers, determine which number has the least number of significant figures. Then round all of the numbers to be multiplied or divided to match the significant figures of the number containing the least number of significant figures in the group.

EXAMPLE 1.7.   Multiply the following numbers:

$$0.0211 \times 25.63 \times 1.05881$$

Answer: Since 0.0211 contains only three significant figures, round the other numbers to three significant figures, then multiply.

$$0.0211 \times 25.6 \times 1.06 = 5.72$$

Rounding-Off Rules

Laboratory measurements should be rounded-off to reflect the precision of the measurement. Rules for rounding off include: (1) if the digit to be dropped is less than 5, the preceding digit is not altered; (2) if the digit to be dropped is more than 5, the preceding figure is increased by one; and (3) if the digit to be dropped is 5, the preceding digit is increased by one if it is an odd number and, it is not altered if it is an even number.

EXAMPLE 1.8.   Round-off the following numbers to two significant figures: 2.64, 2.56, 2.55, and 2.65.

Answer: All numbers round to 2.6.

Scientific Notation

All numbers can be expressed as a number between 1 and 10 with an exponential term (base 10) multiplied by the number to account for the position of the decimal point. Numbers expressed in this manner are in scientific notation.

Numbers Greater than or Equal to 10
To convert a number greater than 10 to scientific notation, count the number of decimal places required to move the decimal point to the left to get a number between 1 and 10. This count will be the power that 10 must be raised in the exponential term when expressing the number in scientific notation.

EXAMPLE 1.9.   Convert the following numbers to scientific notation: 100, 13,500, and 10,000.

Answer:

$$100 = 100 \times \frac{10^2}{100} = 1.0 \times 10^2$$

$$10,000 = 10,000 \times \frac{10^4}{10,000} = 1.0 \times 10^4$$

$$13,500 = 13,500 \times \frac{10^4}{10,000} = 1.35 \times 10^4$$

Numbers Less than 1
When expressing numbers less than 1 in scientific notation, count the number of decimal places required to move the decimal point to the right to get a number

between 1 and 10. This number then will be the correct negative power in the exponential term when expressing the number in scientific notation.

EXAMPLE 1.10.   Convert the following numbers to scientific notation: 0.001, 1.0, and 0.000432.

Answer:

$$0.001 = 0.001 \times \frac{10^{-3}}{0.001} = 1.0 \times 10^{-3}$$

$$1.0 = 1.0 \times 10^0 = 1.0 \text{ (already in scientific notation)}$$

$$0.000432 = 0.000432 \times \frac{10^{-4}}{0.0001} = 4.32 \times 10^{-4}$$

Note that the value of the number does not change, only the position of the decimal point is altered.

UNIT COMPONENT

As mentioned earlier, a measurement in the laboratory consists of a numerical portion and a unit portion (e.g., 8.9 mg/dL). The previous section has concentrated on aspects of the numerical portion. This section will concentrate on the units of the laboratory measurements.

Metric System

The basic units of measurement in the Metric System consist of base units for measurements (e.g., mass–gram) along with prefixes to indicate a multiple of the base unit. Multiples in the Metric System are interconvertible as powers of 10 (i.e., 10^1, 10^2, 10^{-2}, etc.) (see Table 1.3). The basic units of measurement of the metric system are (1) mass–*gram*, (2) length–*meter*, and (3) volume–*liter*.

In clinical chemistry, our measurements on body fluids primarily deal with small amounts of mass and volume, hence we often use fractional prefixes (deci, milli, micro, etc.). Examples of units commonly used in clinical chemistry are listed in Table 1.4.

Older units also used include lambda (λ)—a volume unit equal to a microliter and micron (μ)—a length unit equivalent to a micrometer. These symbols for units have added to the confusion of the reporting of laboratory data and should not be used.

Standard International Units (SI)

The Commission on Clinical Chemistry of the IUPAC and IFCC has proposed a standard international (SI) system of units. Examples of acceptable units follow.

Table 1.3. Prefixes of the Metric System

Multiple Name	Power of 10	Decimal	Prefix	Symbol
Trillion	10^{12}	1,000,000,000,000.0	tera	T
Billion	10^9	1,000,000,000.0	giga	G
Million	10^6	1,000,000.0	mega	M
Thousand	10^3	1,000.0	kilo	k
Hundred	10^2	100.0	hecto	h
Ten	10^1	10.0	deca	da
Basic unit	10^0	1.0	(no prefix)	
One-tenth	10^{-1}	0.1	deci	d
One-hundreth	10^{-2}	0.01	centi	c
One-thousandth	10^{-3}	0.001	milli	m
One-millionth	10^{-6}	0.000001	micro	u
One-billionth	10^{-8}	0.000000001	nano	n
One-trillionth	10^{-12}	0.000000000001	pico	p
One-quadrillionth	10^{-15}	0.0000000000000001	femto	f
One-quillionth	10^{-18}	0.000000000000000001	atto	a

Mass, Length, and Volume

As shown in Tables 1.3 and 1.4, mass units kg, g, mg, etc. are acceptable while grs, gr, oz, kg, mgs, lbs are not acceptable. Length units of m, cm, nm, etc. are acceptable; in. and ft. are not acceptable length units. For volume, cc, cm^3, λ, etc. are no longer acceptable while L, mL, nL are greatly preferred.

Substance Concentration

Although commonly used, M (molarity), N (normality), and mEq/L are not recommended units. Correct concentration units include mol/L, mmol/L, etc. Correct concentration units also include kg/L, g/L, mg/L with unacceptable units being %, g/ml, ppm, and the commonly used mg %.

Table 1.4. Selected Metric Units for Mass, Length, and Volume

Multiple	Mass	Length	Volume
1000	kilogram (kg)	kilometer (km)	kiloliter (kL)
1 (basic standard)	gram (g)	meter (m)	liter (L)
1/10	decigram (dg)	decimeter (dm)	deciliter (dL)
1/100	centigram (cg)	centimeter (cm)	centiliter (cL)
1/1000	milligram (mg)	millimeter (mm)	milliliter (mL)
1/1,000,000	microgram (μg)	micrometer (μm)	microliter (μL)
1/1,000,000,000	nanogram (ng)	nanometer (nm)	nanoliter (nL)
1/10,000,000,000		angstrom (Å)	
1/1,000,000,000,000	picogram (pg)	picometer (pm)	picoliter (pL)

Enzyme Activity

The unit U is used to denote the amount of enzyme activity. One U is based on the conversion of one micromole per minute per milliliter of specimen (1 μmol/min/mL) at specified assay conditions (pH, temperature, etc.). The unit for the concentration of an enzyme is U/L, not IU/L.

Other Units

Area is expressed as a square of a standard length unit (i.e., m^2, cm^2, etc.) Temperature is expressed in degrees centigrade, °C. SI time units are units limited to s, min, h; sec, mins, hrs are not acceptable. Density is expressed in kg/L and not gr/ml or gr/cc. The absorbance unit is A in the SI system. Radioactivity units, as discussed earlier, are based on curies, and include mCi, Ci, etc.

DENSITY

The density of a material is the mass per unit volume and has the units g/mL or kg/L. For example, at 25°C, water has a density of 0.996 g/mL. Density is a physical property of pure material and can be used as a conversion factor to convert mass to volume units and vice versa.

EXAMPLE 1.11.   Compute the density of a pure liquid if 40 mL weighs 51 g.

Answer:

Need: Density or g/mL?

Know: volume = 40 mL; weight = 51 g.

Hence: $\dfrac{51 \text{ g}}{40 \text{ mL}} = 1.3$ g/mL

SPECIFIC GRAVITY

Specific gravity is numerically equal to density; however, specific gravity has no units. Density equals mass per unit volume, while specific gravity is a ratio between the mass of a substance and the mass of an equal volume of pure water at 4°C. At this temperature, 1 mL of water weighs 1 g.

EXAMPLE 1.12.   The label on a bottle of concentrated nitric acid lists sp. gr. = 1.42 and 70% of the solution is HNO_3 by weight. Using these data, prepare 100 mL of a 6 M solution of HNO_3.

Answer:

Need: Volume of concentrated HNO_3 to add to 100 mL volumetric flask or mL?

Know: Specific gravity = 1.42, therefore, density of HNO_3 solution is 1.42 g/mL; 0.70 purity; 6 mol/L, 100 mL total volume; molecular weight (HNO_3) = 63 g/mol:

1H + 1N + 3O = 1 g + 14 g + 48 g = 63 g/mol

Recall: (Density)(% purity)(volume) = wt(HNO_3)
or g/mL × % × mL = g (HNO_3)

$$\frac{wt\ (HNO_3)}{molecular\ weight\ HNO_3} = (molarity \times volume)_{HNO_3}$$

Substituting:

$$\frac{1.42\ g/mL \times 0.70 \times volume\ (HNO_3)}{63\ g/mol} = 6\ mol/L \times 0.1\ L$$

or

$$Vol\ (HNO_3) = \frac{6\ mol/L \times 0.1\ L \times 63\ g/mol}{1.42\ g/mL \times 0.70}$$

$$Vol\ (HNO_3) = \frac{6 \times 0.1 \times 63}{1.42 \times 0.70}\ mL = 38\ mL$$

Hence 38 mL of concentrated HNO_3 must be diluted to 100 mL in order to have a 6 mol/L solution.

DILUTION FACTORS

In clinical chemistry, it is often helpful to treat solutions as a sum of parts. For example, a one-to-ten (1 : 10) dilution of serum would have 1 part (or 1 mL) serum, plus 9 parts (or 9 mL) water giving a total of 10 parts (or 10 mL). It is noteworthy that the concentrations of all substances in the serum would now be reduced by a factor called a dilution factor (DF) (in this case the DF is equal to 10). When performing routine dilutions, it is generally assumed that the diluting solvent contains no detectable amount of analyte.

EXAMPLE 1.13.   Calculate the dilution factor and express the degree of dilution when 3 mL of serum is mixed with 18 mL of saline.

Answer:

Need: DF and parts of serum/total parts of solution?
Know: 3 mL serum added to 18 mL water and each mL equal one part.
Hence: 3 mL + 18 mL = 21 mL or 21 total parts and the degree of solution is 3 parts/21 parts or 1/7. The dilution factor is 7.

Note: Any measurement (excluding Na^+ and Cl^-) performed on this diluted serum must be multiplied by 7 to get the actual serum concentration. Sodium and chloride are, of course, present in the saline diluent, hence the dilution factor cannot be applied to these analytes.

EXAMPLE 1.14.   Prepare 250 mL of a 1 : 10 dilution of serum and saline.

Answer:

Need: Volume of serum needed to dilute to 250 mL or mL (of serum)?
Know: (ratio and proportion)

$$\frac{1 \text{ mL(serum)}}{10 \text{ mL(solution)}} = \frac{X \text{ mL(serum)}}{250 \text{ mL(solution)}}$$

Hence: X mL(serum) $= \dfrac{[1 \text{ mL(serum)}][250 \text{ mL(solution)}]}{10 \text{ mL(solution)}}$

$$X \text{ mL(serum)} = \frac{(1)(250)}{10} \text{ mL(serum)} = 25 \text{ mL(serum)}$$

EXAMPLE 1.15.   How much of a 1 : 16 dilution of urine in distilled water could be made with 3 mL of urine?

Answer:

$$\frac{3 \text{ mL(urine)}}{X \text{ mL(total solution)}} = \frac{1 \text{ part}}{16 \text{ total parts}}$$

$$X \text{ mL(total solution)} = 16 \times 3 \text{ mL} = 48 \text{ mL}$$

Hence: 3 mL of urine diluted to 48 mL retains the 1 : 16 dilution.

EXAMPLE 1.16
(a) A 5 mol/L solution of HCl is diluted 1 : 5. The resulting solution is diluted 1 : 10. Determine the concentration of each of the diluted solutions.

Answer:

First dilution: 5 mol/L $\times$ 1/5 = 5/5 = 1 mol/L

Second dilution: 5 mol/L $\times$ 1/5 $\times$ 1/10 = 5/50 = 0.1 mol/L

(b) A 1 : 10 dilution of a substance is diluted 3/5, rediluted 2/15, and diluted once again 1/2. What is the final concentration?

Answer:

1/10 $\times$ 3/5 $\times$ 2/15 $\times$ 1/2 = 1/250 dilution

(c) A 3% solution is diluted 2/30. What is the resulting concentration?

Answer:

3% $\times$ 2/30 = 6/30 = 0.2%

(Remember: 3% = 3 g/100 mL; 0.2% is 0.2 g/100 mL.)

(d)   A solution that contains 80 mg/dL is diluted 1 : 10 and again 2 : 20. What is the final concentration?

Answer:

$$80 \text{ mg/dL} \times 2/20 \times 1/10 = \frac{160 \text{ mg/dL}}{200} = 0.8 \text{ mg/dL}$$

Note: The final answer has the same units as the original solution because the DF(s) have no units.

CONCENTRATION OF SOLUTIONS IN PERCENT

The three ways of expressing concentration of solutions in percent include (1) weight-solute/weight-solvent (W/W), (2) weight-solute/volume-solution (W/V), and (3) volume-solute/volume-solution (V/V). In the laboratory, W/V and V/V are most commonly used.

EXAMPLE 1.17

(a)   Prepare a 10% (W/W) solution of aqueous NaCl.

   Answer: Mix 10 g of NaCl with 90 g of H_2O.

(b)   Prepare a 10% (W/V) solution of aqueous NaCl.

   Answer: Weigh out 10 g of NaCl, add to a 100 mL volumetric flask, and add enough water to bring total volume to 100 mL.

(c)   Prepare a 10% (V/V) solution of ethyl alcohol in water.

   Answer: To make the 10% (V/V) alcohol solution from two pure liquids, add 10 mL of ethyl alcohol to a volumetric flask; then bring the final volume to 100 mL with water.

Note: As mentioned earlier, percentage units are no longer acceptable.

MATHEMATICAL PROPERTIES OF UNITS

Almost all of our measurements in clinical chemistry are done on body fluids (whole blood, serum, plasma, urine, sweat, etc.) with the analyte (chemical compound being measured) dissolved in the body fluid. We express the concentration of the analyte with units that reflect mass per unit volume. Hence it is useful to look at "concentration units" as unit fractions (8.9 mg/dL) that characterize the preceding number. For example, 8.9 mg/dL indicates there are 8.9 milligrams of Ca^{2+} solute dissolved in one deciliter (or 100 mL) of solution. The number preceding the unit factor always describes the unit (i.e., mg) in the numerator. Unit fractions have the mathematical properties of fractions:

Multiplication:

$$\frac{mg}{dL} \times \frac{mg}{dL} = \frac{mg^2}{dL^2}$$

Division (invert denominator and multiply):

$$\frac{ng/dL}{ng/dL} = \frac{ng}{dL} \times \frac{dL}{ng} = 1$$

Raising to powers:

$$(mg/dL)^2 = mg^2/dL^2$$

UNIT CONVERSION TECHNIQUES

Calculations Involving Concentrations

There are two types of calculations involving concentrations:

1. Calculations in which the concentration values are not changed, but the units of concentration are changed. Use the unit conversion technique described below.
2. Calculations in which the concentration changes as in dilution problems. Use formula $V_1C_1 = V_2C_2$, or a variation thereof, where C (concentration) can be essentially any unit as long as it expresses mass per unit volume. The V (variable) can be in any volume unit. However, if chemical reactions are involved (i.e., neutralization), normality (Eq/L or mEq/mL) units should be used.

Unit Conversion Problems

Since in clinical chemistry, there are many different ways of expressing the mass and volume of our measurements, conversion from one unit system to another using conversion factors is often required. This is accomplished by multiplying by a unit conversion factor (CF). A unit CF is always equal to 1, so the multiplication does not change the value of the original measurement; however, it does change the units.

EXAMPLE 1.18.   Compute the molarity of a H_2SO_4 solution that has a concentration of 49 g/dL.

    Need: Molarity or mol/L?

    Know: Molecular weight (H_2SO_4) = 98 g/mol or $\frac{1}{98}$ mol/g concentration, 49 g/dL, 10 dL = 1 L or 10 dL/L = 1, hence the unit CF = 10 dL/L.

Therefore: $49 \text{ g/dL} \times \frac{1}{98} \text{ mol/g} \times 10 \text{ dL/L} = \frac{(49)(10)}{98} \text{ mol/L}$

        $= 5 \text{ mol/L} = 5 \text{ } M$

Note here that our procedure for working chemistry problems of this sort requires:

1. Interpreting what the problem wants in terms of units.
2. Writing down what is given by the problem with units and noting conversion factors (CFs) needed to get the desired unit (i.e., here g to mol, then dL to L). Note to get unit CF from 98 g/mole, we divide into 1 (i.e., $1/(98 \text{ g/mol})$ = $\frac{1}{98}$ mol/g).
3. Multiplying given information by appropriate CFs, canceling units, then grouping numbers and remaining units.
4. Performing arithmetic and check units against what was needed to answer original question.

EXAMPLE 1.19.   Express 8.9 mg/dL as grams per liter.

   Answer:

   Need: Convert mg to g unit in numerator.
   Recall: 1000 mg = 1 g

$$CF = \frac{1000 \text{ mg}}{1000 \text{ mg}} = \frac{1 \text{ g}}{1000 \text{ mg}} = \frac{1}{1000} \text{ g/mg} = 10^{-3} \text{ g/mg}$$

Hence:

$$8.9 \text{ mg/dL} \times 10^{-3} \text{ g/mg} = 8.9 \times 10^{-3}(10^{-3}) \text{ mg/dL} \times \text{g/mg} = 8.9 \times 10^{-3} \text{ g/dL}$$

Note: We have changed both the numerical portion and unit portion of the calcium level but the laboratory result is still equal to 8.9 mg/dL. Note, also, that during the calculation, numbers are grouped with numbers and units are grouped with units, prior to any arithmetic or unit cancellation.

   Need: Now we need to convert the denominator unit from dL to L.
   Recall: 10 dL = 1 L

$$CF = \frac{10 \text{ dL}}{1 \text{ L}} = \frac{1 \text{ L}}{1 \text{ L}} = 10 \text{ dL/L}$$

Hence:

$$8.9 \times 10^{-3} \text{ g/dL} \times 10 \text{ dL/L} = (8.9 \times 10^{-3})(10) \text{ g/dL} \times \text{dL/L} = 8.9 \times 10^{-2} \text{ g/L}$$

Note: We have converted our laboratory results to the desired SI unit, g/L. Note, also, that the calculations were done in steps. With practice, these CFs can be linked together, as shown below:

$$8.9 \text{ mg/dL} \times 10^{-3} \text{ g/mg} \times 10 \text{ dL/L}$$
$$= (8.9)(10^{-3})(10) \text{ mg/dL} \times \text{g/mg} \times \text{dL/L}$$
$$= 8.9 \times 10^{-2} \text{ g/L}$$

Hence: 8.9 mg/dL = 8.9×10^{-2} g/L.

Calculations Involving Dilutions

Preparation of Standard Solutions

Solutions of known concentration are often required in clinical chemistry. Preparation of a solution of known concentration can be done two ways: (1) dilution of a more concentrated solution to the desired concentration by adding solvent or (2) addition of solvent to a known weight of solute to achieve a specified total volume of solution (usually done in a volumetric flask). For calculations involving solutions, we need to know how much water (or solvent) to add, how much solute to weigh out, and what the final volume of the solution must be. All calculations can be developed from the fact that the amount of solute in solution does not change by the addition of solvent:

$$(\text{grams solute})_{BD} = (\text{grams solute})_{AD}$$

where BD = before dilution and AD = after dilution. If we multiply both sides of the equation by "L solution/L solution", we can develop the equation, $C_1V_1 = C_2V_2$.

$$\left(\frac{\text{g solute}}{\text{L solution}} \times \text{L solution} \right)_{BD} = \left(\frac{\text{g solute}}{\text{L solution}} \times \text{L solution} \right)_{AD}$$

$$(C \times V)_{BD} = (C \times V)_{AD}$$

$$C_1 \times V_1 = C_2 \times V_2$$

This formula is used when a more concentrated solution is diluted to a desired concentration.

EXAMPLE 1.20.   Prepare 100 mL of a 100 mg/L NaCl solution from a solution that has a concentration of 2000 mg/L.

Answer:

Need: mL of concentrated NaCl solution? We need to know how much (or volume) of our concentrated NaCl solution to add to our (100 mL) volumetric flask prior to addition of water to bring the final volume of the solution to 100 mL.

Know: C_1 = 2000 mg/L; V_2 = 100 mL; C_2 = 100 mg/L.

Hence:

$$C_1V_1 = C_2V_2$$

$$2000 \text{ mg/L} \times V_1 = 100 \text{ mg/L} \times 100 \text{ mL}$$

$$V_1 = \frac{100 \text{ mg/L} \times 100 \text{ mL}}{2000 \text{ mg/L}} = \frac{(100)(100)}{2000} \text{ mL}$$

$$V_1 = 5 \text{ mL}$$

Therefore, if 5 mL of a 2000 mg/L NaCl solution are added to a 100 mL flask and diluted to a total volume of 100 mL, the final concentration of the solution will be 100 mg/L.

Dilute Solutions from Solid Material
If we wanted to prepare 100 mL of the same 100 mg/L NaCl solution from solid material, we would need the weight of NaCl required:

$$(\text{g solute})_{BD} = \left(\frac{\text{g solute}}{\text{L solution}} \times \text{L solution} \right)_{AD}$$

$$\text{g solute} = C_2 \times V_2$$

EXAMPLE 1.21. Prepare 100 mL of a 100 mg/L NaCl solution from crystalline NaCl.

Answer:

Need: Grams of solute to be weighed and diluted with water to a volume of 100 mL or g?

Know: $C_2 = 100$ mg/L; $V_2 = 100$ mL or $V_2 = 0.1$ L.

Note: Volume units must match on concentration unit and solution volume unit.

Hence:

$$\text{g solute} = C_2 \times V_2$$

$$\text{g solute} = 100 \text{ mg/L} \times 0.1 \text{ L} = (100)(0.1) \text{ mg/L} \times \text{ L}$$

$$\text{g solute} = 10 \text{ mg}$$

Therefore, 10 mg of NaCl must be weighed and added to a 100 mL volumetric flask before addition of enough water (approximately 100 mL) to bring the volume of the flask to 100 mL.

MOLARITY, NORMALITY, EQUIVALENT WEIGHT, AND MOLECULAR WEIGHT

CHEMICAL REACTIONS

In a chemical reaction, such as the reactions described earlier, we mentioned that the equation was balanced, with the mass of reactants equal to the mass of formed products. We can relate this to an actual gram amount:

$$SO_3 \;+\; H_2O \;\rightarrow\; H_2SO_4$$

before reaction: 80 g   18 g   —
(1 mol) (1 mol)

after reaction: —   —   98 g
(1 mol)

The above chemical reaction states that 1 mol of SO_3 combines with 1 mol of H_2O to give 1 mol of H_2SO_4 (sulfuric acid).

MOLECULAR WEIGHT (MW)

The weight of 1 mole of any compound [i.e., molecular weigh (g/mol)] can be determined by summing the atomic weighs of elements making up the compound. Also, 1 mole of compound consists of 6.023×10^{23} molecules, as defined by Avagadro.

EXAMPLE 1.22.   Calculate the molecular weight (g/mol) of H_2SO_4 (see Table 1.1 for atomic weights).

Answer: Molecular weight is computed by adding the atomic weight of the elements making up the compound.

  Need: g/mol?

         2H = 2 g

         1S = 32 g

         4O = 64 g

  Hence: Molecular weight (H_2SO_4) = 2 g + 32 g + 64 g

           Molecular weight (H_2SO_4) = 98 g/mol

AVAGADRO'S NUMBER

One mole of a given compound is equal to Avagadro's number of molecules (or ions) making up the compound, or there are 6.023×10^{23} molecules/mol. Therefore, 49 g of sulfuric acid would consist of

$$\left(6.023 \times 10^{23} \text{ molecules/mol} \times \frac{1}{98} \text{ mol/g} \times 49 \text{ g}\right)$$

$$= \frac{(6.023 \times 10^{23})(49)}{98} \text{ molecules} = 3.0115 \times 10^{23} \text{ molecules}$$

Note here, we used a conversion factor to get moles from grams:

$$1 \text{ mol} = 98 \text{ g} \quad \text{or} \quad \frac{1 \text{ mol}}{98 \text{ g}} = \frac{98 \text{ g}}{98 \text{ g}} = 1$$

$$CF = \frac{1}{98} \text{ mol/g}$$

EQUIVALENT WEIGHT (EW)

In the equation

$$H_2SO_4 + 2NaOH \rightarrow Na_2SO_4 + 2H_2O$$

	H₂SO₄	2NaOH	Na₂SO₄	2H₂O
before reaction:	1 mol (2 Eq)	2 mol (2 Eq)	—	—
after reaction:	—	—	1 mol (2 Eq)	2 mol (2 Eq)

In this reaction, two equivalents (Eq) of H_2SO_4 combine with two equivalents of NaOH even though there are 1 and 2 moles, respectively, of each compound. Like moles, equivalents are related to a specific weight (called ''combining weight'') of material. The equivalent weight is computed by the following formula:

$$\text{equivalent weight} = \frac{\text{molecular weight}}{n} \quad \text{or} \quad EW = \frac{MW}{n}$$

where n is

1. number of replaceable hydrogens (H^+) in an acid,
2. number of replaceable hydroxides (OH^-) in a base,
3. charge on a cation or anion, or
4. number of electrons lost (or gained) in an oxidation–reduction reaction.

Note: n has the units Eq/mol and is equal to the number of equivalents per mole.

EXAMPLE 1.23.   Calculate the equivalent weight of H_2SO_4. (H_2SO_4 is an acid with two replaceable hydrogens.)

   Answer: The molecular weight of H_2SO_4 is 98 g/mol; H_2SO_4 has two replaceable hydrogens so $n = 2$ Eq/mol.

   Hence:

$$\text{Equivalent Weight}(H_2SO_4) = \frac{98 \text{ g/mol}}{2 \text{ Eq/mol}}$$

$$= \frac{98}{2} \text{ g/mol} \times \text{mol/Eq}$$

$$\text{Equivalent Weight } (H_2SO_4) = 49 \text{ g/Eq}$$

Milliequivalent Weight

The equivalent weight of H_2SO_4 is 49 g/Eq or 49 mg/mEq. Milliequivalents (mEq) are used in clinical chemistry to express concentrations on unicharged cations and anions (i.e., Na^+, K^+, Cl^-). For example, a typical serum concentration of Na^+ is 145 mEq/L, while a typical serum concentration of Cl^- is expressed 98 mEq/L.

Since the charge on the cation (or anion) is equal to 1, the equivalent weight of sodium is equal to the "molecular weight". Therefore, 1 mmol of Na^+ weighs the same as 1 mEq. SI units express concentrations in mmol/L when the molecular weight of the analyte is known. Hence, the serum Na^+ concentration would be converted from 145 mEq/L to 145 mmol/L in SI units. Note here that the term molecular weight implies "molecule" weight; however, we use the term for compounds like NaCl which do not exist as molecules. (Some chemistry texts use "formula weight" for salts.)

MOLARITY AND NORMALITY

In chemical reactions, it is convenient to express concentrations in terms of moles or equivalents since these units conveniently express the amount of material reacting or being formed in a chemical reaction. Molarity (M) is the number of moles per liter of solution (units = mol/L) and normality (N) is the number of equivalents per liter (units = equivalent/L or Eq/L). Molarity and normality are interchangeable by the following formula:

$$\text{molarity} = \frac{\text{normality}}{n} \quad \text{or} \quad M = \frac{N}{n}$$

Note here that n has the same definition as used earlier for the determination of equivalent weight. It equals the number of equivalents per mole.

EXAMPLE 1.24.   Calculate the molarity (M) of a 12 N solution of H_2SO_4.

Answer:

  Need: mol/L?
  Know: H_2SO_4 has a concentration of 12 Eq/L;
         H_2SO_4 has two replaceable hydrogens so $n = 2$ Eq/mol.

  Hence: $M = \dfrac{N}{n}$

$$M = \frac{12\ \text{Eq/L}}{2\ \text{Eq/mol}} = \frac{12}{2}\ \text{Eq/L} \cdot \text{mol/Eq}$$

$$M = 6\ \text{mol/L (or 6 } M \text{ or 6 molar)}$$

Note: Molarity can also be converted to normality by multiplying molarity times (n).

Dilution Problems Based on Molarity and Normality

Preparation of Standard Solutions
As before, the amount of solute material does not change in diluting (or dissolving) the solute:

$$(\text{moles solute})_{BD} = (\text{moles solute})_{AD}$$

where BD = before dilution; AD = after dilution.

$$\left(\frac{\text{moles solute}}{\text{L solution}} \times \text{L solution}\right)_{BD} = \left(\frac{\text{moles solute}}{\text{L solution}} \times \text{L solution}\right)_{AD}$$

$$(\text{molarity} \times \text{volume})_{BD} = (\text{molarity} \times \text{volume})_{AD}$$

$$M_1 \times V_1 = M_2 \times V_2$$

Normality can be substituted in the above equation to give $N_1V_1 = N_2V_2$ and either molarity or normality can be used interchangeably for dilution problems.

EXAMPLE 1.25.   Prepare a 100 mL of 0.1 M H_2SO_4 from a 10 M H_2SO_4 solution.

Answer: We need to know the amount of 10 M H_2SO_4 to add to the 100 mL volumetric flask prior to the addition of diluting solvent (i.e., H_2O). Actually water should never be added to a concentrated acid because of potential for splattering. Hence some initial solvent (i.e., water) should be added to the flask prior to the addition of acid.

Need: V_1 or mL?
Know: $M_1 = 10$ mol/L, $M_2 = 0.1$ mol/L, and $V_2 = 100$ mL.
Hence: $M_1V_1 = M_2V_2$

$$10 \text{ mol/L} \times V_1 = 0.1 \text{ mol/L} \times 100 \text{ mL}$$

$$V_1 = \frac{0.1 \text{ mol/L} \times 100 \text{ mL}}{10 \text{ mol/L}} = \frac{(0.1)(100)}{10} \text{ mol/L} \cdot \text{mL}/1 \cdot \text{L/mol}$$

$$V_1 = 1.0 \text{ mL of 10 } M \text{ } H_2SO_4 \text{ is required}$$

Preparing a Molar or Normal Solution from a Solid

As before, we can develop an equation for diluting a solid material:

$$(\text{moles solute})_{BD} = (\text{mole solute})_{AD}$$

$$\left(\frac{\text{g solute}}{\text{g/mol (solute)}}\right)_{BD} = \left(\frac{\text{mole solute}}{\text{L solution}} \times \text{L solution}\right)_{AD}$$

$$\left(\frac{\text{wt}}{\text{molecular weight}}\right)_{BD} = (\text{molarity} \times \text{volume})_{AD}$$

$$\frac{\text{wt}}{\text{MW}} = MV$$

The above equation can be used for preparing a standard solution based on molarity by diluting (or dissolving) a solid material.

EXAMPLE 1.26.   Prepare 100 mL of a 0.1 M solution of NaOH from solid NaOH pellets.

Answer:

Need: Weight of NaOH pellets or g of NaOH?
Know: "Molecular weight" or formula weight of NaOH:

$$1Na + 1O + 1H = 23\ g + 16\ g + 1\ g = 40\ g/mol$$

$$M = 0.1\ mol/L;\ V = 100\ mL\ or\ 0.1\ L$$

Hence:

$$\frac{wt}{MW} = MV$$

$$\frac{wt}{40\ g/mol} = 0.1\ mol/L \times 0.1\ L$$

$$wt = 0.1\ mol/L \times 0.1\ L \times 40\ g/mol$$

$$wt = (0.1)(0.1)(40)\ mol/L \cdot L \cdot g/mol$$

$$wt = 0.4\ g\ (of\ NaOH\ pellets\ needed)$$

Note that normality can be used to prepare a standard solution from a solid material:

$$\frac{weight}{equivalent\ weight} = normality \times volume$$

$$\frac{wt}{EW} = NV$$

However, normality units (Eq/L, g/Eq) cannot be mixed with molarity units (mol/L, g/mol).

MOLALITY

The molality of a solution is calculated by

$$molality = \frac{moles\ of\ solute}{kilograms\ of\ solvent}$$

Hence a 1 molal solution would contain 1 mol of solute in 1 kg of solvent.

PROBLEMS BASED ON CHEMICAL REACTIONS

The following equations can be used for titration reactions where a solution of known concentration reacts with a solution of unknown concentration:

$$NV = NV$$

and

$$\frac{wt}{EW} = NV$$

Usually a color change indicator is used to indicate when an equivalence point is reached between reactants. In acid–base neutralization, the indicator (e.g., phenolphthalein) will turn from colorless to pink when enough base is added (titrated) to neutralize an acid. When the indicator changes color, the following relationship holds:

$$\text{equivalents acid} = \text{equivalents base}$$

These relationships also develop

$$(N \times V)_{acid} = (N \times V)_{base}$$

$$N_a V_a = N_b V_b$$

For solid acids and solid bases being titrated, the following equations are used for calculations:

$$\left(\frac{wt}{EW}\right)_{\text{solid acid}} = (NV)_{\text{base solution}}$$

$$\left(\frac{wt}{EW}\right)_{\text{solid base}} = (NV)_{\text{acid solution}}$$

Oxidation–Reduction Reactions

All of the preceding equations, used for problem solving for acid–base titrations, can be used for oxidation–reduction reactions. Solid oxidants and reductants can be titrated by solutions of oxidants and reductants, respectively, as long as the indicator changes color at the equivalence point. Because of differences in equivalent weight among reactants and products, it is recommended that normality (not molarity) be used exclusively for calculations involving chemical reaction titrations.

EXAMPLE 1.27.   What volume of 0.1 M NaOH is required to neutralize 10 mL of 6 N HCl?

Answer:

Need: Volume of NaOH or mL?

Know: NaOH: 0.1 mol/L × 1 Eq/mol = 0.1 Eq/L = 0.1 N

HCl: V = 10 mL; N = 6 Eq/L

Hence: $(NV)_{base} = (NV)_{acid}$

0.1 Eq/L × V_b = 6 Eq/L × 10 mL

$$V_b = \frac{6 \text{ Eq/L} \times 10 \text{ mL}}{0.1 \text{ Eq/L}} = 600 \text{ mL}$$

EXAMPLE 1.28.   What weight of HCl is neutralized by 15 mL of 5 N NaOH?

Answer:

Need: Weight of HCl or g?

Know: V (NaOH) = 15 mL

15 mL × 1/100 L/mL = (15)(1/100) mL · L/mL = 0.15 L

5 N(NaOH) = 5 Eq/L

$$\text{Equivalent weight (HCl)} = \frac{\text{molecular weight}}{n}$$

$$= \frac{36.45 \text{ g/mol}}{1 \text{ Eq/mol}} = 36.45 \text{ g/Eq}$$

Hence:

$$\frac{\text{wt}}{\text{EW}} = NV$$

$$\frac{\text{wt}}{36.45 \text{ g/Eq}} = 5 \text{ Eq/L} \times 0.15 \text{ L}$$

wt = 5 Eq/L × 0.15 L × 36.45 g/Eq

wt = (5)(0.15)(36.45) Eq/L · L · g/Eq

wt = 27 g

Preparation of Standard Solutions from Hydrated Salts

Ratio and proportion can be used to compute the weight required when preparing a standard solution from a hydrated salt.

EXAMPLE 1.29.   Prepare a 10% solution (W/V) of $CuSO_4$ if only $CuSO_4 \cdot H_2O$ is available.

Answer:

Need: The weight of the monohydrate equivalent to the weight of 10 g of the anhydrous salt.

Know: Molecular weight ($CuSO_4 \cdot H_2O$) = 178 g/mol

Molecular weight ($CuSO_4$) = 160 g/mol

Hence:

$$\frac{\text{molecular weight (anhydrous)}}{\text{molecular weight (hydrate)}} = \frac{\text{weight anhydrous}}{\text{weight hydrate}}$$

$$\frac{160 \text{ g/mol}}{178 \text{ g/mol}} = \frac{10 \text{ g}}{x} \quad (x = \text{weight of hydrate})$$

$$x = \frac{10 \text{ g} \times 178 \text{ g/mol}}{160 \text{ g/mol}} = 11.1 \text{ g}$$

A weight of 11.1 g of the hydrate is diluted in 100 mL to prepare the 10% solution.

OTHER LABORATORY CALCULATIONS

TEMPERATURE CONVERSION

To convert from degrees Fahrenheit (°F) to centigrade (or Celsius), the following formulas are used:

$$°F = \tfrac{9}{5} °C + 32$$

$$°C = \tfrac{5}{9} (°F - 32)$$

CLEARANCE CALCULATIONS

To calculate clearance of serum analytes by the kidney, we assume the mass of the analyte cleared from the plasma equals the mass detected in the urine:

$$(\text{mass analyte cleared})_{\text{plasma}} = (\text{mass analyte detected})_{\text{urine}}$$

$$(\text{concentration} \times \text{volume})_{\text{plasma}} = (\text{concentration} \times \text{volume})_{\text{urine}}$$

If we divide both sides of the above equation by time (T) and define ''clearance'' (Cr) as Vp/T in mL/min, we have

$$(C_p \times V_p/T) = (C_u \times V_u/T)$$

where the "*p*" subscript denotes plasma and "*u*" indicates urine. Substituting for Cr, we get

$$C_p \times Cr = C_u \times V_u/T$$

$$Cr = \frac{C_u \times V_u}{C_p \times T}$$

where C_u and C_p are both in mg/dL, V_u is in mL, and T is in minutes. The above equation for clearance is commonly used to determine creatinine clearance, a very useful renal function test.

EXAMPLE 1.30. A creatinine clearance test was performed. A 12-hour urine specimen had a volume of 900 mL and a creatinine concentration of 85 mg/dL. The plasma creatinine concentration was 1.2 mg/dL. Assume the body surface area for the patient was normal. Calculate the creatinine clearance in mL/min.

Answer: Calculate V/T or the rate of urine output:

$$\frac{900 \text{ mL}}{12 \text{ h} \times 60 \text{ min/h}} = \frac{900}{(12)(60)} \text{ mL/min} = 1.25 \text{ mL/min}$$

Determine clearance:

$$Cr = \frac{85 \text{ mg/dL}}{1.2 \text{ mg/dL}} \times 1.25 \text{ mL/min}$$

$$Cr = 89 \text{ mL/min}$$

The creatinine clearance is 89 mL/min. This value falls within the normal range.

LOGARITHMS

Occasionally in the laboratory, logarithms (logs) are used for calculations, especially in calculations involving pH. A logarithm is a decimal exponent (or power) between 0.0000 and 1.0000 of a base 10 required to produce a number between 1 and 10. Hence $10^{\log}$ = number (between 1 and 10). Logs then are exponents. Examples of logs and numbers between 1 and 10 are shown below:

$$10^{0.0000} = 1.00; \qquad \log 1 = 0.0000$$

$$10^{0.4471} = 3.00; \qquad \log 3 = 0.4471$$

$$10^{0.8451} = 7.00; \qquad \log 7 = 0.8451$$

$$10^{0.9996} = 9.99; \qquad \log 9.99 = 0.9996$$

Since any number can be expressed in scientific notation (i.e., as a number between 1 and 10), logarithm of any number can be determined.

EXAMPLE 1.31.   Determine the logarithm of 999.

Answer: Convert the number to scientific notation

$$999 = 9.99 \times 10^2 = 10^{0.9996} \times 10^2 = 10^{0.9996 + 2.000} = 10^{2.9996}$$

(note: $10^{0.9996} = 9.99$)

Hence: $\log(999) = \log(10^{2.9996}) = 2.9996$

Note: The decimal portion of a logarithm (i.e., "0.9996") is referred to as a "mantissa"; the digit(s) preceding the decimal point (i.e., "2.") is called the "characteristic" and denotes the final position of the decimal point.

Logarithms and pH

The pH of a solution is defined as the negative logarithm of the hydrogen ion $[H^+]$ concentration, or

$$pH = -\log[H^+]$$

EXAMPLE 1.32.   Compute the pH of a 0.1 M solution of HCl.

Answer:

$$HCl \xrightarrow{\text{H}_2\text{O}} \underset{0.1 \text{ mol/L}}{H^+} + \underset{0.1 \text{ mol/L}}{Cl^-}$$

A 0.1 M solution of HCl has a hydrogen ion concentration of 0.1 mol/L. Hence

$$pH = -\log[H^+]$$

$$pH = -\log[0.1] = -\log[10^{-1}]$$

$$pH = -(-1) = +1$$

BIBLIOGRAPHY

Campbell, J. C. and Campbell, J. C. *Laboratory Mathematics: Medical and Biological Applications,* 2nd ed. St. Louis: C. V. Mosby, 1980.

Johnson, M. D. *Problem Solving and Chemical Calculations.* New York: Harcourt, Brace and World, Inc., 1969.

Loebel, A. B. *Chemical Problem Solving by Dimensional Analysis.* Boston: Houghton Mifflin Company, 1974.

Mortimer, C. E. *Chemistry—A Conceptual Approach,* 2nd ed. New York: Reinhold, 1971.

Remson, S. T. and Ackerman, P. G. *Calculations for the Medical Laboratory.* Boston: Little, Brown and Company, 1977.

Routh, J. I. *Mathematical Preparation for the Health Sciences,* 2nd ed. Philadelphia: Saunders, 1976.

Schaum, D. and Rosenbert, J. L. *Theory and Problems of College Chemistry,* 5th ed. New York: McGraw-Hill, 1966.

POST-TEST

Use the atomic weights listed below when calculating problems:

H-1	C-12
S-32	Cl-35.5
O-16	P-31
N-14	Cu-63.5
Ca-40	Na-23
Mg-24.3	

1. Match the following:

    _____ A. kilo   1. 100
    _____ B. deca   2. 10
    _____ C. hecto   3. 10^{-9}
    _____ D. deci   4. 10^{-6}
    _____ E. nano   5. 10^{-3}
    _____ F. pico   6. 10^{-2}
    _____ G. micro   7. 1000
    _____ H. milli   8. 10^{-1}
    _____ I. centi   9. 10^{-12}

2. Match the following:

    _____ A. 0.0001   1. 3.5×10^{-2}
    _____ B. 0.00000001   2. 10^{-2}
    _____ C. 0.01   3. 0.35×10^{-2}
    _____ D. 10   4. 4.98×10^{3}
    _____ E. 100   5. 10^{1}
    _____ F. 1000   6. 10^{3}
    _____ G. 0.000000001   7. 10^{-4}
    _____ H. 600   8. 10^{-8}
    _____ I. 4980   9. 10^{2}
    _____ J. 0.035   10. 10^{-9}
                               11. 6×10^{2}
                               12. 6×10^{-2}

3. In the metric system, the basic unit for mass is _____ ; the basic unit for length is _____ ; and the basic unit for volume is _____ .

4. How many significant figures do the following numbers contain?

    _____ 0.016
    _____ 1.016
    _____ 0.160

5. Add the following numbers which differ in the number of significant figures and record the answer in the correct number of significant figures:

    $$
    \begin{array}{r}
    0.0131 \\
    22.32 \\
    1.04625 \\
    +\ \ 2.0351 \\
    \hline
    \end{array}
    $$

6. Multiply the following numbers which differ in the number of significant figures and record the answer in the appropriate way.

    $.01115 \times 20.1 \times 1.0154 =$ _____

7. Round off the following numbers to three significant figures.

    _____ a. 6.515
    _____ b. 6.445
    _____ c. 6.116
    _____ d. 6.165

8. A pediatric patient scheduled for a glucose tolerance weighed 44 pounds. Pediatric patients receive 2 g of glucose/kg of body weight. This patient should receive _____ grams of glucose.

9. Determine the molecular weights of the compounds shown.

    a. $H_2N-\overset{\overset{\displaystyle O}{\|}}{C}-NH_2$ _____ g/mol
    b. $CaCl_2$ _____ g/mol
    c. H_3PO_4 _____ g/mol
    d. $Ca_3(PO_4)_2$ _____ g/mol
    e. $CuSO_4 \cdot 5\ H_2O$ _____ g/mol

10. To prepare 100 mL of a 1 M solution of the following, you should weigh:

    a. _____ g $CaCl_2$
    b. _____ g $Ca_3(PO_4)_2$
    c. _____ g $CuSO_4 \cdot 5\ H_2O$

11. Add the following:
    120 mg + 80 μg + 0.01 g = _____ mg.

12. One liter = _____ mL = _____ cc = _____ dL. (Note 1 cc = 1 mL.)

13. Glassware calibration is performed at 20°C. This corresponds to _____ °F.

14. Room temperature was recorded as 70°F. This corresponds to _____ °C or centigrade.

15. A calcium was reported as 10.0 mg/dL. This corresponds to _____ mEq/L or _____ mmol/L.

16. A solution contains 4.5 mmol/mL. The molarity of this solution is _____ .

17. To prepare 1 L of a 1 M solution of H_2SO_4 (specific gravity 1.840, % assay 96), you should use _____ mL of acid and dilute to 1 L with deionized water.

    To prepare 1 L of 1 N solution of this same acid, you should use _____ mL of acid and dilute to 1 L with deionized water.

18. There are _____ g of HNO_3 in a solution with a concentration of 5 M.

    There are _____ g of HNO_3 in a solution with a concentration of 5 N.

19. An assay called for 1 M solution of NaCl. Only 10% NaCl was available. Could you use the 10% NaCl to prepare the 1 M solution? Why? Show calculations.

20. Twenty-five milliliters of 2 M H_2SO_4 diluted to 50 mL gives a final *normality* of:
    a. 1 N
    b. 0.5 N
    c. 2 N
    d. 0.2 N

21. Ten milliliters of 50% ethyl alcohol was diluted to 50 mL to give a final concentration of _____% V/V.

22. Fifty grams of sulfosalicylic acid is dissolved and diluted to 1 L with deionized water. The concentration of this solution is _____% W/V.

23. You are to prepare 100 mL of a standard solution containing 140 mEq/L of sodium. The amount of NaCl required would be _____ mg.

24. In performing a gastric analysis, 10 mL of gastric juice required 2 mL of 0.1 N NaOH. The normality of the gastric juice is _____ .

25. The serum that would be present in 400 mL of a 1 : 50 dilution would be _____ .

26. A urine was diluted in the following manner: 5 mL of urine plus 10 mL of water. This dilution was then used to make a second dilution prepared in the following way: 0.5 mL of the first dilution diluted to a total volume of 5 mL. The dilution factor is _____ .

27. The concentration of a stock creatinine standard contained 100 mg/dL; 500 mL of a working standard with a concentration of 0.006 mg/mL was needed. To prepare this solution, you should use _____ mL of the stock standard and dilute to a volume of 500 mL.

28. Convert 2.3 milliequivalents per liter for magnesium to concentration in mg/dL.

ANSWERS

1. (A) 7; (B) 2; (C) 1; (D) 8; (E) 3; (F) 9; (G) 4; (H) 5; (I) 6
2. (A) 7; (B) 8; (C) 2; (D) 5; (E) 9; (F) 6; (G) 10; (H) 11; (I) 4; (J) 1
3. gram, meter, liter
4. 2, 4, 3
5. $0.01 + 22.32 + 1.05 + 2.04 = 25.42$
6. $0.0111 \times 20.1 \times 1.02 = 0.2275$
7. (a) 6.52; (b) 6.44; (c) 6.12; (d) 6.16
8. $44 \text{ lb} \times \dfrac{1}{2.2} \text{ kg/lb} \times 2 \text{ g/kg} = \dfrac{(44)\,(2)}{2.2} \text{ g} = 40 \text{ g}$
9. (a) 60; (b) 111; (c) 98; (d) 310; (e) 249.5
10. (a) 11.1; (b) 31.0; (c) 25.0
11. 130.08
12. 1000, 1000, 10
13. 68
14. 21
15. $10 \text{ mg/dL} \times 10 \text{ dL/L} \times \dfrac{1}{40} \text{ mmol/mg} = 2.5 \text{ mmol/L}$
16. $4.5 \text{ mmol/mL} \times 1000 \text{ mL/L} \times \dfrac{1}{1000} \text{ mol/mmol} = 4.5 \text{ mol/L}$
17. 27.7 mL
18. $5 \text{ mol/L} \times 63 \text{ g/mol} = 315 \text{ g}$
    $5 \text{ Eq/L} \times 63 \text{ g/Eq} = 315 \text{ g}$
19. $10 \text{ g/dL} \times 10 \text{ dL/L} \times \dfrac{1}{58.5} \text{ mol/g} = 1.7 \text{ mol/L}$

    Yes, the 10% solution is more concentrated and could therefore be diluted to produce the 1 M solution.
20. $(25 \text{ mL})\,(2 \text{ mol/L}) = (50 \text{ mL})\,C$

    $C = \dfrac{(25 \text{ mL})\,(2 \text{ mol/L})}{50 \text{ mL}} = 1 \text{ mol/L}$

    $1 \text{ mol/L} \times 2 \text{ Eq/mol} = 2 \text{ Eq/L} = 2\ N$
21. 10%
22. 5
23. Note! One milliequivalent of sodium is produced for each milliequivalent of NaCl dissolved. The milliequivalent weight of NaCl is 58.5.

    $140 \text{ mEq/L} \times 58.5 \text{ mg/mEq} \times \dfrac{1}{10} \text{ L/dL} = 819 \text{ mg}$
24. $0.02\ N$
25. 8 mL of serum

26. First dilution $= \dfrac{5}{15} = \dfrac{1}{3}$; second dilution $= \dfrac{0.5}{5} = \dfrac{1}{10}$;

   Final dilution $= \dfrac{1}{3} \times \dfrac{1}{10} = \dfrac{1}{30}$

27. Use formula $V_1 C_1 = V_2 C_2$; since C has to be expressed in the same units, change the concentration of the stock to the same units as the working concentration (i.e., mg/mL) before substituting in the formula.

   $(V_1)\,(1) = (500)\,(0.006)$; $V_1 = 3$ mL of stock diluted to 500 mL

28. 2.3 mEq/L $\times \dfrac{1}{10}$ L/dL $\times 12$ mg/mEq $= 2.8$ mg/dL

CHAPTER 2
QUALITY CONTROL AND STATISTICS

OBJECTIVES

On completion of this chapter, you will be able to:

1. Define, illustrate, or explain the following terms:
    a. Histogram
    b. Gaussian distribution
    c. Accuracy
    d. Precision
    e. Predictive value
    f. Mean
    g. Bias
    h. Regression line
    i. "Student's" t test
    j. F test
2. Define and calculate the various statistical parameters commonly used in quality control programs.
3. Discuss the evaluation of plotted quality control data and suggest possible corrective actions when errors are encountered.
4. Discuss the basic principles of a method comparison study and define the meaning of statistical parameters generated by the study.
5. Discuss the predictive value model and associated terms with emphasis on practical applications to the clinical laboratory.
6. Describe how to conduct a "normal range study" and various methods used to compute the normal range.
7. Discuss reagent purity and its relationship to the preparation of standards and diagnostic reagents.

GLOSSARY OF TERMS

Analyte. A substance being measured.

Analytical accuracy. Refers to how closely the measured analyte level approaches the "true" level.

Analytical error. The difference between the measured analyte level and the "true" analyte level. The mean of a number of analyte measurements is the best estimate of the "true" analyte level.

Analytical precision. The agreement between data obtained by repeated analysis of the same specimen or specimen pool. The reproducibility of a measurement.

Constant error. An error that is constant (or fixed) over the entire analytical range.

Control. A solution or specimen, assayed along with patients' specimens, used to ensure the quality and uniformity of the assay.

Diagnostic sensitivity.   Refers to the degree of positivity of a test in association with a specific disease.

Diagnostic specificity.   Refers to the degree of negativity or absence of test positivity in patients not having a particular disease.

Mean.   The average of a group of measurements.

Median.   Refers to the measurement value where the number of observations (or measurements) above the value is equal to the number of observations (or measurements) below.

Mode.   Refers to the measurement value that repeats most often.

Primary standard.   A highly purified chemical that can be weighed directly and used for the preparation of standard solutions to be used for analysis.

Proportional error.   An error that increases as a percentage of the analyte level being measured.

Quality control.   A method used in clinical laboratories to minimize analytical errors.

Standard.   A solution of known concentration used to calculate (or determine) the concentration of other specimens.

True value.   May be approached by the average (or mean) of a number of measurements, on the same sample, using some reference method.

ANALYTICAL ERROR

Every measurement in clinical chemistry has a degree of inaccuracy or error. The actual amount of error is the difference between our measurement and the "true value."

MEAN

Our best estimate of "true value" when using an accurate method is the average of a number of measurements or mean. Accordingly, we will frequently refer to the mean as the "best value." Hence the amount of error in a measurement is estimated as the difference of our measurement from the mean. It is the objective of "quality control" to minimize all levels of error.

ACCURACY

This term refers to how close our measurement is to the true value. Good accuracy on one measurement can be a chance occurrence (or luck).

PRECISION

This term refers to the reproducibility of a measurement.

An example of accuracy and precision is shooting at a target with a rifle: accuracy and precision terms can be used to assess our results (see Figure 2.1).

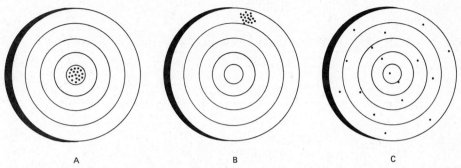

Figure 2.1. Analytical accuracy and precision. (*a*) good accuracy and good precision; (*b*) poor accuracy and good precision; (*c*) good accuracy and poor precision.

GOALS OF QUALITY CONTROL

Our initial goal in quality control in clinical chemistry is to develop methods with high analytical accuracy and precision. Once developed, our second goal is to ensure that the method continues to perform in an accurate and precise manner. We use statistical parameters to evaluate the accuracy and precision of our various analytical methods. In most cases, quality control is based on the evaluation of precision.

MEAN ($\bar{X}$)

The mean of a group of numbers is the computed average of the numbers in the group. To compute the mean of different observations, we must first compute the sum (Σ) of the different observations [($X(1), X(2)$, etc.], then divide the sum by the number of observations (N). The statistical formula to compute the mean is

$$\bar{X} = \frac{\text{sum of observations}}{\text{number of observations}} = \frac{\sum_{I=1}^{N} X(I)}{N} = \frac{X(1) + X(2) + \cdots + X(N)}{N}$$

Note that in the above equation we are using "computer notation" for subscripted variables, with one subscripted variable for each individual observation.

EXAMPLE 2.1. Calculate the mean of the following group of numbers obtained on a calcium control pool: 8.9, 9.1, 8.7, 9.2, 9.9, 8.8, 9.0, 9.3, 9.0, and 8.9 mg/dL.

Answer

Need: Mean or $\bar{X}$?

Know: Number of observations ($N = 10$): $X(1) = 8.9$, $X(2) = 9.1$, $X(3) = 8.7$, $X(4) = 9.2$, $X(5) = 9.9$, $X(6) = 8.8$, $X(7) = 9.0$, $X(8) = 9.3$, $X(9) = 9.0$, $X(10) = 8.9$.

Hence:

$$\bar{X} = \frac{X(1) + X(2) + X(3) + X(4) + \cdots + X(10)}{10}$$

$$\bar{X} = \frac{8.9 + 9.1 + 8.7 + 9.2 + 9.9 + 8.8 + 9.0 + 9.3 + 9.0 + 8.9}{10}$$

$$X = \frac{89.9}{10} = 8.99 \text{ or } 9.0 \text{ mg/dL}$$

As described above, when using an accurate method, the mean of a set of measurements approaches the "best value" of the measurement. Therefore, the mean can be used to calculate the error (or deviation) of each individual measurement from the "best value" by simply subtracting the selected measurement $X(I)$ from the mean ($\bar{X}$):

$$\text{Deviation of single measurement} = \bar{X} - X(I)$$

STANDARD DEVIATION (SD)

This statistic is the average deviation of a set of observations from the mean (or "best value"). To calculate the standard deviation for a set of measurements: (1) determine the mean of the set (as shown in Example 2.1); (2) subtract each measurement from the mean to get the deviation $[\bar{X} - X(I)]$, square the deviation, and add these together $[\Sigma[X - X(I)]^2]$; (3) divide the sum of the (deviation)^2 by the "degrees of freedom" or $N - 1$, giving

$$\frac{\Sigma[\bar{X} - X(I)]^2}{N - 1}$$

and (4) take the square root of the resulting quotient. Hence the complete formula for computing SD is

$$SD = \sqrt{\frac{\Sigma[\bar{X} - X(I)]^2}{N - 1}}$$

EXAMPLE 2.2.   Compute the standard deviation of the following four numbers: 8.9, 9.1, 8.7, and 9.2 mg/dL.

Answer:

Need: Standard deviation or SD?
Know:

$$\bar{X} = \frac{\Sigma X(I)}{N} = \frac{8.9 + 9.1 + 8.7 + 9.2}{4} = 8.97 \text{ or } 9.0$$

Hence:

$$SD = \sqrt{\frac{\Sigma[\bar{X} - X(I)]^2}{N - 1}}$$

$$SD = \sqrt{\frac{(9.0 - 8.9)^2 + (9.0 - 9.1)^2 + (9.0 - 8.7)^2 + (9.0 - 9.2)^2}{4 - 1}}$$

$$SD = \sqrt{\frac{0.01 + 0.01 + 0.09 + 0.04}{3}} = \sqrt{0.05} = \pm 0.22 \text{ mg/dL}$$

DISTRIBUTION OF DATA

FREQUENCY DISTRIBUTIONS

Whenever handling large amounts of quality control data, it is helpful to break the data into subclasses and plot the frequency of occurrence in each subclass.

Histogram

The bar graph plot of a frequency distribution is called a histogram.

EXAMPLE 2.3.   Plot a histogram for the following creatinine kinase (CK) data obtained in 25 separate assays of an aliquoted specimen pool.

Range	Frequency	Values (U/L)
55–59	1	57
60–64	2	62, 61
65–69	5	65, 68, 66, 66, 69
70–74	9	72, 73, 72, 73, 74, 70, 71, 72, 74
75–79	5	76, 78, 75, 78, 75
80–84	2	84, 81
85–90	1	86

The histogram for this CK data is shown in Figure 2.2.

Gaussian Distribution.

The "bell-shaped" curve drawn connecting the histogram cells is called a Gaussian curve (see Figure 2.2). This indicates that the data used to generate the curve constitute a "normal distribution" of data or a Gaussian distribution. Gaussian curves are defined by the equation

$$Y = \frac{1}{SD\sqrt{2\pi}} \, e^{-(1/2)(X - \bar{X})^2/SD^2}$$

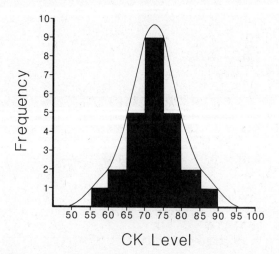

Figure 2.2. A frequency or histogram plot of creatine kinase levels in normal patients. The smooth curve drawn over the histogram depicts a Gaussian distribution.

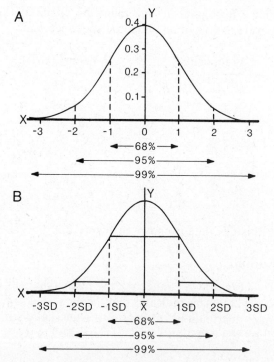

Figure 2.3. Gaussian curves: (a) standard deviation = ±1 and the mean $\bar{X} = 0$; (b) general Gaussian curve used in clinical chemistry.

Using the above relationship, if we plot data where the $\bar{X} = 0$ and the $\underline{SD} = \pm 1$, Y will get larger and approach 0.4 as X approaches zero (or $Y = 1/SD\sqrt{2\pi}$ where SD $= \pm 1$ and $X = 0$). Figure 2.3 shows a normal Gaussian curve (or Gaussian distribution).

Central Tendency

As shown in Figure 2.3, in a Gaussian distribution, data clusters around the mean or has a "central tendency."

$$68\% \text{ of data: } \bar{X} \pm 1\text{SD}$$

$$95\% \text{ of data: } \bar{X} \pm 2\text{SD}$$

$$99\% \text{ of data: } \bar{X} \pm 3\text{SD}$$

Note further from Figures 2.2 and 2.3 that Gaussian data form a symmetric curve around the mean. The width of the symmetric curve is determined by the standard deviation of the data set (see Figure 2.4).

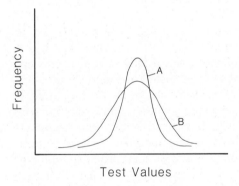

Figure 2.4. Central tendency of Gaussian curves based on standard deviation from the mean.

Median and Mode

Median refers to the central value in a set of measurements. Mode refers to the measurement that has the highest frequency of occurrence in the set of measurements.

EXAMPLE 2.4. What is the mean, median, and mode in the CK measurements listed in Example 2.3?

Answer: The 13th number in the data set is 72, hence the *median* is 72 U/L. The measurement occurring most frequently is 72, hence the mode is 72 u/L as well. The computed mean is 72 u/L. In a true Gaussian distribution, the mean, median, and mode are equal.

CONTROL SPECIMENS

In the chemistry laboratory, quality control is based on data collected by the repeated analysis of a "control" specimen. The control specimen is always assayed along with patient specimens under identical determination conditions. Based on the assay value obtained on the control, the analyst decides whether the assay is "in-control" or "out-of-control." The concomitant patients' data obtained in an out-of-control assay would, of course, not be reported.

REQUIREMENTS FOR A CONTROL

An acceptable control specimen has the following requirements: (1) the control should be stable for the analyte over a period of months or years; (2) the control must have analyte levels suitable to the assay range; (3) the control should closely parallel the constituency of the patients' samples; and (4) the control must be assayed in precisely the same manner as the patients' samples.

QUALITY CONTROL RANGES

COMMERCIAL LYOPHILIZED CONTROLS

Generally, quality control (QC) in the chemistry laboratory can be classified in different categories. Internal laboratory quality control is based on means and standard deviations generated by a control material tested over an extended period of time. Commercially available control materials are lyophilized for stability and have "normal" and "abnormal" target values. These commercial controls can be "assayed" or "unassayed." Most of the constituents found in human serum will be found in commercial lyophilized control material. Assayed controls have target values along with expected ranges based on measurements obtained by the manufacturer's reference laboratories. The assayed values listed are used as a guide in quality control programs; they are not the values to be used on the QC charts. It is best to develop "in-house" quality control ranges whether the commercial control material being used is assayed or unassayed. Liquid control materials are also available commercially.

IN-HOUSE PREPARATION OF QC SERA POOLS

In smaller laboratories, a serum pool can be used for a QC program. This pool can be prepared as follows: (1) pool daily serum samples after analyses are completed (do not use hemolyzed or lipemic samples); (2) pour sera through a plug of glass wool into a plastic bottle; (3) store bottle in a freezer at or below $-20°C$; (4) after at least a six-month supply is collected, thaw and mix thoroughly; (5) adjust concentration of certain constituents (e.g., bilirubin, glucose); (6) adjust pH of sera to pH

7.4 with concentrated sulfuric acid; (7) measure aliquots of sera into test tubes, sufficient for each day's use; and (8) stopper tightly with a soft rubber stopper and store in a freezer at $-20°C$ or preferably $-80°C$.

QC ranges (or limits) for the in-house pool are calculated for each control by (1) assaying the material repeatedly in separate runs; (2) determining the mean of all the control measurements; (3) adding 2SD to the mean to get the upper end of the QC range; and (4) subtracting 2SD to get the lower cut-off of the QC range:

$$\bar{X} + 2SD \qquad \text{Upper QC limit}$$

$$\bar{X} - 2SD \qquad \text{Lower QC limit}$$

Assuming a Gaussian distribution for the control data (see Figure 2.3), 95% of all subsequent control values or 19 out of every 20 results should fall within these QC limits.

EXAMPLE 2.5.   Calculate the 95 percent confidence range for a calcium control given that the mean is 9.0 mg/dL and the SD is 0.22 mg/dL.

Answer:

$$\text{Upper limit} = \bar{X} + 2SD = 9.0 + 2(0.22) = 9.4 \text{ mg/dL}$$

$$\text{Lower limit} = \bar{X} - 2SD = 9.0 - 2(0.22) = 8.6 \text{ mg/dL}$$

Values above or below 9.4 and 8.6, respectively, would indicate that the method may be "out-of-control." Stated another way, there is only a 5 percent chance that one control value could fall beyond these 95 percent limits and the method be "in-control."

PRACTICAL APPLICATIONS OF QC DATA

As mentioned earlier, quality control ensures uniformity of measurement. It is used in the laboratory to ensure the analyst that the results from his or her analyses are accurate and precise.

Quality control can provide the first indication that assay results may not be reliable due to factors such as standard deterioration, an improperly prepared reagent, or other alterations in a procedure. Quality control helps verify the accuracy and precision of reported test values by validating (1) instrument performance, (2) reagent performance, and (3) human performance.

LEVY–JENNINGS CHARTS OR SHEWHART CHARTS

QC data can be readily evaluated visually if plotted on appropriate QC charts. One chart is needed for each method and control, with control results being plotted over a specified period of time (see Figure 2.5).

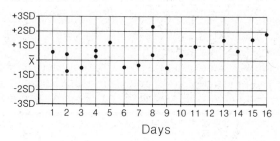

Figure 2.5. A Levy–Jennings chart or Shewhart chart for the evaluation of daily QC data.

Use of Levy–Jennings Charts

Shifts

Generally, controls vary around the mean in a symmetrical Gaussian fashion when plotted on a Levy–Jennings chart. However, if six successive plots fall on the same side of the mean line, the procedure is considered out-of-control. This condition is known as a shift. The assay method should be examined to identify the problem and correct it. Examples of problems causing shifts would include: (1) a deteriorated standard that is maintaining a constant level (i.e., significant evaporation of the standard solvent); (2) a newly prepared standard prepared with a wrong concentration; (3) a reagent that has shifted to a new level of activity; (4) in kinetic enzyme analysis, changes in water bath temperature, timing of the measurements, or changes in concentration or pH of a substrate; (5) a wavelength calibration shifts on the spectrophotometer; and (6) an uncalibrated semiautomated or automated pipette.

Trend

Another common situation seen on the Levy–Jennings chart is referred to as a trend. A trend is indicated by six successive plots being distributed in one general direction. An example of a trend is given in Figure 2.5, days 10 through 16. Trends may be "upward" or "downward" depending on the procedural problem. Trends may be caused by (1) deterioration of one or more reagents, especially the standard (slow downward trends are often due to slow evaporation of the standard); (2) improper wavelength calibration; (3) slight alterations of the pH of an enzyme reagent due to CO_2 absorbances from the air or other contamination; and (4) water bath temperature shifts.

When using control values to calculate statistics for the next month's Levey–Jennings chart, one should not include control values that indicated the shift and the trend. These values were obtained in "out-of-control" situations and should not enter into the calculations for the mean and standard deviations of the next month's chart. However, outliers due to chance occurrence should be included in the next month's cumulative mean and SD.

Random Error

Extreme day-to-day variation of control levels are sometimes seen. An erratic pipette (or pipettor) or a spectrophotometer malfunction is often the cause of the

QC problem. Also, improper reconstitution of the lyophilized control material or inadequate mixing of the reconstituted control causes erratic control results.

Control Outliers
If an occasional control value falls outside of the established mean ±2SD range, the method may be out-of-control. On the other hand, Gaussian theory suggests that 5 of 100 controls tested will be beyond 2SD and 1 control out of 100 tested will be beyond 3SD. Hence, if a single control is beyond the control range, a common practice is to repeat the assay. If the second assay produces a control value beyond the acceptable range, the test procedure should be judged "out-of-control." Patients' values should not be reported. In Figure 2.5, a control was beyond 2SD on day 8. On repeating the assay, the control was in range and near the mean. Therefore, the previous control beyond 2SD was considered an "outlier," hence the "run" was judged to be in-control, and patients results were reported.

WESTGARD RULES

The "Westgard rules" for evaluating QC data involve the simultaneous use of a combination of QC rules. These rules allow for the detection of random analytical errors versus systematic analytical errors. These rules are described below:

"1_{2S}" *Rule.*   This is the "warning rule" where one control exceeds the $\bar{X} \pm$ 2SD range. As discussed earlier, the event stimulates further investigation to determine if the method is "in-control" or "out-of-control."

"1_{3S}" *Rule.*   This rule symbolizes that a control is beyond the $\bar{X} \pm$ 3SD limits. The run should be rejected when this occurs.

"2_{2S}" *Rule.*   This rule states that the run should be rejected when two consecutive controls exceed either the $\bar{X}$ + 2SD or $\bar{X}$ − 2SD limits.

"R_{4S}" *Rule.*   This rule requires rejection of a run when the difference between controls exceeds 4SD. For example, if one control exceeds the $\bar{X}$ + 2SD limit and a second exceeds the $\bar{X}$ − 2SD limit, the difference between the two controls would be 4SD.

"4_{1S}" *Rule.*   This rule requires rejection of a run when four consecutive controls exceed either the $\bar{X}$ + 1SD or $\bar{X}$ − 1SD limit.

"$10\bar{X}$" *Rule.*   This rule states that when 10 consecutive controls fall on the same side of the mean, the run should be rejected.

These Westgard rules can be included in a computer program to evaluate QC data. The programs are currently commercially available. Rules 1_{3S} and R_{4S} detect random analytical error; rules 2_{2S}, 4_{1S}, or $10\bar{X}$ are more sensitive to systematic analytical errors. Systematic errors can be either proportional systematic error or constant systematic error. (These error types will be discussed later in this chapter.)

COMPUTER PROGRAMS FOR QUALITY CONTROL

Microcomputer programs have been developed using QC "rules" to evaluate QC data. However, laboratories have found the manual entry of QC data laborious. Most large laboratory minicomputers collect QC data directly from instruments. In addition, various QC ranges are handled by the computer in order to identify potential analytical and clinical problems. For example, laboratory minicomputers, on entry of any analyte result, perform (1) a technical limits check—where the analyte result is checked to see if it exceeds the upper limit of linearity or the lower limit of analytical sensitivity of the method; (2) a QC check—where a QC result is flagged if it is beyond the $\bar{X} \pm 2SD$ range; (3) a "panic" result check—where a patient's result is flagged to indicate a potential life-threatening condition that needs immediate attention; (4) a delta check—where a patient's current test result is checked against a previous result; if an unreasonable shift or change has occurred, the result is flagged (here the patient is being used as his or her own control); and (5) a "ridiculous" result check—where a test result is flagged if it reflects an impossible (or improbable) analyte result (this check will call attention to data keying errors, in particular, errors of decimal point position). In addition, when the test is initially ordered, the laboratory computer scans the current test file to see if the new test request is a duplicate order error. This error is corrected by notifying the floor or the doctor who ordered the duplicate test, then cancelling the request.

Computerized Laboratory QC Reports

Computer printouts available on laboratory computer systems include (1) a daily printout of all QC data; (2) a printout of all abnormal patients' results; and (3) weekly and monthly QC summary reports including Levy–Jennings plots of QC data. When hard copy reports are not needed, these QC data can be reviewed on a video display terminal. Cumulative means and SD on QC data should be evaluated each month in terms of previous months' means and SDs. Significant shifts in the cumulative monthly means and SDs indicate an unacceptable procedure.

EXTERNAL QUALITY CONTROL

While evaluation of QC data on a day-to-day basis within the laboratory verifies the ongoing precision of the assay, accuracy can be measured by comparing values obtained in the laboratory with values known to be correct. Various organizations such as the College of American Pathologists (CAP, Skokie, IL) conduct proficiency testing for the assessment of accuracy of laboratory methods. "Unknown" specimens are sent to participating laboratories for analysis. Computer analysis of all of the generated data from participating laboratories includes the tabulation of means and standard deviations for each method. Based on these data, CAP will flag those analytes that are beyond acceptable limits for the method and instruments being used for the analyte.

Also, most commercial control vendors furnish computer printouts listing the constituent means and standard deviations of all laboratories using the identical lots of lyophilized control material. Comparisons of internal QC means and ranges can then be made with other laboratories in the group using similar equipment, reagents, and methods.

EVALUATION OF NEW PROCEDURES

Evaluating a new procedure presents a different QC problem. For a new method, an initial literature search should be performed so that the entire method can be understood prior to any testing. Possible reasons for evaluating a new procedure may be that the new procedure is easier to perform, is less costly, or has other advantages excluding greater accuracy and precision over the old procedure. When the latter is true, a Patient Comparison Study (or Correlation Study) is performed involving testing 50 or more patients by both the old and new methods.

Other reasons for evaluating a new test would involve a new method that affords greater accuracy and precision, hence in this case Patient Comparison Studies may not be performed at all. The analyst will concentrate on the comparisons of accuracy and precision data between the old and new method. In this situation, a new normal range study will be required. (Normal range studies will be discussed later in this chapter.)

PATIENTS' COMPARISON STUDY (CORRELATION STUDY)

At least 50 patient specimens should be obtained and arranged in small batches and tested by both the new method (test method) and the old method currently being used (reference method). Specimens collected and tested by both methods should cover a wide range of values (i.e., low, normal, and elevated analyte levels).

Correlation Coefficient (r)

Where paired data are used to compare two methods (a test method and a reference method), the calculated correlation coefficient represents the "linear correlation" between the two methods. The correlation coefficient reveals the significance of the relationship of the two methods being compared. Correlation coefficients are not valid when data over a narrow analytical range are being compared.

The correlation coefficient has values ranging from -1 to $+1$. In perfect positive linear correlation between two methods, $r = 1$; in perfect negative correlation, $r = -1$. When there is no linear correlation between data sets, $r = 0$. The correlation coefficients are computed from the following equation:

$$r = \frac{\Sigma(X - \bar{X})(Y - \bar{Y})}{\sqrt{\Sigma(X - \bar{X})^2(X - \bar{Y})^2}}$$

Evaluating Correlation Coefficients

Correlation coefficients can be evaluated by (1) squaring "r"; (2) subtracting the results from 1.00; and (3) multiplying the remainder by 100 to get the percentage of time when values are not in control.

EXAMPLE 2.6.   Two methods being compared to one reference method had (a) $r = 0.85$ and (b) $r = 0.93$. Which method had fewer results apparently out-of-control?

Answer:

(a) $1.00 - (0.85)^2 = 0.28$ or 28%
(b) $1.00 - (0.93)^2 = 0.14$ or 14%

Method b has r closer to 1.00 and 14% of results apparently out-of-control compared to 28 percent for method a.

Plotting Correlation Data

Data should be compared graphically by plotting the test method on the abscissa (X axis) and reference method on the ordinate (Y axis) with the same units and scale and drawing the "best fit" line.

By plotting the data and connecting points, we get a linear relationship between the test and reference methods as shown in Figure 2.6. The equation for the best fit line is

$$Y = mX + b$$

where Y = test method (dependent variable)

X = reference method (independent variable)

m = slope of the line

b = Y-intercept

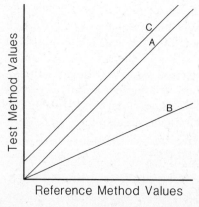

Figure 2.6.   Linear correlation plots between two methods. A-depicts no proportional or constant systematic error; B-indicates systematic proportional error; and C-indicates systematic constant error and no systematic proportional error.

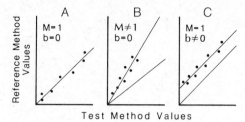

Figure 2.7. Linear correlation plots between two methods. Values of slope and Y-intercepts are shown in A, B, and C.

The slope of the line can be determined graphically by (1) selecting two points on the best fit line; (2) subtracting the two Y coordinates ($Y_2 - Y_1$); and (3) dividing the results by the difference in the X coordinates ($X_2 - X_1$):

$$m = \frac{Y_2 - Y_1}{X_2 - X_1}$$

The Y-intercept can be read directly from the plot (see Figure 2.6) by noting the value on the Y axis that is intercepted by the plotted line.

Appearance of the Comparison Line

In a method comparison study where there is no error indicated between the two methods, the $m = 1$ and $b = 0$. However, the method comparison study usually reveals proportional and constant errors where $m \neq 1$ and $b \neq 0$ (see Figure 2.7).

Least Squares Analysis

Using a statistical calculator or computer, one can calculate the value of the slope and Y-intercept by solving the following regression equations:

$$b = \frac{(\Sigma Y)(\Sigma X^2) - (\Sigma X)(\Sigma XY)}{N \, \Sigma X^2 - (\Sigma X^2)}$$

$$m = \frac{N \, \Sigma XY - (\Sigma X)(\Sigma Y)}{N \, \Sigma X^2 - (\Sigma X)^2}$$

These two equations were developed from the theory of "least squares" used for "fitting" linear data, where N is the total number of paired values; ΣX is the sum of reference procedure values; ΣY is the sum of test procedure values; ΣX^2 is the sum of the squares of the reference procedure; ΣY^2 is the sum of the squares of the test procedure; and ΣXY is the sum of the products of the paired values.

Standard Error (Sy)

The standard error of the estimate can be calculated from the sum terms just described where

$$Sy = \sqrt{\frac{\Sigma d^2}{N}}$$

and

$$\Sigma d^2 = \Sigma Y^2 - (b\, \Sigma Y + m\Sigma XY)$$

Sy is a statistical parameter that indicates the "scatter" of data points about the calculated regression line. A large Sy value indicates that the collective distance between the multiple data points and the line is significant. This observation suggests there is appreciable "random error" between the methods being compared. A large Sy also indicates that the calculated regression line slope and Y intercept may not be valid.

Proportional Error

Proportional error is error between methods that increases as assay values increase. In other words, the difference between results obtained on the test and reference methods gets larger as the level increases (see Figures 2.6 and 2.7, B examples). Proportional error is often due to calibration differences between methods being compared (see Figure 2.7). The percentage proportional error is the absolute value of the difference of the regression line slope and 1.0, times 100.

The proportional error, as indicated by the regression line slope, may not be valid if (1) the spread (or range) of the data is small; (2) if outliers are present in the data; (3) if the data are not linear; or (4) if the standard error of the estimate Sy is large. (This event indicates that random scatter of points around the regression line is large.)

EXAMPLE 2.7.   Compute the proportional error for a comparison study with a slope of (a) 0.93 or (b) 1.21.

   Answer:

(a) $|\, 1.00 - 0.93\, | \times 100 = 7\%$
(b) $|\, 1.00 - 1.21\, | \times 100 = 21\%$

Constant Error

In constant error, unlike proportional error, the difference or error between methods does not vary, regardless of the level of analyte being analyzed. In the absence of proportional error, constant error is equal to the Y-intercept. It is also equal to the bias between the methods.

Bias

The bias between two methods being compared is equal to the test-method mean minus the reference-method mean:

$$Bias = \bar{Y} - \bar{X}$$

Therefore, if the test method gives higher patients' results than the reference method, the method has a "positive bias." In "negative bias," the test method gives lower patients' values than the reference method.

EXAMPLE 2.8.   Calculate the bias of two methods being compared where $\bar{X} = 36$ mg/dL and $\bar{Y} = 21$ mg/dL.

Answer:

Bias $= \bar{Y} - \bar{X} = 21 - 36 = -15$ mg/dL (Negative bias for the test method.)

Student's *t* Test

To evaluate statistically the bias between two methods, the Student's *t* test can be used. The paired *t* value can be calculated from means, frequency (N), and standard deviations (SD) of two methods being compared:

$$t = \frac{\bar{X} - \bar{Y}}{\sqrt{\left(\dfrac{(N_y - 1)SD_x^2 + (N_y - 1)SD_y^2}{N_x + N_y - 2}\right)\left(\dfrac{1}{N_x} + \dfrac{1}{N_y}\right)}}$$

Critical *t* Values

In the statistical tables for the Student's *t* distribution, the critical value of *t* is 2.02 for $n = 40$–60 degrees of freedom at a 95% confidence level. If the calculated *t* is greater than the critical *t* value, the bias is said to be statistically significant (or real). If the computed *t* is less than critical *t* value, we cannot say with 95% confidence that the bias is statistically significant.

Systematic constant error is markedly influenced by proportional error, hence the evaluation of the bias using the Student's *t* test may be invalid in the presence of significant proportional error.

COMPARISON OF METHOD PRECISION

Discussions of regression line slope, bias, Y-intercept, and correlation coefficient are used to compare the test method with the reference method primarily with regard to accuracy. An evaluation of precision can also be made using coefficient of variation (CV), standard error of mean, standard deviation, and F test.

Coefficient of Variation (CV)

The CV statistic is an estimate of the precision of an assay and generally must be below 5.0% for acceptable precision. To calculate CV, the standard deviation is divided by the mean:

$$CV(\%) = \frac{SD}{\bar{X}} \times 100$$

EXAMPLE 2.9.   Calculate the coefficient of variation for a control with a mean of 9.0 mg/dL and a standard deviation of 0.22 mg/dL.

   Answer:

$$CV = \frac{SD}{\bar{X}} \times 100 = \frac{0.22}{9.0} \times 100 = 2.4\%$$

   Interpretation: A CV of 2.4% is well below 5.0%, so the method has "acceptable" precision. If the CV was found to be even lower, the analyst may describe the precision as "good" or even "excellent."

Standard Error of Mean (SEM)

Another statistic for the evaluation of precision (used much less than CV) is "standard error of mean." The SEM is calculated by dividing the standard deviation by the square root of the number of observations.

$$SEM = \frac{SD}{\sqrt{N}}$$

Variance

The variance of a method is equal to the standard deviation squared.

$$Variance = SD^2$$

Variance is related to the "width" of the Gaussian curve. Hence, if the curves in Figure 2.4 represented distribution obtained in a comparison study, method "a" would have a lower variance than method "b". The F test is used to evaluate the variance between two methods. The F statistic is equal to the larger variance divided by the smaller variance:

$$F = \frac{\text{Larger variance}}{\text{Smaller variance}} = \frac{SD^2}{SD^2}$$

The greater the value of F, the more likely the variances differ significantly. On the other hand, as F approaches 1, it is more probable that the two variances do not differ significantly. Like the t test previously described, critical F values are listed relative to degrees of freedom. A separate table is listed for each level of confidence (i.e., a 1 percent chance occurrence). If the computed F is less than the critical F, then the difference in the variance is not statistically significant at the 1 percent level.

EXAMPLE 2.10.   Two methods are compared by testing 24 specimens: method a has a SD of ± 2.3 mg/dL; method b has a SD of ± 2.4 mg/dL. Calculate F and evaluate the variance if the critical F at ($p = 0.01$) is 2.7.

Answer:

$$F = \frac{(2.4)^2}{(2.3)^2} = \frac{5.76}{5.29} = 1.09$$

The computed F of 1.09 is less than 2.7, hence the observed difference in the variance is not significant at a 1% level.

RECOVERY STUDIES

A systematic proportional error is a fixed amount of error, by percentage, and increases as the amount of analyte increases. A recovery experiment is utilized to check the calibration of the test method. Accuracy studies when there is no reference method for comparison often require the use of recovery studies where the analytical balance is used as a reference. The purified analyte is accurately weighed, mixed with a suitable solvent of base pool, and assayed in triplicate by the method being evaluated. The base pool often consists of pooled serum containing a very small amount of the analyte, therefore, addition of the weighed analyte to aliquots of the base pool in increasing amounts can test the useful analytical range of the assay for accuracy. Percentage recovery for each recovery specimen can be calculated:

$$\text{Recovery percent} = \frac{\text{sample level} - \text{base pool level}}{\text{amount added}} \times 100$$

Recovery Experiment

In order not to disturb the protein matrix of the serum specimen, a very small volume of the "spiking" solution should be added to the specimen (or base pool) in recovery experiments. When weighed, powdered materials are dissolved directly in serum, disruption of the protein matrix is usually not a problem.

In an acceptable recovery study, all "spiked" specimens should have recoveries ranging from 90% to 110%.

Linearity

Using recovery studies, a method may not be linear beyond a certain concentration of analyte (see Figure 2.8). Therefore, measurements beyond this upper limit of linearity may be reported as greater than the upper limit or, preferably, diluted and assayed again with diluted analyte results anticipated to fall below the upper limits of linearity.

Interference Studies or Specificity Studies

Whenever systematic constant error is present in the assay, an "interference study" may be indicated. Various substances that may be present in the serum other than the substances being measured are added to the specimen being ana-

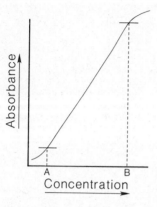

Figure 2.8. The linear assay range. Concentrations above point *B* exceed the upper limits of linearity. Concentrations below point *A* exceed the sensitivity limits of the assay.

lyzed. The assay is then performed. Any increase (or decrease) in analyte levels caused by the addition of the added substance is described as an interference. Methods that are subject to many common interfering substances have low analytical specificity.

Analytical Sensitivity

Analytical sensitivity refers to the least amount of a substance a method can measure or minimum detectable level. It is never appropriate to report a zero level of an analyte. Rather, the results should be reported as less than the sensitivity cut-off for the assay. Sensitivity can often be evaluated with multiple sequential dilutions of the analyte, assaying these dilutions, plotting results, and observing loss of linearity at the low end (see Figure 2.8).

Analytical Specificity

Analytical specificity refers to the method's ability to measure the desired analyte without interference (or crossreactivity) with other substances found in the sample. Early assays in clinical chemistry involved two steps: (1) a separation step and (2) a measurement step. The separation step was often required because of poor method specificity. Also, in urine testing, where many potential interfering substances are present, analytes are extracted in a volatile organic solvent to remove interfering factors. The solvent is then evaporated. The recovered analyte, free of interfering factors and in a more concentrated form, is then reacted with the chromogen. Hence, extraction enhances "specificity," while concentrating the analyte by evaporation of the solvent enhances "sensitivity." Unfortunately, solvent extraction is laborious. Also, extraction efficiency is often quite variable. Hence, extraction methods often have high CVs.

PREDICTIVE VALUE MODEL

Laboratory tests are used primarily by clinicians to rule-in (confirm) or rule-out disease and to monitor therapy. We have previously discussed *analytical* sensitivity and specificity. We will now concentrate on a *diagnostic* sensitivity and specificity.

The ability of a laboratory test to predict the presence or absence of disease is related to the *predictive value* of the test. Evaluation of a new method should involve normal population studies where the distribution of test results for the nondiseased population is established. We should also consider test results on the diseased population and to determine positive results in patients with the disease (disease sensitivity) and negative results in patients with no disease (disease specificity).

The "predictive value model" was first proposed by Bayes in 1763. Later, Vecchio, Gallen, and Gambino applied the predictive value model to laboratory tests (see Table 2.1).

A test has high sensitivity for a disease (*diagnostic sensitivity*) if it is usually positive in patients who have the disease (i.e., the test has high true positives and low false negatives in diseased patients).

A test has high specificity for a disease (*diagnostic specificity*) if it is usually negative in nondiseased patients (i.e., the test has high true negatives and low false positives in nondiseased patients).

POPULATION STUDIES

Population Overlap

The practical use of the Predictive Value Model becomes apparent when we examine Gaussian distributions for healthy (or nondiseased) patients and patients with disease. If, for example, in a patient population where 20% of the patients have disease and 80% do not, a highly predictive test would always be elevated in the diseased patients (or greater than "X," see Figure 2.9a) The test will always be negative (or less than "Y," see Figure 2.9a) in the nondiseased patients. There is no "overlap" between the healthy and diseased populations. The upper end limit for the normal range would obviously set at test level Y.

On the other hand, the more overlap between the diseased and nondiseased population, the lower the overall predictive value of the test (see Figure 2.9b). In this situation, if we set the upper limit at "Y," the test would have many false negatives and low sensitivity because a significant number of diseased patients would have negative results (i.e., fall into the nondiseased range). If we set the test upper cut-off at X (the low value for the diseased population), the test would have many false positives and low specificity because a significant number of nondiseased patients would have positive results.

Disease Prevalence

If the prevalence of disease is quite low in a patient population (1%), the predictive value of a positive result would be only 50% in a test with 99% sensitivity. Since disease prevalence in the general population is low, mass screening programs with laboratory tests to detect occult disease would not be feasible. However, when a physician performs a complete history and physical and orders "indicated" laboratory tests, this increases the prevalence of diseased patients in

Table 2.1. Predictive Value Model

Patients	Positive Test Result	Negative Test Result	Totals
Number with Disease	TP[a]	FN	TP + FN
Number without Disease	FP	TN	TN + FP
Totals	TP + FP	TN + FN	TP + FP + FN + TN

[a] *TP* = True positives; FP = False positives; TN = True negatives; FN = False negaties; and the following formulas apply:

$$\text{Diagnostic sensitivity} = \frac{TP}{TP + FN} \times 100$$

$$\text{Diagnostic specificity} = \frac{TN}{TN + FP} \times 100$$

$$\text{Predictive value of a positive result} = \frac{TP}{TP + FP} \times 100$$

$$\text{Predictive value of a negative result} = \frac{TN}{TN + FN} \times 100$$

$$\text{Predictive efficiency of a test} = \frac{TP + TN}{TP + FP + TN + FN} \times 100$$

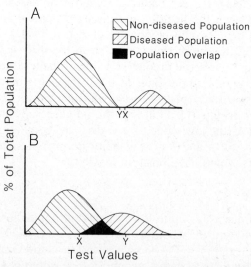

Figure 2.9. Gaussian distributions for the Predictive Value Model for diseased and nondiseased patient populations. (*a*) Gaussian curves show no population overlap; (*b*) Gaussian curves show overlap.

the laboratory test population. Accordingly, the predictive value of a positive laboratory result is increased.

EXAMPLE 2.11.   Fill in the predictive value table for a laboratory test like Figure 2.9a with no false positives and no false negatives. Assume 80 total patients were tested and the prevalence of disease in the group was 37.5%.

Answer:

$$TP = 80 \times 37.5\% = 30$$

$$TN = 80 \times 62.5\% = 50$$

All predictive value parameters are 100%.

Not surprisingly, most laboratory tests have overlap between diseased and non-diseased patients.

EXAMPLE 2.12.   A laboratory test for prostatic cancer had the following results on a total patients' population of 155 where the prevalence of prostatic cancer in the group was 32.2%. Complete the predictive value table and compute the predictive value parameters assuming a predictive value of a positive result = 80% and the predictive value of a negative result = 95%.

	Positive Test	Negative Test
Disease present	40	5
Disease absent	10	100
Test totals	50	105

$$\text{Sensitivity} = \frac{40}{45} \times 100 = 89\%$$

$$\text{Specificity} = \frac{100}{110} \times 100 = 91\%$$

$$\text{Predictive value of a positive result} = \frac{40}{50} \times 100 = 80\%$$

$$\text{Predictive value of a negative result} = \frac{100}{105} \times 100 = 95\%$$

$$\text{Test efficiency} = \frac{140}{155} = 90\%$$

Setting Cut-Off Limits

In diseases that are quite serious and incurable, a false positive test result could be psychologically devastating to a patient. Hence, the positive cut-off for test positivity should be set high for high specificity. On the other hand, in serious diseases that are curable, an infrequent false positive could be tolerated and hence the cut-off for test positivity could be set lower (see Figure 2.9*b*) to maximize sensitivity.

NORMAL RANGE STUDIES

NORMAL RANGES

We have previously described how a 95% confidence range can be used to evaluate whether a method is potentially out-of-control. A similar 95% confidence range is used in most laboratories to classify a patient's laboratory results as being "normal" or "abnormal."

EXAMPLE 2.13.   Fifty overtly healthy people were tested for a calcium normal range study. The histogram plot was Gaussian in appearance, and the mean, median, and mode were equal. The mean of the calcium data set was 9.5 mg/dL; the standard deviation was ±0.45 mg/dL. Compute the normal range.

Answer:

$$\text{Upper end cut-off} = \bar{X} + 2\text{SD} = 9.5 + 0.9 = 10.4 \text{ mg/dL}$$

$$\text{Lower end cut-off} = \bar{X} - 2\text{SD} = 9.5 - 0.9 = 8.6 \text{ mg/dL}$$

Hence the 95% calcium normal range for the healthy subjects was 8.6 to 10.4 mg/dL.

POPULATION SELECTION

A normal range study requires the careful selection of an overtly healthy population for testing. Some of the variables in the selection of a "normal" population that may invalidate the study include race, age, sex, weight, nutritional differences, emotional state, climate and geographical location, sampling time, diurnal variation, and recent meals. A close review of the literature will assist the analyst in deciding which variables to eliminate in the selection of the normal population for a particular test. Also, the selection of a normal population should not be biased by the ease of obtaining volunteer blood samples. If possible, the specimens should be collected and assayed immediately after processing, preferably in the same run to minimize storage and instrument variations. Repeated freezing and thawing of serum will invalidate results in most cases.

After acquisition of the analyte data on the normals, a histogram plot is required to ascertain if the data set is Gaussian. If so, the mean, median, and mode will coincide and a symmetrical distribution of data around the mean will be apparent. For Gaussian data, the normal range is computed by $\bar{X} \pm 2SD$ as described earlier.

NON-GAUSSIAN NORMAL RANGES

For non-Gaussian data, normal ranges are best determined using (1) the inspection–interpolation method or (2) using probability plots. A less accurate method involves the elimination of 2.5% of the normal results data at both the high and low ends of the range, and basing the range on the remaining 95% of the data.

Interpolation Method

The following steps are the interpolation method for non-Gaussian normal ranges (described in *Coulter Currents,* Issue No. 1, Jan. 1979): (1) collect at least 120 data points; (2) write down data lowest to highest; and (3) determine low end cut-off level and upper end cut-off level by eliminating normal values less than 2.5% and greater than 97.5%. This method requires interpolation.

EXAMPLE 2.14.   Compute the normal range for a non-Gaussian set of calcium data using the interpolation method. A total of 140 patients were included, hence $N = 140$.

Answer:

Rank	1	2	3	4	. . .	137	138	139	140
Values	8.2	8.3	8.4	8.5	. . .	10.4	10.5	10.6	10.7

Low end cut-off (L):

1. Determine the rank at the 2.5% level:

$$\text{Rank } (2.5\%) = 0.025 \text{ (number of patients} + 1) = 0.025(140 + 1)$$
$$= 0.025(141) = 3.53$$

2. Interpolating to convert the rank to the low end cut-off

$$\left[\begin{bmatrix} 3.0 & 8.4 \\ 3.53 & x \end{bmatrix} \right]$$
$$\begin{bmatrix} 4.0 & 8.5 \end{bmatrix}$$

$$\frac{0.5}{1.0} = \frac{x}{0.1}$$

$$x = 0.05$$

Low end cut-off $= 8.4 + x = 8.4 + 0.05 = 8.45 = 8.4$ mg/dL

High end cut-off (H):

1. Determine the rank at the 97.5% level:

$$\text{Rank (97.5)} = 0.975 (\text{number of patients} + 1) = 0.975(140 + 1)$$
$$= 0.975(141) = 137.5$$

2. Interpolating to convert the rank to the high end cut-off

$$\left[\begin{bmatrix} 138 & 10.4 \\ 137.5 & x \end{bmatrix} \atop \begin{matrix} 139 & 10.5 \end{matrix} \right]$$

$$\frac{0.5}{1.0} = \frac{x}{0.1}$$

$$x = 0.05$$

$$H = 10.4 + x = 10.4 + 0.05 = 10.45 = 10.4 \text{ mg/dL}$$

Hence the normal range for these non-Gaussian data is 8.4–10.4 mg/dL.

Probability Plots

Another method of determining a normal range using nonparametric statistics involves the use of bracketed histogram data and graphical percentile determinations on probability paper. Our normal population calcium data are listed along with the percentile rank for each calcium level in Table 2.2. To determine the normal range, these percentile data are plotted on normal probability paper.

Table 2.2. Percentile Rankings for Each Calcium Level

Calcium Value	Frequency	Accumulated Frequency	(Accumulated Frequency)/ (Total Number)	Percentile
8.5	4	4	4/72	5.55
8.7				5.55
8.9	9	13	13/72	18.05
9.1	16	29	29/72	40.27
9.3	16	45	45/72	62.50
9.5	15	60	60/72	83.33
9.7	7	67	67/72	93.05
9.9	3	70	70/72	97.22
10.1				
10.3	2	72	72/72	100

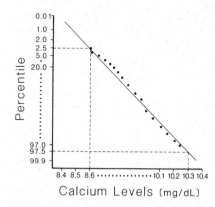

Figure 2.10. A nonparametric plot of normal calcium data on probability paper. Cut-off levels for the upper and lower end of the normal range are indicated.

After plotting the calcium level on the linear abscissa scale along with the percentile rank on the logarithmic ordinate scale and drawing the best line connecting points, the analyst can drop a perpendicular line from 2.5 and 97.5 percentile points and determine the low and high end normal range cut-offs, respectively (see Figure 2.10). For Gaussian normal range data, data points will be essentially linear. Non-Gaussian data will have some points off of the line. For a linear probability plot, the resulting normal range should compare with the $\bar{X} \pm 2SD$ computer normal range.

Concluding, to determine a normal range: (1) select a suitable normal population eliminating undesirable variables (e.g., sex, age, race); (2) assay the analyte in freshly collected specimens; (3) plot a histogram to determine the nature of the distribution and range; (4) if a Gaussian distribution is observed in step 3, calculate the range using $\bar{X} \pm 2SD$; and (5) if non-Gaussian distribution, use the nonparametric approaches to determine the normal range. Also, if possible, selected diseased patients should be tested to ensure that these patients do not have levels overlapping appreciably with the normal range. Based on both the normal and diseased population, the test cut-off for positivity should be refined to maximize either diagnostic sensitivity or specificity as described earlier.

STANDARDS

CONTROLS VERSUS STANDARDS

Up to this point, our discussion of quality control has dealt primarily with controls. As described, controls are used to ensure the uniformity of any measurement. A common error in quality control is the use of a QC specimen for a standard, then using the same material as a control. The control used in this manner is merely another standard and does not monitor the performance of the assay.

In clinical chemistry, the term standard is usually used to indicate a solution used to calculate patients' results. Since most of our measurements in the clinical laboratory are based on spectroscopic measurements and Beer's law, patients' results are calculated from standard data.

EXAMPLE 2.15. Calculate the calcium level obtained on a control that gave absorbance upon addition of cresophthalein complexone of 0.238. The standard (concentration = 10 mg/dL) had an absorbance of 0.260.

Answer:

$$\text{Unknown concentration} = \frac{\text{unknown absorbance}}{\text{standard absorbance}} \times (\text{standard concentration})$$

$$\text{Control Ca} = \frac{0.238}{0.260} \times 10 \text{ mg/dL} = 9.2 \text{ mg/dL}$$

Patients' results for colorimetric methods are computed in a similar manner.

EXAMPLE 2.16. As in Example 2.15, if a patient's sample had an absorbance of 0.280, compute the calcium level.

Answer:

$$\text{Patient's results} = \frac{\text{patient absorbance}}{\text{standard absorbance}} \times (\text{standard concentration})$$

or $$\text{Patient's Ca} = \frac{0.280}{0.260} \times 10 \text{ mg/dL}$$

$$\text{Patient's Ca} = 10.8 \text{ mg/dL}$$

For analysis, it is essential that we accurately weigh standards preferably to the nearest 10^{-5} g (or the nearest one-hundredth of a milligram). A well-maintained analytical balance is required. Aqueous or organic solutions of the standard must be made with certified volumetric glassware that has been adequately cleaned with high purity solvents.

PURITY OF STANDARDS AND REAGENTS

Primary Standards

These standards are highly purified chemicals that can be weighed directly and used for preparation of standard solutions to be used for analysis. The assay purity level of the standard material should be 99.95% or greater. (Note: Hydroscopic materials take on water readily during weighing.)

Secondary Standards

These standards are used commony in clinical chemistry and are often in the form of lyophilized human sera. The concentration of these standards is determined by multiple assays using some reference method calibrated using a primary standard.

Other standards include Certified Standards, certified by various scientific organizations or groups, and Standard Reference Material (SRM), certified by the National Bureau of Standards.

CHEMICAL GRADES

REAGENT AND ANALYTICAL GRADES

Chemicals other than standards (see above) to be used in clinical chemistry analysis must be of the highest degree of purity. Accordingly, "Reagent Grade" or "Analytical Grade" materials must meet American Chemical Society standards for purity with acceptable limits of trace impurities being established and printed on the container label. Analytical grade reagents must be used for routine analytical work.

CHEMICALLY PURE GRADE (CP)

This reagent has no published limits of impurity on its label; hence, the chemist is responsible for establishing purity prior to use. Other grades of chemicals (i.e., USP, NF, Practical, Pure, Technical, or Commercial) should not be used routinely in the clinical chemistry laboratory.

ANTICOAGULANTS (FOR TESTS REQUIRING WHOLE BLOOD OR PLASMA)

Specimen requirements for the laboratory testing are essential for proper quality control of procedures. Coagulation of blood can be prevented by the addition of EDTA (ethylenediaminetetraacetic acid), oxalates, citrates, or heparin. The first three anticoagulants listed remove calcium, an essential component for coagulation, from the blood plasma by precipitation or chelation. Heparin neutralizes thrombin. Tubes with anticoagulant must be handled gently but mixed until all the anticoagulant is dissolved. The type of anticoagulant is indicated by color-coded vacutainter tube stoppers.

CALIBRATION OF PIPETTES

ACCURACY CHECK

Semiautomated and automated pipettes are commonly used in the chemistry laboratory. These pipettes should be checked monthly for accuracy and precision.

Since the volumes dispensed by these pipettes can be quite low (as low as 5 uL or 0.005 mL), pipetting errors introduced by a faulty pipetting device can be significant. A gravimetric pipette calibration procedure is as follows: (1) tare the balance with a weighing vial or boat; (2) pipette water into the vial using the pipette to be calibrated three times, taring and recording the weight for each pipetting; (3) average the three weights and divide the average weight by the density of water at 25°C to get the corresponding average delivery volume; and (4) determine the percentage difference by the following formula:

$$\frac{\text{percentage deviation}}{\text{from stated volume}} = \frac{(\text{aver delivery vol} - \text{stated vol})}{\text{manufacturer's stated volume}} \times 100$$

If the percentage deviation is greater than ±5.0, the pipette should not be used.

PRECISION

The pipette can also be checked for precision by pipetting water into a weighing vial and weighing the amount delivered. The procedure is repeated 10 times. The 10 data points are used to compute the coefficient of variation for the pipette. Acceptable precision should be at least 5% with lower precision levels (i.e., 1–2%) being preferred.

BIBLIOGRAPHY

Bakerman, S. *Review of Clinical Chemistry*, Lecture Notes. Greenville, North Carolina: 1977 and 1983.

Barnett, R. N. *Clinical Laboratory Statistics*, 2nd ed. Boston: Little, Brown, 1979.

Galen, R. S. and Gambino, S. R. *Beyond Normality: The Predictive Value and Efficiency of Medical Diagnosis*. New York: Wiley, 1975.

Henry, J. B. (Ed.): *Clinical Diagnosis and Management by Laboratory Methods*, Vol. I, 16th ed. Philadelphia: Saunders, 1979.

Levey, S. and Jennings, E. R. "The use of control charts in the clinical laboratory," *American Journal of Clinical Pathology*, **20,** 1059 (1950).

Teitz, N. (Ed.) *Fundamentals of Clinical Chemistry*, 2nd ed. Philadelphia: Saunders, 1976.

Westgard, J. O. and Hunt, M. R. "Use and interpretation of common statistical tests," *Clinical Chemistry*, **19**(1), 49 (1973).

Westgard, J. O. and Groth, T. "A multi-rule Shewhart chart for quality control in clinical chemistry," *Clinical Chemistry*, **27**(3), 493 (1981).

Westgard, J. O., deVos, D. J., Hunt, M. R., Quam, E. F., Carey R. N., and Garber, C. C. "Concepts and practices in the evaluation of clinical chemistry methods," *American Journal of Medical Technology*, **44,** 4 (1978).

POST-TEST

1. Match the following:
    ——————— 1. Accuracy
    ——————— 2. Precision
    ——————— 3. Standard
    ——————— 4. Control
    ——————— 5. Standard Deviation
    ——————— 6. Coefficient of Variation
    ——————— 7. 95% Confidence Limit
    ——————— 8. Systematic Shift
    ——————— 9. Systematic Trend
    ——————— 10. Random Error
    a. Encompasses 95.5% of all of the results in a Gaussian distribution.
    b. The deviation expressed as the percentage variance from the mean.
    c. The closeness of the test value to the true mean.
    d. The spread between replicates or variation within the analytic method.
    e. Highly purified material of known composition.
    f. The measure of the scatter around the mean in a Gaussian distribution.
    g. An increasing or decreasing gradual deterioration of the test system.
    h. Caused by intrinsic properties of the method due to chance.
    i. A change in values caused by a sudden sustained test system failure or
        bias.
    j. A material whose physical and/or chemical composition closely resem-
        bles the unknown test specimen.

2. The reagent grade considered most pure is:
    a. Chemically Pure (CP)
    b. Purified or Pure Grade
    c. Analytical Reagent Grade (AR)
    d. Technical or Commercial Grade

3. A new test for ionized calcium was evaluated. The results are listed below:

    Twenty patient samples with primary hyperparathyroid disease (known posi-
    tives) and 20 normal patient samples (no disease known to affect calcium)
    were tested with the following results:

20 Known Positive		20 Known Negative	
16 positive results (TP)	4 negative results (FN)	18 negative results (TN)	2 positive results (FP)

    These data may be interpreted in the following way:

    Out of 100 samples from known hyperparathyroid patients, this test proce-
    dure will be positive in —————(number) samples and negative in
    —————(number) samples.

Out of 100 samples from normal patients (known negatives), this test will be negative in _____ (number) samples and positive in _____ (number) samples.

The sensitivity of this test would be _____ %.

The specificity of this test would be _____ %.

4. Give the formula for the determination of a standard deviation unit and describe what each symbol in the formula represents.

5. A normal range study was conducted. The concentration value of the samples versus number were plotted on a bar graph producing the symetrical bell-shaped curve in Figure 2.11. Answer the questions that follow.
   a. The patient population that falls in this area refers to _____ SD units, which would include _____ % of all values.
   b. The patient population that falls in this area refers to _____ SD units, which would include _____ % of all values.
   c. The patient population that falls in this area refers to _____ SD units, which would include _____ % of all values.

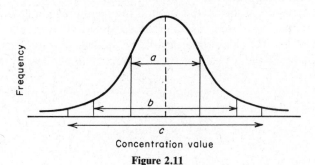

Figure 2.11

6. The data from 200 samples for a BUN normal range study is as follows:

   Mean value of all results = 15 mg/dL

   One standard deviation unit = 2.5 mg/dL

   The BUN reference range (normal range) based on these results would be _____ (give units).

7. Below are the QC results for the low (normal) quality control sera for the *o*-toluidine procedure for glucose. Use the data from Figure 2.12 and calculate:
   a. Mean   _____
   b. Median   _____
   c. Mode   _____
   d. The number of values that fall outside of the ±2SD limits is _____ .
   e. One standard deviation (1SD) is _____ .
   f. Two standard deviations (2SD) is _____ .

g. The range for allowable limits for the next month's quality control chart is _____ .

h. The coefficient of variation is _____ .

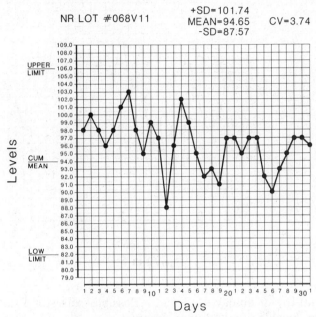

NR LOT #068V11   +SD=101.74
MEAN=94.65   CV=3.74
-SD=87.57

Figure 2.12

8. Shown in Figure 2.13 is the QC chart for the high control sera for glucose. The control limits for this laboratory are ±2SD marked on the chart. Use Figure 2.13 to answer the questions that follow:

1. On day 10, the QC sera value exceeds the −2SD limits. The *best* course of action would be:

    a. Repeat the control and if the value falls back within QC limits, report patient values.

    b. Report the patient results, since this represents a random analytical error.

    c. Repeat the control plus several patients. If the control falls back within ±2SD limit and patient values are similar, report results.

    d. Make up a new bottle of control sera and repeat the control since the first bottle may have deteriorated in storage.

    e. Report the patient results, since in using 95% confidence limits, 5% or 1 in 20 results are expected to fall outside the ±2SD limits.

2. A trend begins on day _____ . Possible causes of a trend in QC programs include (mark all that apply):

    a. Gradual deterioration of a standard.

    b. Continuous deterioration of a reagent

    c. Deteriorating radiant energy light source

    d. A newly prepared standard prepared with a wrong concentration

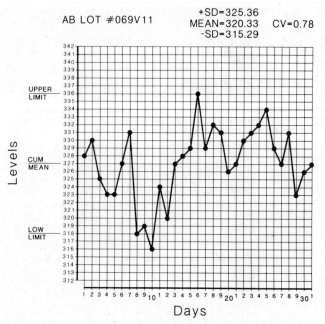

AB LOT #069V11   +SD=325.36
                        MEAN=320.33   CV=0.78
                        -SD=315.29

Days

Figure 2.13

3. A shift begins on day _____ . Possible causes of a shift in QC pro-
    grams include (mark all that apply):
    a. A deteriorated standard that is maintaining a new concentration
        level
    b. A reagent that has shifted to a new level of activity
    c. Change in reagent lot number
    d. Improper calibration or standardization
    e. Contamination of reagents

9. In QC programs, control sera are used to monitor _____ and are known
    as _____ quality control; participation in check sample programs is used
    to monitor _____ and is known as _____ quality control. Use the
    following terms in the blanks:
    a. precision or reproducibility
    b. accuracy
    c. internal
    d. external

10. Some correlation studies were performed. The results are plotted below. The
    heavy line represents test result plots. Match the graphic picture with the
    proper interpretation. (Note: The dotted line represents perfect correlation.)

    _____ a. These results show constant
                  bias

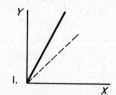

_____ b. These results show proportional bias

_____ c. These results show perfect
correlation

_____ d. These results show both constant and proportional bias

11. The slope describes _____ error.
    a. constant
    b. proportional
    The y-intercept describes _____ error.
    a. constant
    b. proportional

12. The least squares analysis is a statistical tool that is used to:
    a. calculate slope and y-intercept of a comparison study
    b. determine if the bias between two test methods is significant
    c. estimates the precision of an assay
    d. estimates the degree of varience between two methods

13. The coefficient of variation (CV) estimates the precision of an assay proce-
    dure. In general, if the CV is no greater than _____ , the precision of the
    procedure is considered acceptable.
    a. 2%
    b. 10%
    c. 5%
    d. 8%

14. A sample pool was analyzed and the mean concentration of an analyte was
    obtained. A measured amount of this pure analyte was then added to the
    base pool. This spiked pool was then assayed in triplicate. These results
    were compared to the expected concentration. The type of study performed
    is:
    a. interference
    b. linearity
    c. precision
    d. recovery

15. A base pool is analyzed to obtain an accurate concentration of an analyte. The base pool is then spiked with a high concentration of this pure analyte. Multiple dilutions of the spiked pool are assayed. You plot expected values (x-axis) against assayed values (y-axis). The type of study done is:
    a. precision
    b. linearity
    c. recovery
    d. accuracy

16. The following data were obtained by performing one type of BUN assay on a single sample: 50, 48, 52, 49, 55, 52, 53, 56, 54, 50, 47. The type of study performed is:
    a. accuracy
    b. precision
    c. recovery
    d. linearity

17. Analytical sensitivity refers to _____; analytical specificity refers to _____. Diagnostic sensitivity refers to _____; diagnostic specificity refers to _____.
    a. The ability of the test system to measure low concentrations of the analyte.
    b. Test positivity when disease is present.
    c. The ability of the test system to measure only the substance under assay.
    d. Test negativity when disease is absent.

18. A sample was analyzed for a particular analyte. Lysed red blood cells were then added, and the sample was reanalyzed. The results were compared. The type of study performed is:
    a. recovery
    b. precision
    c. accuracy
    d. interference

19. In the clinical laboratory, standards are used to _____; controls are used to _____.
    a. Calculate concentration values
    b. Validate results
    c. Calibrate an instrument

20. All of the anticoagulants inhibit coagulation by binding calcium except:
    a. EDTA
    b. potassium oxalate
    c. sodium citrate
    d. heparin

21. The reference method for pipette calibration is:
    a. dye dilution method
    b. gravimetric method
    c. radioisotope method
    d. repeated delivery into volumetric glassware

ANSWERS

1.   1. c
     2. d
     3. e
     4. j
     5. f
     6. b
     7. a
     8. i
     9. g
    10. h

2. c

3. 80, 20, 90, 10, 80, 90

4. $$SD = \sqrt{\frac{\Sigma(\bar{X} - X)^2}{N - 1}}$$
    SD = Standard Deviation
    Σ = Sum of
    $\bar{X}$ = Mean
    X = Patient value
    N = Number of values

5. a. ± 1, 68
    b. ± 2, 95
    c. ± 3, 99+

6. 10–20 mg/dL
    $15 \pm 2SD$ = 10–20 mg/dL

7. a. 96
    b. 97
    c. 97
    d. none
    e. 3.4 mg/dL
    f. 7 mg/dL
    g. 96 ± 7 mg/dL = 89–103 mg/dL
    h. 3.5%

8. 1. c
    2. 20, a, b, c
    3. 13 and 21, a, b, c, d, e

9. a, c, b, d

10. a. 2
     b. 1
     c. 3
     d. 4

11. b, a

12. a

13. c

14. d
15. b
16. b
17. a, c, b, d
18. d
19. a, c; b
20. d
21. b

CHAPTER 3
LABORATORY MANAGEMENT

VII.   SPECIMEN TESTING
    A.   Work Station Worklists
    B.   Interfaced Equipment
    C.   Results Verification
    D.   Results Reporting

VIII. OTHER MANAGEMENT FUNCTIONS
    A.   Billing
    B.   Workload
    C.   Interpreting Productivity

IX.   COST ANALYSIS

X.   LABORATORY INFORMATION SYSTEMS (LIS)
    A.   Functions of Computers
    B.   Computer Hardware
        1.   CPU
        2.   Computer Software
    C.   Functions of a LIS
        1.   Computer Programs
        2.   Computer Prompts
        3.   Sample Clinical Computer Programs
        4.   Computer Files
        5.   Laboratory Microcomputers

XI.   LABORATORY DESIGN AND WORKFLOW

XII.   MANAGEMENT STYLE
    A.   Scheduling of Work

OBJECTIVES

On completion of this chapter, you will be able to:

1. Define, illustrate, or explain the following terms or concepts:
    a.   batch testing
    b.   CAP workloads
    c.   lipemic specimen
    d.   TAT
    e.   diurnal variation
    f.   hemolysis
    g.   CPU
    h.   prompt
    i.   computer hardware
    j.   computer program

2. Discuss the different areas of management involved in the "testing process," beginning with the physician's order.

3. Describe the difference between stat and batch work stations.

4. Discuss the general and special considerations involved in collecting a blood specimen.

5. Describe the integration of the laboratory computer in the function and management of the laboratory.

6. Discuss the use of CAP workload recording in the assessment of laboratory productivity.

7. Describe the procedure involved in determining the cost of a laboratory test.

8. Describe a simple computer program in Basic to add numbers, compute the average, and print results.

9. Describe a rotating scheduling system for a chemistry section.

GLOSSARY OF TERMS

Batch testing. Lower priority testing or routine testing where specimens are not tested immediately. Rather, testing is delayed until a "batch" of specimens accumulate.

CAP workload. The sum of all of the CAP units (in minutes) for all procedures performed in a laboratory section.

CPU. Central processing unit. The "brains" of the computer consisting of an arithmetic logic unit (ALU), memory, and an input/output controller.

Computer data file. The arrangement of related data or fields of data into records with the collection of records being called a file. The data file is usually stored on a computer disk or magnetic tape. A data file may be loaded into computer memory for processing.

Computer program. A set of precise computer instructions which are executed by the computer in a step-by-step fashion in order to perform a well-defined task.

Float work station. A laboratory work station with variable task assignments.

Hemolysis. Term used to describe the red color of a serum or plasma specimen that has an increased hemoglobin concentration due to destruction (lysis) of the cell walls of erythrocytes.

Instrument interface. The computer hardware and programs to allow the transfer of data from instruments directly to a computer.

Lipemic. Term used to describe a milky or turbid specimen due to an increased concentration of lipid.

LIS. Laboratory information system or laboratory computer system.

Productivity. The total workload (in minutes) divided by the total hours worked. Hence, it is the number of minutes of each hour the section is producing test results.

Stat testing. High priority testing where laboratory data are urgently needed to assist in patients' care.

TAT. Turn-around time for a laboratory test or time required from specimen receipt in the laboratory to the reporting of test results.

Workload unit. In the CAP workload system, it is the number of minutes of technical, clerical, and aid time required to produce a test result.

Work station. A laboratory area (instruments, bench area, etc.) with associated tasks and a daily workload sufficient for one technologist.

SPECIFIC AREAS OF MANAGEMENT

The management of the clinical laboratory must encompass all phases of the "testing process," which begins with the physician's order and is completed after the reporting of the test results. Hence, management of a modern chemistry laboratory can be divided into five major areas; these are as follows:

1. *Specimen Acquisition.* Test ordering, transport of specimens, and phlebotomy services.
2. *Specimen Processing.* Receipt and labeling of specimens, specimen logging, separation of serum or plasma, aliquoting specimens, delivery of specimens to work stations, and referral testing.
3. *Specimen Testing.* Arranging specimens into batches, assaying specimens, quality control, instrument maintenance, evaluating patient's test results, telephone reporting, research, and methods development.
4. *Results Reporting.* Filing reports, telephone inquiry, sorting, and sending results to reporting locations.
5. *Billing and Workloads.* Generating bills and determining workload from daily test talleys.

In addition to the specific "test-related" tasks listed above, other activities are involved in the management of a laboratory; these may include (1) teaching or staff development; (2) ordering, and stocking of laboratory supplies; (3) budgeting; (4) organization and planning; (5) laboratory design; (6) scheduling of staff; (7) recruitment of new staff; (8) data processing; (9) conforming with governmental regulations and requirements for accreditation; and (10) coordination of laboratory services with the needs (goals, objectives, etc.) of the hospital. There are of course many other management activities not listed. It follows, then, that the successful management of the clinical laboratory requires a "team effort" from a broad spectrum of talented and dedicated professionals. The style of management required to achieve this "team effort" will be discussed in this chapter.

GOALS OF THE CLINICAL LABORATORY

There are many goals and objectives in a complex operation like a clinical laboratory. However, in a service-oriented hospital or private clinical laboratory, a major goal of management must be to supply a variety of reliable test results for patients' care in a predictable expeditious manner, appropriate to the condition of the patients being served. Moreover, this high level of quality laboratory services must be provided in a cost-efficient manner with effective use of laboratory resources. Generally, laboratory resources include personnel, equipment, space, reagents, and other supplies. Also, the testing service in hospitals must often be available 24 hours a day, seven days a week, including holidays. Other goals relative to teaching and research, especially in university laboratories, are significant but should not supercede the patient care objectives described above.

SUBDIVISIONS OF TESTING

STAT VERSUS BATCH TESTING

There are two major types of testing in the clinical laboratory: (1) "stat" testing or assays that must be performed immediately and (2) "batch" testing where specimens are tested in groups. A considerable amount of stat testing is expected from emergency rooms, intensive care areas, and clinic areas (where patients wait for laboratory results). Batch testing is often required by nonautomated, labor-intensive methods (i.e., radioimmunoassay). Physicians should be made aware of available stat tests and the time required to obtain stat results. Also, physicians should be notified as to when batch assays are performed (i.e., the "batch cutoff time") and time required for completion of patient's test results (see Table 3.1). Stat tests must be available at all times on all shifts; batch tests may only be available once daily or even once a week.

TURN-AROUND TIME

The turn-around time of a laboratory test is the total time required from receipt of a specimen in the laboratory to reporting of test results.

Table 3.1. Batch Scheduling and Cutoff Times

Test	Specific Requirements for Serum	Days Done	Batch Cutoff (a.m.)	Analysis Completed By
ACTH	0.3 mL (plasma)	M	9	4:00 p.m. (Thurs.)
Aldosterone	1.5 mL	M	9	4:00 p.m. (Thurs.)
B12 and folate	0.5 mL	Tu	9	10:00 a.m. following day
β-2 microglobulin	0.5 mL (25 mL urine)	M,W,F	9	5:00 p.m.
CEA	0.7 mL	M	9	5:00 p.m.
Cortisol	0.1 mL (5 mL urine)	M,W	9	4:30 p.m.
Digoxin	0.3 mL	M–F	9	3:00 p.m.
Digitoxin	0.2 mL	M–F	9	3:00 p.m.
Ferritin	0.5 mL	W	9	10:00 a.m. following day

EXAMPLE 3.1.   Typical turn-around times for the following tests may be acceptable: stat blood gases, 10 minutes; stat potassium (serum), 40 minutes; and batch digoxin assay, 5 hours.

Serum and plasma must be centrifuged and separated (10–20 minutes required). For serum, complete clotting of the specimen is required (20–30 minutes). To eliminate clotting time delays, certain tests like glucose are often performed on plasma.

LABORATORY WORK STATIONS

A laboratory work station is an instrument(s) or laboratory bench area(s) with specific testing assignments.

STAT AND BATCH WORK STATIONS

Certain work stations in the laboratory are called "stat work stations," where primarily stat testing or higher priority testing is performed. These work stations may be grouped in a stat laboratory section. A batch work station is generally associated with lower priority testing. In many laboratories, during the evening, nights, weekends, and holidays, only stat work stations are scheduled with minimal batch work station coverage. Generally, technologists are assigned to a par-

Table 3.2.   Sample Work Station Assignments for a Chemistry Section with Nine Technologists

Work station 1.	Stat Lab 1 (Team Leader)—blood gases, immunoenzymatic drug assays, and rate nephelometer assays
Work station 2.	Stat Lab 2—stat (and batch) discrete electrolyte, glucose and blood urea nitrogen analyzer
Work station 3.	Stat Lab 3—stat (and batch) discrete chemistry analyzer
Work station 4.	Automated chemistry profile
Work station 5.	Electrophoresis and toxicology
Work station 6.	Radioimmunoassay

Lower priority work stations are also established; these are listed below:

Float #1.	Special test development assignments, pipette calibration, temperature checks, special instrument maintenance, quality control review, workloads, etc.
Float #2.	Same as Float #1
Float #3.	Same as Float #1

ticular work station with associated time schedules and duties. Some selected chemistry work station assignments are listed in Table 3.2. Note that in addition to stat and batch work stations, lower priority "float" work stations are also assigned for methods development, teaching, etc.

THE TESTING PROCESS

SPECIMEN ACQUISITION

Since most of the testing in the chemistry laboratory is done on serum or plasma, blood collection is required.

Test Request

Prior to the collection of a blood specimen, the test must be ordered by the physician whereupon a test requisition is completed by the nurse or ward clerk and sent to the laboratory. In computerized hospitals, the test order is usually placed on the hospital computer [or hospital information system (HIS)], which then prints out the test request in the laboratory. If a laboratory computer [laboratory information system (LIS)] is in place and interfaced to the HIS, the test request is transmitted directly to the laboratory computer files prior to printing.

In some computerized laboratories, the test request is sent to the laboratory where it is keyed directly into the laboratory computer, a process called "computer accessioning." Generally, during the accessioning process, the computer assigns a unique accession code (or number) to the sample. If a batch or routine test is requested, a laboratory phlebotomist is sent to collect the required blood sample (at a designated time or rounds). For a request of a stat blood test, the phlebotomist will collect the specimen as soon as possible.

Sample Collection

Considerations Prior to Phlebotomy
Once the patient is located, the phlebotomist should verify the identity of the patient by (1) checking the patient's hospital armband and (2) asking the patient to identify him- or herself. Prior to collecting the blood, the phlebotomist should know (1) the type of blood specimen required for the test (i.e., whole blood, plasma, or serum); (2) if whole blood or plasma is required, the type of anticoagulant; (3) the volume of blood required; (4) special handling requirements of specimens such as collection on ice for renin assays; (5) special requirements for the patient such as posture, fasting, or unfasting; and (6) special requirements for the time of collection for stimulation and suppression studies in endocrinology and "peak" and "trough" drug levels in therapeutic drug monitoring. The patient may also be on medications or special diets that may interfere with the laboratory testing procedure.

Intravenous Fluid Contamination
Also, when a patient is receiving an intravenous fluid, drawing fluid from a vein above the fluid line will result in a blood specimen contaminated with the intraven-

ous fluid mixture. Alternatively, the phlebotomist should select another arm or foot vein for sampling. Also, the fact that the patient is receiving intravenous fluid should be indicated on the laboratory request slip or requisition along with the type of fluid being administered.

Patient Preparation for Phlebotomy

Assuming the phlebotomist is now prepared to collect the specimen, the patient should be in bed or sitting with his or her arm supported by a pillow or table. A tourniquet is applied to the upper arm in the bicep area to expose veins. The patient should clench and unclench his or her hand to expose veins. However, clenching of the fist is contraindicated for some tests, such as venous pH, ionized calcium, and electrolytes. Venous insertion of the needle, bevel-up, at 15 degree angle is recommended.

Special Considerations for Phlebotomy

Special considerations should also be given to (1) the patient who is difficult to stick; (2) a fainting patient; (3) finger puncture; (4) preparation of blood cultures; (5) specimen collections in restricted areas (e.g., surgery, psychiatry); (6) collections from infants; and (7) patients in isolation. For patients in isolation, the phlebotomist must observe all procedures and precautions indicated on a color coded card placed on the door of the patient's room usually by the infection control nurse.

Once collected, the specimen should be properly labeled and delivered to the laboratory as rapidly as possible for testing. In computerized laboratories, the phlebotomist must verify that the specimen has been collected and note the time of collection on the laboratory computer. In noncomputerized laboratories, the phlebotomist may log the specimen onto a "master log."

Other Specimen Collection Variables

Early morning blood specimens are preferred for chemistry analysis in order to avoid (1) diurnal variation; (2) postural effects; (3) postprandial changes; and (4) exercise effects.

Diurnal Effects

Certain blood analytes show changes in concentrations throughout the day, hence it is essential that specimens for these analytes be collected at a standardized time of day. For example, cortisol concentration is highest in the early morning and falls throughout the day to a low level in the afternoon. Iron levels are highest in the morning, while cholesterol levels are highest in the afternoon.

Postural Effects

Some blood constituents vary in concentration in an upright (erect) position as opposed to a supine (lying down) position because a shift in body fluid occurs. Upon standing, water shifts from the intravascular space to the interstitial compartment. Nonfilterable substances such as proteins and protein-bound compounds become more concentrated since they are held in the vascular tree. Examples of constituents showing postural effects are (1) protein—the total protein level may increase by as much as 9% while erect as opposed to the supine position; and (2) substances bound to protein—analytes normally bound to protein (e.g., iron, calcium, and lipids) are increased in the erect position.

Postprandial Changes

Fasting specimens are preferred for chemical analysis. The patient should not eat 10 hours prior to sampling. Most analytes are not affected directly by a recent meal; however, the turbidity due to recently digested fat (chylomicrons) interferes with many colorimetric methods of analysis. Serum blanking may correct this type of interference. Constituents directly affected include glucose and triglycerides, which are increased in nonfasting specimens.

Phosphorus is decreased in nonfasting blood owing to cellular absorption and utilization in glucose phosphorylation, the initial step of the glycolytic pathway. Lipemia of nonfasting specimens can also reduce the activity of certain enzymes to give falsely low results. Gross lipemia also falsely lowers sodium levels obtained by flame photometry. While a 10 hour fast is recommended, a prolonged fast alters the level of many blood constituents and is not recommended.

Exercise

Exercise prior to blood sampling can alter various analyte components including (1) enzymes—with increases in creatine kinase, lactate dehydrogenase, aspartate transaminase, and aldolase; and (2) blood gases—with increases in pCO_2 and decreases in pH due to lactate accumulation.

Hemolysis

Disruption of the cell wall of erythrocytes causes the contents of these cells to spill into the plasma. The hemoglobin of erythrocytes imparts a red appearance to serum or plasma when the hemoglobin concentration exceeds 200 mg/L. Erythrocytes contain higher concentrations of certain analytes than plasma; these include potassium, magnesium, phosphorus, acid phosphatase, lactate dehydrogenase isoenzymes 1 and 2, and the transaminases AST (SGOT) and ALT (SGPT). Therefore, in a hemolyzed serum or plasma specimen, these constituents would be falsely elevated.

Hemolysis can also (1) inhibit the activity of lipase; (2) interfere with the diazo reaction in bilirubin measurements; and (3) interfere with most colorimetric methods by adding color to the reaction medium.

Bilirubin Interference

Patients with jaundice have elevated serum bilirubin levels imparting a yellow color to the specimen. The specimen from a jaundice patient is sometimes called "icteric." The color of bilirubin can interfere with colorimetric measurements of glucose and protein; this interference can be eliminated with a specimen blank.

Tourniquet Problems

When collecting a venous blood specimen, a tourniquet is used to expose veins. However, if the patient opens and closes his hand to fill veins and the tourniquet is released prior to drawing the sample, laboratory results consistent with "metabolic acidosis" will be obtained. The blood sample will have an elevated CO_2, decreased pH and an elevated lactic acid. Hence, upon premature release of the tourniquet, the collection site is flushed with blood with falsely acidotic analytes.

Sunlight

Light exposure of certain analytes can facilitate decomposition. For example, unconjugated bilirubin and porphyrins are particularly susceptible to degradation by sunlight. Hence specimens for these analytes should be protected from light exposure.

Stability of Specimens

Most analytes are stable at room temperature for 1–2 hours after separation from cells. Specimen for the analysis of gases (NH_3, pO_2, pCO_2) and other analytes (lactic acid, pH, renin, and aldosterone) are not stable at room temperature. After separation from cells, most analytes are stable for several days at 4°C. Many analytes are stable for weeks and even months when specimens are stored frozen at −20 to −40°C.

Blood Gases

Exposing a blood gas sample to air contamination will decrease the carbon dioxide and increase the oxygen level. The loss of CO_2 increases the pH of the specimen. This event increases the binding of ionized calcium to protein; hence, specimens for ionized calcium measurements should be collected under anaerobic conditions.

Glucose Stability

Samples for glucose analysis should be separated from cells promptly because of the rapid cellular metabolism of glucose. If separated from cells, glucose in serum is stable for several hours at 25°C (room temperature).

Analyte Degradation Due to Refrigeration

Refrigeration should not be used for specimen storage of samples for lactate dehydrogenase (LDH) measurements and lactate dehydrogenase isoenzyme studies. LDH isoenzymes 4 and 5 are labile at 4°C. Refrigeration also decreases the resolution of lipoprotein fractions in lipoprotein electrophoresis.

Capillary Versus Arterial Blood

Capillary blood is more similar in constituency to arterial blood than venous blood. Glucose is higher in capillary blood than venous blood, but the difference is usually minimal (+5 mg/dL).

SPECIMEN PROCESSING

CENTRAL RECEIVING AND PROCESSING AREA

Many modern clinical laboratories have a central receiving and processing area for the receipt of all specimens. Blood specimens for chemistry testing usually require serum or plasma, hence they must be processed prior to delivery to work stations for testing.

COMPONENTS OF BLOOD PROCESSING

Processing of blood specimens may include (1) logging of the specimens (or accessioning on the laboratory computer); (2) allowing the specimen to clot; (3) spinning the specimen in a centrifuge to separate serum and cells; (4) pouring off specimens into labeled aliquot tubes for various testing work stations; and (5) delivery of the specimen to the work station (or arranging specimens in batches for later deliver to batch work stations). For anticoagulated blood, urine, and other fluid specimens not requiring blood processing, after logging or laboratory computer accessioning, the specimen should be delivered directly to the appropriate work station by the collecting phlebotomist or laboratory transporter. Examples of specimens not requiring blood processing include (1) whole blood (EDTA) for hematology; (2) whole blood (heparin) for blood gases; (3) urine for urinalysis; (4) blood specimens for coagulation testing; and (5) specimens for microbiological studies.

REFERRAL TESTING

The central receiving and processing area usually handles all referral testing (or testing referred to reference laboratories). Most referred tests involve serum testing (hormones, drug levels, and specific proteins), and usually one member of the processing staff is assigned primary responsibility for referral testing. Most reference laboratories provide an on-line, computer-generated report from the referral laboratory via telephone lines.

SPECIMEN TESTING

As described earlier, once the specimens have been logged into the laboratory or accessioned into the laboratory computer, they can be tested at a laboratory work station. Various types of laboratory testing equipment will be described in Chapter IV; various testing procedures are described throughout this text.

WORK STATION WORKLISTS

In general, for batch testing, the laboratory computer will generate a work station worklist for specimens needing testing. In noncomputerized laboratories, the list is prepared manually. Testing is performed on the specimens and results are recorded onto the worklist and are either keyed into the computer or manually transferred to the reporting document.

INTERFACED EQUIPMENT

Some laboratory computers are "interfaced" to laboratory testing equipment and receive patients' data (or test results) directly from the instrument, obviating the need for the analyst to key or transcribe the patients' results onto the worklist or reporting document.

RESULTS VERIFICATION

In computer systems, the analyst must "verify" the patients' results by reviewing the results and associated QC data. Only after verification will the patients' results be released for reporting.

RESULTS REPORTING

Laboratory results can be reported in a variety of ways. For stat or emergency testing, results can be called or printed on a computer printing terminal in the patient care area. Stat results are often called to the patient care area or ordering doctor. For routine testing, cumulative computer reports are generated, containing all the patients' laboratory data. Noncomputerized laboratories generally deliver a copy of the reporting slip to the floor for charting. To facilitate telephone inquiry concerning patients' laboratory results, a file is usually maintained in the central processing and receiving area with a copy of each laboratory slip. In computerized laboratories, laboratory results can be retrieved throughout the hospital using terminals connected to the laboratory computer. (Laboratory results can be obtained from the HIS when the HIS–LIS computer interface supports "results inquiry.")

OTHER MANAGEMENT FUNCTIONS

BILLING

Many laboratories have a manual system for billing, workloads, and other management-related activities. A laboratory computer often collects information for billing on test request or on completion of the laboratory test.

The actual generation of patient bills, which requires associated accounts receivable files and programs, usually takes place on the hospital computer.

WORKLOAD

A laboratory computer will also count and store information concerning workloads, a management tool for the allocation of laboratory resources. Using a workload recording system (College of American Pathologist, Skokie, IL), the productivity of a laboratory section can be determined based on the amount of tests performed compared to the hours worked. Paid hours differs from actual hours worked because of "nonproductive time" such as vacations and sick leave.

Each test has an assigned unit value in minutes/test (e.g., digoxin = 7 minutes/test). At the end of the month, the total "raw count" or number of patients' digoxin assays (plus controls, standards, repeats, dilutions) are multiplied times the unit value for each test to get the "total unit value" for the test. The unit for total unit value is minutes:

$$\text{Total unit value (minutes)} = 1000 \text{ digoxin tests} \times 7 \text{ minutes/test}$$

$$= 7000 \text{ minutes}$$

Each test in the chemistry section is treated in a similar manner, and the total unit value for each test are summed. This would give the total "predicted" time (in minutes) required to perform the test workload. This total time in minutes is called the "workload":

$$\text{grand total CAP unit values for chemistry (November)} = 181,421 \text{ min}$$

Dividing this grand total by the actual hours worked by all of the staff in the section gives the number of minutes worked per hour (or productivity):

$$\text{Productivity} = \frac{\text{Grand total unit values (minutes)}}{\text{Hours worked}}$$

If the total hours worked for the month were 4,774 hours,

$$\text{Productivity (Chemistry, November)} = \frac{181,421 \text{ minutes}}{4,774 \text{ hours}}$$

$$\text{Productivity} = 38 \text{ minutes/hour}$$

Hence, based on these figures, the chemistry section is productive 38 minutes of each hour.

INTERPRETING PRODUCTIVITY

Time spent in activities that do not generate workload units is called "non-specified hours." Teaching hospitals often have lower productivity because of teaching programs for medical technologists, residents, and medical students. Research and development activities are also non-specified hours that do not generate workload units. Therefore, in interpreting productivity data, non-specified hours should be considered along with the productivity of the section compared with similar sections in other hospitals.

COST ANALYSIS

Prior to the introduction of any new test procedure, a cost analysis should be performed to determine an appropriate amount to charge for the test. A procedure cost analysis will arrive at the total cost per patient of performing the test.

EXAMPLE 3.2.   Determine the cost per patient of a digoxin assay with a patients' batch size of 10, seven standards, and two controls. Two hours of technologist time is required to perform the assay. A technologist salary including benefits is $12.00 per hour and reagent cost is $0.60 per tube. All assays are performed in duplicate. Note: Phlebotomy and processing charges are included in a 100% overhead figure. Overhead costs also include a professional component for management expense, result interpretation, space, utilities, maintenance, capital equipment, etc.

Answer:

1. Determine the cost of the standard curve and controls for each patient assuming a 10 patient batch:

    Cost for standard curve and controls (cost/patient)

    $$= 18 \text{ tubes/run} \times \$0.60/\text{tube} \times 1 \text{ run/10 patients}$$

    $$= \frac{(18)(0.60)}{10} \text{ \$/patient} = \$1.08/\text{patient}$$

2. Determine the technologists' labor costs for the run:

    Technologist time cost per patient

    $$= 2 \text{ hours/run} \times \$12.00/\text{hours} \times 1 \text{ run/10 patients}$$

    $$= \frac{(2)(12.00)}{10} \text{ \$/patient} = \$2.40/\text{patient}$$

3. Add costs in parts 1 and 2 to get totals costs:

    Total cost = $1.08 + $2.40 = $3.48/patient

    Add Overhead Cost: Assume a 100% overhead cost

    Cost plus overhead = $3.48/patient × 2 = $6.96/patient

4. Add profit to determine a suitable test charge (assume a 20% percent profit margin):

    Profit = $6.96/patient × 0.20 = $1.39

Therefore, a reasonable total patient's charge for performing a digoxin assay would be $6.96 plus $1.39, giving a total of $8.35.

LABORATORY INFORMATION SYSTEMS (LIS)

FUNCTIONS OF COMPUTERS

Computers perform three types of tasks rapidly and accurately. These tasks include (1) multiple and repetitive calculations; (2) the acquisition, storage, and retrieval of information; and (3) the monitoring and control of laboratory equipment. Since the laboratory is confronted with tasks of these types, many modern clinical laboratories have computerized their operation by installing a laboratory information system (LIS) dedicated to a variety of laboratory functions.

Computerization of the clinical chemistry laboratory began with the development of automated analyzers under minicomputer and later microprocessor control. In recent years, a major goal has been total computerization of the laboratory to include functions such as: (1) test ordering; (2) specimen collection lists or labels; (3) worklist generation; (4) on-line data acquisition from instruments; (5) quality control; (6) cumulative patients' reports; (7) laboratory billing; (8) filing and retrieval of patients' data; and (9) workload computations and other management reports. The laboratory computer also maintains a "trail" or "log" for inquiry as the specimen proceeds through the various stages of the testing process such as (1) specimen acquisition; (2) specimen processing; (3) specimen analysis; and (4) results reporting. The laboratory computer system also maintains valuable information regarding patient census and patients' demographic and clinical information. In addition, extensive technical information regarding various laboratory procedures is maintained by the LIS.

Based on clerical staff savings (see Table 3.3) and improved billing data, a laboratory computer system can be cost-justified and represents a good investment for a hospital with a possible payback period for the computer system of four to five years.

Table 3.3. Clerical Activities Conveniently Handled or Supported by a Laboratory Computer

Specimen collection and distribution

Logging of test requests

Production of test worksheets

Recording of test results

Inquiry for test results

Report distribution

Management control
  workloads
  billing

COMPUTER HARDWARE

CPU

The CPU (central processing unit) functions as the "brains" of the LIS and has many functions similar to the human brain: (1) memory areas for computer programs and data; (2) arithmetic logic component; and (3) an input/output central controller.

The LIS computer hardware also includes equipment located near the CPU: (1) disk drives for storage and retrieval of information and computer programs; (2) a CPU console for communications to the CPU; and (3) the input/output controller, which allows peripheral access to the CPU. Peripheral computer hardware in a LIS includes: (1) video display terminals for input and output of information to the CPU; (2) printers for output of information; (3) instrument interfaces; and (4) card readers and other special devices for computer input.

Computer Software

Software for a LIS includes all computer programs that are executed by the CPU to perform useful tasks in the laboratory. Software may also include certain data files or tables.

FUNCTIONS OF A LIS

Computer Programs

A computer program is a set of precise instructions that the computer executes in a step-by-step fashion in order to perform a well-defined task. Generally, the LIS is in a "watch-and-wait" mode until an operator asks for a program, whereupon (1) the computer selects the desired program from the disk; (2) loads all or part of the program into CPU memory; and (3) executes the instructions in the program beginning with the first instruction.

Computer Prompts

Invariably, the executing computer program requires additional information from the operator, so at various steps of the program, execution stops with a messages (called computer prompts) to the operator requesting additional information. For example, a computer program that collects information for a patient's census file would prompt the operator for (1) patient's identification number; (2) patient's name; (3) bed location; (4) billing number; (5) doctor's name; (6) admitting diagnosis, etc. After all of the needed information is keyed by the operator, the CPU program creates a "record" of the information in a "file" on the computer disk for later use. The program then cycles back (or loops) to the beginning of the program and prompts for the next patient's name. Hence, most laboratory computer programs (1) require initial operator input; (2) process information by performing manipulations and calculations; (3) output the processed information; and (4) repeat the cycle (see Figure 3.1). In LIS systems, most of the processing involves storage and retrieval of laboratory information (data) with less emphasis on complex calculations.

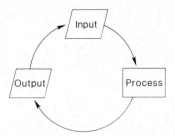

Figure 3.1. Computer cycle for input, processing, and output of data.

Sample Clinical Computer Programs

Sample programs in an LIS are ADM—admission of patient; TR—test request; EDIT—alter patient's data; BILLS—generate billing report; PCR—generate patient's cumulative report; and LIST—list the records in a file.

EXAMPLE 3.3. Write a computer program in the Basic computer language to (1) **input**—computer input of two numbers stored in variables X and Y; (2) **process**—sum the contents of X and Y and compute the average; (3) **output**—print the average of the two numbers; and (4) **cycle**—return program to "input" for additional number sets.

Answer:

1. A computer "input" statement consists of a line number, an input command, a screen prompt, and variable name. Hence, for input:

    10 INPUT "X ="; X

    20 INPUT "Y="; Y

2. Process instructions are analogous to algebraic equations. Hence, for process after input:

    30 A = (X + Y)/2

    (Note that the variable "A" will contain the "average" of the numbers stored in computer variables X and Y.)
3. A computer output statement includes a line number, a print command, a result label, and a variable name. Hence, for output:

    40 PRINT "AVERAGE ="; A

4. A "cycle" command returns the program to computer input. It consists of a line number, a "GOTO" command, and a target line number:

    50 GOTO 10

Our complete program then is

```
10 INPUT "X="; X
20 INPUT "Y="; Y
30 A = (X + Y)/2
40 PRINT "AVERAGE ="; A
50 GOTO 10
```

When the operator types the command "RUN", execution of the program when stored in computer memory will stop on each "INPUT" statement and wait for a keyed response from the operator (e.g., operator enters the desired number followed by a carriage return). The "carriage return" key on the terminal restarts the program. (All prompts and output will be displayed on the video display screen.)

X = ?

Note, the program stops with the above prompt at line 10. Entering a number and depressing the carriage return will result in

X = 3
Y = ?

Entering the value for Y will result in a screen display as follows:

```
X = 3             (carriage return by operator)
Y = 6             (carriage return by operator)
Average = 4.5     (output on screen)
X = ?             (cycle to line 10 for input)
```

Computer Files

In addition to processing information by arithmetic manipulation, the computer also arranges specific "fields" of data (e.g., patient name field, room number field, etc.) into records of a similar type (e.g., a patient admitting record). Each record is taken from computer memory and stored in a "data file" on the computer disk for later use (e.g., patients' file, test file). Computer programs are also stored on the disk. Data files can be created, updated, and listed by computer programs. Data files on the disk created as output from one computer program can serve as input data to another.

Laboratory Microcomputers

Smaller, more limited microcomputers or "personal computers" are used in most clinical laboratories, but, at this writing, microcomputers have not replaced the larger LIS minicomputer systems that control overall laboratory functions. Rather, microcomputers have been dedicated with enormous success to specific tasks on instruments. Microcomputers are often much easier to program than the LIS. Recently, microcomputers have been connected together (networked) so they can share disk drives for information storage. Very soon, it will be possible to computerize even large complex laboratories using this approach.

LABORATORY DESIGN AND WORKFLOW

The design of clinical laboratories has changed in recent years to reflect changes in automation and computerization. In the past, laboratories have generally been designed inefficiently with separate rooms or areas for the various departments (i.e., chemistry, hematology, etc.). Although this arrangement is somewhat functional during the day when areas are completely staffed, it is inadequate on evenings, nights, weekends, and holidays when staffing levels are at a minimum. A more feasible approach to laboratory design concentrates on testing priority (i.e., stat versus batch testing) and generally minimizes the separation of testing into separate rooms for each department (except when appropriate, for example, microbiology, radioassay, etc.).

An acceptable arrangement or design of the modern computerized laboratory involves (1) the centralization of receiving and processing as the "hub" of the laboratory; (2) the arrangement of a phlebotomy and transport staging area in close proximity to the central receiving area; and (3) the grouping of emergency testing equipment or stat work stations in close proximity to the central receiving area (no more than 10–20 steps away) with good visual and voice communication emphasized. Larger batch analyzers and departments (or subdepartments) that perform primarily batch testing are then grouped farther back from the central receiving "hub."

Cross-training of staffing, especially on evening and night shifts in chemistry, hematology, and urinalysis, etc., is encouraged in more efficiently operated laboratories.

MANAGEMENT STYLE

The newer, more automated laboratories require a highly trained technologist to operate the equipment and perform tests. The old dictatorial style management, common several years ago in clinical laboratories, has given way to a team approach where the manager functions best as a team-leader or coach. It is the manager's responsibility to provide the resources (staff, equipment, supplies, etc.) required to meet the major goals of the department. However, the staff in the department should have major input into how the goals are to be met. Major emphasis should be placed on the professional growth of the staff members. A good manager should place emphasis on the quality of work and avoid emphasis on quantity. This is especially required in the clinical laboratory, where laboratory errors greatly effect patients' care.

SCHEDULING OF WORK

It is most efficient for a manager to set up a rotating scheduling system where the employees can actually schedule themselves. Some preliminary management

Month: June–July															
Date:		27	28	29	30	1	2	3	4	5	6	7	8	9	10
Technologist	Rotating assignment	Mon	Tue	Wed	Thu	Fri	Sat	Sun	Mon	Tue	Wed	Thu	Fri	Sat	Sun
Jones, B.	Work station 2	V	V	V	V	V	Off	Off	7–3:30	7–3:30	7–3:30	7–3:30	7–3:30	Off	Off
Smith, S.	Float 2	7–3:30 W2	7–3:30 W2	7–3:30 W2	7–3:30 W2	7–3:30 W2	Off	Off	7–3:30	H	7–3:30 P2	7–3:30 P2	7–3:30 P2	Off	Off
Baker, B.	Work station 5	8–4:30	8–4:30	8–4:30	8–4:30	H	Off	Off	7–3:30	8–4:30	8–4:30	8–4:30	8–4:30	Off	Off
Sanford, B.	Work station 1	Off	Off	7–3:30	7–3:30	7–3:30	7–3:30	7–3:30	H	7–3:30	7–3:30	7–3:30	7–3:30	Off	Off
Lee, J.	Work station 6	8–4:30	8–4:30	8–4:30	8–4:30	8–4:30	7–3:30	7–3:30	H	Off	Off	8–4:30	8–4:30	Off	Off
Todd, C.	Float 1	7–3:30 W1	7–3:30 W1	7–3:30 P1	Off	Off	7–3:30	7–3:30	H	8–4:30 W6	8–4:30 W6	7–3:30 P3	7–3:30 P3	Off	Off
Haskins, F.	Work station 3	Off	Off	7–3:30	7–3:30	7–3:30	Off	Off	H	7–3:30	7–3:30	7–3:30	7–3:30	7–3:30	7–3:30
Martin, D.	Work station 4	8–4:30	8–4:30	Off	Off	8–4:30	Off	Off	H	8–4:30	8–4:30	8–4:30	8–4:30	7–3:30	7–3:30
Bush, T.	Float 3	7–3:30 W3	7–3:30 W3	8–4:30 W4	8–4:30 W4	8–4:30 W5	Off	Off	H	7–3:30 P4	7–3:30 P4	Off	Off	7–3:30	7–3:30

This two-week working timetable is the final stage in a scheduling process that begins when technologists block out their vacations at the beginning of the year. Every six weeks, technologists schedule their own compensatory weekdays off for the next three pay periods. Next the department supervisor schedules rotations through six regular and three floating work stations; "floats" fill in for absentees or work on special projects. Three teams, indicated by different colors, cover weekends and holidays in rotating order. Each two-week schedule is posted a week in advance of its effective date.

V: Vacation
H: Holiday
W1, W2, etc.: Work station 1, work station 2
P1: Project 1 (pipette calibration and workload recording)
P2: Project 2 (evaluation of new cortisol assay)
P3: Project 3 (blood gas lectures to medical technology students)
P4: Project 4 (monthly maintenance on automatic chemistry profile analyzer)

Figure 3.2. A sample two-week working schedule.

changes are required. Initially, specific work stations (stat and batch) are established as described earlier (see Table 3.2). Also, all staff members of the chemistry section must be cross-trained on all workstations. Weekend and holiday coverage of the section is based on a rotating schedule.

The scheduling process can involve three different scheduling documents: (1) a vacation schedule book (consisting of a calendar book); (2) a six-week tentative schedule; and (3) a two-week working schedule. Each year during the first week of January, the vacation book is circulated through the section by seniority allowing technologists to reserve vacation time with some restrictions.

Since most employees are only allowed to work 10 days during a two week pay-period, off days (or compensatory days) are required for individuals working weekends and holidays. To schedule these days, a tentative, incomplete six-week schedule is rotated through the section, which indicates the weekend and holiday rotating assignments along with vacations (and associated "off days") reserved in the vacation book. Technologists mark their desired off days to compensate for their weekend and holiday assignments. The one major restriction for reserving off days on the six-week schedule is that only a certain number of people can be scheduled off for any given weekday. This ensures that the major work stations can be covered. The final two-week working schedule is prepared from the six-week schedule by the manager (see Figure 3.2).

Note that the nine staff technologist names in Figure 3.2 are listed on the schedule with work stations rotated from top to bottom each two weeks. When one of the priority work station technologists was scheduled off (vacation, etc.), a "float" technologist is scheduled for coverage with the abbreviation of the work-station job assignment indicated along with the scheduled time to be at work. Workstations 4, 5, and 6 are all from 8:00 a.m. to 4:30 p.m. because they involved batch testing. The Stat Lab assignments are from 7:00 a.m. to 3:30 p.m. The working two-week schedule should be posted at least one week prior to the work period covered. This gives employees ample time to review their work station assignments for the coming week along with special developmental assignments or other delegated tasks.

BIBLIOGRAPHY

Bennington, J. L., Westlake, G. E., and Louvan, G. E. *Financial Management of the Clinical Laboratory*. Baltimore: University Park Press, 1974.

Bennington, J. L., Boer, G. B., Louvan, G. E., and Westlake, G. E. *Management and Cost Control Techniques for the Clinical Laboratory,* Baltimore: University Park Press, 1977.

Blanchard, K. and Johnson, S. *The One Minute Manager*. New York: William Morrow and Company, 1972.

Blick, K. E. "A do-it-yourself approach to scheduling," *Medical Laboratory Observer,* **15**(12), 99–104 (1983).

Chaney, W. H. and Beech, T. R. *The Union Epidemic—A Prescription for Supervisors*. Germantown, MD: Aspen Systems Corporation, 1976.

Deegan, A. X. *Management by Objective for Hospitals*. Germantown, MD: Aspen Systems Corporation, 1977.

Elevitch, F. R. and Aller, R. D. *Planning for a Computer in Your Laboratory*. American Society of Clinical Pathology, Commission on Special Education, 1980.

Henry, J. B. (ed.). *Clinical Diagnosis and Management,* 16th ed. Philadelphia: Saunders, 1979.

Johnson, J. L. *Achieving the Optimum Information System for the Laboratory*. Northbrook, IL: J. Lloyd Johnson Associates, 1976.

Lundberg, G. D. *Managing the Patient-Focused Laboratory*. Oradell, NJ: Medical Economics Company, 1975.

Metzger, N. *The Health Care Supervisor's Handbook*. Germantown, MD: Aspen Systems Corporation, 1978.

Newell, J. E. *Laboratory Management*. Boston: Little, Brown, 1972.

Teitz, N. (ed.). *Fundamentals of Clinical Chemistry*, 2nd ed. Philadelphia: Saunders, 1976.

Westlake, G. E. and Bennington, J. L. *Automation and Management in the Clinical Laboratory*. Baltimoe, MD: University Park Press, 1972.

Workload Recording Committee. *Manual for Laboratory Workload Recording Method*. Skokie, IL: College of American Pathologists, 1984.

Yanda, R. L. *Doctors as Managers of Health Care Teams*. New York: American Management Association, 1977.

POST-TEST

1. A serum iron was ordered and the blood was collected with the routine morning collection. That afternoon, the physician requested a repeat to verify test results. Since the morning sample was insufficient, a second afternoon sample was collected and assayed. You would expect the morning versus afternoon assay results to (choose the *best* answer):
   a. remain essentially the same since iron is not directly affected by nonfasting conditions.
   b. show variation since iron exhibits diurnal variation.
   c. show variation due to postural effects of protein bound substances.
   d. show variation because iron concentration is directly affected by exercise.

2. An ambulatory patient, who was to be admitted to the hospital, stopped by the laboratory to have some admission laboratory work drawn before proceeding to the hospital ward. The next day, additional work was collected with the morning routine fasting collection. The physician noted a discrepancy in the total protein assay results on day 1 and day 2. You could explain the discrepancy in the following way(s):
   a. The patient was nonfasting on admission and the resulting lipemia interfered with test methodology.
   b. Total protein exhibits postural effects, hence the specimen collected in the supine position will be lower.
   c. The presence of hemolysis in one of the samples may alter test results since Hb absorbs at the wavelength used in the Biuret methodology.
   d. The stoppered tube of separated serum from the first sample was left overnight at room temperature before analysis and owing to the instability of protein, test results on sample one would be altered.

3. A hemolyzed sample was received in the laboratory. The tests requested included BUN, glucose, calcium, iron, potassium, magnesium, acid phosphatase, and inorganic phophorus. Since the concentration of some of these components is higher inside the RBC's than in the plasma, falsely elevated test results would occur for:
   a. K^+, glucose, BUN, acid phosphatase, iron
   b. K^+, inorganic phosphorus, glucose, BUN

    c. Mg^{2+}, Ca^{2+}, glucose, acid phosphatase, iron

    d. K^+, Mg^{2+}, inorganic phosphorus, acid phosphatase, iron

4. A tourniquet was used to collect a venous blood sample for pH measurement. The technologist applied the tourniquet and asked the patient to open and close his hand in order to fill the vein. After entering the vein, the tourniquet was released and the sample was drawn. You would expect the pH results to be:

    a. unaffected

    b. falsely alkaline

    c. falsely acidotic

    Justify your answer.

5. A morning microcapillary sample from pediatrics was received in the laboratory for bilirubin assay. The physician phoned the laboratory in the afternoon and requested that the assay be repeated. The morning sample had been left by the window in the collection rack. Precluding any assay errors, do you think the results would correlate? Justify your answer.

6. The fasting and one hour specimens for a glucose tolerance test were collected by venipuncture. The technologist was unable to collect the third sample by venipuncture, so she resorted to capillary collection for the remainder of the test. The samples were assayed on the Beckman Glucose/BUN analyzer. Mark all answers below that apply to this situation.

    a. The test would be unaffected since the glucose concentration in venous and capillary blood is the same.

    b. This change in the collection site at mid-test is unacceptable since the glucose content in capillary blood is higher than in venous blood.

    c. It is possible that the patient test results may be interpreted as glucose intolerant even though the patient may be normal.

    d. Since the oxygen content would be higher in a capillary blood than in venous blood, the test results would be affected using the Beckman system for glucose assay.

7. A sample was collected for blood gas analysis. The sample was exposed to air before measurement. Test results would be affected in the following way:

    a. pH $\uparrow$, pCO_2 $\uparrow$, pO_2 $\downarrow$

    b. pH $\downarrow$, pCO_2 $\uparrow$, pO_2 $\downarrow$

    c. pH $\uparrow$, pCO_2 $\downarrow$, pO_2 $\uparrow$

    d. pH $\downarrow$, pCO_2 $\downarrow$, pO_2 $\uparrow$

8. In the same person, the concentration of protein bound components in the plasma will be _____ in an upright position as opposed to a supine position.

    a. unaffected

    b. increased

    c. decreased

9. Computer programs are generally designed for execution in the following order:

    a. process–input–cycle–output

    b. output–input–process–cycle

     c. input–process–cycle–output

     d. input–process–output–cycle

10. Computer software consists of the following:
    a. computer programs and data files
    b. CPU and computer programs
    c. terminals
    d. disk drives and magnetic tape drives

11. The correct relationship of data stored on a computer data file is
    a. fields → records → files
    b. records → fields → files
    c. files → records → fields
    d. records → files → fields

12. Computer programs are _____ on the disk and _____ in memory.

13. Productivity of a laboratory section is computed by:
    a. CAP workload/specified time
    b. paid time/CAP workload
    c. CAP workload/hours worked
    d. none of the above

ANSWERS

1. b
2. a, b, and c
3. d
4. c—when the tourniquet was released the collection site was flushed with blood containing CO_2 and lactic acid—both will lower pH.
5. no—unconjugated bilirubin is broken down by light, especially UV light which is present in sunlight, therefore the afternoon value will be falsely decreased
6. b, c
7. c
8. b
9. d
10. a
11. a
12. stored; executed
13. c

CHAPTER 4
INSTRUMENTATION

115

XII.   GAS CHROMATOGRAPHY
    A.   Gas–Liquid Chromatography
        1. Principle
        2. Sample Requirements
        3. Stationary Phase
        4. Mobile Phase
    B.   Basic Components
        1. Carrier Gas
        2. Gas FLow Regulator
        3. Sample Injection
        4. Column Oven
        5. Column
        6. Detectors
        7. Recorder
    C. Uses

XIII.   IMMUNOASSAY TECHNIQUES
    A. Radioimmunoassay
        1. RIA Competitive Binding Assay
            a. Principle
            b. Test
            c. Interpretation
        2. Immunoradiometric Analysis (IRMA)
        3. Uses
    B.   Enzyme Immunoassay
        1. Enzyme-Multiplied Immunoassay Technique (EMIT)
            a. Principle
            b. Test
            c. Interpretation
            d. Uses
        2. Enzyme-Linked Immunoabsorbent Assays (ELISA)
            a. ELISA Sandwich Technique
            b. ELISA Immunoenzymometric Technique
            c. ELISA Competitive Binding Technique
            d. ELISA Antibody Quantitation
            e. Uses
    C. Fluorescent Immunoassay (FIA)
        1. Uses

OBJECTIVES

On completion of this chapter, you will be able to:

1.   Define the following terms:
    a. visible light
    b. ultraviolet light

    c.   infrared light

    d.   wavelength

    e.   photon

    f.   nanometer

    g.   transmittance

    h.   absorbance

    i.   bandpass

2. Discuss what is meant when a procedure is described as following Beer's law.

3. Calculate concentration values using Beer's law.

4. Draw the component arrangement and describe the function of each component of a photometer or spectrophotometer.

5. Discuss the methods used to check for:

    a.   wavelength calibration

    b.   linearity

    c.   photometric accuracy

    d.   stray light

6. For each instrument listed below, (1) discuss principle of operation, (2) draw component arrangement and discuss the function of each component, (3) list the potential sources of error, and (4) list analytes measured:

    a.   nephelometer

    b.   fluorometer

    c.   flame photometer

    d.   atomic absorption spectrophotometer

    e.   scintillation counter

7. Define the following terms:

    a.   nuclide

    b.   beta emission

    c.   atomic number

    d.   mass number

    e.   alpha emission

    f.   gamma emission

    g.   fluor

8. List the instruments that measure light absorption; list the instruments that measure light emission.

9. List the four general groups of automated systems and discuss the principle of operation for each system.

10. In regard to chromatography, define the following terms:

    a.   stationary phase

    b.   mobile phase

    c.   support phase

    d.   normal phase

    e.   reverse phase

11. Draw the component arrangement and describe the function of each component of a liquid chromatograph system.

12. Discuss the principle of separation for:
     a. adsorption chromatography
     b. partition chromatography
     c. ion exchange chromatography
     d. molecular exclusion chromatography

13. Calculate R_f values.

14. Draw the component arrangement and describe the function of each component of a gas chromatograph.

15. In regard to immunoassay procedures, define the following terms:
     a. homogeneous methods
     b. heterogeneous methods
     c. antigen
     d. antibody
     e. radioimmunoassay
     f. enzyme immunoassay
     g. fluorescent immunoassay
     h. EMIT
     i. ELISA

16. Discuss the principle of competitive binding immunoassay procedures.

GLOSSARY OF TERMS

Abscissa. The horizontal scale or x axis on a chart or graph.

Absorption. To incorporate the material by taking it in.

Adsorption. To attach a substance to a surface without incorporating it inside.

Anode. The positive electrode in electron tubes and electrophoretic cells. The negative electrode in a voltaic cell.

Antibody. A glycoprotein whose synthesis is stimulated in man and animals by the presence of an antigen. The antibody binds to the stimulating antigen with specificity.

Antigen. A molecule or chemical substance that when introduced into an animal stimulates the synthesis of specific antibodies capable of binding specifically with the antigen.

Bandpass (bandwidth). The width of the spectrum that is allowed to pass through the exit slit of the wavelength-selecting device.

Band spectra. Groups of emission lines that merge forming a band of spectral energy.

Cathode. The negative electrode in electron tubes and electrophoretic cells. The positive electrode in a voltaic cell.

Chromatogram. The pattern of solute separation.

Continuous spectra. A spectrum that contains light energy of about the same intensity over a wide range of wavelengths.

Detector. A device that converts light energy into electrical energy.

Deuterium lamp. A UV light source that is used to make measurements in the ultraviolet portion of the spectrum.

Excited state. The energy state produced when orbital electrons of an atom or molecule absorb sufficient energy to raise electrons to higher energy levels.

Fluor. Substances that are capable of converting the kinetic energy of particles or photons into flashes of light.

Ground state. The unexcited state of an atom or molecule in which orbital electrons are in their normal orbital planes.

Harmonic wavelength. A light wave whose frequency is some integral multiple of a fundamental frequency.

Heterogeneous methods. Immunoassay methods that require the separation of the bound and unbound (free) fractions.

Homogeneous methods. Immunoassay methods that do not require the separation of the bound and unbound (free) fractions.

Infrared light. Electromagnetic energy, less energetic than visible light, defined as those wavelengths that fall in the 700–2000 nm (approximate) range.

Ionization. The removal of one of the orbital electrons from an atom that results in the production of a positive ion and an electron. Ionization can occur when the energy supplied to the atom is of sufficient nature that an orbital electron is removed rather than only raised to a higher energy level.

Line spectra. Emitted energy that is monochromatic (consisting of a single wavelength). The emission spectra would appear as well defined lines on a spectral emission plot.

Linearity. The relationship existing between absorbance and concentration that produces direct proportionality.

Mobile phase. The solvent or gas that transports the material to be separated through a column.

Monochromator. A device that is used to isolate a selected portion of the electromagnetic spectrum.

Nanometer. A unit of measure (10^{-9} m) that describes the frequency interval of wave cycles in the ultraviolet and visible portions of the spectrum.

Ordinate. The vertical scale or y axis on a chart or graph.

Phase. The fractional part of a cycle through which a periodic wave (light wave) has advanced at any instant with reference to a standard position.

Rotational energy. The energy with which the entire molecule rotates around some axis.

Stationary phase. The separating material that remains in a fixed position in the chromatographic column. The stationary phase may be a solid or liquid.

Stray light. All radiant energy that reaches the detector and falls outside of the spectral bandpass selected by the monochromator.

Support phase. An inert material to which the stationary phase is attached in a chromatographic column.

Tungsten lamp.   Light source that emits a continuous spectrum in the upper UV and visible portion of the spectrum.

Ultraviolet light.   Electromagnetic energy, more energetic than visible light, defined as those wavelengths that fall in the 200–400 nm (approximate) range.

Vibrational energy.   The energy with which interatomic bonds vibrate within the molecule.

Visible light.   Electromagnetic energy visible to the eye, defined as those wavelengths that fall in the 400–700 nm (approximate) range.

Wavelength.   The distance between successive wave crests of an electromagnetic wave cycle, which is measured in nanometers.

Many instruments commonly used in clinical laboratories measure electromagnetic radiation. Photometers, spectrophotometers, and atomic absorption instruments measure light absorption; flame photometers, fluorometers, and scintillation (gamma) counters measure light emission; nephelometers measure scattered light.

LIGHT

DEFINITION

Light is a form of radiant energy. Physicists have described radiant energy as behaving as though it possessed an electrical field and a magnetic field, hence the term electromagnetic radiation. Light is a visible form of electromagnetic radiation. Other nonvisible forms of electromagnetic radiation include gamma radiation, X-rays, ultraviolet, infrared, microwaves, and radio waves.

There is proof that under suitable conditions light and all electromagnetic radiation behave as waves in motion. These waves can be pictured as those that would be produced when a pebble is thrown into water. If a portion of the wave cycle could be viewed horizontally, it would appear as a series of peaks and troughs (Figure 4.1). The distance between successive wave crests is a measure of wavelength (λ), and this distance is measured in nanometers (10^{-9} m). In the past, wavelength was measured in angstroms (10^{-10} m) or millimicron units. A millimicron is the same as a nanometer. Angstrom and millimicron are considered obsolete terminology. Wavelength is the factor that determines color in the visible spectrum.

In addition to possessing wavelike properties, light is also described as possessing energy packets called photons, that is, tiny particles that have energy but no

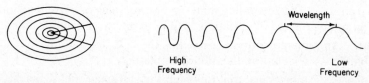

Figure 4.1.   The wave theory describes light as traveling in a wavelike pattern.

Table 4.1. The Electromagnetic Spectrum

Energy and Frequency Relationship	Type of Radiation	Approximate Range of Wavelengths Included
More energy	Gamma	
Greater frequency	X-rays	
Short wavelength	Ultraviolet	200–400 nm
	Visible	400–700 nm
Less energy	Infrared	700–2000 nm
Lesser frequency	Microwaves	
Long wavelength	Radio waves	

mass. Frequency is the term used to describe the repetition of wave cycles per second. Light energy is directly related to frequency, that is, light of greater frequency possesses more energy than light of lesser frequency (Figure 4.1). The greater the frequency, the shorter the wavelength and vice versa. Light energy and wavelength are, therefore, inversely related. The range of the electromagnetic spectrum is indicated in Table 4.1.

Color description is not precise because the range of wavelengths that comprise the visible portion of the spectrum varies from person to person owing to differences in individual eyesight. When the visible portion of the spectrum is arranged by wavelength (400–700 nm), a continuous spectrum is formed. Each color is visible in a specified range of wavelengths. Beginning at the shortest wavelength, the colors in order are violet → blue → green → yellow → orange → red. The visible color of a solution, that is, the color you actually see, is determined by the wavelengths of light that are transmitted, not absorbed. In photometry and spectrophotometry, the color that is absorbed is measured. The color absorbed is the complementary color of that transmitted. For example, a solution that appears purple to the eye (color transmitted) absorbs principally in the green portion of the spectrum. The wavelength chosen to measure the absorbance of the purple solution would be in the green portion of the spectrum.

SPECTRAL SCAN

In actual practice, the wavelength selected in a test procedure is determined by performing a spectral scan. Every colored solution has a characteristic absorption spectrum. This spectrum is determined by recording the absorbance of a solution at various wavelengths. The values, when plotted on linear graph paper, wavelength on the abscissa and absorbance on the ordinate, form a spectral scan (Figure 4.2).

The wavelengths at which the substance absorbs most strongly are called absorption maxima; the wavelengths at which the substance transmits best are called absorption minima. The wavelength chosen to read a specific procedure is usually that wavelength where the absorbance is maximal because the sensitivity (ability to detect small changes in concentration) of the procedure is greatest at

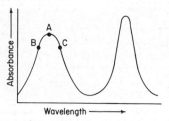

Figure 4.2. Spectral absorbance scan of green food coloring.

this wavelength. However, other variables must be considered in wavelength selection.

Peak Characteristics

It is best to choose a wavelength that is on a flat portion of the absorption peak, that is, point *A*, Figure 4.2. If point *B* or *C* were chosen, a small error in wavelength setting may result in a large absorbance error because the slope of the absorbance peak at *B* and *C* is quite steep. A small error in wavelength setting would not result in a large change in absorbance with point *A* because the peak is wide and relatively flat. In this absorbance scan, more than one absorbance peak exists. The second peak is sharper than the first peak, therefore the wider peak would be the better choice for measurements.

Presence of Impurities

If there is a choice of wavelengths, and an impurity absorbs at one wavelength but not another, the choice wavelength would be that at which the impurity does not absorb. The impurities often encountered in blood samples include hemoglobin and bilirubin. Hemoglobin absorbs at 575 nm and 540 nm and bilirubin absorbs at 460 nm; therefore, if a choice is available, a wavelength in another portion of the spectrum would be more desirable. If an alternative wavelength is not available, serum blanks are needed to correct for the presence of these interfering impurities.

Test Solvent

The solvent used in the test procedure must not absorb at the wavelength chosen to read the test. In the Jaffe procedure for creatinine, the reagents show maximal interference at the wavelength of maximal absorbance of creatinine; therefore, an alternate wavelength must be selected for this procedure.

Extended Linearity

By choosing a wavelength where the substance does not absorb maximally, the linearity of the procedure may be extended. The reduction in sensitivity may change the slope of the calibration line and extend the linear portion of the standard curve.

BEER'S LAW

Beer's law states that absorbance (optical density) is directly proportional to the concentration of the substance absorbing light or inversely proportional to the logarithm of the transmitted light. The law also states that the light source must be monochromatic (light of one specific wavelength) and the distance traveled by the light through the cuvette must be fixed. Beer's law may be expressed in the following way:

$$A = abc$$

where A = absorbance

a = absorptivity of the substance, which is characteristic for each substance depending on its chemical content, molecular structure, and properties of solvent/solution used

b = light path length, which is determined by cuvette size

c = concentration of the absorbing substance

Since absorbtivity (a) or component being assayed remains constant for each test, and light path (b) remains constant for a particular cuvette size, absorbance (A) becomes directly proportional to concentration (c), that is, $A \propto c$.

Absorbance values have no units and cannot be measured directly. The light transmitted through a solution can be measured, and the transmittance (T) is defined as the ratio of transmitted light (or light emerging from the cuvette, I_s) to incident light (or light entering the cuvette, I_0) or I_s/I_0. The incident light is always expressed as 1, or in percent (i.e., 100%), since the instrument is always adjusted to 100% transmittance or zero absorbance with a reference solution (blank) prior to reading test solutions. Since absorbance and transmittance are logarithmically and reciprocally related, absorbance may be expressed as

$$A = - \log \frac{I_s}{I_0} = - \log T = \log \frac{1}{T}$$

To convert T to %T, multiply numerator and denominator by 100:

$$A = \log \frac{1}{T} = \log \frac{100}{\%T} = \log 100 - \log \%T = 2 - \log \%T$$

All photometers and spectrophotometers measure light transmitted; absorbance is derived from the transmittance readings by using the formula, $A = 2 - \log \%T$.

To determine if a procedure follows Beer's law on a particular instrument, a standard curve is needed. To set up a standard curve, a concentrated standard is diluted so that at least three, preferably four or more, dilutions are obtained. Each dilution is assayed by the method under evaluation. The results are plotted on graph paper. To plot absorbance (ordinate) versus concentration (abscissa), use coordinate graph paper (both scales in linear units). To plot percent transmittance (log or ordinate) versus concentration (abscissa), one-cycle semi-log graph paper should be used. When Beer's law is followed, the plotted points, when connected, will form a straight line and will show direct proportionality between concentration and absorbance.

These concepts are illustrated in Table 4.2 and Figure 4.3. A standard curve was constructed for the creatinine procedure using the Jaffe reaction. The standards were read against a water blank and against a reagent blank (Table 4.2). In the creatinine procedure, the reagents also absorb at the test wavelength. Therefore, it is necessary to read the reagent blank against water to determine the y intercept at zero concentration for the plot read against water.

Note that both plots produce straight lines when the standard points are connected (Figure 4.3). However, this procedure follows Beer's law only when read against a reagent blank. For the reagent blank curve, note the direct proportionality between standard values and absorbance units. In a Beer's law curve, each successive concentration and its absorbance will increase in the same multiple units. In the example given, note that the 1.5 mg/dL standard has an absorbance of 0.113; therefore, the 3.0 mg/dL standard will have an absorbance of 0.226, and so on.

This direct proportionality does not occur with those standards read against the water blank, because the slope of the line does not produce this direct proportionality relationship.

When Beer's law is followed, a calculated *factor* for each standard point is the same. This factor is derived by dividing the concentration of each standard by its absorbance (c/A). In Table 4.2, the factor (13.3) is the same for all standards read against a reagent blank, but the factor varies widely for those standards read against a water blank.

Table 4.2. Standard Curve Assay Results, Creatinine Procedure, Jaffe Reaction

Concentration	%T	Absorbance	Factor c/A
Standards Read Against Water Blank			
Water blank	100	0.000	
Reagent blank	73	0.136	
1.5 mg/dL	57	0.244	6.1
3.0 mg/dL	43.5	0.362	8.3
4.5 mg/dL	33.5	0.475	9.5
6.0 mg/dL	26	0.585	10.2
Standards Read Against Reagent Blank			
Reagent blank	100	0.000	
1.5 mg/dL	77	0.113	13.3
3.0 mg/dL	59.5	0.226	13.3
4.5 mg/dL	46	0.339	13.3
6.0 mg/dL	35.5	0.452	13.3

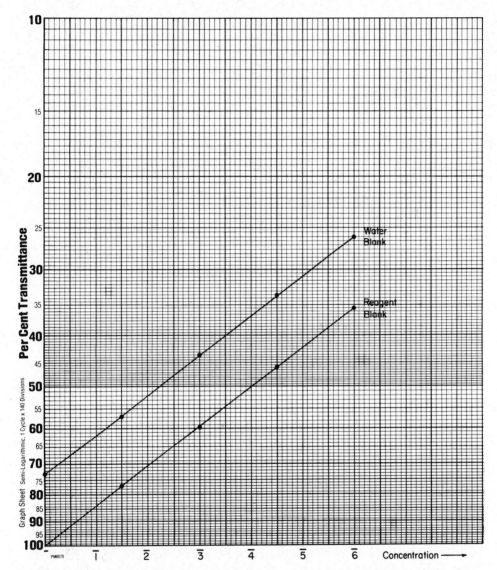

Figure 4.3. Standard curve, creatinine procedure, Jaffe reaction. The standards were read against a reagent blank and against a water blank and the results were plotted on one-cycle, semi-log paper, % T, ordinate, concentration, abscissa.

When Beer's law is followed, unknown concentrations may be determined by a single standard (because the factor is the same throughout the linear area of the curve) using Beer's law formula, which is derived from the direct proportionality statement:

$$\frac{\text{absorbance of standard } (A_s)}{\text{concentration of standard } (c_s)} = \frac{\text{absorbance of unknown } (A_u)}{\text{concentration of unknown } (c_u)}$$

When the direct proportionality statement is solved for the concentration of the unknown, Beer's law formula is derived:

$$(A_s)(c_u) = (c_s)(A_u)$$

$$c_u = \frac{A_u}{A_s} \times c_s$$

The following example illustrates the use of Beer's law to determine unknown concentrations.

EXAMPLE 4.1. Serum total protein was determined by the Biuret procedure (which follows Beer's law). The following absorbance values were obtained:

Standard (7.1 g/dL) = 0.162
Control (acceptable range, 4.3–4.7 g/dL) = 0.102
Patient #1 = 0.154
Patient #2 = 0.216

The concentration values calculated by using Beer's law formula are:

Control: $\dfrac{0.102}{0.162} \times 7.1$ g/dL = 4.5 g/dL

Patient #1: $\dfrac{0.154}{0.162} \times 7.1$ g/dL = 6.7 g/dL

Patient #2: $\dfrac{0.216}{0.162} \times 7.1$ g/dL = 9.5 g/dL

Beer's law may only be valid over a narrow concentration range. When Beer's law formula is used to calculate unknown concentrations, all measurements must be confined to the linear portion of the curve. If sample concentrations exceed the linear limits of the procedure, dilution of the sample with reassay will be necessary. The answer is then corrected by the dilution factor.

Some procedures do not follow Beer's law. For these procedures, unknown concentrations can only be determined by reading concentration values directly from a standard curve plot.

PHOTOMETERS AND SPECTROPHOTOMETERS

PRINCIPLE

Photometers and spectrophotometers measure light transmitted by a solution in order to determine the concentration of a light-absorbing substance present in the solution.

BASIC COMPONENTS

The major components of photometers and spectrophotometers are illustrated in Figure 4.4. The lamp or light source provides the radiant energy; the monochromator selects the band of light that passes to the cuvette; the cuvette holds the sample and provides a constant light path; the detector senses the radiant energy that is not absorbed by the sample; and the meter provides a readout of the transmitted light, either in $\%T$ or in absorbance units (which are calculated by the formula, $A = 2 - \log \%T$). Each individual component will be discussed in detail.

Light Source

The function of the light is to provide a source of radiant energy. Tungsten filament lamps emit light of many wavelengths in a continuous pattern and can be used for measurements at wavelengths in the visible and longer ultraviolet portion of the spectrum (from 350 to 700 nm). At wavelengths below 340 nm, the glass envelope surrounding the filament begins to absorb, therefore the tungsten lamp is not recommended for measurements below 350 nm. Some authors recommend a cut-off point of 360 nm. At operating temperatures, some of the tungsten metal may vaporize and then condense on the glass envelope of the bulb. The darkened surface reduces the intensity of the radiant energy, which can change the spectrum emitted. When darkening occurs, the light source must be replaced to avoid this introduction of error.

The deuterium lamp, which provides a continuous spectrum in the ultraviolet region, is recommended for measurements in the wavelength range of 190–360 nm. At longer wavelengths, the emission is no longer continuous. Hydrogen, mercury, and xenon arc lamps may also be used for measurements in the ultraviolet region of the spectrum.

The spectrum emitted from the light source will vary with changes in temperature; therefore, the voltage must be carefully regulated to ensure that the light spectrum remains unchanged.

Monochromator

The function of the monochromator is to isolate the desired wavelength of light and to exclude unwanted wavelengths.

Photometers use filters to select desired wavelengths. Glass filters function to absorb light of wavelengths other than the desired wavelength. Filters are not "true monochromators" since they transmit light of more than one wavelength. Filters may be made of one or more layers of colored glass or may consist of colored gelatin between clear glass plates (Wratten filters). Interference filters are able to provide light of greater spectral purity than is possible with glass filters. Interference filters are constructed of two pieces of mirrored glass separated by a spacer. The thickness of the spacer determines the wavelength selectivity (the distance must be precisely one-half the desired wavelength). Interference filters transmit

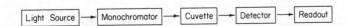

Figure 4.4. Arrangement of components in a single beam spectrophotometer.

only wavelengths in phase (those wavelengths two times the spacer distance or multiples of these). The combined energies of those wavelengths in phase is sufficient to break through the mirrored surface and pass on to the cuvette. Other wavelengths will cancel due to phase differences and will not be transmitted. Since interference filters also pass multiples of the desired wavelength (i.e., those $\frac{1}{2}$ or $2\times$ the desired wavelength), accessory glass filters are required to eliminate these harmonic wavelengths. Interference filters remove unwanted wavelengths by destructive interference and not by absorption.

In spectrophotometers, the light is dispersed into a spectrum from which the desired wavelength (or band) of light is isolated by mechanical slits. Usually, a designated narrow band is isolated. The slit width determines the width of the light band (bandpass) that is allowed to pass to the cuvette. Simple spectrophotometers have a fixed slit width. More complicated instruments have an adjustable slit width that can allow manual selection of the bandpass.

The light may be dispersed into a wavelength spectrum by either a prism or a diffraction grating. When prisms are used for this dispersion, the shorter wavelengths are diverted or bent more strongly than the longer wavelengths. The spectrum emitted is nonparallel with the longer wavelengths more crowded than the shorter wavelengths (Figure 4.5). Ordinary glass prisms are desirable because their dispersion qualities are better than quartz. However, shorter wavelengths are mainly absorbed by glass. To include spectra below 350 nm, the prism material must be quartz or fused silica.

A diffraction grating produces a parallel or linear spectrum. It consists of a highly polished plate with many etched parallel sharp cornered grooves spaced at close intervals. Since rays of radiant energy will bend around a sharp corner with the degree of bending dependent on wavelength, as light strikes the grating, many tiny spectra are formed, one for each line of the grating. As the light moves past grooved corners, wave fronts are formed. Those in phase reinforce one another, those not in phase cancel one another and disappear. A linear spectrum results. Quality diffraction gratings generally provide better spectral isolation than do prisms. Because of imperfections in the ruling of the grooves, stray light error is more likely to occur in those instruments that use diffraction gratings than in those using prisms. *Stray light* is defined as light striking the detector at wavelengths other than that for which the wavelength dial is set.

To select the desired wavelength, the prism or diffraction grating is rotated directing the desired wavelength through the exit slit.

Half-bandpass or half-bandwidth are terms used to describe the spectral purity of a filter or other monochromator. Bandpass or bandwidth describes the width of the band of light that is allowed to pass to the sample and is measured in nanometers. The half-bandpass of a monochromator is obtained by plotting the intensity of the light emerging from the monochromator against the wavelength and measuring the peak width at one-half peak height (Figure 4.6).

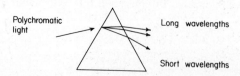

Polychromatic light

Long wavelengths

Short wavelengths

Figure 4.5. Diffraction of light through a prism.

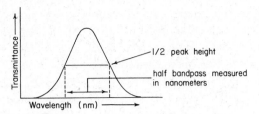

Figure 4.6. Spectral transmission scan and measurement of half-bandpass of a monochromator.

A narrow bandpass instrument will be able to isolate a narrow portion of the spectrum, hence the light striking the cuvette will be more monochromatic than in a wide bandpass instrument. The narrower the bandpass, the more sensitive the instrument is to concentration changes. The bandpass in state-of-the art instruments ranges from 2 nm to fractions of a nanometer. The bandpass can be manually adjusted to different widths. In wide bandpass instruments, the bandpass is generally fixed and may be in the range of 10 nm or greater.

In order to provide an accurate absorbance value, the spectral bandpass of a spectrophotometer must be one-tenth or less the width of the absorption peak (of the substance assayed) measured at half-peak height. For example, at 340 nm, nicotinamide adenine dinucleotide (NADH) has a half-peak height width of 60 nm. In order to measure NADH within desired accuracy limits, the spectrophotometer should have a bandpass of 6 nm or less. Many chromophores, or substances measured in the visible portion of the spectrum, have relatively wide absorbance bands (many, 100 nm or more), therefore wide bandpass instruments may be used for the measurement of these substances. Wide bandpass instruments would be unacceptable for measurements in the ultraviolet portion of the spectrum because absorbance bands in this region are much narrower (60 nm or less). A procedure that follows Beer's law on narrow bandpass instruments may not adhere to Beer's law on wide bandpass instruments.

Cuvettes

The function of the cuvette is to hold the solution in the instrument when measuring absorbance and to provide a constant light path length.

Cuvettes may be round or square; square cuvettes are preferred to round cuvettes. When light is directed onto round surfaces, some light is lost by reflection and refraction. Since the surfaces of square cuvettes are flat, loss of light by these processes is minimal. Square cuvettes also provide a constant light path (usually 1 cm); the light path may vary in round cuvettes, since they are not necessarily round but oval. Round cuvettes should always be matched and then positioned in the spectrophotometer in exactly the same manner each time they are used. Purchased cuvettes are matched at the manufacturing site and are marked with a label. The label should always be placed in the same position when measurements are made. To match round cuvettes, the following procedure may be used:

1. Prepare a stable solution of a colored substance in a concentration that will read approximately 50% T at the appropriate wavelength.

2. Set the instrument wavelength to correspond to the appropriate wavelength for measurement. Adjust the instrument to 100% T with the reference blank.
3. Place the solution into a cuvette and note the % T reading. Etch the top of the tube to indicate the placement position.
4. Add the solution to each dry cuvette to be tested and note the % T readings. The cuvettes may be rotated to obtain the best position to match the % T reading of the beginning or reference cuvette. Permanently mark the top so that the tubes may be correctly positioned in the spectrophotometer when future readings are taken. Discard all tubes that are outside selected tolerance limits (i.e., 0.5% T or other selected limits).

When measurements are made in the visible portion of the spectrum (approximately 350–700 nm), glass or disposable plastic cuvettes are preferred. Glass absorbs ultraviolet light; therefore, this type of cuvette cannot be used when measurements are made at wavelengths below 350 nm. Quartz or fused silica cuvettes are used in the ultraviolet range (approximately 200–400 nm).

Cuvettes should be cleaned immediately after use and should never be left to soak in detergent solution. They may be washed with a mild detergent, rinsed well with tap water and then in deionized water, and placed upside down in plastic-coated racks to dry.

Detectors

The function of the detector is to convert light energy into electrical energy in proportion to the intensity of the light striking the sensing surface.

A barrier layer cell (photocell) is composed of photosensitive material that will release electrons when exposed to light energy. The flow of electrons is in proportion to the intensity of the light that strikes the photosensitive material. No external voltage source is required to operate a barrier layer cell. It relies on internal electron transfer to produce a current in an external circuit. This type of detector is used with wide bandpass instruments or those with higher levels of illumination. The output of energy is seldom amplified. Barrier layer cells may become "fatigued"; with continuous high light intensity, the electrical output decreases with time. Barrier layer cells are slow to respond to changes in light intensity. For this reason, they cannot be used in instruments which "scan" the wavelength spectrum or in those utilizing "choppers". The electrical output is also very temperature dependent. The response of barrier layer cells becomes nonlinear at very high or very low levels of illumination.

A second type of detector is the phototube, which requires an outside voltage source for operation. In the presence of light, electrons are emitted from the cathode. The electrons are attracted to a positive anode where they are collected and returned through an external measurable circuit. The type of material used in the cathode determines the range of wavelengths at which the tube will give its highest response.

Photomultiplier tubes operate on the same principle as phototubes; however, the electrons emitted from the cathode are attracted to a series of anodes called dynodes. These dynodes will give off secondary electrons when struck by other electrons, thus multiplying (or amplifying) the signal. Since photomultiplier tubes *amplify* the initial signal, low levels of illumination can be measured. Any stray light can introduce significant error, since the signal from the stray light is also amplified. Photomultiplier tubes are used as detectors in narrow bandpass instruments. They respond rapidly to changes in light intensity. Therefore, they may be used in instruments with wavelength scanners or in double beam spectrophotometers where a fast response time is necessary.

Readout Device

The readout device functions to read the electrical output from the detector and then to present the data in some interpretable form. The data are usually presented as either transmittance or absorbance. Some instruments are capable of reporting the data directly in concentration units. Readout devices may include meters, digital displays, printed readouts, or "strip-chart" recorders that provide line tracings of transmittance or absorbance.

Power Source

The intensity of the light emitted from the light source must be constant. If voltage changes occur, the output from the lamp varies. Voltage regulators are built into instruments to keep the voltage applied to the detector and lamp constant. Errors associated with voltage fluctuations can be avoided with the use of double beam spectrophotometers. Schematic representations of a double beam-in-space and a double beam-in-time spectrophotometer are illustrated in Figure 4.7.

In the double beam-in-space spectrophotometer, a single light source is split by a mirror. The light is passed through two monochromators. Half of the light is directed to the reference cuvette and half is directed to the sample cuvette. The signal output from each sample compartment is detected by two photomultiplier tubes. The readout device compares the signal output from each detector. A ratio of the two signals is an indication of voltage output. Any voltage fluctuation that would affect the output intensity of the radiant energy source would affect both

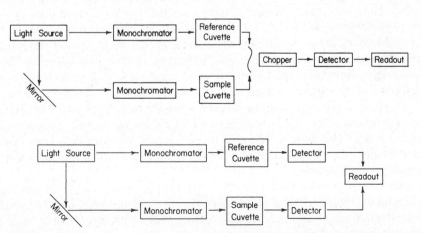

Figure 4.7. Double beam-in-time spectrophotometer (top); double beam-in-space spectrophotometer (bottom).

signals equally. Since both are affected equally, the ratio would remain the same and no error would be introduced.

Double beam-in-time spectrophotometers use only one detector. A chopper, which is a rotating disk that presents alternately a mirror and an opening is placed distal to the cuvette compartments. The chopper alternatively focuses the sample and reference beams onto the detector, thus allowing the detector to discriminate between the source of each signal. This arrangement of components would eliminate error that might be introduced due to unequal response of dual photomultiplier tubes.

PROCEDURES TO CHECK PHOTOMETRIC FUNCTIONS

Wavelength Calibration

Wavelength calibration is performed to verify that the wavelength indicated by the instrument wavelength dial corresponds to that selected by the instrument. It is advisable to check wavelength calibration daily. If the light source is changed, wavelength calibration is mandatory.

One method for checking wavelength calibration involves the use of rare-earth glass filters such as didymium or holium oxide. The didymium filter has several broad absorption bands and may be used to verify wavelength calibration with wide bandpass instruments. The holium oxide filter has a number of sharp absorption bands that occur at precisely known wavelengths in the visible and ultraviolet (280–650 nm range) region of the spectrum and may be used to verify wavelength calibration with narrow bandpass instruments. These filters, available from the National Bureau of Standards (NBS), can be inserted directly into the sample compartment of the instrument. Wavelength calibration is verified by comparing absorption peaks as indicated by the instrument with established values provided by the manufacturer of the calibration filter. Since filters may deteriorate with time, the wavelength accuracy should be checked periodically.

A more accurate method to verify wavelength calibration involves replacing the source lamp with a radiant energy lamp that has strong emission lines at well-defined wavelengths. Mercury vapor, hydrogen, and deuterium lamps have this property. Mercury lamps have many emission lines between 300 and 600 nm and are therefore used to verify wavelength calibration in the ultraviolet and visible spectral range. Deuterium and hydrogen lamps have intense emission lines at 656 and 486 nm and hence can be used to verify wavelength calibration in the visible range. To verify wavelength calibration, the wavelength selector of the instrument is adjusted several nanometers to either side of the known strong emission lines provided by the light source. The absorbance output is observed. The point of maximal absorbance should correspond to the known emission lines characteristic of the source lamp in use.

A third method to verify wavelength calibration involves the use of pure solutions that have characteristic absorption peaks. For example, holium chloride has sharp absorption peaks and may provide a two-point check at 241.1 and 361.5 nm in narrow bandpass (2 nm) instruments.

All calibration procedures compare instrument response at designated wavelengths to known absorption peaks of a reference material. The operator's manual provided with each instrument should give detailed instructions of wavelength calibration.

Linearity Check

A linearity check is performed to certify that the detector response is linear over the range of wavelengths used to make measurements (i.e., the useful analytical range).

The most common method for certification of linearity uses solutions that are known to follow Beer's law. A solution of known concentration is diluted so that at least three concentrations are obtained. The absorbance of each dilution is read at the appropriate wavelength against the reference solution. Absorbance (ordinate) versus concentration (abscissa) is plotted on linear graph paper. If the detector response is linear, the plotted points form a straight line. Compounds that have been used for linearity check procedures include potassium chromate (370 nm), copper sulfate (650 nm), cyanmethemoglobin (450 nm), and cobalt ammonium sulfate (512 nm). The most convenient solution is green food coloring, which has absorption maxima at 257, 410, and 630 nm, allowing linearity checks at each end of the visible portion of the spectrum and an additional check in the ultraviolet range.

It is recommended that linearity certification be performed weekly. Absorbance plots can be compared each week to determine whether the slope of the line is changing with time. The effect of an aging source lamp may be detected by a change in linearity response. Stray light may also cause a gradual deviation from linearity at the high concentration end of the linearity plot.

Photometric Accuracy

Photometric accuracy is performed to certify that an instrument will produce an absorbance reading on a standard solution that compares to a reference value. It is a method that checks the ability of the spectrophotometer to indicate correctly the energy level presented to the detector. Photometric accuracy should be performed each week on routinely used spectrophotometers.

A set of three neutral density filters (SMR 930b) may be purchased from the National Bureau of Standards. These neutral density filters have defined absorbance values at specified wavelengths. Solutions prepared from high-purity chemicals have also been used as standards.

Stray Light

A stray light check is performed to verify that no radiant energy other than that indicated by the monochromator setting reaches the detector.

Stray light may originate from radiation that is scattered by dust or corrosion of optical surfaces or by light reflection from gratings or mirrors or through misalignment of the sample cell or cuvette. Stray light causes deviation from Beer's law because it appears that the solution absorbs less light than the actual amount of light-absorbing substance present (more light is detected by the detector). This nonlinearity becomes more apparant at increasing concentrations, hence there will be a gradual departure from linearity at the high end of a standard curve when stray light is present.

Stray light is measured by selecting a solution or filter that has a very high absorbance at the test wavelength; indeed so high that any measured transmittance must be due to stray light. Stray light checks should be performed when warranted by quality control data.

TURBIDIMETRY

PRINCIPLE

Substances assayed by turbidimetry form a fine precipitate with an added reagent. The precipitate is so finely divided that it remains in a stable suspension. When the cloudy solution is then placed in a spectrophotometer, the light from the radiant energy source is either absorbed or scattered (greater portion) by the particles, or passes through and is detected by the photocell. The reduction in transmitted light caused by the suspension is an indication of the concentration of the unknown. Beer's law is not followed; therefore, calibration curves must be used to calculate concentration values.

DISADVANTAGES

The decrease in light transmittance is related to the number and size of the particles in the light beam. Therefore, the size of the particles for standard and unknown samples have to be similar for accurate results. Particles may settle out upon standing or the turbidity may change with time; therefore, the length of time between sample preparation and measurement has to be kept constant for reproducible results.

USES

Turbidimetry methods are used to measure total proteins in cerebrospinal fluid and urine. The standard biuret total protein methodology is insensitive to the low protein levels encountered in these body fluids.

The activity of the enzymes, amylase and lipase, may also be determined by a turbidimetric methodology. The spectrophotometer measures a change in turbidity as substrates are hydrolyzed. The substrates (starch for amylase and triglycerides for lipase) are turbid. As hydrolysis occurs, the turbidity decreases. The degree of decrease in turbidity (decrease in absorbance) is related to the concentration of the enzyme.

NEPHELOMETRY

PRINCIPLE

Nephelometry is similar in principle to turbidimetric methods. There are two major differences: 1) in nephelometry, the antibody specific to the unknown to be measured (antigen) forms a complex with the antigen and 2) the light that is *scattered* by antigen–antibody complexes is measured. The amount of light scattered by unknown samples is compared to known calibrator material (which has been treated in a similar manner) to arrive at a concentration value.

BASIC COMPONENTS

The arrangement of components of a nephelometer is illustrated in Figure 4.8.

The light source may be either tungsten–halogen or laser. Laser sources provide monochromatic light. The tungsten–halogen source provides light in a broad spectral range and is preferred when particle size varies, which would be the situation in a clinical laboratory.

The detector is a photomultiplier tube located at an angle to the light beam. The size and shape of the antigen–antibody molecule may cause variation in the scattering intensity (scattering may be asymmetrical); therefore, the angle of the detector to the light beam is important. The detector, in relation to the light beam, is placed at an angle that will allow the greatest amount of forward scattered light to be detected, because this angle will provide the greatest sensitivity. The detector in some nephelometers is located at a right angle to the incident beam; in other instruments the observation angle is less (usually 37° or 70° forward scatter).

Endpoint nephelometers measure scattered light generated by reactions that have reached equilibrium. Rate nephelometers measure light scattering as the reaction is proceeding in the cuvette compartment (kinetic). Rate nephelometers are generally preferred because 1) kinetic procedures require no blanks and 2) they measure a change in light scattering per unit of time. Results are generated rapidly, generally in 10–60 seconds, in contrast to endpoint measurements which require steady-state equilibrium before measurements can be made. Also, rate nephelometers are generally more analytically sensitive than endpoint nephelometers.

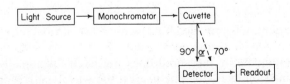

Figure 4.8.   Arrangement of components in a nephelometer.

USES

Nephelometry is ideal for measuring individual proteins. The sensitivity and specificity provided by antibody reagents allow the determination of a single protein species in the presence of other proteins without the need to isolate the protein prior to the assay. Protein species assayed by nephelometric techniques include immunoglobulins, CSF albumin, fibrinogen, haptoglobin, C-reactive protein, complement components, rheumatoid factor, alpha-1-antitrypsin, transferrin, ceruloplasmin, and alpha-2-macroglobulin. Nephelometric inhibition techniques are also used to measure drug concentrations.

FLUORESCENCE AND PHOSPHORESCENCE

PRINCIPLE

As molecules absorb discrete amounts of energy, they are described as reaching an *excited state*. An excited state is one in which an electron or electrons are at an energy level higher than the ground state. As the electrons return to the ground state or lowest molecular energy level, they emit the absorbed energy in the form of light. In the return to the ground state, some of the energy may be lost through collision with other molecules and through other energy modes such as vibration and rotation; therefore, the wavelength of light emitted is longer than that absorbed for excitation. Also, when electrons return to the ground state indirectly through other "less excited" states, the length of time required for emission is extended. When the length of time between absorption and emission of energy is in the range of 10^{-8} to 10^{-4} seconds, the process is called fluorescence; if the length of time is longer than 10^{-4} second, the process is call phosphorescence. The absorption and emission phenomena are highly specific, since both are dependent on electron molecular orbital structure and hence will be unique for a particular molecule. Some compounds may absorb light however electrons return "directly" to the ground state hence emitted light has precisely the same wavelength as the exciting light; therefore, not all compounds are fluorescent.

BASIC COMPONENTS

The basic components of a fluorometer are illustrated in Figure 4.9.

Light Source

The excitation radiation is provided by a high-energy ultraviolet light source. The mercury arc lamp, which is commonly used, emits line spectra from 200 to 1000 nm. If the compound to be measured does not excite maximally at one of the mercury lamp emission lines, the line nearest the wavelength of maximal excitation is chosen.

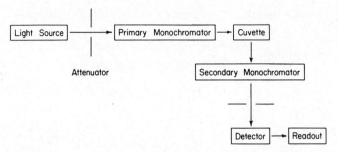

Figure 4.9.   Arrangement of components in a fluorometer.

The xenon lamp emits a continuous spectrum from 200 to 800 nm. Since the spectrum is continuous, it is possible to select any desired wavelength by incorporating a prism or diffraction grating in the instrument components. This type of instrument would be called a spectrofluorometer as opposed to a fluorometer, which uses only filters for wavelength selection.

Attenuator

The attenuator is a device that is used to control the intensity of the light that is directed to the primary monochromator. The sensitivity, or the ability to detect changes in concentration, can be increased or decreased by adjusting the intensity of the exciting light.

Monochromator

Filters or a combination of filters select wavelengths in a fluorometer. The primary filter or monochromator isolates the desired wavelength for excitation. The secondary filter or monochromator selectively removes unwanted wavelengths of light and allows only selected light to fall on the detector. The secondary monochromator is located at a right angle to the beam of the exciting light source. Spectrofluorometers are equipped with prisms or gratings that allow selection of primary and secondary wavelengths by adjusting the wavelength selector dial.

Detector

The detector is usually a photomultiplier tube, since these tubes allow light to be measured at very low intensities. If both visible and ultraviolet light are measured, two photomultiplier tubes may be needed, since the cathode material of the photomultiplier tube determines the range of wavelength detection.

Readout

The readout device reflects the light that is emitted by the molecules in the sample compartment in numerical terms. In dilute solutions where measurements are in the linear portion of a standard curve, the amount of fluorescence is directly proportional to the concentration of the fluorescent substance. Under these conditions, the unknown may be calculated by using a single standard:

$$\text{unknown concentration} = \frac{\text{fluorescent reading unknown}}{\text{fluorescent reading standard}}$$

$$\times \text{ concentration of standard}$$

ADVANTAGES

Fluorometry is relatively sensitive and can measure concentrations to μg levels.

DISADVANTAGES

At high concentrations, *quenching* occurs. Quenching is any process taking place in the sample container that results in a decrease in the intensity of fluorescence. When the concentration of the fluorophores in the sample is too high, that portion nearest to the exciting light absorbs so much of the light that less and less is available for the remainder of the solution. The fluorescence becomes concentrated in that portion of the solution nearest the light source, which then results in a nonlinear curve.

At high concentrations, self-absorption may also occur. The emitted light must pass through the solution before passing to the detector. This emitted light is at the energy optimum to be absorbed by other molecules of the substance under assay, resulting in decreased fluorescence. This process is negligible at low concentrations.

Quenching can also occur when foreign material forms nonfluorescent complexes with the substances undergoing analysis. All of these phenomena result in false low values.

Fluorescence is subject to environmental influences. Temperature affects the probable loss of energy by collision. At cold temperatures, less energy is lost by collision and more energy is lost by fluorescence; at warm temperatures, more energy is lost by collision and less energy is lost by fluorescence. The standard and samples should both be at the same constant temperature when readings are taken in order to avoid introduction of error.

Fluorescence is also sensitive to pH changes. The pH affects the molecular structure of the compound and hence the electrons and their associated molecular orbitals. A substance may fluoresce only at a narrow pH range or fluoresce maximally at a specific pH.

Errors may also be introduced due to the presence of undefined fluorescent material in reagents. New reagents should always be checked for fluorescence before use.

USES

Fluorometry is used to measure catecholamines, porphyrins, and some drugs (quinidine, salicylates, procainamide, and acetylprocainamide). Fluorometers are also used to detect and quantitate substances separated by chromatographic procedures.

FLAME PHOTOMETRY

PRINCIPLE

When a substance (as an aerosol) is exposed to a flame, thermal energy is absorbed by orbital electrons of neutral or ground-state atoms; these electrons then "jump" into higher, more energetic but less stable orbitals. Being unstable, the electron returns almost instantly to its previous base or ground state energy level and *emits* the absorbed energy in a photon of light of a particular wavelength characteristic of the element.

An element may have more than one emission line. The wavelength selected for the determination of each element is that at which the emission line is the most intense. For example, sodium emits light most strongly at a wavelength of 589 nm. Heat from the flame causes the electron in the M shell, s orbital ($3s$ level), to be displaced to the $3p$, $3d$, or $4p$ level (Figure 4.10). These excited states are unstable and the electron falls back to the ground state. The line spectra emitted is dependent on the orbital energy level of electron displacement. The most intense band for each metal is the transition from the lowest excited state to the ground level, which for Na would be the $3p \rightarrow 3s$ transition. The strongest signal occurs at 589 nm for Na, 767 nm for K, 671 nm for Li, and 852 nm for Cs. If the emitted light is isolated by means of a monochromator that allows only the selected wavelength of interest to pass, the intensity of the light will be proportional to the concentration of the element.

The intensity of the light emitted will be dependent on the number of neutral atoms in the flame and the temperature of the flame. The temperature of the flame is dependent on the fuel to air ratio. Since it is difficult to keep these factors constant for extended periods of time, an internal standard is used to correct for changes in temperature or aspiration rate that would affect the emission signal.

For each reading, two signals are utilized, one for the substance to be analyzed and one for the internal standard (reference). A ratio of the two signals (sample beam/reference beam) is used to determine concentration values. If the flame temperature changes or the aspiration rate varies, both signals will be similarly affected but the ratio will remain unaffected and the signal will remain constant.

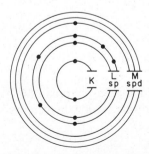

Figure 4.10. Sodium atom with electron shells indicated by K, L, and M. The letters s, p, and d refer to orbitals with progressively increasing energy levels.

In choosing an internal standard, the following must be considered:

1. The internal standard must consist of a substance that is not normally found in biological fluids to ensure that the reference emission is constant.
2. The amount of energy required to excite the internal standard must be close to that required to excite the element being measured (i.e., excitation potential is similar), because if the flame temperature changes, the emissions of sample and reference will be equally affected.
3. The emission lines of samples and internal standard must be far enough apart so that they can be resolved by a monochromator and measurement can be accomplished.
4. The concentration of the internal standard must be precisely the same in all samples and standards so that the reference beam is constant.
5. The concentration of the internal standard should be similar in magnitude to the unknowns so that a reasonable ratio is obtained.

Either cesium or lithium meets the above criteria and may be used as the internal standard for the determination of sodium and potassium. Although lithium is used in the treatment of depression, the therapeutic range is so low that usually no interference is introduced. Flame emission instruments (which use lithium as an internal standard) can also measure the serum concentration of lithium using the potassium signal as an internal standard.

The internal standard also acts as a radiation buffer. When measuring potassium, the internal standard tends to buffer the change in signal response due to variations in sodium concentration. In serum, sodium is present in much higher concentrations than potassium. Hence, in the flame, the number of excited sodium atoms is far greater than the number of excited potassium atoms. These excited sodium atoms are able to transfer their energy to potassium atoms. Therefore, the number of excited potassium atoms (and the signal produced) would be influenced by the sodium concentration. To overcome this source of error, samples are all diluted with high concentrations of internal standard so that the potassium signal is not influenced by the sodium concentration.

BASIC COMPONENTS

The components of a flame photometer utilizing an internal standard are illustrated in Figure 4.11.

Aspirator

The function of the aspirator is to pull sample into the atomizer. To accomplish aspiration, a jet of air is directed across the upper tip of a capillary tube. The lower tip of the capillary tube is immersed in sample. The negative pressure created by the air stream causes sample to be drawn into the capillary tube, which then enters the atomizer.

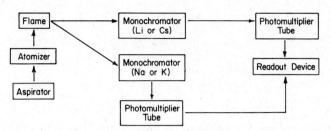

Figure 4.11. Arrangement of components in a flame photometer utilizing an internal standard.

Atomizer

The function of the atomizer is to disperse the solution into fine droplets for delivery to the flame. As the sample enters the atomizer, the force of the air stream blowing across the capillary tip converts the sample into a fine mist, which is then introduced into the flame. If the mist particles are too large, they do not enter the flame but drain to the bottom of the atomizer and to waste. The sample must be introduced into the flame in a stable, reproducible rate.

Sample viscosity, density, and surface tension affect droplet size and the rate of atomization. High sample dilutions are used to minimize these effects.

Flame

As the sample enters the flame, the solvent is evaporated, leaving dry salts that are then vaporized. The function of the flame is to provide energy to break chemical bonds and to excite neutral atoms. If any atoms exist in the ionized form, the reducing gases in the flame (i.e., carbon monoxide and free radicals) reduce metal ions to neutral, ground-state atoms. These reactions are summarized below:

$$Na^+ \text{ (ionized)} \xrightarrow[\text{reducing gas}]{+1e^-} Na^0 \text{ (ground state)}$$

$$Na^0 \text{ (ground state)} \xrightarrow{\text{flame}} Na^* \text{ (excited state)}$$

$$Na^* \text{ (excited state)} \to Na^0 \text{ (ground state)} + \text{emitted light (589 nm)}$$

Fuels for the flame may consist of propane (most common), acetylene, or ordinary gas, which is mostly methane.

Monochromator

The function of the monochromator is to isolate the emission wavelength characteristic for the substance being assayed and to exclude all other spectra from reaching the detector. Interference filters are frequently used as monochromators. However, cutoff filters, prisms, or gratings are also used. The number of interference or cutoff filters needed is dependent on the number of line spectra measured. For example, if Na, K, Li, and Cs are measured, four filters are needed.

Detector

The function of the detector is to measure the intensity of the emission signal by converting light energy to a proportional electrical current. The detector system usually consists of photomultiplier tubes.

Readout

The readout device usually displays the concentration of the element being assayed. When a straight line standard curve is obtained, concentration is directly proportional to the reading ratio, and unknown concentrations may be calculated by the following formula:

$$\text{unknown concentration} = \frac{\text{emission ratio of unknown}}{\text{emission ratio of standard}} \times \text{standard concentration}$$

SOURCES OF ERROR

The fuel to oxidant ratio can raise or lower flame temperature. Temperature affects the number of excited atoms present in the flame. If the flame temperature is too low, insufficient atoms will be excited to provide the sensitivity needed for accurate measurements. Even in an average flame, only about 1–5% of the atoms achieve an excited state. If the flame temperature is too hot, ionization occurs. Ionic metals cannot emit light at the necessary line spectrum; therefore, emission is lost and a nonlinear calibration curve results. The amount of air is regulated so that it is just sufficient to burn the fuel.

The aspiration rate has to be adjusted so that sufficient sample is delivered to the flame for proper sensitivity. If the atomization rate is too high, the flame temperature will be lowered and salts will become encrusted on the burner tip.

Protein may coat atomizer walls, affecting the ability of the atomizer to produce the fine mist needed for introduction into the flame. The atomizer is periodically flushed with a cleaning solution to prevent protein buildup. Protein may also coat the sample capillary tube, which affects the amount of sample drawn into the atomizer. To minimize protein buildup, the sample capillary tube is cleaned with a wire probe of correct diameter. If the size of the probe is incorrect, the bore of the capillary tube may become distorted or scratched, leading to error.

If the sample, introduced into the flame, is too highly concentrated, self-absorption occurs. A ground-state metal atom absorbs energy, becomes excited, and emits a photon of light. However, this emitted light is absorbed by a second ground-state atom; thus, less light is emitted than the sample concentration should produce. The result is a nonlinear curve, which flattens at high concentrations.

USES

Flame emission photometry is commonly used to measure sodium, potassium, and lithium. It is possible to measure calcium and magnesium by flame photometry, but the sensitivity is poor.

ATOMIC ABSORPTION SPECTROPHOTOMETRY

PRINCIPLE

Atomic absorption is similar in principle to flame photometry; however, light *absorbed* by ground-state atoms is measured in atomic absorption spectroscopy, while flame photometers measure light emitted by excited atoms. All elements are capable of absorbing electromagnetic energy of the same wavelength as that emitted by the element. In atomic absorption spectrophotometry, the electromagnetic energy is supplied by a hollow cathode lamp, which is composed of the same metal being assayed. Atoms of the hollow cathode lamp are excited and as they return to the ground state, emit light of a wavelength characteristic for the metal cathode of the hollow cathode lamp. This light energy from the hollow cathode lamp may then be absorbed by neutral atoms of the same kind in the sample. The detector measures a decrease in the intensity of a beam of light from the hollow cathode lamp; the decrease in intensity is proportional to the concentration of the metal in the sample.

BASIC COMPONENTS

The components of an atomic absorption spectrophotometer are illustrated in Figure 4.12.

Hollow Cathode Lamp

The function of the hollow cathode lamp is to supply electromagnetic radiation of a specific wavelength. The lamp consists of a cylindrical cathode enclosed in a gas-tight chamber. The chamber is filled with either helium or argon gas. When voltage is applied, electrons, given off by the cathode, collide with the gas and produce ionized gas atoms, which are then attracted to the cathode. The collision of the gas with the cathode causes excited metal atoms to be dislodged into the lamp chamber. As the excited metal atoms return to the ground state, light is emitted that is characteristic for the metal.

The cathode of the hollow cathode lamp is always made of the same metal as that being assayed (i.e., to determine calcium, a calcium hollow cathode lamp is used, etc.).

Chopper

The chopper (a rotating disk that alternatively passes and blocks the light beam) is used to pulsate the light emitted from the hollow cathode lamp. When the sample

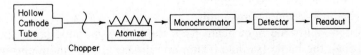

Figure 4.12.   Arrangement of components in a single beam atomic absorption spectrophotometer.

is vaporized, both ground-state and excited atoms are produced. The light emitted by the excited atoms will be of the same wavelength as that absorbed and will therefore be transmitted by the monochromator. In order to measure only absorbed light, the detector must be able to differentiate between light originating from the hollow cathode lamp and that originating from flame spectrum and sample emission. The detector detects all light, but the detector amplifier is electrically tuned to accept only pulsed signals.

When a chopper is not used, pulsation of the hollow cathode lamp emission can also be achieved by chopping the current supplied to the lamp.

Atomizer

The function of the atomizer is to break chemical bonds and to form ground-state atoms. Atomization can be achieved by the flame or other thermal processes. The flame resides in a long narrow burner head and can only accommodate liquid samples. In the premix type burner, sample, air, and fuel are premixed and converted to an aerosol before introduction into the flame. Large size droplets drain off so that only fine droplets are introduced into the flame. In the total consumption burner, all sample that is aspirated is presented to the flame. The atomizer functions as a cuvette. Hence, it is necessary to control flame size in order to provide a constant light path.

High-temperature furnace atomizers can allow analysis of solid samples as well as liquid samples. Furnace thermal atomization provides a greater number of ground-state atoms than flame atomization, which therefore increases test sensitivity and provides lower detection limits.

Monochromator

The monochromator selects the wavelengths that are allowed to pass to the detector. Prisms or gratings with narrow bandpass are used as monochromators.

Detector/Readout

The detector is a photomultiplier tube. The amount of light absorbed is measured and Beer's law is used to calculate unknown concentrations. The readout device may present the data as either concentration, absorbance, percent transmittance, or percent absorption. The formula used to convert percent absorption to absorbance is

$$\text{absorbance} = 2 - \log(100 - \text{percent absorption})$$

DOUBLE BEAM ATOMIC ABSORPTION SPECTROPHOTOMETERS

The hollow cathode lamp may require a long warm-up time before the exciting line is stable. A double beam arrangement eliminates the relatively long warm-up time needed in a single-beam arrangement.

This system splits the light beam from the hollow cathode lamp so that half passes through the flame (sample beam) and directly to the monochromator and the other half passes around the flame (reference beam) and is recombined in the monochromator.

The photomultiplier tube then compares the ratio of the two beams as an indication of sample absorption. Any error introduced by changes in the output from the hollow cathode lamp is then eliminated, since both beams would be affected in the same manner producing no change in the ratio. This component arrangement compensates only for changes in output from the hollow cathode lamp but will not eliminate error introduced due to changes in flame temperature or sample aspiration rate.

ADVANTAGES

Atomic absorption spectrophotometry provides good sensitivity and specificity. In order to absorb energy the atoms need to exist only in the ground state, while flame emission methodology requires excited atoms. The number of ground-state atoms available in the flame is far greater than the number of excited atoms, hence the sensitivity of atomic absorption spectrophotometry is generally greater than flame emission spectroscopy.

Since the band of light absorbed by neutral atoms is very narrow, atomic absorption spectrophotometry is almost free from spectral interference, therefore providing good specificity.

INTERFERENCES

Chemical Interference

Chemical interference refers to the situation incurred when the flame is unable to dissociate the sample into ground-state atoms so that absorption can occur.

Calcium, assayed by atomic absorption, provides an example of chemical interference. Calcium phosphate complexes cannot be readily dissociated by the flame unless a special high-temperature burner is used. Strontium or lanthanum (preferred) is added in excess, which preferentially binds to phosphate thus liberating the calcium. The calcium is then able to form ground-state atoms:

$$LaCl_3 + Ca_3(PO_4)_2 \rightarrow LaPO_4 + 3Ca^{2+} + 3Cl^-$$

Ionization Interference

If the atomization temperature is too high, ionization may occur. Ionization occurs because orbital electrons are lost, and the resulting ion is no longer able to absorb energy from the hollow cathode lamp. Ionization interference can be corrected by using lower flame temperatures or by the addition of a substance that will readily ionize. The added substance will then absorb most of the flame energy so that the metal of interest will not become ionized.

Matrix Interference

Matrix interference may occur in those situations where the solvent contains high concentrations of salt. Much of the flame energy is utilized to decompose the salt and the atomizing effect of the flame is decreased. Fewer ground-state atoms will be formed. This interference can be overcome by dilution.

USES

Atomic absorption assays are generally limited to metallic elements. Metals routinely determined by atomic absorption include calcium, magnesium, copper, lead, mercury, zinc, and chromium. Arsenic and other nonmetals can be measured by atomic absorption as well. Atomic absorption is the reference method for calcium.

GAMMA (SCINTILLATION) COUNTER

Radioactive emission is usually discussed in terms of energy rather than wavelength. Radiation comes from an unstable atomic nucleus that rearranges to become more stable and, in the process, the atom emits energy, particles, or both.

An atom characterized by its nucleus is called a nuclide. The mass number describes the number of protons plus neutrons present in the nucleus of the element and is indicated as a superscript. The atomic number describes the number of protons present in the nucleus of the element and is indicated as a subscript. The nuclei of isotopes contain the same number of protons as the original or native element but differ in the number of neutrons. For example, $^{131}_{53}I$ is heavier than $^{125}_{53}I$. The ^{131}I isotope contains more neutrons (78) than are present in ^{125}I (72). Isotopes with unstable nuclei disintegrate to change the neutron/proton ratio and, in the process, emit radiation. Hence these isotopes are termed radioactive. Not all isotopes are radioactive. Chemically, isotopes behave identically, since they contain the same number and arrangement of orbital electrons surrounding the nucleus.

There are three types of nuclear radiation: (1) alpha, (2) beta, and (3) gamma. In alpha emission, alpha particles, which consist of two protons and two neutrons, are ejected from the nucleus. With alpha particle emission, both the atomic number and mass number of the atom decrease. Alpha particles are weakly penetrating and will penetrate only the most superficial layers of the skin. Alpha emitters have little application in the clinical laboratory.

In beta emission, a neutron disintegrates to form a proton and a highly energetic electron (a beta particle). Beta emission is characteristic for nuclei with high neutron/proton ratios that are too high for stability. Beta emission results in an increase in the atomic number but produces no change in the mass number of the atom.

In gamma radiation, no particles are emitted, only energy. When rearrangements occur within the nuclei, they are left with excess energy, which is emitted as gamma rays. Gamma rays have extremely short wavelengths and are therefore of very high energy. The emitted gamma rays have discrete energy levels characteristic for a particular atom. The emission of gamma rays results only in a change in the energy state of the nucleus; the atomic number and mass number are not altered. Internally, gamma rays are less hazardous than alpha and beta radiation, but present a greater external hazard because of their high penetrating power.

PRINCIPLE

Certain materials, when exposed to ionizing radiation, will convert the kinetic energy of the particles or photons into flashes of light (scintillations). Materials capable of this conversion property are called fluors or scintillators. The excitation of the fluor by nuclear radiation leads to light emission.

BASIC COMPONENTS

The components of a solid scintillation counter are illustrated in Figure 4.13.

Fluor

The fluor serves as the detector for the radiation. The most common fluor for gramma radiation is a crystal of sodium iodide that is activated by thallium. The activator is needed for maximal production of scintillations. A hole or well is drilled into the crystal so that the sample can be inserted into the crystal. The gamma rays will be emitted in all directions and insertion of the sample into the crystal allows detection of a greater number of gamma rays.

Detector

The detector is a photomultiplier tube, which is placed against the surface of the crystal. The crystal and detector are then sealed to keep all radiation other than the scintillations from reaching the detector. The signal to the detector is in the form of light pulses. The size of the light pulse is in proportion to the energy of the gamma ray striking the fluor.

Amplifier

Each signal from the detector is amplified in direct proportion to the energy of the gamma ray that produced the scintillation.

Spectrometer or Pulse Height Analyzer

The spectrometer is the component that is used to select the pulse sizes that will be counted. This is determined by electronic circuits that allow only designated voltages to pass to be counted by the scaler. Each type of radionuclide emits a different gamma ray energy; therefore, by choosing the pulse size that will be counted, it is possible to count the gamma emissions from one type of isotope and to exclude all others.

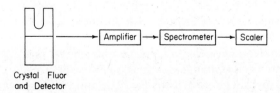

Crystal Fluor
and Detector

Figure 4.13.   Arrangement of components in a scintillation counter.

Scaler

The function of the scaler is to count the pulses that are passed by the spectrometer. The number of pulses per unit of time (generally, counts per minute, or cpm) is counted. The counting time is determined by a timer mechanism, which can be manually set by the operator.

A standard curve is needed in order to determine sample concentrations.

USES

Radioisotopes are used as tags in radioassay or radioimmunoassay (RIA) procedures. By using antibodies and other protein binders as reagents in assay procedures, it is possible to measure substances by specific binding without purification or separation steps. Radioassay procedures are also very sensitive and have made possible the measurement of substances whose concentrations are so low that quantitation is impossible by standard assay methodology.

Substances routinely measured in clinical laboratories by radioassay techniques include thyroid hormones (T_3, T_4), thyroid stimulating hormone (TSH), cortisol, vitamin B12, folate, digoxin, and digitoxin. RIA methodology is also available for many other compounds.

Isotopes are also used in diagnostic studies. Since the isotopes of an element react the same chemically, radioactive substances will be utilized by an individual in the same manner as the parent or native element. The radioactivity provides a method to trace the components via a scintillation counter. For example, radioactive iodine can be used to determine the ability of the thyroid gland to trap iodine; radioactive vitamin B12 is used to determine the presence or absence of intrinsic factor (see Schilling test).

AUTOMATED SYSTEMS

Automated systems may be classed into four general groups: (1) continuous flow systems, (2) discrete analyzers, (3) centrifugal analyzers, and (4) dry chemical methods.

CONTINUOUS FLOW SYSTEMS

Continuous flow analyzers were first introduced into the clinical laboratory in 1957 and represented the first attempt to automate clinical chemistry analysis. Although the term AutoAnalyzer is applied to many chemistry systems, it is the registered trade mark of the Technicon Instrument Corporation and should be reserved for Technicon instruments. The AutoAnalyzer instrument, when first introduced, was able to perform only one automated test procedure at a time (one channel) while later two simultaneous analyses were possible (two channel). Now multiple channel instruments are available. The abbreviation SMA means *sequential multiple analyzer*. The terminology SMA 12/60 means that 12 analyses per sample are performed, and 60 samples per hour are processed. The SMAC, a sequential multiple analyzer with a computer, is able to analyze up to 150 samples per hour.

With the AutoAnalyzer, it is not possible to select tests from a test menu; a sample processed on the AutoAnalyzer is analyzed automatically for all tests available on the system. Therefore, continuous flow systems may use excess reagents to produce unrequested test results. The AutoAnalyzer serves best as an instrument to perform screening profiles.

Continuous flow equipment is based on the following principles:

1. The internal diameter of the tubing determines the amount of reagents and sample used; the greater the diameter, the greater the volume aspirated.
2. Air bubbles are introduced at intervals. Air bubbles are used to separate sample and reagent streams and also serve to wipe the tubing walls to reduce interaction between samples.
3. In those samples that require deproteinization before analysis, dialysis is used to remove protein.
4. Glass coils are used for mixing; the tumbling action of the reagent stream causes heavier liquids to fall through the lighter liquid until the liquid bubble is uniform.

Theory of Operation

Continuous flow systems are based on the principle that the standards and samples are tested in precisely the same way. Therefore, it is not necessary for the reaction to reach completion before measurements are taken.

Basic Components

Sampler

The sampler consists of a disk that holds sample cups. As the disk rotates, a sample probe moves automatically into cups containing standards, controls, and samples. Between samples, the probe moves into a water wash to minimize sample carry over. The sample probe is driven by a cam, which is a notched wheel, which determines the rate of analysis and the dwell time in specimens (sample to wash ratio). Air bubbles are used to separate samples. In the SMAC, the rapidity of samples processed per hour increases the probability of sample carry over. To circumvent this problem, the probe enters and leaves the specimen several times during the aspiration process, which then introduces air segments and reduces the probability of contamination between specimens.

Peristalic Pump

The function of the pump is to aspirate and dispense portions of sample, reagents, and air into the AutoAnalyzer system in a reproducible manner through a series of polyvinyl tubes. The assembly of polyvinyl tubes and fittings is called a manifold. The manifold is mounted on a platen, a table that supports the manifold. A roller assembly is used to compress the tubes of the manifold. One manifold tube is attached to the sample probe, and the ends of the other tubes are submerged in appropriate reagents or left exposed so that air can be drawn into the tubing. As the rollers close over the tubing, compression on the platen forces the solutions and air along the tubes on the up-stream side of the pump, which then causes aspiration on the down-stream side of the pump. By changing the diameter of the tubing, the volume of sample and reagents may be changed.

Dialyzer

The dialyzer separates larger molecular weight substances like proteins automatically from the substances to be analyzed. The dialyzer consists of two flat plates that contain mirror image grooves on the face of the plates. A semipermeable membrane is placed between the two plates with the grooves facing each other, thus creating a channel that is separated by the membrane. The diluted sample flows on one side of the membrane while the reagent stream flows on the other side of the membrane. The small molecular weight molecules pass through the membrane into the reagent stream, while the large molecules pass into the waste. Equilibrium is not achieved and much of the substance to be measured is lost in the waste. The actual amount of material that diffuses across the membrane is estimated to be only about 10–20% of the total. The percent of material that diffuses across the membrane depends on the temperature, the area of contact, the thickness and porosity of the membrane, and the concentration gradient across the membrane. The temperature and membrane characteristics must remain the same for sample and standards to obtain valid results. Protein may deposit on the membrane and decrease diffusion rates; therefore, membranes should be changed periodically as part of the maintenance program.

Not all samples need to be deproteinized before analysis; these samples would circumvent the dialyzer module.

Heating Bath

Heating baths are used to heat the recipient stream for a specified time in order to promote color development or to allow enzyme reactions to proceed. The glass coils are submerged in oil to maintain constant temperature. The retention time in the heating bath can be varied by using multiple coils connected in series or by using coils of larger diameter.

Detector

The most common detector is a colorimeter (photometer). The function of the colorimeter is to measure continuously the intensity of the color that is present in the solution flowing through the flow cell (cuvette) and to produce an electrical signal proportional to the increase or decrease in light transmitted. The colorimeter is a double beam instrument. The light from the lamp is split, half is directed through the monochromator then through the flow cell to the photocell (sample beam), and half is directed through a second monochromator to a second photocell (reference beam). The ratio of voltage from the sample photocell to that of the reference photocell is measured. By using a double beam arrangement, errors due to voltage fluctuation are reduced. Interference filters are used as monochromators. Since the stream has bubbles (which interfere with spectral measurements), the bubbles and waste are removed by a F-tube before passing to the colorimeter. The SMAC system does not remove bubbles. The noise caused by sudden changes in voltage (due to bubbles) is eliminated electronically. The computer is able to recognize the large voltage change due to bubbles from the gradual voltage change associated with differences in color intensity.

Recorder

Early AutoAnalyzers used a strip chart recorder, which produced a continuous tracing of the difference in voltage between test and reference photocells. To determine concentration values, the sample peaks were compared with a calibration curve derived from standards that were assayed simultaneously with the samples. In newer instruments, the raw data are converted directly into concentration units.

DISCRETE ANALYZERS

Discrete analyzers have the capability to select one or more tests from a group of available tests. Only those tests programmed by the operator will be performed. There are many instruments on the market today that offer multiple discrete analyses. In principle, these systems mimic manual methods, but the procedural steps are performed with minimal operator intervention.

The steps common to chemical analysis include:

1. Sample pick-up is accomplished by a sample probe that enters the sample cup and aspirates sample from the cup and then dispenses it into a reaction chamber. The sample may be washed out of the probe by diluent or reagent.
2. Reagents are dispensed by the instrument, mixing occurs, and the reaction is allowed to proceed. The reaction chamber may be heated to hasten the reaction.
3. The reaction is then detected by visible or UV photometry, nephelometry, or fluorometry.
4. The test results are presented on a digital display, printed tape, computer terminal, or printed report.

Many automated instruments do not eliminate the protein matrix prior to analysis. Instead, protein interferences are minimized by using calibrator material, which is dissolved in a protein base. In the determination of unknown concentrations, sample values are compared to calibrator values. Assuming both sample and calibrator are affected in a similar manner, the error introduced is minimized. However, in those individuals with abnormal albumin/globulin ratios (i.e., renal disease, multiple myeloma), significant error may be introduced from protein interference.

Automated systems also minimize protein error by the use of kinetic rather than end-point methodology.

CENTRIFUGAL ANALYZERS

Centrifugal analysis was first described in 1969. Since that time, applications of the centrifugal analyzer have increased each year.

A unique feature of the system consists of a disk that contains cavities for sample and reagents. When the disk is placed on the rotor of the instrument and rotation begins, the centrifugal force moves both the reagent(s) and sample outward so that they mix. This reaction mixture then moves through a channel into a cuvette where the reaction is monitored. The cuvettes are located either on a separate rotor that fits around the disk or are incorporated into the body of the disk. In some instruments, the disks are automatically washed and dried for reuse. Disposal disks are also available. By eliminating the washing components, smaller and more compact instruments are possible. The light source is located so that the light passes through the cuvette in a perpendicular pathway to the rotor. The detector can be a spectrophotometer, fluorometer, or laser nephelometer.

The centrifugal analyzer is capable of performing only one assay procedure on multiple samples at a time. Multiple tests on a single specimen require each sample to be processed several times since each test has different reagent and wavelength requirements.

Initial rate, kinetic, and end-point tests can all be performed by centrifugal analyzers. Since standards and samples are all processed under identical conditions and test readings can be taken at specified time intervals, some test procedures normally run in an end-point mode can be run as initial rate tests. In initial rate tests, the change in absorbance at the beginning of the reaction is proportional to the concentration. This is advantageous because the absorbance reading can be taken before interferring substances alter test results, since interferring components frequently react at different rates. This reading time interval must be determined for each individual test.

The instrument is also very efficient in kinetic methodology. In kinetic assays, the reaction is monitored as it proceeds in the cuvette with the change in absorbance per unit of time used to calculate concentration values or enzyme activity.

For end-point tests, multiple readings are taken on each sample as the disk rotates past the detector system. A series of consecutive readings are averaged to arrive at an absorbance value.

Advantages of centrifugal analyzers include rapid throughput (when compared to single channel analyzers), the ability to use small sample and reagent volumes, the capability to adapt to different batch sizes, and the capability of adapting many different assay methods.

DRY CHEMICAL METHODS

Dry chemical methods utilize reagent slides that are composed of several layers which may include: (1) spreading layer; (2) scavenger layer; (3) reagent layer(s); and (4) plastic or support layer. The reagent layer(s) contains enzymes, dye precursor, and buffers necessary for the analysis of a specific component. Sample, control, or standard is deposited on the spreading layer. Selected components are allowed to penetrate to the reaction layer(s), which in turn activate the dehydrated reagents. A chemical reaction is initiated to produce a color. Incident light is passed from beneath the support or plastic layer and is directed through the reagent layer(s). As the light hits the white spreading layer, some of the light reflects back through the reagent layer(s) to a photocell. The amount of reflected light, which is in proportion to color intensity, is used to determine the concentration of the analyte.

Assay by reflectance photometry offers several advantages. The storage requirements for reagents is minimal since no wet reagents are required. No pipetting steps are needed as slides are prepared by the manufacturing company. Since no wet reagents are used, reflectance instruments require no plumbing. No sample dilutions are required and 10 or 11 μL of sample per test is used.

At the present time, analytes that can be measured by reflectance photometry include glucose, BUN, ammonia, bilirubin, uric acid, cholesterol, triglycerides, total calcium, total protein, albumin, creatinine, phosphorus, and serum enzymes.

A dry chemical method to determine sodium, potassium, chloride, and carbon dioxide (electrolytes) has been introduced which employs ion-selective electrode methodology. The slide consists of single-use side-by-side ion-selective electrodes that are joined by a paper bridge. A drop of reference and a drop of patient sample are deposited, each on its respective electrode. The samples interact with coated reagent layers to create a pair of electrochemical half-cells. The drops spread toward one another across the paper bridge, meeting at the center and forming a stable liquid junction. A voltmeter measures the potential difference of the two half-cells, which is then used to determine concentration values.

LIQUID CHROMATOGRAPHY

PRINCIPLE

Chromatography provides a method for the separation of compounds in a mixture. The separation is achieved because compounds differ in distribution between two distinct phases (i.e., mobile and stationary phases). The distribution varies because molecules differ in the degree of polarity, charge or size.

The word *chromato* means color. The name chromatography was derived from the original work performed by a Russian botanist, Mikhail Tswett, who was working with pigments extracted from plants. He could see definite bands in his column, hence he suggested the name chromatography. However, the materials in the column are usually not visible.

There are two major divisions of liquid chromatography: (1) column chromatography and (2) thin-layer chromatography, which includes paper chromatography.

CHROMATOGRAPHIC PROCEDURE

The basic components of a high-pressure liquid chromatograph are illustrated in Figure 4.14.

The compounds to be separated (the sample is a liquid or can be dissolved in a liquid) are applied to the top of a column, which contains the stationary phase (Figure 4.15). Solvent (liquid phase) is allowed to flow through the column, either by gravity or by pressure applied to the top of the column [high-pressure liquid

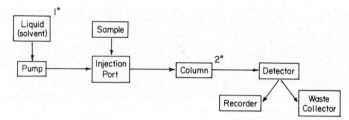

Figure 4.14. Arrangement of components in a high-pressure liquid chromatograph. 1* represents the mobile phase; 2* represents the stationary phase.

chromatography (HPLC)]. As the solvent and sample flow through the column, components of the sample are separated by four physicochemical processes to be described. If the solvent flow is allowed to continue, the components may be eluted or washed out of the column and collected or detected as separate fractions. The separation is based on different affinities of the molecules for the two phases, the stationary phase and the mobile phase. In Figure 4.15, component #1 has greater affinity for the stationary phase than components #2 and #3 and will therefore elute last. Component #3 has greater affinity for the solvent than components #1 and #2, therefore it will elute first.

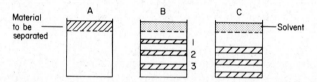

Figure 4.15. Separation by chromatography. The material is applied to the top of column A. As solvent is added, separation occurs. The separated components may eventually elute as separate fractions if solvent flow is continued.

DETECTOR TYPES

The choice of detector system in a liquid chromatograph depends on which type of chemical or physical characteristic is to be detected. Some detectors are very selective and measure a specific characteristic possessed by few molecules, while others measure a characteristic displayed by many molecules (universal detector).

There are four basic types of detectors: (1) refractive index detector; (2) spectrophotometric detector; (3) fluorometric detector; and (4) electrochemical detector.

Refractive Index (RI) Detector

The refractive index detector is a universal detector and measures the difference between the refractive index of the column effluent and a reference solution, which is usually the mobile phase. The refractive index of a component changes with changes in temperature, therefore the temperature has to be carefully controlled when a RI detector is used.

Spectrophotometric Detector

This type of detector is more selective than a RI detector. A spectrophotometric detector measures the absorption of light, either visible or ultraviolet, of the column effluent. Simple fixed wavelength detectors or those with a multitude of wavelength selections are available.

Fluorometric Detector

This type of detector uses a fluorometer to detect fluorescent compounds. The fluorometric detector is highly sensitive to fluorescent compounds and possesses moderate specificity.

Electrochemical Detector

Electrochemical detectors are very sensitive and exhibit good specificity. The amperometric detector is the most widely used electrochemical detector. In this detector, a voltage is applied to an electrode. As the column effluent flows past the electrode, the analyte molecules at the interface between the electrode and solution can either be reduced (accept electrons) or oxidized (give up electrons). The net movement of electrons causes a current flow that is in proportion to the concentration of the analyte at the interface.

SEPARATION PROCESSES

There are four basic types of liquid chromatography or physicochemical processes involved in separation. They are: (1) adsorption chromatography, (2) partition chromatography, (3) ion exchange chromatography, and (4) molecular exclusion chromatography.

Adsorption Chromatography (Liquid–Solid)

Principle
In adsorption chromatography, components of a mixture are separated on a column by adsorption (surface adsorption) of the solute on the packing material. There is competition between the solvent and the adsorbent for the solute molecules. Those molecules with high affinity for the column material will be retained longer on the column than those with little affinity.

Stationary Phase
The stationary phase is a solid. Common examples of stationary phases include silica and alumina.

Mobile Phase
The mobile phase is usually a relatively nonpolar solvent mixture.

Uses
At one time, this was the most widely used separation approach, but adsorption chromatography lacks speed and reproducibility which limits its usefulness. It is still used for class separations.

Partition Chromatography (Liquid–Liquid)

Principle
In partition chromatography, both the mobile phase and stationary phase are liquids. A thin film of solvent is adsorbed onto column particles. The solvent, coating the packing material, is the stationary phase. The packing material is the inert or support phase. The mobile phase, another solvent, is immisible with the stationary phase. Separation is achieved because components interact with the stationary phase in different ways, the more interactive components will be held in the column longer (retention time longer) than the less interactive components. Partition separation, the most widely used type of separation, is ideal for the separation of compounds based on chemical polarity differences. There are two types of partition chromatography, normal phase and reverse phase.

Normal Phase. In normal phase chromatography, the stationary phase is polar. Examples of normal phase stationary phases include water and polyethylene glycol. The mobile phase is a nonpolar solvent. Examples of normal phase mobile phases include hexane and chloroform.

Normal phase chromatography is used to separate components that differ most in polar functionalities. A molecule with greater polarity will react with the column with greater affinity (retention time longer) than one which is less polar, thus allowing separation.

Reverse Phase. In reverse phase chromatography, a nonpolar stationary phase is used. A common reverse phase stationary phase is hydrocarbon chains bonded to a solid support. The mobile phase is a polar solvent. Examples of reverse phase mobile phases are water, acetonitrile, and methanol.

Reverse phase chromatography is used to separate nonpolar components of a mixture. The separation is based on the interaction of the nonpolar groups with the column packing material. The degree of affinity of nonpolar components with the nonpolar packing material will determine the order of elution.

Ion Exchange Chromatography

Principle
In ion exchange chromatography, the column is packed with an organic resin which contains ionic groups as part of its structure. These ionic groups are exchanged for either anions or cations present in the solution undergoing separation. In cation exchange resins, cations of the solution are exchanged for cationic groups on the resin, and in anion exchange resins, anions of the solution are exchanged for anionic groups on the resin. Only charged components (ions) can be separated by ion exchange chromatography.

Stationary Phase
The stationary phase consists of an ion exchange resin. These resins are polmerized cross-linked hydrocarbons that contain ionized functional groups.

Mobile Phase
The mobile phase is a liquid. Those ionic solutes that interact weakly with the resin are eluted rapidly, while those reacting more strongly are retained longer on the column.

Uses
Ion exchange chromatography is ideal for the separation of species on the basis of charge differences. It is used to separate mixtures of amino acids, heme break-down products, small hormones (catecholamines), and isoenzymes, and to remove interfering substances.

Molecular Exclusion Chromatography

Principle
In molecular exclusion chromatography, components of a mixture are separated based on molecular size. The separation is achieved because the gel contains pores of controlled size. Small molecules are trapped in the pores, while the larger molecules cannot enter the pores and pass around the gel. Large molecules then are eluted before small molecules.

Stationary Phase
The column is made of a polysaccharide polymer (dextran gel), which contains pores of controlled size.

Mobile Phase
The mobile phase is either an aqueous or organic solvent. Some stationary phases are compatible only with organic solvents.

Uses
Gel permeation chromatography is ideal for the separation of particles that differ in size such as proteins from lower molecular weight molecules and ions.

Choice of a Stationary Phase

The function of the stationary phase is to selectively retain components of a mixture to effect separation. Retention is accomplished by affinity between the stationary phase and the various components. In choosing the stationary phase, the general rule to remember is "like attracts like." Nonpolar molecules have high affinity for nonpolar stationary phases while polar compounds have high affinity for polar stationary phases.

THIN-LAYER CHROMATOGRAPHY

In thin layer chromatography a thin layer of a stationary phase is applied to a supporting material such as glass plates, aluminum foil, plastic, or other inert material. The material to be separated and known substances are applied as spots near one end of the plate and then dried. The sample application end of the plate is then dipped into the solvent (mobile phase), which is in a closed container. The level of the solvent must not cover the applied spots. The solvent rises to the top of the plate by capillary action. Separation of components is achieved because molecules differ in affinity for the solvent (mobile phase) and stationary phase. By choice of solvents and stationary phases, any of the four basic types of chromatography can achieve separation.

Identification is achieved by comparing the migration distance of knowns and unknowns, by spraying with reagent solutions, and by using UV light to detect fluorescent molecules. R_f values of knowns and unknowns may be compared for identification purposes.

$$R_f \text{ value} = \frac{\text{distance unknown traveled from the point of application}}{\text{total distance solvent front moved}}$$

For example,

Spot traveled 8.1 cm

Solvent front moved 12.0 cm

$$R_f \text{ value} = \frac{8.1}{12.0} = 0.68$$

For quantitative purposes, the stained material may be cut from the plate, eluted and read colorimetrically or the chromatogram may be scanned with a densitometer. The densitometer consists of a light source, an optical filter, a slit, and a photocell in an arrangement similar to a photometer (Figure 4.16). The stained chromatogram is placed on the carriage. A slit in the carriage allows the light source to scan a selected portion of the chromatogram. A recorder, connected to the photocell, traces the light transmitted by each separated spot and provides a tracing of this pattern as an indication of band absorbance.

Two-Dimensional Chromatography

For greater separation purposes, two-dimensional chromatography may be used. The unknown is applied at the edge of the plate and allowed to migrate. The plate is removed and the solvent evaporated. A second edge of the plate, at right angles to the original line of migration, is then dipped into a second solvent and a second migration of components occurs (Figure 4.17). Two-dimensional chromatography can be used to separate two substances with similar R_f values.

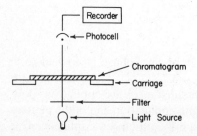

Figure 4.16. Arrangement of components in a densitometer.

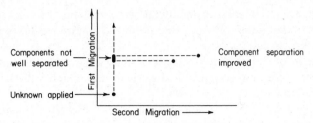

Figure 4.17.   Two-dimensional chromatography used to separate components with similar R_f values.

Uses

Thin-layer and paper chromatography are used to separate and identify drugs, sugars, and amino acids. The phospholipids lecithin and sphingomyelin (L/S ratio), which are used to determine fetal lung maturity, can be quantitated using thin-layer chromatography.

GAS CHROMATOGRAPHY

There are two major categories of gas chromatography: (1) gas–solid chromatography (GSC) and (2) gas–liquid chromatography (GLC).

In GSC, the stationary phase is a solid of large surface area. Common packings include silica gel, molecular seives, and charcoal. The mobile phase is an inert gas.

In GLC, the stationary phase is a nonvolatile, high-boiling-point liquid, which is coated on an inert solid support. The mobile phase is an inert gas. GLC is the type commonly found in the clinical laboratory.

GAS–LIQUID CHROMATOGRAPHY

Principle

In GLC the substances are separated in the gaseous phase. Separation is achieved because the volatile substances vary in vapor pressure, which affects their solubility in a high-boiling-point liquid (stationary phase). Those substances that are present to a greater extent in the gaseous phase (mobile phase) than in the stationary phase will be eluted first. The purpose of the column is to retard the progress of the sample through the column.

Sample Requirements

The sample to be introduced into the column may be a solid, liquid, or gas; however, liquids and solids (in solution) must be able to be vaporized. Only rarely can biological samples be injected without prior purification. The number of purification steps is dependent on the type of compound undergoing analysis. For example, the determination of blood alcohol (by some methods) may require no purification steps, while multiple purification steps are needed for determining urinary hormones. The number and concentration of compounds present in the sample, the presence of interfering substances, and the type of column packing used are factors that have to be considered in the determination of the number of purification steps needed.

After purification, some substances in a family of compounds may not separate well unless they are first converted to derivatives, which will then allow more efficient separation.

After sample preparation, a suitable solvent is selected for sample injection. The solvent is generally an organic solvent such as acetone, alcohol, chloroform, or hexane.

After injection, the sample solvent mixture is converted into a discrete plug of gas, which then enters the column.

Stationary Phase

The stationary phase is usually a high-boiling-point liquid, which coats the surface of an inert solid support (crushed fire brick, glass beads, or diatomaceous earth). Sometimes the high-boiling-point liquid is adsorbed on the walls of the column (capillary column).

The columns are made of glass or metal and vary in length and diameter. Common analytical columns are 3–10 ft in length. They are often coiled to conserve space. Since the material to be analyzed has to be in vapor form, the column is housed in a temperature-controlled oven.

Mobile Phase

The carrier gas (mobile phase) usually consists of nitrogen gas, but helium and argon have been used. The type of carrier gas used is dependent on the type of detector system present in the gas chromatograph system. Nitrogen gas can be used with flame ionizer, electron capture, or thermal conductivity detectors. The gas has to be dry and the flow rate consistent, otherwise retention times will vary.

BASIC COMPONENTS

The components of a gas chromatograph are illustrated in Figure 4.18.

Carrier Gas

The carrier gas (mobile phase) is the system that carries the vaporized sample forward. The gas has to be inert and may be nitrogen, helium, or argon.

Gas Flow Regulator

The rate of flow of the carrier gas affects separation and therefore must be carefully controlled.

Sample Injection

The sample to be analyzed is usually dissolved in a volatile solvent and then is injected into the column. The sample inlet area is heated to a higher temperature

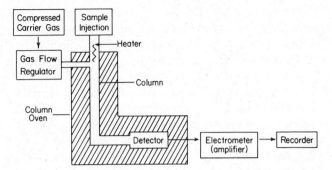

Figure 4.18. Arrangement of components in a gas chromatograph.

than the column in order to obtain rapid vaporization of the sample. Temperatures that are too high may cause thermal decomposition of the sample.

Column Oven

The temperature of the column has a marked effect on the retention times of the compounds. Retention times are increased with lower temperatures, and separations are usually better. However, the temperature in the column has to be high enough to accomplish the separation in a reasonable time. If the column temperature is too high, the stationary phase may bleed from the column and the column will lose its separatory properties. High temperatures may also decompose the material being analyzed.

Column

The column is packed with a liquid-solid support that will selectively impede the forward movement of the sample components. In general, small particle size packing material is desired because this will produce a more efficient column. However, the packing material must not be so dense that the flow rate is impeded.

The ratio of liquid to support phase is one factor that determines the speed of sample movement through the column. Low concentrations of liquid phase increase the efficiency and speed of analysis. However, if active sites on the solid support become exposed, the sample may be adsorbed onto its surface.

Detectors

The detector functions to detect the sample components as they pass out of the column.

The thermal conductivity detector (TCD) is the most universal and least sensitive of the three types to be discussed. This type of detector is based on the principle that the molecular weight of the gas affects its ability to conduct heat from a filament. The TCD consists of two sets of wire filaments, a reference set and a sample set. A constant current is passed through the wires, which generates heat in a constant amount. As the gas passes over the wire, heat is dissipated. The amount of heat dissipated is a function of the molecular weight of the gas. When the carrier gas contains sample, the rate of heat loss is increased and the electrical resistance of the sample filament changes, while the reference filament resistance remains unchanged. The resistance change is measured and this signal is fed to a strip chart recorder where it appears as a peak.

For most analytical work, a flame ionizer detector (FID) is needed. This detector will respond only to components that produce charged ions when burned in a hydrogen flame. As the carrier gas leaves the column, it is mixed with hydrogen and burned at the tip of the jet. Any ions produced are collected by two electrodes. This current flow is then amplified and fed to a recorder. The current flow is directly proportional to the amount of sample burned in the flame. The FID will detect essentially all organic substances but will not detect inorganic gases and water vapor. The insensitivity to water is an asset since water is present in most clinical samples. The FID will not respond to oxygen, nitrogen, carbon monoxide, and carbon dioxide and, therefore, cannot be used for respiratory gas analysis. A TCD is needed for these analytes.

The electron-capture device (ECD) is very selective and very sensitive. This detector will respond only to materials capable of adding an electron to form a negative ion. As the sample enters the detector system, electrons strike the vaporized molecules. These electrons have just enough energy to be captured by the molecule but not enough energy to cause ionization. Only halogenated substances and a few other classes can be measured by ECDs. The ECD is used to measure pesticide residues.

Recorder

The recorder measures retention time. Retention time is defined as the time from sample injection into the column to the maximum peak. Gas chromatography offers good sensitivity with the ability to provide rapid separation of components with quantitation.

USES

GLC is used in toxicology laboratories to screen and quantitate drugs. It may be used to measure steroids, fatty acids, alcohols, blood gases, anesthetics in blood, intermediates of metabolism, vitamins, pesticide residues, and various amines including serotonin, histamine, and tryptamine.

IMMUNOASSAY TECHNIQUES

Immunoassays offer two distinct advantages over standard assay methods. They provide a high degree of specificity (detects only the substance under assay) and offer good sensitivity (ability to detect low levels). In immunoassay procedures, a label is needed in order to monitor the antigen–antibody reaction. Three labels commonly used in immunoassays are radioisotope, fluorescent, and enzyme.

Nonisotopic labels offer three distinct advantages over isotopic labels. The shelf life of kits using enzyme and fluorescent labels is longer than the shelf life of kits using radioisotope labels. The detector system for enzyme and fluorescent labels is a spectrophotometer or fluorometer; radioisotopes require expensive scintillation counters as detectors. The nature of the reagents in nonisotopic kits is considered less hazardous than those in isotopic kits. Because of these advantages, nonisotopic procedures will probably dominate the immunoassay market in the future.

Immunoassay procedures may be divided into two categories, homogeneous and heterogeneous. Homogeneous methods do not require separation of free and antibody-bound fractions, while in heterogeneous methods, a separation step is required. Immunoassay methods may be divided as follows:

I.   Homogeneous Immunoassay Methods
     A.   Enzyme-Multiplied Immunoassay Technique (EMIT)
     B.   Fluorescent Immunoassay (Ames Co.)
II. Heterogeneous Immunoassay Methods
     A.   Radioimmunoassay (RIA)
          1.   Competitive binding technique
          2.   Immunoradiometric technique (IRMA)
     B.   Enzyme-Linked Immunosorbent Assay (ELISA)
          1.   Sandwich technique
          2.   Immunoenzymometric technique
          3.   Competitive binding technique
          4.   Antibody quantitation
     C.   Fluorescent immunoassay

RADIOIMMUNOASSAY

RIA Competitive Binding Assay

The use of radiolabeled ligands was first introduced by Yalow and Berson in 1959.

Principle
Radioactive-labeled antigen and unlabeled antigen compete for limited binding sites on antibody molecules. The extent of binding of the labeled antigen is dependent on the concentration of the unlabeled antigen present. That is, the higher the amount of unlabeled antigen present, the lower the binding of the labeled antigen to the antibody. Free and antibody-bound antigen fractions are separated. The amount of radioactivity in the separated fractions determines the extent of binding that has occurred.

Test
1.   To unlabeled antigen (unknown, control, or standard), a precise amount of labeled antigen is added. The labeled antigen is like the antigen of interest; that is, if cortisol is being determined, labeled cortisol is used.
2.   To this mixture, a precise amount of specific antibody is added. The sample is incubated to allow competitive binding to occur.

3. The bound and unbound antigen fractions are separated.
4. The radioactivity in the separated fractions is determined by means of a scintillation counter.

Several methods are available that will allow separation of free from antibody-bound antigen. In solid phase affinity techniques, the antibody is bonded to some solid support such as a test tube wall, glass beads, or cellulose fibers. The Ag–Ab complexes (bound fraction) can then be separated by decantation or washing.

In exclusion chromatography, separation is achieved because the unbound antigen can easily diffuse in and out of a substance such as dextran-coated charcoal. The bound fractions are too large to penetrate into the charcoal and will be excluded.

In precipitation techniques, such as by zirconium gel or ammonium sulfate, one fraction, usually antibody-bound antigen, will precipitate. Unbound antigen will not precipitate and will remain in the supernatant thus allowing the components to be separated by filtration.

Separation may also be achieved by using a second antibody. The second antibody is directed to the primary antibody. The primary antibody binds the antigen. The precipitate that forms can be separated by centrifugation and represents the bound fraction. The second antibody may also be bound to a solid phase, such as a test tube wall. Separation can then be achieved by decantation, the bound fraction remaining in the tube.

Interpretation

The interpretation is based on the fraction counted, that is whether bound or unbound (free) labeled fraction is counted. The most common procedure is to count the bound fraction. When the bound fraction is counted, a high count means low sample concentration. This can be illustrated hypothetically as follows:

$$\begin{array}{ll} 0 \text{ Ag} \\ 4 \text{ Ag}^* \end{array} + 4 \text{ Ab} \rightarrow 4 \text{ Ag}^*\text{Ab} \qquad 4000 \text{ cpm}$$

$$\begin{array}{ll} 4 \text{ Ag} \\ 4 \text{ Ag}^* \end{array} + 4 \text{ Ab} \rightarrow \begin{array}{l} 2 \text{ AgAb} \\ 2 \text{ Ag}^*\text{Ab} \end{array} \qquad 2000 \text{ cpm}$$

$$\begin{array}{ll} 12 \text{ Ag} \\ 4 \text{ Ag}^* \end{array} + 4 \text{ Ab} \rightarrow \begin{array}{l} 3 \text{ AgAb} \\ 1 \text{ Ag}^*\text{Ab} \end{array} \qquad 1000 \text{ cpm}$$

Note that the same amount of labeled antigen (*) and specific antibody is always added to each tube. The only variable factor is the amount of unlabeled antigen present. The degree of binding of the labeled antigen fraction to the antibody is dependent on the amount of unlabeled antigen present. The law of mass action states that labeled and unlabeled fractions will compete for binding sites on a one-to-one basis; therefore, the ratio of labeled and unlabeled antigen bound by the antibody will be the same as the ratio of labeled and unlabeled fractions present in the unbound form. When bound fractions are counted (pictured), low counts per minute (cpm) mean high sample concentrations. Unknown concentration values are determined by comparing the cpm of samples to that of standards treated in the same manner (i.e., from a calibration curve).

In those procedures that count the free or unbound fraction, the interpretation would be exactly opposite, that is, high cpm means high sample concentration.

Immunoradiometric Analysis (IRMA)

In radiometric analysis, the *antibody* is labeled instead of the antigen. The antigen of interest and the antibody react to form Ag–Ab* complexes. After the bound and unbound fractions are separated, the radioactivity of the complex would directly produce measurement of the concentration of the antigen.

In radiometric analysis, extremely pure antibodies with high association constants are required so that most of the antigen reacts with the antibody. These conditions have not previously been met, currently limiting the usefulness of radiometric analysis. With the availability of monoclonal antibodies, immunoradiometric assay methods are now available for some analytes (e.g., CK–MB sandwich method, HCG).

Uses

Hormone measurements available by RIA techniques include the thyroid hormones (T_3, T_4), thyroid stimulating hormone (TSH), parathyroid hormone (PTH), gastrin, cortisol, aldosterone, adrenocorticotropic hormone (ACTH), calcitonin, beta-human chorionic gonadotropin (β-HCG), estrogens, follicle stimulating hormone (FSH), growth hormone (GH), insulin, luteinizing hormone (LH), and prolactin. Other substances measured by radioisotope methods include prostatic acid phosphatase (PAP), digoxin, digitoxin, renin, vitamin B12, folate, myoglobin, and several hepatitis antigens and antibodies.

ENZYME IMMUNOASSAY

Enzyme-Multiplied Immunoassay Techniques (EMIT)

EMIT procedures were first described in 1972.

Principle
EMIT procedures are similar to RIA competitive binding procedures except the antigen label is an enzyme rather than a radioisotope. Enzymes make suitable labels because their catalytic property allows them to act as amplifiers. Enzymes are not consumed as reactions proceed but continue to catalyze reactions as long as substrate and coenzymes (if needed) are present. In EMIT procedures, competitive binding occurs between a antigen (unknown, control, or standard) and an enzyme-labeled antigen (like the unknown) for limited binding sites on antibody molecules. In order to determine the extent of Ag–Ab binding, the activty of the enzyme, which was used as the label, is determined.

Test
1. The antigen (unknown, control, or standard) is diluted with buffer.
2. A precise amount of specific antibody is added.

3. A precise amount of an enzyme-labeled antigen is added. This antigen is like the antigen of interest; that is, if phenobarbital is to be assayed, enzyme-labeled phenobarbital is used. There is competition between labeled and unlabeled antigen for the limited number of binding sites on the antibody molecules. The extent of binding of the labeled antigen is dependent on the concentration of the unlabeled antigen present. The enzyme-labeled antigen is prepared so that if binding occurs, enzyme activity is inhibited.

4. Substrate is added. Any enzyme-labeled antigen not bound to the antibody is free to react with the substrate.

5. The enzyme activity is monitored kinetically using a spectrophotometer. The concentration of the sample is computed by comparing the absorbance of the sample to the absorbance values of a set of standards.

Interpretation
The enzymatic activity is proportional to the concentration of the antigen present in the sample, that is, high enzyme activity means high sample concentration. If patient antigen is high, more of the antibody-binding sites will be occupied by unlabeled antigen (hence less labeled antigen), thus making more enzyme available to react with the substrate.

Uses
EMIT procedures are available for the assay of over 30 drugs with many more drug analyses under development. It is used to monitor antiepileptic drugs, anti-asthmatic drugs, cardioactive drugs, antineoplastic drugs, antimicrobials, and drugs of abuse. EMIT technology can also be applied to some hormone assays. EMIT procedures are capable of automation and can be used in continuous flow systems, centrifugal analyzers, and discrete analyzers.

Enzyme-Linked Immunoabsorbent Assays (ELISA)

Enzyme-linked immunoabsorbent assays (ELISA) methods were first described in 1971. ELISA methodology can be used to assay for either an antigen or an antibody. The sandwich, immunoenzymometric, and competitive binding techniques all assay for an antigen. An ELISA technique is available which measures the concentration of an antibody. Two enzyme labels, commonly used in ELISA procedures, are alkaline phosphatase and horseradish peroxidase. They are chosen because they have high activity, they are inexpensive, and the reaction product is highly visible.

ELISA Sandwich Technique
The sandwich technique compares the arrangement of components to that of a sandwich. Two antibodies are used and are analogous to the slices of bread, the antigen is the filler in the sandwich (i.e., Ab–Ag–Ab).
   Principle. The antigen (unknown, control, or standard) must contain two antibody-binding sites. A solid phase antibody binds the antigen of interest. Excess enzyme-labeled antibody is added which reacts with the second antibody-binding site on the antigen. The excess labeled antibody is washed away. Substrate is added to determine enzyme activity which determines the extent of Ag–Ab binding.

Test.

1. A test receptacle is used that contains specific antibody bound to the surface of the tube, or the antibody may be bound to glass beads.
2. The tube is washed with nonionic detergent in order to minimize nonspecific binding.
3. The antigen to be assayed (unknown, control, or standard) is added. Any antigen present will bind to its specific immobilized antibody.
4. Excess components are washed away.
5. Excess enzyme-labeled antibody is added that binds to the second binding site of the immobilized antigen.
6. Excess enzyme-labeled antibody is washed away.
7. Substrate is added. Any enzyme-labeled antibody present is free to react with the substrate.
8. Absorbance measurements are recorded on a spectrophotometer.

Interpretation.   The enzymatic activity is proportional to the amount of antigen present in the sample, that is, high enzyme activity means high sample concentration.

In the clinical chemistry area, the sandwich ELISA method may be used to assay for carcinoembryonic antigen.

ELISA Immunoenzymometric Technique

Principle.   Antigen (unknown, control, or standard) and excess enzyme-labeled specific *antibody* are allowed to react. The excess unreacted antibody is removed. Substrate is added in order to determine the extent of Ag–Ab binding.

Test.

1. Antigen (unknown, control, or standard) and excess enzyme-labeled specific antibody are incubated.
2. After binding has occurred, the unreacted enzyme-labeled antibody is removed with excess solid phase antigen.
3. Substrate is added to the supernatant. Enzymatic activity is measured in order to determine the extent of Ag–Ab binding.

Interpretation.   The enzymatic activity is proportional to the concentration of the antigen present, that is, high enzyme activity means high sample concentration.

ELISA Competitive Binding Technique

Principle.   As in other competitive binding assays, labeled and unlabeled antigen compete for limited binding sites on specific antibody molecules. The degree of binding by the labeled antigen to the antibody is dependent on the amount of unlabeled antigen present. In EIA techniques, the indicator system consists of an enzyme.

Test.

1. Antigen (unknown, control, or standard), and a specific amount of enzyme-labeled antigen (this antigen is of the same type as that being assayed), are added to a tube which has solid phase antibodies.
2. This mixture is incubated to allow competitive binding to occur.
3. Separation is achieved by decantation and washing, which removes unreacted antigen.
4. Substrate is added. Any enzyme-labeled antigen present is free to react with the substrate.

Interpretation.   High enzyme activity means low sample concentration. When sample concentration is high, more binding sites will be occupied by unlabeled antigen, hence less enzyme-labeled antigen. Following separation, less enzyme will be available to react with the added substrate.

ELISA Antibody Quantitation

Principle.   This method measures the amount of antibody present. The wall of a plastic tube or microtitration plate is coated with antigen. Sample is added. If the sample contains specific antibody, binding of the antibody to the antigen will occur. Excess enzyme-labeled anti-human globulin is added. This binds with the Ag–Ab complexes (i.e., Ag–Ab–Ab). The addition of substrate allows the extent of binding to be determined.

Test.

1. A plastic tube or microtitration plate is coated with an antigen; the antigen is usually bound by simple adsorption.
2. The tube or plate is washed thoroughly.
3. The sample (unknown, control, or standard), which contains antibodies, is added. Binding is allowed to occur.
4. The tube or plate is washed to remove excess components.
5. Enzyme-labeled anti-human globulin is added, which will attach to the antibody bound to the antigen.
6. The tube or plate is washed to remove excess components.
7. Substrate is added to determine the enzyme activity of the bound enzyme-labeled antibody.

Interpretation.   The enzymatic activity is proportional to the amount of antibody present, that is, high enzyme activity means high sample antibody titers.

Uses
ELISA procedures, to determine antibody titers, are used in screening programs to determine congential–perinatal infections of toxoplasma, rubella, cytomegalovirus, and herpes (TORCH).

ELISA procedures, to determine antigen concentrations, are available for alpha-fetoprotein, CEA, ferritin, rheumatoid factor, digoxin, insulin, placental lactogen, estrogens, thyroid hormones (T_3, T_4), herpes virus, and hepatitus B surface antigen.

FLUORESCENT IMMUNOASSAY (FIA)

In fluorescent immunoassay, a fluorescent molecule is used as a label. The fluorescent label allows measurement of substances in the micro-, nano-, and picogram range. Formerly, sensitive RIA procedures were necessary to measure substances at these low concentrations.

FIA methods have several advantages when compared to RIA procedures. The reagents used in FIA have a longer shelf life than does RIA reagents. Reagent disposal in FIA requires no special handling precautions, which is in contrast to radioisotope disposal. A special license must also be obtained when radioisotopes are handled.

A homogeneous fluorescent immunoassay method is available from the Ames Company, Elkhart, IN, that allows measurement of gentamicin, tobramycin, and amikacin (aminoglycosides). In this procedure, substrate-labeled drug and drug compete for limited binding sites on antibody molecules. When the substrate-labeled drug is bound to the antibody, the substrate is unavailable for reaction. Unbound substrate is free to react with an enzyme (added), which forms a fluorescent product. The amount of fluorescence is proportional to the drug concentration, that is, high fluorescence means high sample concentration.

Heterogeneous fluoroimmunometric tests are also available. The FIAX system (International Diagnostic Technology, Inc., Santa Clara, CA) employs a dip stick (StiQ Sampler) that carries the solid phase reagents. The StiQ Sampler contains either adsorbed pure antigen or antibody (depending upon the assay). The solid StiQ sampler is then reacted with diluted patient sample. The specifically bound analyte is then reacted with fluorescein labeled antihuman antiglobulin (either IgG or IgM). The StiQ Sampler is inserted into a fluorometer (available from the company) to measure fluorescence. The fluorescence is directly proportional to the concentration of the analyte in the sample.

Uses

Fluorescent methodology is available to measure therapeutic drug levels and specific proteins (e.g., low level albumin, immunoglobulins, complement components).

BIBLIOGRAPHY

Alexander, L. R. and Barnhart, E. R. *Clinical Chemistry Photometric Quality Assurance Instrument Check Procedures.* U.S. Dept. of Health and Human Services Publication, Centers for Disease Control, 1980.

Bakerman, S. "Enzyme immunoassays," *Laboratory Management,* 21–29 (Aug. 1980).

Bender, G. *Clinical Instrumentation: A Laboratory Manual Based on Clinical Chemistry.* Philadelphia: Saunders, 1972.

Buffone, G. J. "Improved nephelometric instrumentation," *Laboratory Management*, 14–99 (April 1977).

Cawley, L. P. *Principles of Radioimmunoassay, Part I and II*. Division of Education Media Services of ASCP, 1975.

Cawley, L. P. "Comparing competitive immunoassays," *Lab World*, 47–51 (Jan. 1981).

Chamran, M. and Keiser, R. "Maintaining optimum spectrophotometric performance. Part I," *Laboratory Medicine*, 9(4), 33 (1978).

Chamran, M. and Keiser, R. "Maintaining Optimum spectrophotometric performance. Part II." *Laboratory Medicine*, 9(5), 35 (1978).

Coiner, D. *Basic Concepts in Laboratory Instrumentation*. Bellaire, TX: American Society of Medical Technology Education and Research Division, 1977.

Finley, P. R. "Nephelometry: Principles and clinical laboratory applications," *Laboratory Management*, 34–44 (Sept. 1982).

Frings, C. S. and Broussard, L. A. "Calibration and monitoring of spectrometers and spectrophotometers," *Clin Chem.* 25, 1013 (1979).

Gerson, B. "HPLC monitoring improves drug therapy" *Lab World*, 43–46 (Jan. 1981).

Helman, E. Z. and Manning, R. J. *Basic Spectrophotometry in the Clinical Laboratory*. Fullerton, CA: Beckman Instruments, 1972.

Henry, J. B. (ed.). *Clinical Diagnosis and Management by Laboratory Methods*, 16th ed. Philadelphia: Saunders, 1979, Vol. I.

Hicks, R., Schenken, J. R., and Steinrauf, M. A. *Laboratory Instrumentation*. 2nd ed. Hagerstown: Harper and Row, 1980.

Miller, J. G. "Enzyme immunoassay," *Lab 78*, 45–49 (Sept./Oct. 1978).

Schram, S. B. *The LDC Basic Book on Liquid Chromatography*. Petersburg, FL: Milton Roy Co., 1981.

Tietz, N. (ed.). *Fundamentals of Clinical Chemistry*. 2nd ed. Philadelphia: Saunders, 1976.

Walls, K. "Pigs, parasites and ELISA," *Lab World*, 55–57 (Jan. 1981).

POST-TEST

1. Match the following:

   _____ 1. visible spectral region   a. 700–2000 nm
   _____ 2. ultraviolet spectral region   b. 200–400 nm
   _____ 3. infrared spectral region   c. 400–700 nm

2. One nanometer is equal to _____ meters.

3. Radiant energy has wavelength and frequency characteristics. The shorter the wavelength, the:
   a. greater the frequency, the greater the energy
   b. greater the frequency, the lesser the energy
   c. lesser the frequency, the greater the energy
   d. lesser the frequency, the lesser the energy

4. If the wavelength dial selector is turned slowly from 400 to 700 nm, the colors you would expect to see are:
   a. blue, violet, green, yellow, red, orange
   b. red, orange, yellow, green, blue, violet

c. yellow, green, blue, red, orange, violet
d. violet, blue, green, yellow, orange, red

5. Mathematically, absorbance may be written in all of the following ways *except:*
    a. $A = 2 - \log \% T$
    b. $A = \log 100 - \log \% T$
    c. $A = -\log T$
    d. $A = \dfrac{\% T}{2}$

6. If Beer's law applies, the following combination will give a straight line on semi-log paper:
    a. $\% T$ versus concentration
    b. absorbance versus concentration
    c. optical density versus concentration
    d. wavelength versus concentration
    e. both a and b

7. If Beer's law applies, the following combination will give a straight line on linear graph paper:
    a. $\% T$ versus concentration
    b. absorbance versus concentration
    c. optical density versus concentration
    d. wavelength versus concentration
    e. both b and c

8. Percent transmittance may be converted to absorbance by using the formula:
    a. $2 - \log \% T$
    b. $\% T - \log 2$
    c. $2 + \log \% T$
    d. $\log 2/\% T$

9. Assuming Beer's law is followed, answer the following questions:
    a. The absorbance of a solution read in a cuvette with a 1.5 cm light path was 0.660. The absorbance of this same solution read in a cuvette with a 1.0 cm light path should read _____ .
    b. The absorbance of a solution containing 5.0 mg/dL calcium was 0.300. The absorbance of this same solution containing 10.0 mg/dL calcium should read _____ .

10. The absorbance will be equal to _____ when the $\% T$ is equal to 100.

11. The percent transmittance of a solution is 10%. The absorbance of this solution is _____ .

12. The absorbance of a calcium standard containing 9.3 mg/dL of calcium was 0.494. The absorbance of the control sera was .504. The procedure is a Beer's law procedure. The concentration of the control is _____ .

13. Below is a list of the basic components of a spectrophotometer. Match the component with its function.

    _____   1. Light source   a. holds the solution in the instru-
    _____   2. Monochromator   ment and provides a constant

    _____ 3. Readout device   light path

    _____ 4. Cuvette   b. isolates specific wavelengths

    _____ 5. Detector   c. provides radiant energy

                                       d. reads the output from the detector directly in %T or absorbance units

                                       e. changes light energy to electrical energy

14. Use the words listed below to fill in the blanks.

A test procedure is to be read at a wavelength of 310 nm. The correct light source would be a _____. The correct cuvette to use should be made of

_____.

A test procedure is to be read at a wavelength of 510 nm. The correct light source would be a _____. The correct cuvette to use should be made of

_____.

    a. deuterium lamp
    b. tungsten lamp
    c. glass
    d. quartz or fused silica

15. A photometer will exhibit maximum sensitivity at the wavelength where the change in absorbance per unit of concentration is:
    a. greatest
    b. least
    c. the same

16. A substance to be assayed has a half peak-height width of 40 nm. The spectral bandwidth of the instrument used to measure absorbance should be _____ or less.

17. Figure 4.19 shows a spectral absorbance scan of a solution. Several selected wavelengths were chosen marked A, B, and C on the graph. The solution was then diluted to give different concentrations and read at each of these selected wavelengths. Absorbance versus concentration of these solutions was then plotted as shown on graph 2.

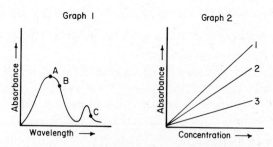

Figure 4.19

Match the wavelength on graph 1 with the expected slope of the line on graph 2.

_____ Wavelength *A*
_____ Wavelength *B*
_____ Wavelength *C*

18. The deviation of a wavelength reading from an accepted recognized standard is a measure of:
    a. wavelength range
    b. absorbance accuracy
    c. photometric linearity
    d. wavelength accuracy

19. Each blank may have one or more answers.
    On wide bandpass (10 nm or greater) instruments, wavelength calibration may be verified by using _____ ; on narrow bandpass instruments, wavelength calibration may be verified by using _____ .
    a. mercury lamp
    b. holium oxide filter
    c. didymium filter
    d. holium chloride solution

20. The procedure that is used to confirm photometric linearity requires all *except:*
    a. use of a substance that conforms to Beer's law
    b. serial dilutions and preparation of a calibration curve
    c. straight line relationship of plotted points
    d. use of neutral density filters with defined absorbance values

21. It is recommended that wavelength calibration should be performed _____ ; linearity verification should be performed _____ ; photometeric accuracy should be performed _____ ; stray-light check should be performed _____ .
    a. daily
    b. weekly
    c. monthly
    d. when data suggest a problem

22. Match the optical arrangement with the correct instrument.
    _____ 1. Single beam spectrophotometer
    _____ 2. Flame photometer
    _____ 3. Fluorometer
    _____ 4. Double beam spectrophotometer
    _____ 5. Nephelometer
    _____ 6. Atomic absorption
    _____ 7. Scintillation counter
    _____ 8. Liquid chromatograph
    _____ 9. Gas chromatograph

A. UV light source → Primary filter → Cuvette
↓
Secondary
filter
↓
Detector
↓
Readout device

B. Light Source → Monochromator → Cuvette → Detector → Readout device

C. Atomizer → Flame → Interference filter → Detector → Readout device

D. Liquid   Sample   Recorder
   ↓   ↓   ↑
   Pump → Injection → Column → Detector
                                 ↓
                                 Waste

E. Light source → Monochromator → Cuvette
↘
Detector
↓
Readout device

F. Hollow cathode tube → Chopper → Atomizer → Monochromator → Detector → Readout device

G. Sample → Crystal → Detector → Amplifier → Spectrometer → Readout device

H. Light source → Monochromator → Cuvette → Detector ⌐
                  ↓   Readout device
                  ↓   → Monochromator → Cuvette → Detector ⌐
   I.   Sample

   Gas ⌐

   Column → Detector → Readout device

23. Match the instruments with the principle of operation.

    _____ 1. Nephelometer   a. measures light emitted by ex-
    _____ 2. Flame photometer   cited atoms
    _____ 3. Atomic absorption   b. measures a decrease in light
    _____ 4. Turbidmetry   transmittance by a cloudy so-
    _____ 5. Gamma counter   lution
                                      c. measures scintillations pro-
                                         duced by radioactive material
                                      d. measures light absorbed by
                                         ground-state atoms
                                      e. measures light scattered by
                                         Ag–Ab complexes

24. In fluorometry, the wavelength of light emitted is _____ than the wave-
    length absorbed for excitation.
    a. longer
    b. shorter

25. Sodium was assayed by flame photometry. The light emitted is from:
    a. sodium atoms (Na^0)
    b. sodium ions (Na^+)
    c. both
    d. neither

26. An internal standard used in flame photometry serves all of the purposes listed below except:
    a. corrects for changes in aspiration rate
    b. corrects for changes in flame temperature
    c. serves as a radiation buffer
    d. corrects for difference in sample viscosity

27. The light source in an atomic absorption instrument is a:
    a. mercury lamp
    b. deuterium lamp
    c. tungsten lamp
    d. hollow cathode lamp

28. In the determination of calcium by atomic absorption spectrophotometry, lanthanum is added to overcome:
    a. ionization interference
    b. matrix interference
    c. phosphate interference
    d. magnesium interference

29. An atom of iodine contains 53 protons. Therefore, ^{125}I contains _____ neutrons and ^{131}I contains _____ neutrons.

30. In a nuclide, the mass number describes _____; the atomic number describes _____ .

31. Some instruments measure light absorbed and some instruments measure light emitted. Classify each instrument based on the type of measurement performed.

    _____ 1. Spectrophotometer   a. measures light emission
    _____ 2. Atomic absorption   b. measures light absorption
    _____ 3. Fluorometer
    _____ 4. Scintillation counter
    _____ 5. Flame photometer

32. A substance was to be analyzed by a continuous flow system. The substance under assay required protein removal before color development. The components of a continuous flow system are listed below. Mark sequential numbers in the blanks in the order of sample processing.

    _____ manifold
    _____ sampler
    _____ colorimeter
    _____ readout device
    _____ dialyzer
    _____ heating bath

33. AutoAnalyzers use air to:
     a. segment samples and cleanse tubes
     b. segment samples and slow down flow
     c. cleanse tubes and absorb gases
     d. dilute samples and slow down flow

34. Match the following:
     _____ 1. Initial rate reactions   a. reaction monitored at timed
     _____ 2. Kinetic reactions   intervals
     _____ 3. End-point reactions   b. reaction has reached equi-
                                                    librium
                                                c. reaction read before equilib-
                                                    rium is reached and before
                                                    interfering substances react

35. In an attempt to identify an unknown sugar, thin-layer chromatography was
     performed. The distance traveled by the unknown was 6.4 cm. The distance
     traveled by the solvent front measured 8.0 cm. The R_f value for the unknown
     is _____ .

36. In the first blank, match the type of chromatography with the principle (items
     a–g). In the second blank, use items h–l and select the type of chromatogra-
     phy that could be used to *best* achieve separation. The second blank may or
     may not have a correct answer given in the list. If no answer is given, mark
     "X" in the blank.
     ____ ____ 1. Ion exchange   a. Separation is achieved by
     ____ ____ 2. Molecular exclusion   differences in vapor pres-
     ____ ____ 3. Gas   sure which affects the solu-
     ____ ____ 4. Adsorption   bility of the substance in a
     ____   5. Partition   high-boiling-point liquid.
           ____   a. Normal phase   b. Separation is achieved by
           ____   b. Reverse phase   the exchange of a cation
                                                for an anion or vice versa.
                                            c. Separation is achieved due
                                                to differences in surface
                                                adsorption of the solute on
                                                packing material.
                                            d. Separation is achieved due
                                                to differences in solubility
                                                of solute molecules in two
                                                immisible liquids.
                                            e. Separation is achieved on
                                                the basis of differences in
                                                molecular size, shape and
                                                hydration; the small parti-
                                                cles pass out of the column
                                                first.
                                            f. Separation is achieved on
                                                the basis of differences in
                                                molecular size, shape, and

hydration; the smaller particles are retained in the column longer than the larger particles.

g. Separation is achieved by the exchange of a cation for another cation or an anion for another anion.

h. Separation of polar substance

i. Separation of nonpolar substances

j. Separation of high molecular weight particles such as proteins from lower molecular weight substances

k. Separation of volatile substances

l. Separation of charged particles

37. Which instrument would be the best choice to quantitate the components listed below:

_____	1.	Sodium	a. Flame photometer
_____	2.	Trace metals	b. Nephelometer
_____	3.	Porphyrins	c. Fluorometer
_____	4.	Total protein	d. Spectrophotometer
_____	5.	Glucose	e. Atomic absorption
_____	6.	Potassium	f. Gas chromatograph
_____	7.	IgG	
_____	8.	C_3	
_____	9.	Alcohols	

38. The EMIT procedure is an example of a _____ method; the ELISA procedures are examples of a _____ method.
   a. homogeneous
   b. heterogeneous

39. The most sensitive immunoassay methods are those which use _____ labels.
   a. fluorescent
   b. radioisotope
   c. enzyme

40. Two samples for a plasma cortisol were determined by a RIA procedure using a specific antibody. The count on the tagged antigen–antibody fraction in sample #1 was 8000 cpm. The count on the tagged antigen–antibody fraction in sample #2 was 6000 cpm. From this information, you can conclude (two answers are required):
   a. The cortisol concentration in sample #1 was higher than in sample #2.

   b.   The cortisol concentration in sample #2 was higher than in sample #1.

   c.   In the above example, there is an inverse relationship between radioactivity and cortisol concentration.

   d.   In the above example, there is a direct relationship between radioactivity and cortisol concentration.

41.   In an RIA competitive binding procedure, the unbound or free fraction was counted. High radioactivity indicates:

   a.   high patient concentration

   b.   low patient concentration

42.   Two samples were assayed for theophylline using the EMIT procedure. The enzyme activity was found to be higher in sample #1 than in sample #2. From this information, you can conclude (two answers are required):

   a.   The theophylline concentration in sample #1 is higher than in sample #2.

   b.   The theophylline concentration in sample #2 is higher than in sample #1.

   c.   Enzyme activity and drug concentration are inversely related.

   d.   Enzyme activity and drug concentration are directly related.

43.   In an ELISA sandwich procedure, the enzyme activity was higher in sample #1 than in sample #2. From this information, you can conclude (two answers are required):

   a.   The concentration of sample #1 is higher than sample #2.

   b.   The concentration of sample #2 is higher than sample #1.

   c.   Enzyme activity is inversely related to concentration.

   d.   Enzyme activity is directly related to concentration.

ANSWERS

---

1.   (1) c; (2) b; (3) a

2. 10^{-9}

3.   a

4.   d

5.   d

6.   a

7.   e

8.   a

9.   (a) 0.440; (b) 0.600

10.   zero

11.   1.0

12.   9.5 mg/dL

13.   (1) c; (2) b; (3) d; (4) a; (5) e

14.   a, d, b, c

15.   a

16. 4 nm
17. 1, 2, 3
18. d
19. c; a, b, d
20. d
21. a, b, b, d
22. (1) B; (2) C; (3) A; (4) H; (5) E; (6) F; (7) G; (8) D; (9) I
23. (1) e; (2) a; (3) d; (4) b; (5) c
24. a
25. d
26. d
27. d
28. c
29. 72, 78
30. number of protons and neutrons; number of protons
31. (1) b; (2) b; (3) a; (4) a; (5) a
32. 2, 1, 5, 6, 3, 4
33. a
34. (1) c; (2) a; (3) b
35. 0.8
36. (1) g, l; (2) f, j; (3) a, k; (4) c, x; (5) d, (a) h, (b) i
37. (1) a; (2) e; (3) c; (4) d; (5) d; (6) a; (7) b; (8) b; (9) f
38. a; b
39. b
40. b, c
41. a
42. a, d
43. a, d

CHAPTER 5
CARBOHYDRATES

OBJECTIVES

On completion of this chapter, you will be able to:

1. Describe the chemical composition of carbohydrates.
2. Define the following words as they relate to carbohydrate nomenclature:
   a. aldoses
   b. ketoses

    c. D and L sugars

    d. alpha and beta configuration

3. List the three monosaccharides of biological importance.

4. For the three disaccharides of biological importance:

    a. List the monosaccharide units of each disaccharide.

    b. Describe the reaction that occurs in the formation of a glycosidic bond.

    c. Define the characteristic that classifies the sugar as a reducing or nonreducing sugar.

5. List three common polysaccharides. Describe the type of glycosidic bond that can be hydrolyzed by the enzyme amylase.

6. List examples of biologically important compounds of the body that are derivatives of carbohydrates.

7. Define the terms:

    a. glycogenesis

    b. glycogenolysis

    c. gluconeogenesis

    d. glycolysis

    e. lipogenesis

    f. lipolysis

8. Trace the biochemical pathways associated with carbohydrate metabolism. Discuss:

    a. digestion and absorption processes

    b. the glycolytic or Embden–Meyerhof pathway

    c. the pentose phosphate or shunt pathway

    d. common pathways

9. List the hormones that regulate blood glucose levels. Describe the mechanism of glucose regulation for each hormone.

10. Discuss specimen requirements for glucose assay. Describe the variability of glucose concentration in the following situations:

    a. whole blood sample versus serum or plasma sample

    b. fasting sample versus nonfasting sample

    c. capillary sample versus venous sample

11. For each glucose assay procedure, discuss:

    a. procedure principle

    b. sources of error

12. List normal and panic blood glucose values for adults and neonates.

13. Describe the correct procedure to be followed in the preparation of an individual for an oral glucose tolerance test. Include:

    a. dietary requirements preceding the test

    b. medication precautions

    c. fasting requirements

    d. glucose load

14. Discuss oral glucose tolerance test results associated with:
    a. normal individuals
    b. individuals with decreased glucose tolerance
    c. individuals with increased glucose tolerance
15. Discuss the glycosylated hemoglobin test. Include:
    a. origin of glycosylated fractions
    b. proposed test applications
    c. mechanisms of measurement
    d. sources of error
16. List four types of diabetes mellitus. For each type describe:
    a. associated symptoms
    b. laboratory data
    c. diagnostic criteria for group classification
17. Correlate laboratory findings associated with reactive versus pathological hypoglycemia.
18. Discuss the application of the C-peptide assay in the differentiation of hypoglycemia associated with insulinomas versus self-administration of insulin.
19. Discuss test performance and interpretation of results for the following tests:
    a. tolbutamide test
    b. epinephrine test
    c. lactose tolerance test.

GLOSSARY OF TERMS

Aldehyde. A derivative of hydrocarbons that contains a carbonyl ($C{=}O$) group at the end of the carbon chain.

Anomeric carbon. Refers to the carbon of the carbonyl group that participates in the formation of the ring structure; this carbon then becomes asymmetric.

Asymmetric carbon. A carbon atom bonded to four different atoms or groups of atoms.

Carbohydrate. Polyhydroxy aldehydes and ketones and their derivatives.

Carbonyl group. The group represented by the structural formula $\diagdown C{=}O$.

Condensation reaction. The reaction that occurs as molecules are joined together; water is formed in the process. A synonym for this reaction is dehydration synthesis. Polysaccharides, proteins, and triglycerides are formed by condensation reactions.

Dextrins. Products of partial hydrolysis of starch; a complex mixture of molecules of different sizes.

Dextrorotary. Describes a monosaccharide that rotates plane polarized light to the right.

Disaccharide. A carbohydrate that is composed of two monosaccharide units joined together by a glycosidic bond.

Empirical formula.   The formula that indicates the elements present in the compound and the number of atoms of each element necessary to form the compound.

Glycosidic linkage.   The bond that is formed when monosaccharide units are joined together.

Ketone.   A derivative of hydrocarbons that contains the carbonyl group (C=O) at an intermediate position in the chain.

Levorotary.   Describes a monosaccharide that rotates plane polarized light to the left.

Monosaccharide.   Simple sugars that cannot be hydrolyzed into more simple forms.

Oxidation.   Increase in oxidation state of an atom due to a loss of electrons.

Polyhydroxy.   Refers to many hydroxyl (—OH) groups.

Reduction.   Decrease in oxidation state of an atom due to a gain in electrons.

Structural formula.   The formula that indicates the arrangement of the atoms and groups which compose a molecule.

Structural isomer.   Molecules that have the same empirical formula but differ in their structural formula.

The typical western diet consists of about 45% carbohydrate, 45% fat, and 10% protein. Carbohydrate sources include grains, starchy vegetables, fruits, cane and beet sugars, and milk products. Animal sources contain minimal carbohydrate (less than 1%) in the form of glycogen.

CHEMICAL COMPOSITION

Carbohydrates are composed of carbon, hydrogen, and oxygen, with the ratio of hydrogen to oxygen the same as water (2 : 1), hence the name carbohydrates or hydrates of carbon. The general molecular formula for a carbohydrate is $C_n(H_2O)_n$ ($n = 3$ or some larger number). However, there are carbohydrates that do not contain this H : O ratio, and some carbohydrate derivatives contain phosphorus, nitrogen, and sulfur in addition to carbon, hydrogen, and oxygen.

Carbohydrates may be defined as polyhydroxy (more than one hydroxyl group) aldehydes or ketones and their derivatives. An aldehyde has a carbonyl (C=O) group at the end of the carbon chain; a ketone has the carbonyl group on an internal carbon atom.

CLASSIFICATION AND NOMENCLATURE OF CARBOHYDRATES

MONOSACCHARIDES

Monosaccharides may be defined as simple sugars that cannot be hydrolyzed into more simple carbohydrates. Most contain from three to seven carbon atoms. The group names of monosaccharides are derived from the number of carbon atoms in the chains. Sugars that contain three carbon atoms are called trioses; with four

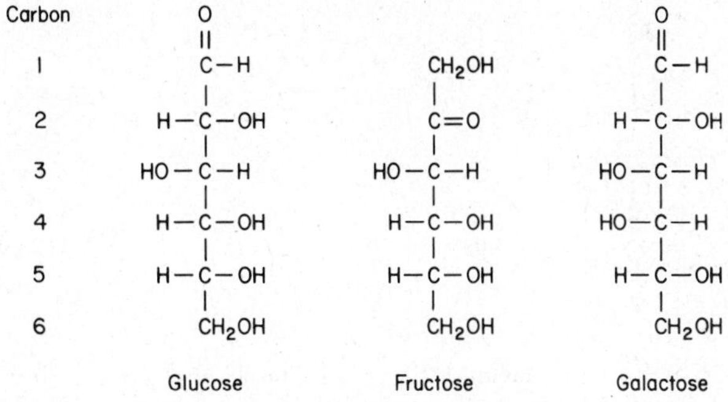

Carbon			
1	$\overset{\overset{\text{O}}{\|\|}}{\text{C}-\text{H}}$	CH_2OH	$\overset{\overset{\text{O}}{\|\|}}{\text{C}-\text{H}}$
2	$\text{H}-\text{C}-\text{OH}$	$\text{C}=\text{O}$	$\text{H}-\text{C}-\text{OH}$
3	$\text{HO}-\text{C}-\text{H}$	$\text{HO}-\text{C}-\text{H}$	$\text{HO}-\text{C}-\text{H}$
4	$\text{H}-\text{C}-\text{OH}$	$\text{H}-\text{C}-\text{OH}$	$\text{HO}-\text{C}-\text{H}$
5	$\text{H}-\text{C}-\text{OH}$	$\text{H}-\text{C}-\text{OH}$	$\text{H}-\text{C}-\text{OH}$
6	CH_2OH	CH_2OH	CH_2OH
	Glucose	Fructose	Galactose

Figure 5.1. Biologically important hexoses.

carbon atoms, tetroses; with five carbon atoms, pentoses; with six carbon atoms, hexoses; and with seven carbon atoms, heptoses. The hexoses are the predominate type of monosaccharide found in the body. Two pentose sugars, ribose and deoxyribose, are important constituents of nucleic acids.

The three hexoses of biological importance—glucose, fructose, and galactose—all have the same empirical formula, $C_6H_{12}O_6$ (Figure 5.1). Glucose and galactose are aldehyde sugars because the carbonyl group is on the end carbon. Glucose and galactose are known as aldoses. Fructose is a ketone sugar, because the carbonyl group is on an internal carbon. It is called a ketose. The aldehyde and keto groups (carbonyls) are called reducing groups, because they will chemically reduce certain metal ions (copper). This property has been used in the laboratory for the analysis of sugars.

Structural Isomers and Steroisomers

Glucose, fructose, and galactose are *structural isomers*. They have the same number of atoms and kinds of groups (same empirical formula) but differ in the arrangement of the groups around the carbon atoms (different structural formula).

The number of possible optical isomers or *steroisomers* is related to the number of asymmetric carbon atoms present. Asymmetric carbon atoms are those bonded to four different atoms or groups of atoms. In Figure 5.1, each of the internal carbons, 2 through 5 for glucose and galactose and 3 through 5 for fructose, are asymmetric carbons. Sugars with asymmetric carbon atoms can possess optical activity and rotate polarized light. Mirror-image isomers rotate light in opposite directions.

D and L Sugars

For sugars having two or more asymmetric carbon atoms, D and L refer to the arrangement of groups around the asymmetric carbon located at the greatest distance from the carbonyl carbon atom. By convention, if the —OH group on the most distal asymmetric carbon is on the right, it is called a D sugar (dextro, right); if the —OH group is on the left, it is called an L sugar (levo, left) (Figure 5.2).

The D and L forms of glucose are mirror-image forms, will rotate light in opposite directions, and are called optical isomers. The majority of sugars in the body are of the D configuration.

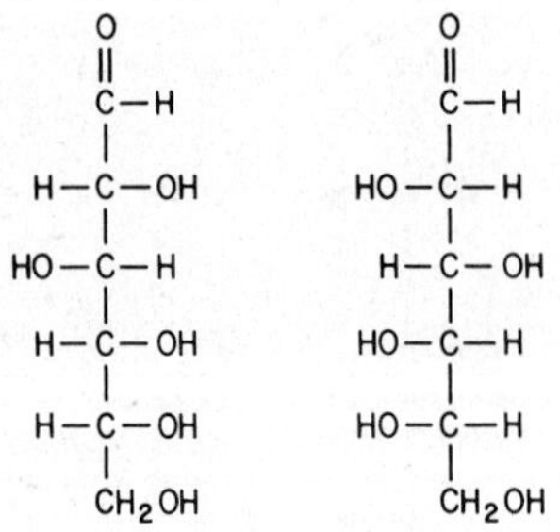

```
        O                    O
        ||                   ||
        C—H                  C—H
        |                    |
   H—C—OH              HO—C—H
        |                    |
  HO—C—H                H—C—OH
        |                    |
   H—C—OH              HO—C—H
        |                    |
   H—C—OH              HO—C—H
        |                    |
     CH2OH                CH2OH

  D – Glucose          L – Glucose
```

Figure 5.2. D and L forms of glucose.

Hemiacetyl or Ring Formation

Ring and chain forms of carbohydrates exist in chemical equilibrium. Only a small amount of monosaccharides are present in the open chain; most exist as five- or six-membered heterocyclic rings. The rings result from intramolecular reactions between the active carbonyl group (C=O) and a hydroxyl group located on the next to the last carbon atom. When the bond is formed, the carbon of the carbonyl group becomes asymmetric and is called an *anomeric* carbon. In the ring structure, if the hydroxyl group on the anomeric carbon is written to the right, the sugar is called alpha (α); if the hydroxyl group is on the left, it is of the beta (β) configuration (Figure 5.3).

When using Haworth projection formulas, it is customary to not show the carbon atoms. The ring is considered to be perpendicular to the plane of the paper; the heavy line indicates the plane of the ring to be projecting toward the reader. When the ring closes, the hydroxyl group on the anomeric carbon may lie either above or below the plane of the ring. In the alpha form, the hydroxyl group lies below the plane of the ring; in the beta form, the hydroxyl group lies above the plane of the ring. In the D series, the terminal primary alcohol group (—CH$_2$OH) projects above the plane of the ring; in the L series the terminal primary alcohol group projects below the plane of the ring (Figure 5.4).

DISACCHARIDES

Disaccharides yield two molecules of the same or different monosaccharide upon hydrolysis. The general formula for disaccharides is $C_n(H_2O)_{n-1}$. In formation, the

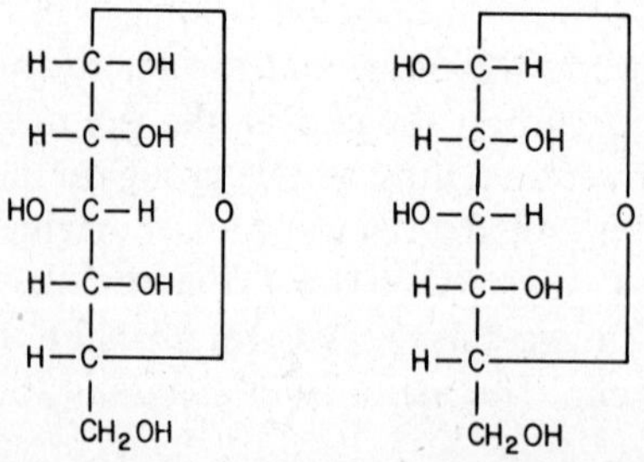

```
     H—C—OH                   HO—C—H
     |                        |
     H—C—OH                   H—C—OH
     |                        |
   HO—C—H      O            HO—C—H      O
     |                        |
     H—C—OH                   H—C—OH
     |                        |
     H—C                      H—C
     |                        |
     CH2OH                    CH2OH

  α–D–Glucose              β–D–Glucose
```

Figure 5.3. Alpha and beta forms of glucose.

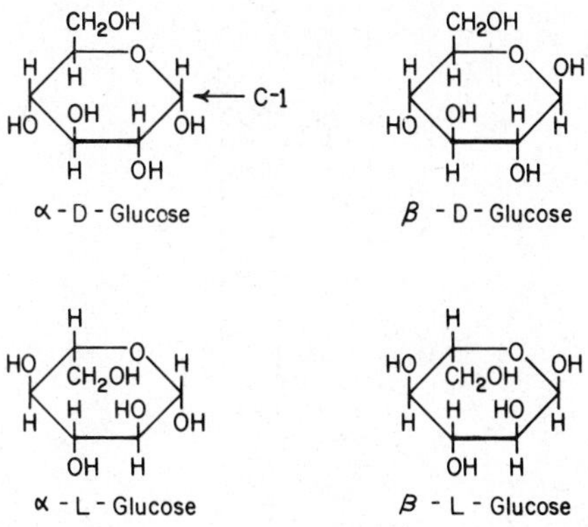

CH2OH
H H O H ← C-1
HO OH H OH
H OH
α - D - Glucose

CH2OH
H H O OH
HO OH H H
H OH
β - D - Glucose

H
HO CH2OH O H
H H HO OH
OH H
α - L - Glucose

H
HO CH2OH O OH
H H HO H
OH H
β - L - Glucose

Figure 5.4. Haworth formulas, alpha and beta, D and L nomenclature.

monosaccharide units are covalently linked together by a condensation reaction (dehydration synthesis) in which one molecule of water is removed. This *glycosidic bond* is usually formed between a hydroxyl group of one sugar and a hydroxyl group located on an anomeric carbon of an adjacent sugar molecule. If hydroxyl groups on both anomeric carbons are involved in the formation of the glycosidic bond, a nonreducing sugar is formed. If the bond is formed involving the hydroxyl group of only one anomeric carbon, the sugar is a reducing sugar due to the presence of a potentially free carbonyl grouping. In naming the glycosidic bond, the projection of the hydroxyl group (i.e., α or β) on the anomeric carbon forming the bond is designated, and the numbers of those carbon atoms that participate in the bond are indicated (Figure 5.5).

Three common disaccharides ($C_{22}H_{22}O_{11}$) are maltose, lactose, and sucrose (Figure 5.6). Maltose is composed of two glucose units and is a reducing sugar. Lactose (milk sugar) is composed of galactose and glucose and is a reducing sugar. Sucrose (cane sugar, table sugar) is composed of glucose and fructose and is not a reducing sugar.

OLIGOSACCHARIDES AND POLYSACCHARIDES

Oligo means few. Oligosaccharides are those carbohydrates that yield 2–10 monosaccharide units on hydrolysis.

Poly means many. Polysaccharides yield more than 10 molecules of monosaccharides on hydrolysis. They are formed by the condensation of large numbers of

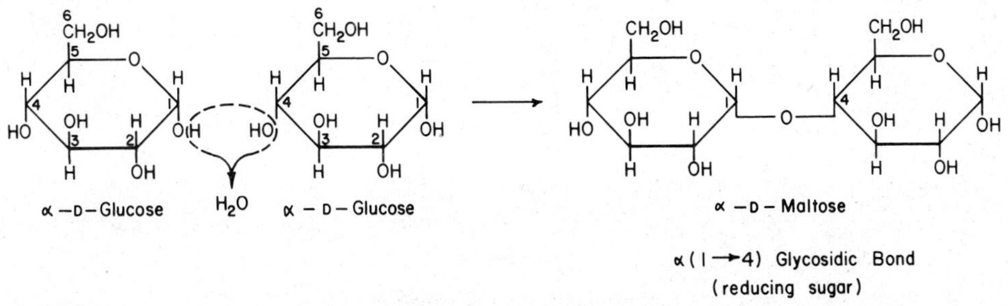

6
CH2OH
5 O
H H 4 H
HO OH 3 2 OH H HO
H OH
α —D—Glucose   H2O

6
CH2OH
5 O
H H 4 H
HO OH 3 2 OH
H OH
α − D − Glucose

CH2OH
O
H H H
HO OH H O 4
H OH

CH2OH
O
H H H
OH H OH
H OH

α —D— Maltose

α (I →4) Glycosidic Bond
(reducing sugar)

Figure 5.5. Condensation reaction with glycosidic bond formation.

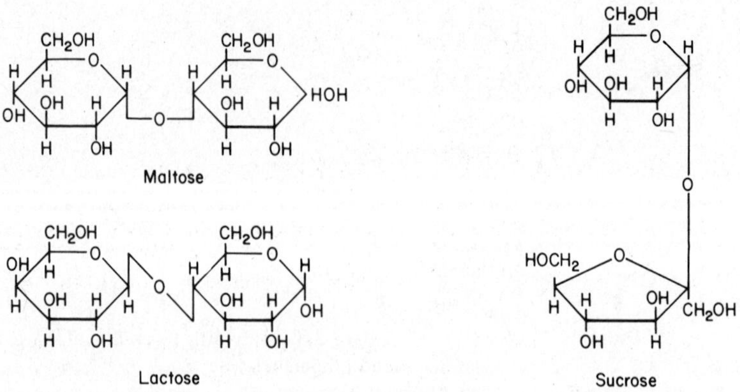

Figure 5.6. Common disaccharides indicating α (1 → 4) bond (maltose), β (1 → 4) bond (lactose), and α (1 → 2) bond (sucrose).

monosaccharide units. Starch, cellulose, and glycogen are built entirely from glucose units.

Starch

Starch is a high-molecular-weight storage carbohydrate of plant cells. It is composed of amylose, a linear polymer, and amylopectin, a branched polymer. It is readily digested by man. Products of partial hydrolysis of starch are called dextrins, which consist of a mixture of carbohydrate molecules of different sizes.

The glucose units in amylose are joined by α(1 → 4) glycosidic bonds. The

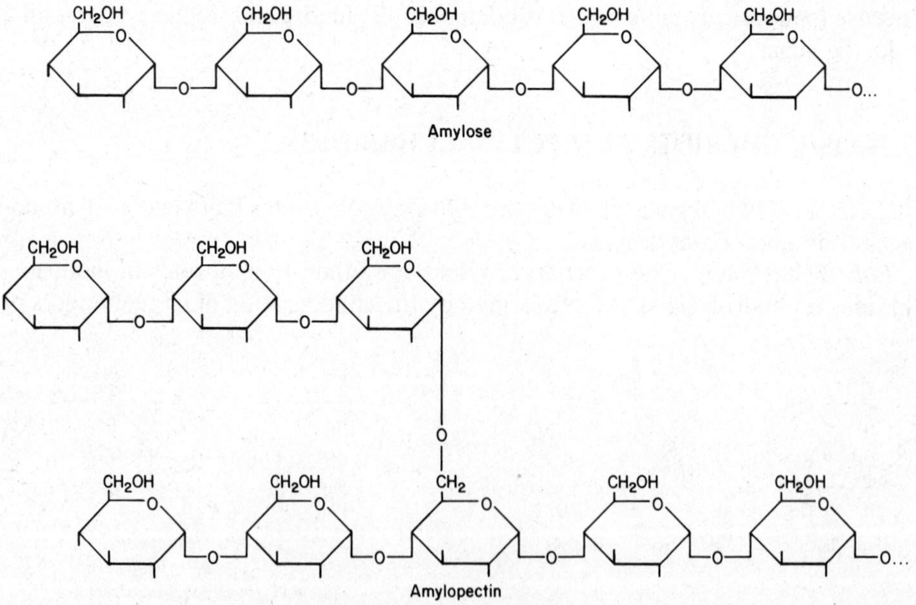

Figure 5.7. Starch structure.

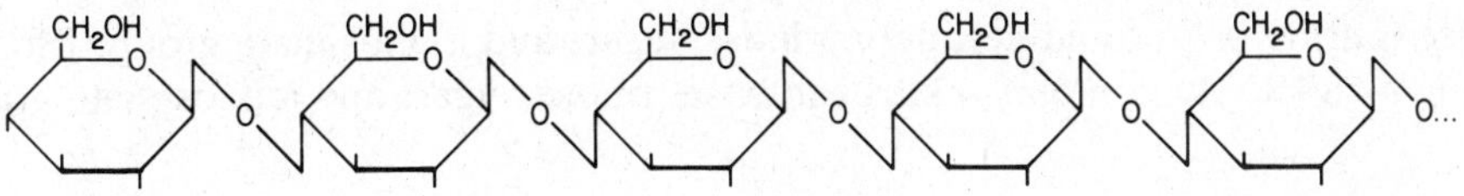

Figure 5.8. Cellulose structure.

glucose units in amylopectin are joined by $\alpha(1 \rightarrow 4)$ glycosidic bonds, with $\alpha(1 \rightarrow 6)$ glycosidic bonds forming the branch points (Figure 5.7).

Cellulose

Cellulose (Figure 5.8) is a structural carbohydrate that forms cell walls and other supporting tissues of plants. The linkage between the glucose units is composed of $\beta(1 \rightarrow 4)$ glycosidic bonds and is not digested to any extent by man, because amylase cannot hydrolyze these bonds. Certain bacteria and protoza contain enzymes that can hydrolyze these bonds. These organisms are present in the rumen or colon of herbivorous animals, allowing these animals to utilize cellulose as food.

Glycogen

Glycogen is the storage carbohydrate of animal tissue and is found principally in the liver, where it may comprise up to 10% of the net weight, and to a lesser extent in skeletal muscle (1–2%). It is a branched polymer of thousands of glucose units like amylopectin, but the branch points in glycogen occur more frequently. It has been described as having a treelike shape. The liver is able to convert glucose to glycogen; glycogen is stored then broken down to glucose as needed. Glycogen stores are used by the body to maintain the blood sugar level between meals. After a 12–18 hour fast, the liver becomes almost totally depleted of glycogen.

DERIVATIVES OF MONOSACCHARIDES FOUND IN THE LIVING ORGANISM

A number of biologically important compounds in the body are derivatives of carbohydrates.

PHOSPHORIC ACID ESTERS OF PENTOSE

Adenosine triphosphate (ATP), adenosine diphosphate (ADP), and cyclic-adenosine monophosphate (cAMP) are all examples of compounds containing ribose sugar. They are nucleotides formed by the nitrogen base, adenine, the sugar, ribose, and one or more phosphate groups. When ATP forms ADP, energy is released for muscular contraction and other types of cellular work; cAMP serves as the second messenger in one of the hormonal mechanisms of action.

The chromosomal material, DNA (deoxyribonucleic acid), contains a nitrogen

base (purine or pyrimidine), deoxyribose sugar, and a phosphate group. Ribonu-
cleic acid (RNA) contains a nitrogen base, ribose sugar, and a phosphate group.

AMINO SUGARS

Glucosamine and galactosamine are two amino sugars found in many tissues of
the body. Structurally, an amino group (NH_2) replaces a hydroxyl group on carbon 2 of either glucose or galactose.

SUGAR ACIDS

An important six-carbon sugar acid is glucuronic acid. Glucuronic acid is formed
when carbon 6 in glucose is oxidized to a carboxyl group (—COOH). Many drugs,
toxic substances, and steroid hormones are conjugated with glucuronic acid in the
liver to facilitate excretion.

COMPLEX MACROMOLECULES CONTAINING CARBOHYDRATE

ACID MUCOPOLYSACCHARIDES

Acid mucopolysaccharides are complex macromolecules; many contain monosac-
charide-derived units such as amino sugars and sugar acids. The most abundant
acid mucopolysaccharide is hyaluronic acid, which occurs in synovial fluid, in the
vitreous humor of the eye, and in the extracellular ground substances of connective tissue (intercellular cement). Heparin, an anticoagulant synthesized in the
liver, and chondroitin sulfate, which forms the structure of cartilage and bone, are
other examples of acid mucopolysaccharides.

GLYCOPROTEINS

Glycoproteins are complex macromolecules containing carbohydrate groups attached covalently to a polypeptide chain. Glycoproteins contain many different
monosaccharides and monosaccharide derivatives. The carbohydrate content of
glycoproteins may vary from less than 1% to as high as 80%.

Most glycoproteins occur as extracellular components. Extracellular glycoproteins include the blood group substances (antigens A and B on cells), the circulating form of some protein hormones (chorionic gonadotropin, follicle-stimulating
hormone, thyroid stimulating hormone), the immune globulins, substances
present in saliva (secretors), and some of the alpha globulins.

CARBOHYDRATE METABOLISM

The following terminology is used to describe carbohydrate metabolism:

1. *Glycogenesis.* The conversion of glucose to glycogen.
2. *Glycogenolysis.* The breakdown of glycogen to form glucose and other intermediate products.
3. *Gluconeogenesis.* The formation of glucose from noncarbohydrate sources (e.g., amino acids).
4. *Glycolysis.* The conversion of glucose or other hexoses into lactate or pyruvate.
5. *Lipogenesis.* The formation of fat from carbohydrate.
6. *Lipolysis.* The breakdown of fat and the release of fatty acids from fat cells into the blood.

DIGESTION AND ABSORPTION

The digestion of carbohydrate begins in the mouth. The enzyme, ptyalin or salivary amylase, attacks plant starches and glycogen and changes them to dextrins and maltose.

In the stomach, the salivary amylase is inactivated by the HCl of the gastric juice. As the stomach contents enter the duodenum, small intestinal hormones are secreted that stimulate the pancreas to produce alkaline pancreatic fluid. Pancreatic amylase (amylopsin) is more enzymatically active than salivary amylase. Further hydrolysis occurs, producing disaccharides and small amounts of monosaccharides.

The intestinal wall enzymes—maltase, sucrase, lactase, and isomaltase (disaccharidases)—then catalyze the hydrolysis of the disaccharide units into monosaccharides. The monosaccharides are absorbed by both active transport and simple diffusion through the intestinal wall (jejunum) into the blood stream. If the capacity of the jejunum to absorb is exceeded, absorption takes place in the ileum. The monosaccharides are then transported to the liver via the portal vein. Any disaccharides that are absorbed, pass through the liver, enter the blood, and are excreted in the urine. In the liver, the monosaccharides may then be used for energy purposes, converted to glycogen for storage purposes, converted to carbohydrate derivatives or pentose sugars, converted to fat, or converted to amino acids. Any glucose that escapes hepatic uptake becomes available for use by peripheral tissue.

Glucose is the only monosaccharide that is used by the body for energy. The liver has the necessary enzymes to convert other monosaccharides (galactose and fructose) to glucose. Glucose, in the form of glycogen, can be stored in the liver and muscle. The hepatic glycogen is available at any time the blood glucose level is decreased, but the muscle glycogen is not available to the blood; it is available for muscle energy only. The liver contains the enzyme, glucose-6-phosphatase, required to convert glycogen to glucose. Muscle lacks this enzyme. The digestion of carbohydrate may be summarized as follows:

$$\text{polysaccharides} \xrightarrow[\text{amylase}]{\text{salivary}} \text{dextrins and maltose} \xrightarrow[\text{amylase}]{\text{pancreatic}} \text{disaccharides}$$

$$\xrightarrow[\text{disaccharidases}]{\text{intestinal}} \text{monosaccharides} \rightarrow \text{portal blood} \rightarrow \text{liver}$$

PATHWAYS OF GLUCOSE METABOLISM

Glucose is needed to produce adenosine triphosphate (ATP). ATP is the primary compound for energy storage in the cell. ATP cannot cross cell membranes. Each cell must therefore produce its own supply of ATP. When ATP is hydrolyzed to ADP, energy is released. A portion of this energy is utilized for contraction of muscles, nerve conduction, maintenance of body temperature, active transport, and synthesis processes. A simple schematic representation of glucose metabolism is presented in Figure 5.9. The key below is used to interpret each numbered step of Figure 5.9.

1. In the first step of glucose utilization, energy is used. Glucose enters the cell under the influence of insulin. ATP reacts with glucose (intracellular process) in the presence of hexokinase to form glucose-6-phosphate and ADP:

$$\text{glucose} + \text{ATP} \xrightarrow{\text{hexokinase}} \text{glucose-6-phosphate} + \text{ADP}$$

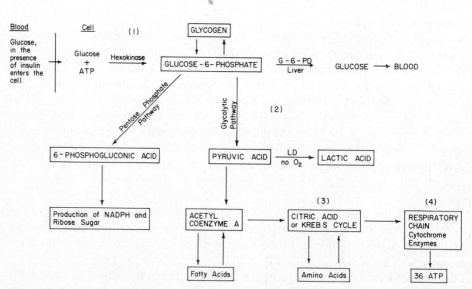

Figure 5.9. Pathways of glucose metabolism. The numbers in parentheses refer to reactions described in the text.

Glucose-6-phosphate is a common intermediate for a number of routes by which glucose is metabolized. There are two main pathways for its breakdown: (1) glycolytic or Embden–Meyerhof Pathway and (2) pentose phosphate or shunt pathway.

Glucose-6-phosphate may also be converted to glycogen. This process requires a complex enzyme system.

2.   The glycolytic pathway consists of a series of reactions by which one molecule of glucose (six carbons) is converted to two molecules of pyruvic acid (three carbons). This process does not require oxygen (anaerobic) and takes place in the cytoplasm of the cell. The coenzyme, nicotinamide adenine dinucleotide (NAD^+), is the carrier that accepts the hydrogen removed during glycolysis. The glycolytic pathway provides only two ATPs per molecule of glucose. Glycolysis can be represented as

$$C_6H_{12}O_6 + 2NAD^+ + 2ADP + 2Pi \rightarrow 2C_3H_4O_3 + 2NADH_2 + 2ATP$$

The fate of pyruvic acid depends on whether or not oxygen is present. If there is no oxygen present, the pyruvic acid is converted to lactic acid.

3.   If there is abundant oxygen present, the pyruvic acid enters the citric acid cycle, also called the Krebs Cycle or the tricarboxylic acid cycle. The major portion of the pyruvic acid undergoes oxidative decarboxylation to form a two-carbon (acetyl) fragment. The acetyl fragment in combination with coenzyme A (a combination of ATP and one of the B vitamins) forms acetyl coenzyme A (acetyl CoA). The citric acid cycle refers to the chemical processes involved in the complete oxidation of acetyl CoA. These processes require the presence of oxygen (aerobic) and take place in the mitochondria in the cell. NAD^+ acts as the hydrogen acceptor forming $NADH_2$. In this stage, the electrons are transported in the form of hydrogen (electron + proton) rather than as free electrons.

4.   The electron transport and oxidative phosphorylation cycle is the final pathway for all electrons removed from the substrate during respiration. The hydrogens from $NADH_2$ are channeled into the electron transport system, a chain of electron carriers, most of which are the cytochrome enzyme group. (The cytochromes are iron-containing enzymes similar to hemoglobin.) These carriers are reduced and oxidized as they transport the free electrons one step at a time to oxygen. During this sequential oxidation–reduction of electron carriers, the free energy of the electrons is used to convert ADP to ATP (oxidative phosphorylation). Thirty-six ATPs are generated from the electron transport system. One molecule of glucose completely oxidized produces 38 ATP molecules (2 ATPs produced from glycolysis and 36 ATPs from the electron transport system):

$$C_6H_{12}O_6 + 6O_2 \rightarrow 6CO_2 + 6H_2O + Energy$$

$$where: Energy\ is\ 38\ ATP$$

About 40% of the energy released from the complete oxidation of glucose is stored in ATP; the rest is lost as heat.

The pentose phosphate or hexose monophosphate shunt pathway is secondary in most tissues and does not generate ATP. This pathway provides a mechanism for synthesis of ribose sugar, the carbohydrate component of nucleotides (DNA, RNA). The shunt pathway also generates NADPH, which is utilized in the formation of lipids and helps to maintain the integrity of the red blood cell membrane.

COMMON PATHWAYS

Gluconeogenesis

Energy may be derived from noncarbohydrate sources (gluconeogenesis). Fragments from all classes of foodstuffs can eventually be channeled into the citric acid cycle for the production of energy.

A common junction point is through acetyl CoA. Certain amino acids (derived from the digestion of proteins) and fatty acids (derived from the breakdown of fats) can be converted to acetyl CoA. Amino acids may also be converted into several intermediates of the citric acid cycle. Glycerol (derived from the breakdown of fat) can be converted into an intermediate in the glycolytic cycle.

Lipids and Protein Synthesized From Carbohydrates

Metabolites of carbohydrates can form fatty acids. Any carbohydrate, taken in excess of that required for energy, is converted to and stored as fat (fatty acid + glycerol). Metabolites of carbohydrates can also form the carbon skeleton of nonessential amino acids, providing a mechanism for protein synthesis from carbohydrate.

Lactic Acid

NAD^+ is present in small amounts in the cell. When the six-carbon glucose molecule is split into two three-carbon fragments, NAD^+ serves as the hydrogen acceptor. If NAD^+ could not be reoxidized after reduction, glycolysis would become stalled after all available NAD^+ was reduced. The formation of lactic acid from pyruvic acid provides a mechanism for recycling NAD^+ when oxygen is absent. The reaction is catalyzed by the enzyme lactate dehydrogenase (LD) (Figure 5.10). The formation of lactic acid from pyruvic acid frees sufficient NAD^+ so that glycolysis can proceed when oxygen supplies are insufficient or absent. The lactic acid that accumulates in the muscles can be converted back to pyruvic acid when sufficient oxygen is present. Any lactic acid that diffuses into the blood stream is removed by the liver and converted back to glucose.

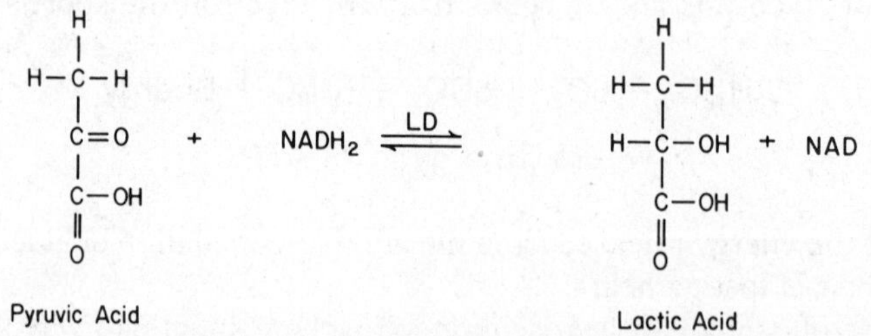

Figure 5.10. Mechanism for recycling NAD.

Glycogenolysis

Glucose may be derived from glycogen through glycogenolysis. The enzymes necessary for glycogen conversion are activated by various hormones (epinephrine and glucagon in the liver; epinephrine in muscle; epinephrine and norepinephrine in adipose tissue). Muscle glycogen cannot add to the blood glucose, because muscle lacks the glucose-6-phosphatase enzyme necessary to convert glycogen back to glucose. Muscle glycogen can be used only for energy purposes in muscle cells. The glucose-6-phosphatase enzyme is present in the liver. Liver glycogen can therefore form glucose, which can then enter the blood to provide an energy source for many body cells:

$$\text{glycogen} \rightarrow \text{glucose-6-phosphate} \xrightarrow{\text{glucose-6-phosphatase}}$$

$$\text{glucose} \rightarrow \text{released to blood}$$

REGULATION OF BLOOD GLUCOSE

Since glucose is not stored to any extent within the cell, it has to be obtained from the blood. The brain is almost completely dependent on the blood for its supply of glucose and is extremely sensitive to changes in blood glucose.

At least six hormones are involved in the control of blood glucose levels. Of these, only insulin acts to lower blood glucose; all others raise the blood glucose level.

INSULIN

Insulin is synthesized in the beta cells of the islets of Langerhans of the pancreas and is needed for the entry of glucose into cells. However, certain tissues (kidney, liver, brain, and intestinal mucosa) do not need insulin for transport of glucose across cell membranes. In skeletal muscle and heart cells, the glucose is metabolized to produce energy; any glucose entering adipose tissue is metabolized to acetyl CoA for fat synthesis.

Insulin may also increase or decrease the activity of enzymes in the liver. Insulin increases the formation of glycogen (glycogenesis) and decreases the release of glucose to the blood. Insulin inhibits gluconeogenesis.

GLUCAGON

Glucagon is synthesized in the alpha cells of the islets of Langerhans of the pancreas. Glucagon increases the breakdown of hepatic glycogen (glycogenolysis) but has no effect on muscle glycogen.

Glucagon and insulin act as the main control mechanisms in the regulation of blood glucose and operate via negative feedback systems. An increase in blood glucose stimulates the release of insulin. Insulin lowers the blood glucose, which then supresses insulin secretion. A decrease in blood glucose or ingestion of a protein meal stimulates glucagon release, which raises blood glucose levels. As the blood glucose rises, glucagon synthesis is suppressed.

EPINEPHRINE

Epinephrine, a catecholamine, is produced in the adrenal medulla. Epinephrine causes glycogenolysis of both hepatic and muscle glycogen; however, since muscle does not contain the glucose-6-phosphatase enzyme needed to catalyze the liberation of free glucose, muscle glycogen does not contribute to blood glucose. Epinephrine also inhibits the release of insulin from the pancreas and provokes glucagon release. A tumor of the adrenal medulla (pheochromocytoma, which produces epinephrine) may produce moderate hyperglycemia as long as glycogen stores in the liver are available.

Stress causes an increase in the release of epinephrine and may lead to hyperglycemia.

GLUCOCORTICOIDS

The glucocorticoids (cortisol, cortisone), produced in the adrenal cortex, increase gluconeogenesis. Glucocorticoids also raise blood sugar by decreasing the response of muscle and adipose tissue to insulin-stimulated glucose uptake (i.e., antagonistic to the effect of insulin). Hyperglycemia may be found in individuals with hyperplasia of the adrenal cortex or in the presence of an adrenal cortex tumor (Cushing's syndrome).

THYROXINE (T$_4$)

Thyroxine, produced in the thyroid gland, increases glycogenolysis. Thyrotoxic individuals may show symptoms of mild diabetes and almost complete absence of liver glycogen. Thyroxine also increases the rate of glucose absorption from the gastrointestinal tract.

GROWTH HORMONE (GH)

Growth hormone, produced in the anterior pituitary, is antagonistic to the action of insulin. GH inhibits glycolysis and glucose uptake by muscle cells. Individuals with acromegaly (a disease caused by increased GH) may be hyperglycemic. GH secretion is stimulated by hypoglycemia.

GLUCOSE METHODOLOGY

SPECIMEN REQUIREMENTS

The preferred specimen for glucose assay is fasting serum or plasma. Whole blood can be used in some assays but the glucose value depends on the hematocrit. If the hematocrit is low, the glucose value will increase; if the hematocrit is high, the value will decrease. Whole blood glucose values will be lower than serum or plasma values. Whole blood represents a convenient sample source in pediatric patients when limited specimen is available.

When glucose remains in contact with cells at room temperature, the glucose will decrease at the rate of about 7% in 1 hour as a result of cellular glucose metabolism. In nonhemolyzed serum separated from cells glucose is generally stable for up to 72 hours at 4°C (refrigerator temperature) or up to 8 hours at room temperature (25°C); however, at room temperature, bacterial utilization will cause glucose loss in an unsterile sample. Sodium fluoride, a preservative, has been used in collection tubes, but preservatives may inhibit enzymatic reagents, rendering the specimen unsuitable for analysis by some procedures.

The glucose content in arterial and capillary blood is higher than in venous blood. In the fasting state, the difference is small (only about 2–3 mg/dL). However, when a glucose tolerance test is performed, the difference may be much greater (20–30 mg/dL). In collecting specimens for a glucose tolerance test, the blood should always be collected from the same source throughout the test; hence it is inappropriate to collect venous blood then change to capillary blood during the test. If capillary blood is used for the test, the source of collection should be noted on the slip so that the physician can interpret the results correctly.

In summary, for valid results, the glucose assay should be performed within 30 minutes of collection. If this is impossible, separate the nonhemolyzed serum from the cells and store the serum in the refrigerator. The analysis should be performed within 72 hours. Whole blood should be analyzed as soon as possible after collection.

OXIDATION–REDUCTION METHODS

Folin–Wu Copper Reduction

Principle
In hot alkaline solutions, glucose reduces cupric (Cu^{2+}) to cuprous (Cu^{+}) ion with the formation of cuprous oxide. The cuprous ion reduces phosphomolybdic acid to phosphomolybdous acid, which is blue in color and can be measured spectrophotometrically. A Folin–Wu filtrate (type of protein free filtrate) is used:

$$\text{glucose} + Cu^{2+} \xrightarrow{\text{hot, alkaline}} Cu_2O$$

$$Cu^{+} + \text{phosphomolybdic acid} \xrightarrow{\text{reduced } (+e^{-})} \text{blue color}$$

Sources of Error
This method measures all reducing substances present. Uric acid, creatinine, and sulfhydryl compounds may cause a positive error (or interference) by reacting with Cu^{2+}.

Ferricyanide Method

Principle
The yellow ferricyanide ion is reduced by glucose to the colorless ferrocyanide ion. The decrease in color is measured with a spectrophotometer (inverse colorimetry) or the ferrocyanide may be measured directly by reacting it with excess ferric ions to form a blue ferric ferrocyanide:

$$\text{glucose} + \text{Fe(CN)}_6^{3-} \text{ (yellow)} \xrightarrow{\text{heat}} \text{Fe(CN)}_6^{4-} \text{ (colorless)}$$

Sources of Error
Other reducing substances (uric acid, creatinine) can cause falsely elevated glucose values, therefore, this methodology is not recommended for glucose determinations in uremic patients.

Neocuproine Method

Principle
Glucose reduces cupric (Cu^{2+}) to cuprous (Cu^+) ion. The cuprous ion forms a stable orange-colored complex with neocuproine, which can be measured spectrophotometrically:

$$\text{glucose} + \text{Cu}^{2+} \xrightarrow[\text{hot, alkaline}]{\text{reduced } (+e^-)} \text{Cu}^+$$

$$\text{Cu}^+ + \text{neocuproine} \rightarrow \text{cuprous–neocuproine (orange-colored complex)}$$

Sources of Error
Other reducing substances (uric acid, creatinine, ascorbic acid) interfere positively and, therefore, this method is relatively nonspecific for glucose.

ENZYMATIC METHODS

Glucose Oxidase

Principle
The enzyme glucose oxidase catalyzes the oxidation of glucose to gluconic acid and hydrogen peroxide. The product, hydrogen peroxide, is used as a substrate for a coupled reaction using a second enzyme, peroxidase, and a chromogenic oxygen acceptor (*O*-dianisidine), which results in the formation of a color (oxidized chromogen) that can be measured spectrophotometrically:

$$\beta\text{-D-glucose} + O_2 + H_2O \xrightarrow{\text{glucose oxidase}} \text{gluconic acid} + H_2O_2$$

$$H_2O_2 + \text{reduced chromogen (not colored)} \xrightarrow{\text{peroxidase}}$$
$$\text{oxidized chromogen (colored)} + H_2O$$

Sources of Error
Glucose in solution exists as 36% alpha and 64% beta forms. Glucose oxidase is highly specific for β-D-glucose. Standard solutions prepared from dry glucose should be allowed to stand at least 2 hours prior to use to ensure that mutarotation has reached the equilibrium state. Some commercial enzyme preparations contain the mutarotase enzyme, which accelerates this reaction.

Other reducing substances (uric acid, ascorbic acid, glutathione, bilirubin) may inhibit the reaction, presumed through competition with the chromogen for hydrogen peroxide resulting in a negative bias or compounds may be present that oxidize the indicator dye resulting in a positive bias.

Glucose Oxidase–Oxygen Electrode Method

A second glucose oxidase procedure is available that does not require the coupled peroxidase reaction and, therefore, interferences associated with the second reaction are eliminated.

Principle
Oxygen is consumed at the same rate as glucose reacts to form gluconic acid. The oxygen consumption, as measured by an oxygen electrode, is directly proportional to the glucose concentration. Hydrogen peroxide, a product of the reaction, is destroyed by ethanol in the presence of catalase, a pathway that does not lead to the generation of oxygen:

$$\beta\text{-D-glucose} + O_2 \xrightarrow{\text{glucose oxidase}} \text{gluconic acid} + H_2O_2$$

$$H_2O_2 + \text{ethanol} \xrightarrow{\text{catalase}} \text{acetaldehyde} + H_2O$$

Hydrogen peroxide may also be removed by reaction with iodides.

Sources of Error
This procedure is not appreciably sensitive to elevated levels of bilirubin, uric acid, creatinine, lipids, or hemoglobin (up to 1 g/dL). The method is not subject to interference from agents used to prevent coagulation (ammonium, sodium or lithium heparin, sodium oxalate, sodium citrate, or EDTA) or glycolysis (sodium fluoride and iodoacetate).

Two known substances that cause interference are hydroxyethyl starch, a plasma expander that produces a slightly positive bias (at the concentration *in vivo*, the bias is less than 3%) and 2-deoxyglucose.

This methodology may be used to determine glucose in serum, plasma, urine, or cerebrospinal fluid. Since oxygen is bound by hemoglobin, whole blood cannot be used with this method.

Hexokinase Method (Reference Method)

Principle
The enzyme hexokinase transfers a phosphate group from ATP to glucose to form glucose-6-phosphate (G-6-P) and ADP. The product glucose-6-phosphate is the substrate for the second or coupled reaction. The enzyme glucose-6-phosphate dehydrogenase (G-6-PD) catalyzes the oxidation of G-6-P by nicotinamide adenine dinucleotide phosphate (NADP⁺) with the formation of NADPH. The absorbance of NADPH, measured at 340 nm, is directly proportional to the glucose concentration.

$$\text{glucose} + \text{ATP} \xrightarrow{\text{hexokinase, Mg}^{2+}} \text{glucose-6-phosphate} + \text{ADP}$$

$$\text{glucose-6-phosphate} + \text{NADP}^+ \xrightarrow{\text{G-6-PD}} \text{6-phosphogluconate} + \text{NADPH} + \text{H}^+$$

Utilizing an additional coupled reaction, a colorimetric method has been developed using the electron transport system, INT (iodonitrotetrazolium salt), and PMS (phenazine methosulfate). The NADPH (formed in the above reaction) is reacted with INT. PMS functions as an intermediate electron carrier. Reduced INT produces a color.

Sources of Error
Phosphate esters and enzymes released from red blood cells may react to produce changes in the NADP⁺ concentration, therefore, hemolyzed specimens (more than 0.5 mg/dL hemoglobin) cannot be used. Ascorbic acid or uric acid do not interfere. The hexokinase method may be used to determine glucose in urine.

OTHER METHODS

Ortho-Toluidine Method

Principle
Aldehyde sugars will react with *o*-toluidine in hot acetic acid solution to produce colored (green) derivatives. The reaction is not specific for glucose, since other aldehyde sugars (galactose and mannose), if present, will produce variable amounts of color.

Sources of Error
Bilirubin may interfere because of the conversion of bilirubin to the green pigment biliverdin. This procedure may yield falsely elevated glucose values in neonates with high serum bilirubin.

REFERENCE RANGE

Adult Glucose Values

Serum or plasma "true glucose"	70–110 mg/dL
Whole blood "true glucose"	60–100 mg/dL
Serum or plasma (Folin–Wu)	80–120 mg/dL
Panic values:	
Low	equal to or less than 40 mg/dL
(Low blood glucose may cause brain damage)	
High	equal to or greater than 700 mg/dL
(High blood glucose may cause diabetic coma)	

The glucose level in cerebrospinal fluid is about two-thirds or 60–70% of the plasma levels.

Neonatal Glucose Values

Fetal glucose is normally about two-thirds to three-fourths of the mother's value. In stressed infants (as in respiratory distress), the neonates may be hypoglycemic.

Premature, serum or plasma	20–80 mg/dL
Premature, whole blood	20–65 mg/dL
Full term, serum, plasma, whole blood	20–90 mg/dL

(Although the lower figure is frequently noted in neonates, interpretation of this value as normal is debatable)

Panic values:	
Low	equal to or less than 30 mg/dL
High	equal to or greater than 300 mg/dL

ORAL GLUCOSE TOLERANCE TEST

PRINCIPLE

The patient is given a standard glucose load, and the ability of the patient to handle the glucose load is monitored by measuring blood glucose levels versus time.

PATIENT PREPARATION

The oral glucose tolerance test (OGTT) should be administered only to ambulatory, otherwise healthy people and should not be performed while the patient is hospitalized. At least three days prior to the test, the patient should be placed on a carbohydrate diet that contains at least 150 g of carbohydrate per day. If carbohydrate intake has been too low preceding the test, a false diabetic-type curve may be obtained. Drugs known to affect plasma glucose levels should be avoided at least three days prior to the test. Oral contraceptives should be omitted for one complete cycle prior to the test. Before the test is performed, the patient should have fasted for at least 10 hours, but no more that 16 hours; no smoking is permitted during the test, and the patient should remain seated throughout the test, although limited walking is permitted. During the test, the patient may drink water, but no coffee, tea, or gum is permitted.

GLUCOSE LOAD

The load of glucose administered should be 75 g dissolved in water, with a concentration no greater than 25 g/dL. (The 100 g glucose load often used in the United States is not the dose recommended by the American Diabetes Association.) For children, the load is based on body weight (1.75 g/kg ideal body weight up to a maximum of 75 g). Commercially prepared carbohydrate preparations equivalent to this glucose load are also acceptable. A 100 g oral glucose load is recommended to evaluate gestational diabetes.

TEST

A fasting blood sample is obtained, after which the glucose load is given. The glucose should be consumed in a period of approximately 5 minutes. Blood samples are collected at 30 minute intervals for two hours (zero time is at the beginning of the drink). In pregnant females (to evaluate gestational diabetes), the Diabetes Association recommends sampling at fasting, 1-, 2-, and 3-hour intervals. If possible, venous blood should be collected. Some laboratories collect urine samples (to be tested for glucose) at the same timed intervals as the blood samples are collected, but this is not necessary for test evaluation.

INTERPRETATION OF RESULTS

Normal Tolerance Curve

(See Figure 5.11.) The fasting level of glucose is within normal limits (serum or plasma, 70–110 mg/dL.) The peak concentration of blood glucose is reached between 30 and 60 minutes and should not exceed the renal threshold (160 ± 20 mg/dL). No glucose should be excreted in the urine. By about two hours, the

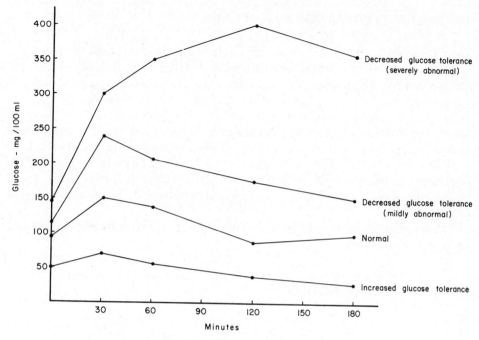

Figure 5.11.   Glucose tolerance curves.

blood level tends to drop slightly below fasting levels and then returns to normal in approximately three hours.

When glucose is first given, the absorption occurs rapidly, stimulating the pancreas to secrete insulin. After 30–60 minutes, the blood glucose level begins to decrease. More insulin is present than is necessary, and the blood glucose level drops below the fasting level after approximately $1\frac{1}{2}$–2 hours. Endocrine factors are stimulated by hypoglycemia, and the blood sugar returns to the fasting level by approximately three hours.

Decreased Glucose Tolerance

In persons with decreased glucose tolerance, blood glucose levels may be elevated for the entire postprandial period. The peak blood level frequently occurs late and may exceed the renal threshold.

Increased Glucose Tolerance

In individuals with increased glucose tolerance, the peak glucose level remains low and may decrease to hypoglycemic levels for the remainder of the test. In patients with adrenal insufficiency, the response to the oral glucose load may appear to be normal during the first several hours of the test but the glucose falls to hypoglycemic levels later. When the OGTT is used to evaluate hypoglycemia, the test is extended to four or five hours.

TWO-HOUR POSTPRANDIAL GLUCOSE

The two-hour postprandial (2 hr pp) glucose test has been recommended as a test to screen for diabetes mellitus, to diagnose diabetes mellitus, and to monitor glucose control. The patient is given either a glucose load or eats a meal containing carbohydrate (75–100 g carbohydrate). A two-hour postprandial specimen is drawn and the glucose level is determined. In normal persons, the 2 hr pp value should approximate the fasting glucose level. In older individuals, the delay in insulin release may yield values slightly higher than the fasting glucose level.

Diabetic individuals will have significantly elevated 2 hr pp levels (200 mg/dL or greater). In monitoring diabetic control, a value greater than 150 mg/dL should be considered poor control, while values less than 130 mg/dL are considered good control.

The significance of the 2 hr pp test is limited because no controlled conditions have been established.

GLYCOSYLATED HEMOGLOBIN

ORIGIN

The hemoglobin found in red blood cells of normal adults and children consists of hemoglobin A (90–95%), hemoglobin A_2 (less than 2.5%), hemoglobin F (less than 0.5%), hemoglobin A_{1a} (1.6%), hemoglobin A_{1b} (0.8%), and hemoglobin A_{1c} (3–6%). HbA_{1a}, HbA_{1b}, and HbA_{1c} are collectively known as "fast" hemoglobin, glycosylated hemoglobin, or glycohemoglobin. These minor hemoglobins are formed by the addition of sugars to HbA. The glycosylation occurs after the hemoglobin has been synthesized and is a slow, continuous, nonenzymatic process that occurs throughout the life of the red blood cell. Once the stable ketamine structure is formed, it is relatively irreversible. The carbohydrate attaches to the N-terminal valine of the beta globin chains. The rate of glycohemoglobin formation is believed to reflect the concentration of glucose in the blood over an extended period of time (4–10 weeks); the higher the blood glucose level, the greater the percentage of glycohemoglobin present. Diabetic patients who are poorly controlled will have glycohemoglobin levels two to three times higher than normal individuals.

USES OF GLYCOSYLATED HEMOGLOBIN

The measurement of glycosylated hemoglobin has been proposed as a test that could be used to measure long-term glycemic control in the diabetic patient. The complications found in diabetics, that is, accelerated vascular disease (peripheral and cardiovascular), neuropathy, nephropathy, and retinopathy, have all been associated with uncontrolled diabetics who have high levels of glycohemoglobin. The current belief is that tight control of blood glucose levels may prevent or arrest the development of these complications.

It is postulated that the mechanism of injury results from the nonenzymatic glycosylation of cellular proteins in different organs, resulting in changes in biochemical properties leading to the development of secondary complications (microvascular, kidney, nerve, and eye disease). Since the glycosylation process is slow, only those proteins that are slowly replaced are significantly altered by glycosylation. These proteins include those in blood vessel linings, those that insulate nerves, and those within the lens of the eye.

The two methods of assessing glucose control, namely, the measurement of plasma glucose or the measurement of urinary glucose, measure only the control achieved at one specific time. The measurement of glycosylated hemoglobin reflects the degree of glycemic control during the preceding 8–10 weeks and, therefore, is a useful index in assessing the degree of glycemic control achieved over a long period. Thus, the glycohemoglobin test may be very helpful in the management of diabetic patients.

A proposed second application of the glycosylated hemoglobin test is as a screening procedure to detect impaired glucose tolerance. Preliminary data suggest that the glycosylated hemoglobin test may be used to detect severe glucose intolerance but is not a sensitive indicator of mild or chemical diabetes. Further studies are needed to determine the value of this test in achieving this goal.

The third potential application of the glycosylated hemoglobin test is in the management of glucose levels in diabetics during pregnancy. Careful control of blood glucose levels during pregnancy significantly lowers the fetal risk factors associated with diabetic mothers.

METHODOLOGY

Specimen Requirements

To measure glycosylated hemoglogin, a nonfasting sample taken any time during the day may be used. EDTA as the anticoagulant is the specimen of choice; samples collected in sodium fluoride are also acceptable. The samples may be stored up to five days at 2–6°C. Stability for longer periods of time is presently controversial.

Hemoglobin A_{1c} has been studied most extensively. The term glycohemoglobin or fast hemoglobin can refer to hemoglobin A_{1c} or to the other hemoglobins that contain the sugar moiety. Most laboratories use the microcolumn ion-exchange technique, which isolate the glycohemoglobins as a group. This method correlates favorably with the macrocolumn method, which is considered the reference method.

Test Principle

A cation exchange column is used to separate the glycosylated hemoglobin from the other hemoglobins. Depending on their cation character, the hemoglobin fractions vary in their affinities for attachment to the column resin. The glycosylated hemoglobin fractions may be eluted separately, based on the pH of the elution buffer.

Test

1. A hemolysate is made from the sample and applied to a column using a Pasteur pipette. The sample is allowed to absorb onto the resin surface.
2. The glycohemoglobins (called fast fraction) are eluted with a phosphate buffer. The eluted fraction is collected and diluted to a specified volume with pure water.
3. To determine total hemoglobin, another portion of the hemolysate is added to a second tube and diluted to a specified volume with pure water.
4. The absorbance of each fraction is determined using a spectrophotometer.
5. The glycohemoglobin level is determined as a percentage of the total.

Sources of Error

1. Any disorder that results in a decreased erythrocyte life span in the circulation (hemolytic anemias, bleeding) will result in decreased glycosylated hemoglobin, while postsplenectomy patients (increased RBC survival) will have elevated glycosylated hemoglobin levels.
2. Hemoglobin F is not bound by the column resins and will coelute with the fast fraction, causing false elevations of glycosylated hemoglobin.
3. Individuals with HbS or HbC will show false low glycosylated hemoglobin values, because glycosylated HbS and HbC do not elute with HbA_1.
4. Chromatographic methods also yield increased levels of glycosylated hemoglobin in individuals with chronic renal diseases and should be used with caution to measure diabetic control in this group of individuals.
5. The elution rates are affected by buffer pH and ionic strength and are extremely temperature dependent. A temperature–glycohemoglobin conversion chart is used to correct for ambient room temperature.

Glycosylated hemoglobin may also be determined by electrophoresis, RIA, HPLC, isoelectric focusing and colorimetric techniques.

REFERENCE RANGE

Normal glycosylated hemoglobin: 5–8%
Insulin requiring diabetes mellitus: 10–22%

HYPERGLYCEMIA

DIABETES MELLITUS

Diabetes mellitus may be defined as a genetically and clinically heterogeneous group of disorders associated with an absolute or relative lack of insulin and characterized by hyperglycemia. It affects about 5% of the American population and is the third leading cause of death in the United States.

Diabetes mellitus may be divided into three groups: (1) insulin-dependent diabetes mellitus (IDDM), (2) noninsulin-dependent diabetes mellitus (NIDDM), and (3) other types.

Insulin-Dependent Diabetes Mellitus (IDDM)

Insulin-dependent diabetes mellitus was formerly called juvenile or juvenile-onset diabetes mellitus, brittle diabetes, or ketosis-prone diabetes. Approximately 10% of all persons with diabetes mellitus have this form of the disease. IDDM is characterized clinically by the abrupt onset of symptoms, lack of insulin requiring dependence on injected insulin to sustain life, with tendency toward ketosis. Most of the cases occur in juveniles, but IDDM may be present at any age. The associated symptoms include polyuria, polydipsia, polyphagia, and rapid weight loss. Laboratory findings include hyperglycemia, glycosuria, ketouria, and metabolic acidosis.

Evidence regarding etiology suggests genetic and environmental or acquired factors, association with certain histocompatibility antigens (HLA) on chromosome 6, and abnormal immune responses, including autoimmune reactions. Islet cell antibodies are usually present. Viral infections have also been implicated but the data for this theory are inconclusive.

Case History, Nonketotic Hyperosmolar Diabetes Mellitus

A three-year-old white male was seen in the emergency room. The patient had no history of previous illness other than the expected childhood diseases. Three days prior to this admission, the child developed flulike symptoms with malaise, nausea, and vomiting. The patient developed polydipsia and continually requested fluids. The vomiting continued. The admission laboratory data are listed in Table 5.1.

Discussion
The diabetes developed suddenly following a flulike illness. Although viruses have been suspected as an etiological agent of juvenile diabetes, no conclusive evidence exists to support this theory. History revealed diabetes on the maternal side of the family, both insulin-dependent diabetes mellitus and noninsulin-dependent diabetes mellitus. The blood gases were within normal limits. In ketosis, metabolic acidosis with Kussmaul respiration may be an associated clinical finding. Fluids and insulin were administered, and the vomiting was controlled. The elec-

Table 5.1. Diabetes Mellitus Case History Laboratory Results[a]

Serum Glucose (70–110) mg/dL	Serum Acetone (negative)	BUN (7–21) mg/dL	Na (138–145) mmol/L	K (3.5–5.0) mmol/L	Cl (98–106) mmol/L	pH (7.34–7.45)	pCO2 (32–45) mm Hg
1470	Small amount	11	119	5.3	79	7.36	36

[a] The reference range for components is indicated in parentheses.

trolytes and blood sugar were returned to normal values. The patient was placed on dietary and insulin therapy.

Noninsulin-Dependent Diabetes Mellitus (NIDDM)

Noninsulin dependent diabetes mellitus was formerly called adult-onset diabetes, maturity-onset diabetes, ketosis-resistant diabetes, or stable diabetes. NIDDM is divided into two categories: one category includes those individuals who are obese, and the second includes those who are not obese. The age of onset is usually after 40 years, although NIDDM may occur at an earlier age. Persons in this group are not insulin dependent or ketosis prone, although there are those who may require insulin for the control of hyperglycemia and can develop ketosis under special circumstances (stress, infections, trauma). The serum insulin levels may be depressed, normal, or elevated. Many persons in this group show an exaggerated rise in insulin after eating and are described as insulin resistant. The insulin resistance is thought to be due to a reduction in the number of insulin receptors on cell membranes or to decreased affinity of the receptor for insulin. Compromised insulin sensitivity leads to hyperglycemia. About 60–90% of NIDDM subjects are obese, and weight reduction and physical activity will decrease the insulin levels, increase the number of insulin receptors, and improve the insulin effectiveness.

There are probably multiple etiologies for NIDDM. Environmental factors superimposed on genetic susceptibility are probably involved. Obesity is suspected as an etiological factor. Islet cell antibodies have not been found in this group.

Other

Diabetes mellitus is also associated with other conditions and syndromes, and the hyperglycemia is secondary to such disorders as pancreatic disease (hemochromatosis, pancreatic insufficiency), hormonal abnormalities (acromegaly, Cushings, pheochromocytoma, glucagonoma, somatostatinoma, primary aldosteronism), certain genetic syndromes, and insulin-receptor abnormalities, or it can be induced by drugs or chemicals.

Diagnosis

In nonpregnant adults, The National Diabetes Data Group recommends that the diagnosis of diabetes mellitus be restricted to those individuals who exhibit any one of the following:

1. The presence of the classic symptoms of diabetes together with unequivocal hyperglycemia.
2. The presence of a fasting venous plasma glucose concentration equal to or greater than 140 mg/dL on more than one occasion. (If this criterion is met, an OGTT is not needed to make the diagnosis of diabetes mellitus.)
3. The presence of a fasting venous plasma glucose concentration less than 140 mg/dL but sustained hyperglycemia during an OGTT (75 g load) on more than one occasion. The OGTT must yield two or more sample values equal to or

greater than 200 mg/dL (both the two-hour sample and another sample taken after the ingestion of the glucose and within the two hour period).

The diagnosis of diabetes mellitus can be made in children who exhibit either of the following:

1. The presence of the classic symptoms of diabetes together with a random plasma glucose greater than 200 mg/dL.
2. In asymptomatic individuals, both an elevated fasting glucose concentration (equal to or greater than 140 mg/dL venous plasma or 120 mg/dL capillary whole blood) and sustained hyperglycemia during an OGTT (1.75 g/kg not to exceed 75 g load) on more than one occasion. The OGTT must show two or more sample values to be equal to or greater than 200 mg/dL for venous plasma or capillary whole blood (both the two-hour sample and another sample taken after the ingestion of the glucose and within the two-hour period).

IMPAIRED GLUCOSE TOLERANCE

Definition

Those individuals who have elevated plasma glucose levels and abnormal OGTT (but do not meet the criterion established for diabetes mellitus) are classified into a group called impaired glucose tolerance (IGT). This group was formerly called asymptomatic diabetes, chemical diabetes, subclinical diabetes, borderline diabetes, or latent diabetes.

Diagnosis

The National Diabetes Data Group recommends that the criterion necessary to classify a nonpregnant adult in the IGT group is as follows:

1. The fasting glucose concentration must be below the value that is diagnostic for diabetes (less than 140 mg/dL venous plasma) *and*
2. Following an OGTT (75 g load), the two-hour venous plasma sample value must be between 140 and 200 mg/dL, and an additional sample obtained between the $\frac{1}{2}$ hour and the $1\frac{1}{2}$ hour interval must be unequivocally elevated (equal to or greater than 200 mg/dL).

For children, the criterion that must be met are:

1. The fasting glucose concentration must be below the value that is diagnostic for diabetes (less than 140 mg/dL venous plasma or 120 mg/dL capillary whole blood) *and*
2. The glucose concentration two hours after an oral glucose challenge must be elevated (greater than 140 mg/dL venous plasma or 120 mg/dL capillary whole blood).

GESTATIONAL DIABETES MELLITUS

Gestational diabetes mellitus (GDM) is restricted to pregnant women who develop impaired glucose tolerance or diabetes during pregnancy. The diagnosis of GDM is important so that effective management of blood glucose levels during pregnancy can be initiated to prevent infant morbidity and mortality.

MANAGEMENT OF HYPERGLYCEMIA

The management of individuals with elevated blood glucose consists of dietary measures and the administration of insulin or oral hypoglycemic agents when indicated. Diabetes is a life-long disease. The complications associated with diabetes mellitus involve both the small vessels (microangiopathy) and the large vessels (macroangiopathy). Small vessel disease causes injury to the kidney (nephropathy) and eyes (retinopathy). Neuropathy may cause impotence in males. Foot lesions are common and may occur secondary to the neuropathy, which causes loss of feeling in the feet, or secondary to insufficient blood supply associated with the peripheral vascular disease. Blood lipids are elevated, and the risk for developing cardiovascular disease is markedly increased.

There is evidence that careful control of blood glucose levels will probably lower the incidence of small vessel injury, but there is no scientific evidence to suggest a decrease in large vessel injury with careful control of blood glucose levels. Seventy-five percent of diabetic patients will die of large vessel disease, and most of these individuals succumb to myocardial infarctions.

HYPOGLYCEMIA

DEFINITION

The symptoms associated with hypoglycemia are dependent on the rate of onset of the hypoglycemic state. If the hypoglycemia occurs rapidly, symptoms include sweating, shakiness, trembling, weakness, and anxiety. If the hypoglycemia occurs slowly, symptoms include headache, irritability, and lethargy.

Hypoglycemia may be defined as a fasting glucose value below 50 mg/dL; however, not all investigators agree on this definition.

The hypoglycemia that occurs following a meal is called *reactive hypoglycemia* and can be divided into three types: (1) functional, (2) prediabetic or diabetic, and (3) alimentary hypoglycemia.

Functional hypoglycemia is quite common in adults and is characterized by hypoglycemia occurring two to four hours following a meal. A five-hour OGTT gives a pattern of low glucose levels at about three hours (Note: a sample should be obtained at any time symptoms of hypoglycemia develop). In functional hypoglycemia, insulin levels throughout the test are appropriate for the glucose levels. These individuals do not develop diabetes at a rate greater than the general population.

Following an oral glucose tolerance test, hypoglycemia in the prediabetic or diabetic type occurs later than in the functional type. In the prediabetic or diabetic type, the insulin response is delayed and exaggerated, and these individuals may develop diabetes mellitus.

The alimentary type of hypoglycemia occurs in those individuals who have had gastrointestinal surgery. The food enters the intestine at an accelerated rate. The absorbance of glucose is rapid, with a corresponding increase in insulin secretion from the pancreatic beta cells. The response to the accelerated insulin release leads to hypoglycemia.

Ethanol and other drugs may also cause hypoglycemia. The hypoglycemia associated with ethanol ingestion occurs following prolonged ingestion of alcohol with concurrent depletion of liver glycogen stores. Other drugs associated with hypoglycemia include phenformin, salicylates, and sulfonylurea. *Factitious hypoglycemia* is self-induced by exogenous insulin injections or sulfonylurea ingestion.

The hypoglycemia that has no association with meals is probably pathological and may occur at night. Tumors that secrete insulin or insulinlike substances produce hypoglycemia that is generally more profound and unremitting than any of the reactive types. The hypoglycemia does not induce a fall in insulin.

An insulinoma, a tumor of the pancreatic beta islet cells, may occur at any age but is most common in the fourth to the sixth decade. These tumors may be either benign (more common) or malignant. The symptoms may mimic many disorders. The criteria known as Whipple's triad are used as a diagnostic tool. These include (1) hypoglycemia attacks that are precipitated by fasting, (2) plasma glucose less than 40 mg/dL during an attack, and (3) symptoms promptly relieved by the administration of glucose. To confirm the diagnosis, serum insulin levels (available by RIA) are measured. In normal persons, fasting will produce low insulin levels. In individuals with insulinomas, the insulin level is inappropriately high for the degree of hypoglycemia. Insulin/glucose ratios are useful for diagnostic purposes. An OGTT is generally not helpful in the diagnosis of islet cell tumors.

C-PEPTIDE

Hypoglycemia may also be self-induced. This type of hypoglycemia is called factitious hypoglycemia and occurs in response to the self-administration of insulin (exogenous insulin). Endogenous insulin (insulin produced within the body) is produced in the pancreatic beta islet cells as the proinsulin form, which is subsequently cleaved in the pancreas to the active form. The secretory products from the pancreatic gland include equimolar concentrations of the active insulin plus the portion of the molecule cleaved from the proinsulin called the C-peptide or connecting peptide. Exogenous insulin or commercial insulin does not contain the C-peptide, because it is removed from the product when manufactured. In individuals with factitious hypoglycemia, the insulin level will be high but the C-peptide will be decreased or absent; in individuals with insulinomas, the insulin and C-peptide levels will both be elevated.

Because the C-peptide measurement gives an indication of pancreatic insulin production, the measurement of C-peptide may also be useful in the evaluation of insulin production in those individuals who develop antibodies in response to previous insulin treatment. Insulin antibodies interfere with the measurement of insulin by RIA techniques.

C-peptide appears to be metabolized by the kidney; therefore, C-peptide may be elevated in those individuals with chronic renal failure.

OTHER TOLERANCE TESTS

TOLBUTAMIDE TEST

Principle

Tolbutamide (Orinase) is a substance that stimulates the pancreas to produce insulin. The tolbutamide test may be used to differentiate insulinomas from other hyperinsulinemic states. Throughout the test, the patient has to be monitored for signs of severe hypoglycemia, a potentially life-threatening situation.

Test

1. After an overnight fast, blood samples for glucose and insulin are drawn.
2. Tolbutamide is given intravenously.
3. Blood is drawn at intervals. At least three samples are obtained at 5 minute intervals after the tolbutamide is given. Next, additional samples are obtained at timed intervals, including one sample drawn at 60 minutes. If necessary, the test may be extended to 2 hours.
4. The plasma glucose and insulin levels are measured in all samples.

Interpretation

In normal individuals, the blood level of glucose will decrease (lowest level in 30 minutes) as insulin is secreted and then return to normal in about 2 hours.

Individuals with insulinomas will have a profound decrease of blood glucose levels, which remain low to the end of the test.

In diabetics, the pancreatic insulin reserve is decreased. The blood glucose will not fall to levels as low as those seen in normal individuals because of the insufficiency in insulin secretion.

Obesity may cause a false positive intravenous tolbutamide tolerance test.

EPINEPHRINE TEST

Principle

Individuals with glycogen storage disease (Type I, von Gierke's) are deficient in or lack the liver glucose-6-phosphatase enzyme, which is necessary for converting liver glycogen to glucose. Glycogen stores in the liver are increased, and blood glucose levels are low. Epinephrine induces glycogenolysis.

Test

1. A blood sample is drawn to provide a baseline glucose value.
2. Epinephrine is given intramuscularly.
3. Blood is drawn at 35, 45, 60, 90, and 120 minutes following the epinephrine injection.
4. The glucose level is measured in all samples collected.

Interpretation

In normal individuals, the epinephrine induces glycogenolysis, increasing the blood glucose level within 40–60 minutes.

Individuals with von Gierke's glycogen storage disease show absent or impaired increase in blood glucose following the administration of epinephrine, as glycogen stores cannot be utilized.

LACTOSE TOLERANCE TEST

Principle

The lactose tolerance test is used to evaluate lactase deficiency. The lactase enzyme is a gastrointestinal enzyme that hydrolyzes lactose (disaccharide) to glucose and galactose (monosaccharides), which may then be absorbed. When lactose is not absorbed, large amounts remain in the gastrointestinal tract. Bacteria metabolize the carbohydrate, leading to gas production. The sugar causes an osmotic load, and watery diarrhea follows.

Test

1. An oral glucose tolerance test is performed to be used as a basis for comparison.
2. The next day, the patient is given an oral 100 g load of lactose.
3. Blood samples are collected at timed intervals.
4. All samples are assayed for glucose.

Interpretation

If lactase activity is present, the lactose is hydrolyzed to glucose and galactose, and the tolerance curve following the lactose load is similar to the baseline glucose tolerance curve. With lactase deficiency, the lactose tolerance curve is flat.

BIBLIOGRAPHY

Bates, H. M. "Glycohemoglobin determinations in blood glucose monitoring," *Laboratory Management,* 21–24 (May 1980).

Boehm, T. M. and Lebovitz, H. E. "Statistical analysis of glucose and insulin responses to intravenous tolbutamide: Evaluation of hypoglycemic and hyperinsulinemic states," *Diabetes Care,* **2,** 479 (1979).

Clark, C. M. "Hypoglycemia in the diabetic." *Journal of the Indiana State Medical Association,* 103–108 (Feb. 1979).

Cohen, A. *Handbook of Cellular Chemistry.* St. Louis: Mosby, 1975.

Felig, P., Baxter, J., Broadus, A., and Frohman, L. (eds.). *Endocrinology and Metabolism.* New York: McGraw-Hill, 1981.

Fletcher, J. "For the assessment of beta cell function C-peptide measurement," *BioScience Reports,* 4–5 (Spring, 1982).

Hammons, G. T. "Glycosylated hemoglobin and diabetes mellitus," *Laboratory Medicine,* **12**(4), 213 (1981).

Henry, J. B. (ed.). *Clinical Diagnosis and Management by Laboratory Methods,* 16th ed. Philadelphia: Saunders, 1979.

Meites, S. (ed.). *Pediatric Clinical Chemistry,* 2nd ed. Washington, D.C.: American Association of Clinical Chemistry, 1981.

National Diabetes Data Group. "Classification and diagnosis of diabetes mellitus and other categories of glucose intolerance," *Diabetes* **28**(12), 1039 (1979).

Olefsky, J. M. "The insulin receptor: Its role in insulin resistance of obesity and diabetes. *Diabetes,* **25**(12), 1154 (1976).

Peterson, C. M. "Renewed interest sparked in glycosylated hemoglobin," *Laboratory Management,* 37–44 (June 1979).

Tietz, N. (ed.). *Fundamentals of Clinical Chemistry,* 2nd ed. Philadelphia: Saunders, 1976.

Watts, N. "Diabetes mellitus: Diagnostic and monitoring techniques," *Laboratory Management* 43–55 (May 1982).

Watts, N. "Oral glucose tolerance test may be done too fast," *Lab World,* 68–72 (June 1981).

POST-TEST

1. Carbohydrates are composed primarily of:
   a. carbon, nitrogen, and oxygen
   b. carbon, nitrogen, and hydrogen
   c. carbon, oxygen, and phosphorus
   d. carbon, hydrogen, and oxygen

2. Match the following
   Carbohydrate Nomenclature

   _____ 1. Aldehyde sugar
   _____ 2. Ketone sugar

   a.
   ```
       CH2OH
        |
       C=O
        |
   HO - C - H
        |
   H - C - OH
        |
   H - C - OH
        |
       CH2OH
   ```
   b.
   ```
        O
        ||
        C - H
        |
   H - C - OH
        |
   HO - C - H
        |
   H - C - OH
        |
   H - C - OH
        |
       CH2OH
   ```

   _____ 3. L-glucose
   _____ 4. D-glucose

   a.
   ```
        O
        ||
        C - H
        |
   HO - C - H
        |
   H - C - OH
        |
   HO - C - H
        |
   HO - C - H
        |
       CH2OH
   ```
   b.
   ```
        O
        ||
        C - H
        |
   H - C - OH
        |
   HO - C - H
        |
   H - C - OH
        |
   H - C - OH
        |
       CH2OH
   ```

_____ 5. α-D-glucose
_____ 6. β-D-glucose

a. H–C–OH
    H–C–OH
    HO–C–H   O
    H–C–OH
    H–C
    CH₂OH

b. HO–C–H
    H–C–OH
    HO–C–H   O
    H–C–OH
    H–C
    CH₂OH

3. The three hexoses of biological importance include:
   a. sucrose, glucose, fructose
   b. glucose, fructose, galactose
   c. fructose, maltose, sucrose
   d. lactose, glucose, fructose

4. Match the following (each has three answers):

Monosaccharide composition		Reducing properties			
_____	_____	_____	1. Maltose	a.	glucose
_____	_____	_____	2. Sucrose	b.	galactose
_____	_____	_____	3. Lactose	c.	fructose
				d.	reducing sugar
				e.	nonreducing sugar

5. All of the following are examples of polysaccarides *except:*
   a. starch
   b. ribose
   c. glycogen
   d. cellulose

6. Amylase can hydrolyze _____ bonds.
   a. alpha glycosidic bonds
   b. beta glycosidic bonds
   c. both

7. Because of the dependence on enzymes, only _____ glucose can be utilized by the body.
   a. D-glucose
   b. L-glucose

8. Match the following:
   Carbohydrate derivatives

   _____ 1. DNA
   _____ 2. ATP
   _____ 3. Glucuronic acid
   _____ 4. Glycoproteins
   _____ 5. Mucopolysacchrides

   a. hyaluronic acid
   b. A and B blood group substances
   c. purine or pyrimidine, deoxyribose sugar, phosphate group
   d. adenine, ribose sugar, phosphate groups
   e. sugar acid

9. Match the following:
    _____ 1. Glycolysis   a. refers to the breakdown of glyco-
    _____ 2. Glycogenesis   gen to form glucose and other in-
    _____ 3. Glycogenolysis   termediate products
    _____ 4. Gluconeogensis   b. refers to the conversion of glu-
                                         cose to glycogen
                                      c. refers to the formation of glucose
                                         from noncarbohydrate sources
                                      d. refers to the conversion of glu-
                                         cose to lactate or pyruvate

10. Because muscle lacks the enzyme, _____ , muscle glycogen cannot con-
    tribute to blood glucose.
    a. hexokinase
    b. glucose oxidase
    c. glucose-6-phosphatase
    d. glucose-6-dehydrogenase

11. Hormones regulate blood glucose. Match the following:
    _____ 1. Insulin   a. raises blood glucose levels by
    _____ 2. Glucagon   increasing absorption from the
    _____ 3. Epinephrine   GI tract and promotes glyco-
    _____ 4. Cortisol   genolyis
    _____ 5. Thyroxine   b. stimulates gluconeogenesis
    _____ 6. Growth hormone   c. increased in stress with in-
                                         crease in glycogenolysis
                                      d. produced in beta cells of the
                                         pancreas; acts to lower blood
                                         glucose
                                      e. produced in alpha cells of the
                                         pancreas; acts to raise blood
                                         glucose
                                      f. produced in the anterior pitui-
                                         tary and is antagonistic to the
                                         action of insulin

12. In the evaluation of blood glucose levels, all of the following are true *except:*
    a. whole blood will give lower values than serum or plasma
    b. capillary blood values will be lower than venous values
    c. glycolysis decreases glucose values if the specimen is allowed to sit
        before analysis
    d. serum or plasma is the preferred specimen for glucose assay

13. Match the following:
    _____ 1. Oxidation   a. loss of electrons
    _____ 2. Reduction   b. gain of electrons

14. The reference method for glucose is:
    a. hexokinase
    b. ortho-toluidine
    c. ferricyanide
    d. glucose oxidase

15. Match the test principle with the assay procedure:

    _____ 1. Hexokinase   a. glucose reduces metal ions; the
    _____ 2. Glucose oxidase   metal ion reduces a reagent to
    _____ 3. Ortho-toluidine   form a colored compound
    _____ 4. Copper reduction b. method is specific for glu-
    _____ 5. Ferricyanide   cose; requires the coenzyme,
                                      NADP$^+$
                                   c. only aldehyde sugars react to
                                      produce colored derivatives
                                   d. glucose reduces this metal ion
                                      to a colorless form; the reduc-
                                      tion in color is measured (in-
                                      verse colorimetry)
                                   e. a colorimetric method specific
                                      for glucose; hydrogen peroxide
                                      formed in the first reaction is
                                      used as a substrate for the sec-
                                      ond reaction

16. Reducing substances that interfere with nonspecific glucose procedures in-
    clude all *except:*
    a. ascorbic acid
    b. uric acid
    c. creatinine
    d. sucrose

17. Using the NADP$^+$ → NADPH reaction, the form that absorbs at 340 nm is
    _____ .

18. Factors that can alter glucose tolerance test results include all of the follow-
    ing *except:*
    a. smoking during the test
    b. drinking black coffee during the test
    c. drinking water during the test
    d. diet deficient in carbohydrate content preceding the test

19. For nonpregnant adults, the glucose load recommended in the standardized
    OGTT is:
    a. 100 g oral glucose load
    b. 75 g oral glucose load
    c. 1.75 g/kg body weight not to exceed 100 g oral dose
    d. 75 g given intravenously to avoid vomiting

20. Match the adult nonpregnant patient with the expected laboratory findings:

Fasting	2 hr pp 75 g load	Glyco-sylated Hb		Fasting venous blood values

				Fasting
_____	_____	_____	1. Normal	a. 80 mg/dL
_____	_____	_____	2. Diabetes mellitus	b. 150 mg/dL
				c. 120 mg/dL
_____	_____	_____	3. Impaired glucose tolerance	d. 50 mg/dL

2 hr pp

e. 50 mg/dL
f. 175 mg/dL
g. 240 mg/dL
h. 90 mg/dL

Glycosylated Hb

i. 6%
j. 18%
k. 10%
l. zero

21. Glycosylated hemoglobin refers to:
    a. HbA, HbA_2, HbA_{1c}
    b. HbA_{1a}, HbA_{1b}, HbA_{1c}
    c. HbA_{1c} only
    d. HbA_{1c}, HbF, HbC

22. Juvenile diabetes mellitus may be manifested by all of the following *except:*
    a. polyuria
    b. polydipsia
    c. polyphagia
    d. hyperglycemia
    e. hyperinsulinemia

23. Match the following:
    _____   1. Formerly called juvenile diabetes   a. IDDM
    _____   2. Formerly called adult-onset diabetes   b. NIDDM
    _____   3. Sudden onset and usually occurs at an
                 early age
    _____   4. Onset usually after 40 years of age
    _____   5. Onset associated with obesity
    _____   6. Associated with islet cell antibodies
    _____   7. Not associated with islet cell antibodies
    _____   8. Most common type of diabetes mellitus

      _____   9. Usually no measurable insulin present
      _____ 10. Associated with decreased number of insulin receptor sites on cells
      _____ 11. Can usually be controlled by dietary measures

24. A patient was brought into the emergency room in a coma. To differentiate between diabetic coma and hypoglycemic shock, the most helpful test would be:
    a. blood pH
    b. measurement of C-peptide
    c. blood glucose
    d. urinary ketone bodies

25. A patient with diabetic keto acidosis may present with all of the following *except:*
    a. metabolic alkalosis
    b. elevated blood sugar and positive urine sugar
    c. Kussmaul respiration
    d. dehydration with polyuria
    e. positive ketone bodies in the urine

26. The most common cause of death associated with diabetes mellitus is:
    a. atherosclerotic-cardiovascular disease
    b. diabetes acidosis
    c. Kimmelsteil–Wilson's lesions
    d. insulin shock

27. Secondary diabetes mellitus may be associated with all of the following *except:*
    a. Cushing's
    b. hemochromatosis
    c. acromegaly
    d. Addison's disease

28. The lactose tolerance test is used to evaluate:
    a. pancreatic deficiency
    b. hormone deficiency
    c. bile acid deficiency
    d. disaccharidase deficiency

29. The tolbutamide test is used to evaluate:
    a. insulin reserve
    b. glucagon production
    c. exocrine pancreatic function
    d. C-peptide production

30. Reactive hypoglycemia refers to _____ .
    a. hypoglycemia associated with an insulinoma
    b. hypoglycemia associated with the ingestion of sulfonylurea agents
    c. hypoglycemia associated with self-administration of insulin
    d. hypoglycemia following meals

31. The C-peptide is found in:
    a. endogenous human insulin
    b. commercial animal insulin
    c. both
    d. neither

32. The C-peptide will be _____ in factitious hypoglycemia; in insulinoma, the C-peptide will be _____ .
    a. increased
    b. decreased
    c. normal

33. An insulinoma is characterized by all of the following *except:*
    a. hypoglycemia
    b. presence of insulin antibodies
    c. symptoms precipitated by fasting
    d. hyperinsulinemia

ANSWERS

1. d
2. (1) b; (2) a; (3) a; (4) b; (5) a; (6) b
3. b
4. (1) a, a, d; (2) a, c, e; (3) a, b, d
5. b
6. a
7. a
8. (1) c; (2) d; (3) e; (4) b; (5) a
9. (1) d; (2) b; (3) a; (4) c
10. c
11. (1) d; (2) e; (3) c; (4) b; (5) a; (6) f
12. b
13. (1) a; (2) b
14. a
15. (1) b; (2) e; (3) c; (4) a; (5) d
16. d
17. NADPH
18. c
19. b
20. (1) a, h, i; (2) b, g, j; (3) c, f, k
21. b
22. e
23. (1) a; (2) b; (3) a; (4) b; (5) b; (6) a; (7) b; (8) b; (9) a; (10) b; (11) b
24. c

25. a
26. a
27. d
28. d
29. a
30. d
31. a
32. b, a
33. b

CHAPTER 6
PROTEINS

a. Free and Bound Light Chains
b. Kappa/Lambda Ratio
c. Immunosuppression

B. Waldenströms Macroglobulinemia
C. Amyloid Disease
D. Cryoglobulinemia
E. Associated Renal Disease

XIII. SERUM ELECTROPHORETIC PATTERNS IN DISEASE

A. Stress Pattern
B. Delayed Response Pattern
C. Hepatic Cirrhosis Pattern
D. Hypogammaglobulinemia Pattern
E. Monoclonal Gammopathy
F. Nephrotic Syndrome Pattern
G. Protein-Losing Enteropathy Pattern
H. Other Disease Patterns

XIV. GLYCOPROTEINS AND MUCOPROTEINS

A. Tumor Markers

XV. IMMUNODIFFUSION METHODS FOR PROTEIN MEASUREMENT

A. Single Diffusion (Radial Immunodiffusion) (RID)
B. Double Diffusion
C. Immunoelectrophoresis (IEP)

XVI. INTERPRETATION OF LABORATORY DATA

OBJECTIVES

On completion of this chapter, you will be able to:

1. Define, illustrate, or explain the following terms or concepts:
   a. protein
   b. amino acid
   c. amphoteric
   d. zwitterion
   e. peptide bond
   f. nitrogen balance
   g. globulins
   h. immunoglobulin
   i. A/G ratio
   j. amyloid
   k. Bence Jones protein
   l. multiple myeloma

2. Discuss the relationship between pH of a solution and charges on amino acids and proteins.
3. Discuss proteins in terms of structure, functional groups, elemental and molecular constituents, and function.
4. Discuss the basic principles and components of electrophoresis.
5. List the five electrophoretic zones and describe proteins found in each including function, clinical significance, and expected levels.
6. Describe the five classes of immunoglobulins and the unique features of each class.
7. Describe the various proteins identified in myeloproliferative and lymphoproliferative malignancies.
8. Describe various immunodiffusion methods for protein measurements.
9. Identify serum electrophoretic patterns of various diseases.

GLOSSARY OF TERMS

Amino acids. Organic compounds containing a carboxylic acid group, an amine basic group, and an organic R group all bonded to the same alpha carbon. Amino acids serve as building blocks of proteins through polymerization via peptide bond formation.

Amino acids, essential. Amino acids that cannot be synthesized by the body and hence must be included in the diet.

Amphoteric. Chemical compounds in solution that can serve as either an acid (proton donor) or base (proton acceptor) depending on the pH of the solution.

Antigen. Any foreign material that can elicit an immunological response in a host.

Globulins. Refers to serum proteins other than albumin and prealbumin.

Isoelectric point. Refers to the pH at which an amphoteric amino acid (or protein) has the same number of positive and negative charges.

Nitrogen balance. Means that the net gain of protein in the diet is sufficient to meet the needs of growth and repair of the body.

Polypeptide. Chemical polymeric compounds containing 6–30 amino acids linked together by peptide bonds.

Protein. A polymeric compound consisting of 40 or more amino acids held together by peptide bonds.

Proteins are complex, chemical compounds, which have major structural and biochemical functions in all living cells. Plasma proteins, of primary concern in clinical chemistry, serve as antibodies, hormones, enzymes, clotting factors, and plasma transporters of various substances. In addition, proteins function (1) as the complement system; (2) as a nutrition source; (3) as part of the plasma buffering system; and (4) to control water distribution between the intra- and extravascular compartments.

ELEMENTAL COMPOSITION OF PROTEINS

Proteins are large organic molecules (called macromolecules) with molecular weights ranging from 10,000 to 10,000,000 g/mol or higher. In addition to elements carbon, hydrogen, oxygen, and nitrogen, proteins contain about 1% sulfur. Specialized proteins contain other elements such as phosphorus in casein (a milk protein), iron in hemoglobin, iodine in thyroglobulin (a hormone-storage protein of the thyroid), copper in ceruloplasmin, and zinc. The presence of nitrogen in proteins makes them unique organic molecules especially when compared to lipids and carbohydrates. Lipids and carbohydrates are primarily composed of elements carbon, hydrogen, and oxygen. Most proteins contain from 12 to 18% nitrogen with albumin, major plasma protein, containing about 16% nitrogen.

AMINO ACIDS

PROTEIN FORMATION

As described in Chapter 1, chemical elements serve as building blocks for compounds. Similarly, amino acids serve as building blocks for proteins. Proteins are formed by condensation reactions of the various amino acid monomers, forming a much larger molecular structure (or polymer).

There are at least 21 amino acids commonly found in proteins of the body. These amino acids are shown in Figure 6.1.

Essential Amino Acids

Ten of the amino acids are called "essential amino acids" because they cannot be synthesized by the body and therefore must be supplied in the diet. Proteins that contain all the essential amino acids are called "complete proteins;" these proteins are usually of animal origin (e.g., meat, milk, eggs, and cheese). Plant proteins are often lacking in one or more of the essential amino acids. Obviously, diet plays the significant role in diseases caused by essential amino acid deficiencies.

FUNCTIONAL GROUPS

Amino acids, as the name implies, are organic carboxylic acids (—COOH) containing a basic amino group (—NH$_2$) attached to the same carbon atom. Because of the close bonding proximity of these two different functional groups, amino acids are unique chemical compounds in nature. The bonding of functional groups of the "alpha" carbon is shown in Figure 6.2.

All naturally occurring amino acids have the amino grouping attached to the alpha carbon, hence these are termed alpha amino acids. Also, with the exception of glycine, which has two hydrogens on the alpha carbon, the alpha carbon on amino acids is asymmetrical, giving rise to at least two optically active enantio-

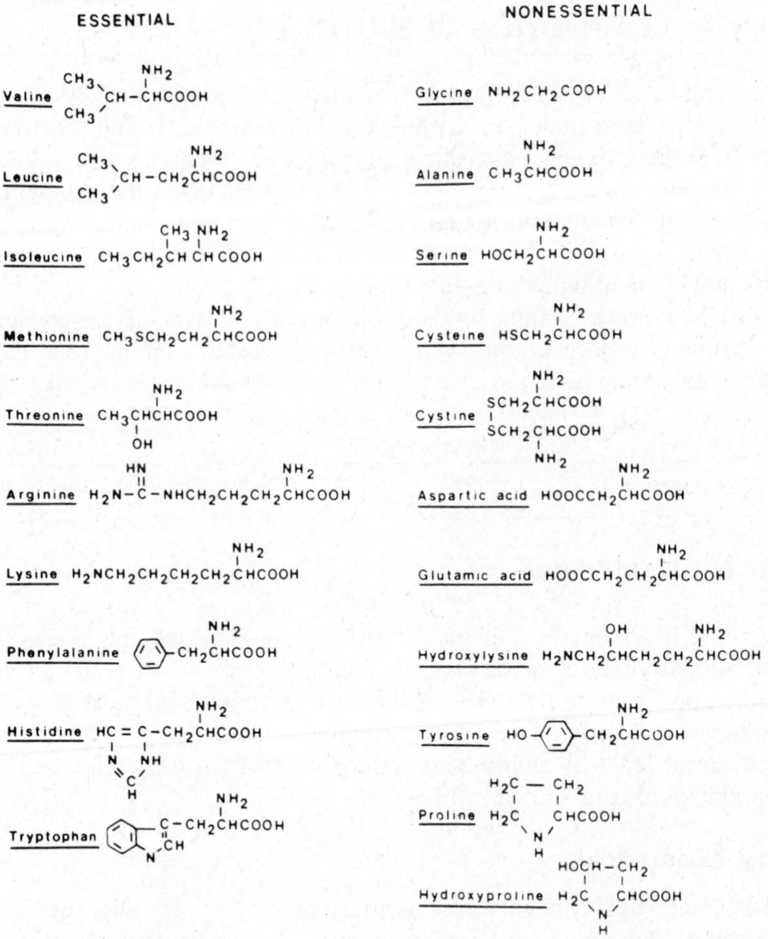

ESSENTIAL   NONESSENTIAL

Valine CH_3 $CH-CHCOOH$ with NH_2 Glycine NH_2CH_2COOH
        CH_3

Leucine CH_3 $CH-CH_2CHCOOH$ with NH_2 Alanine $CH_3CHCOOH$ with NH_2
         CH_3

Isoleucine $CH_3CH_2CH\ CHCOOH$ with CH_3 NH_2 Serine $HOCH_2CHCOOH$ with NH_2

Methionine $CH_3SCH_2CH_2CHCOOH$ with NH_2 Cysteine $HSCH_2CHCOOH$ with NH_2

Threonine $CH_3CHCHCOOH$ with NH_2 and OH Cystine $SCH_2CHCOOH$ / $SCH_2CHCOOH$ with NH_2 / NH_2

Arginine $H_2N-C-NHCH_2CH_2CH_2CHCOOH$ with HN (double bond) and NH_2 Aspartic acid $HOOCCH_2CHCOOH$ with NH_2

Lysine $H_2NCH_2CH_2CH_2CH_2CHCOOH$ with NH_2 Glutamic acid $HOOCCH_2CH_2CHCOOH$ with NH_2

Phenylalanine ⬡$-CH_2CHCOOH$ with NH_2 Hydroxylysine $H_2NCH_2CHCH_2CH_2CHCOOH$ with OH and NH_2

Histidine $HC=C-CH_2CHCOOH$ with NH_2, ring N, NH, C, H Tyrosine $HO-⬡-CH_2CHCOOH$ with NH_2

Tryptophan ⬡$C-CH_2CHCOOH$ with NH_2, ring N, CH Proline H_2C-CH_2 / H_2C $CHCOOH$ / N / H

Hydroxyproline $HOCH-CH_2$ / H_2C $CHCOOH$ / N / H

Figure 6.1.   The essential and nonessential amino acids. (Reproduced with permission from Henry, J. B. *Clinical Diagnosis and Management by Laboratory Methods,* 16th ed. Philadelphia: Saunders, 1979.)

meric forms. Because of the asymmetric alpha carbon, amino acids can have two nonsuperimposable isomeric structures, both of which are optically active. The isomer (or enantiomer) that rotates plane polarized light to the right is called D (or dextrorotary); the enantiomer that rotates the polarized light to the left is called L (or levorotary). All amino acids in human proteins are of the L configuration.

PROPERTIES OF AMINO ACIDS

Amphoteric

Compounds that can serve as both acids and bases in water solutions are called amphoteric. All amino acids and proteins are amphoteric because (1) the amino group can accept a proton to become $-NH_3^+$ (bases are proton acceptors) and (2) the carboxyl group can lose a proton to form $-COO^-$ (acids are proton donors).

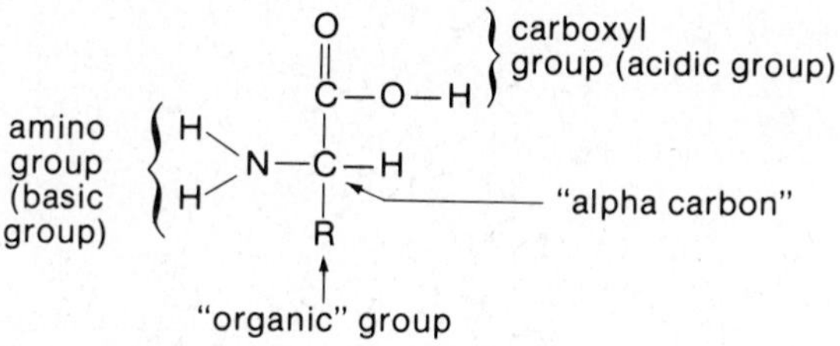

Figure 6.2. General formula of amino acids.

Isoelectric Point

The isoelectric point (PI) refers to the pH at which an amphoteric molecule has the same number of positive and negative charges. Amino acids can form a *zwitterion* ion at the isoelectric point (i.e., a polyatomic ion with one positive and one negative charge). In electrophoresis experiments, if the buffer pH is the same as PI for the amino acid or protein, the compound will not migrate in the electric field because the net charge is zero (see Figure 6.3).

If the pH is above the PI, the excess hydroxide groups in solution (OH^-) will remove the proton from the $-NH_3^+$ group, hence the net charge on the remaining ion will be -1 (see Figure 6.4). On the other hand, if the pH is below the PI, excess protons (H^+) will protonate the COO^- group with the net charge on the resulting ion being $+1$ (see Figure 6.4).

R Groups of Amino Acids

The "R" organic group of aspartic and glutamic acids contains an additional carboxylic acid group (see Figure 6.1). Hence these "acidic" amino acids lose two protons in solutions at high pHs, forming an amino acid anion with a -2 charge.

The "R" organic groups of lysine, arginine, and histidine contain additional amino groups. These "basic" amino acids can gain an additional proton in solutions at low pHs, forming a dipositive amino acid cation.

The "R" groups of other amino acids are aliphatic carbon chains, aromatic carbon rings, heterocyclic imino acid rings, and amides. Generally, proteins exhibit acid–base properties similar to amino acids in aqueous solutions. In solutions with pHs from 7.6 to 8.4 and above, proteins are negatively charged; with pHs ranging from 3.0 to 3.2 and below, proteins are positively charged.

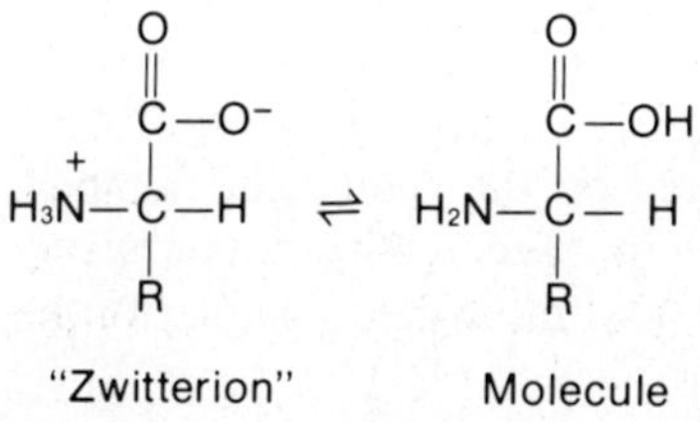

"Zwitterion"   Molecule

Figure 6.3. A zwitterion structure, which has the same net charge as the parent molecule.

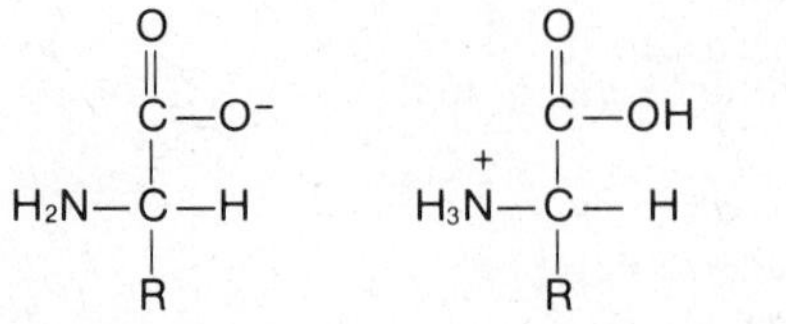

$$
\begin{array}{cc}
\underset{\text{H}_2\text{N}-\underset{\overset{|}{\text{R}}}{\overset{|}{\text{C}}}-\text{H}}{\overset{\displaystyle \overset{\text{O}}{\underset{\|}{\text{C}}-\text{O}^-}}{}} &
\underset{\overset{+}{\text{H}_3\text{N}}-\underset{\overset{|}{\text{R}}}{\overset{|}{\text{C}}}-\text{H}}{\overset{\displaystyle \overset{\text{O}}{\underset{\|}{\text{C}}-\text{OH}}}{}}
\end{array}
$$

(Anionic form)   (Cationic form)
pH > PI   pH < PI
pH = 9.1 to 10.8   pH = 1.7 to 2.6

Figure 6.4. Different net charges on amino acids depending on the pH of the solution.

FORMATION OF PROTEINS

PEPTIDE BONDS

Amino acids link together forming proteins by forming peptide bonds. Peptide bond formation is a condensation reaction (loss of water) between the nucleophilic "lone electron pair" on the amino group with the electrophilic carbonyl carbon on the other reacting amino acid. A water molecule is lost with the resulting peptide bond being formed. A possible mechanism for this reaction is shown in Fig. 6.5.

A more simplified representation of the formation of a peptide bond is given in Figure 6.6. As shown in Figure 6.6, this dipeptide has an N-terminal end (on the left) and a C-terminal end (on the right). If three amino acids had reacted, a tripeptide would be formed; if four a tetrapeptide would be formed; etc. Amino acid "residues" are the remaining components of the original amino acid as it resides in the protein structure.

POLYPEPTIDES

Polypeptides contain 6–30 peptide bonds, while polypeptides with 40 or more peptide bonds are often called proteins.

STRUCTURE OF PROTEINS

PRIMARY STRUCTURE

The primary structure of a protein molecule is based on the sequence, number, and type of amino acid residues making up the protein. Hence the primary structure is based on the peptide bonds linking the 30–50,000 amino acid residues of the various proteins.

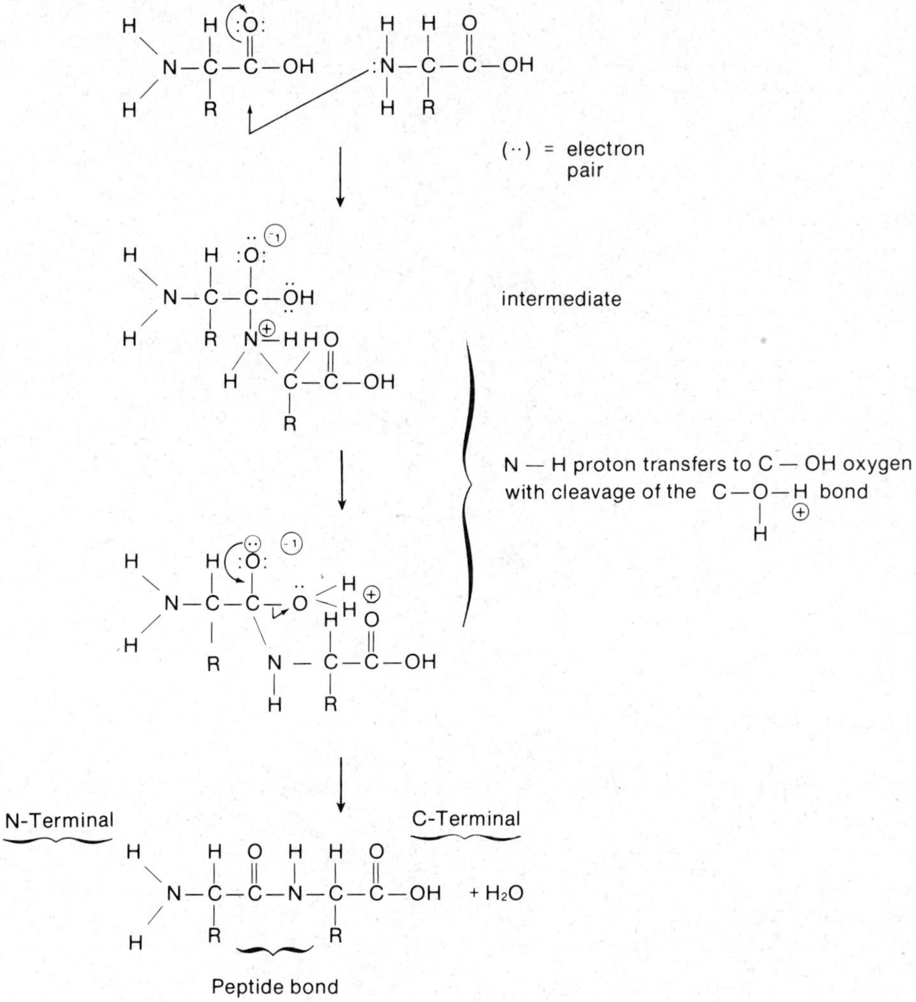

Figure 6.5. A mechanism for the formation of the peptide bond.

SECONDARY STRUCTURE

The secondary structure of proteins is related to the helix formation that occurs due to hydrogen bonding between amino acid residues on a single peptide chain.

TERTIARY STRUCTURE

Weak intramolecular forces between amino acid residues on the peptide chain cause the molecule to become folded, looped, and twisted about itself. These forces are due to weak interaction between R side chain groups and are disrupted during a process called "denaturation." Properties of proteins such as water solubility and enzyme activity are lost when "native" proteins are denatured.

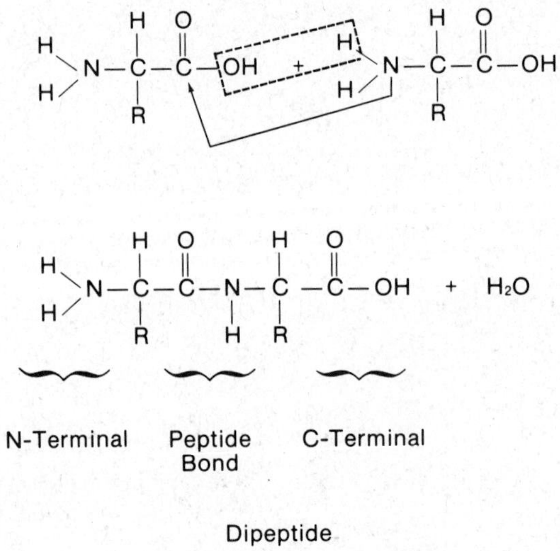

Figure 6.6. The peptide bond.

QUATERNARY STRUCTURE

Proteins have quaternary structure when two or more peptide chains are linked together to form the protein.

PROTEIN CLASSIFICATIONS

Many different methods of classifying proteins are used including classifications based on structure, chemical composition, solubility, biological functions, and electrophoretic migration.

FIBROUS PROTEINS

Fibrous proteins include (1) collagen and elastin or proteins of connective tissue such as bone, cartilage, ligament, and tendons; (2) keratin or sulfur-containing proteins of the skin, hair, and nails; (3) fibrin, the protein formed when blood clots; and (4) myosin, the contractile protein of muscle fiber.

GLOBULAR PROTEINS

Highly coiled proteins that form colloidal spheres in aqueous solutions are called globular proteins and include (1) most body enzymes; (2) plasma proteins (excluding albumin); and (3) hemoglobins.

CHEMICAL COMPOSITION OF PROTEINS

Simple proteins contain only alpha amino acids (as described above) and include albumin, globulins, and albuminoids of supportive tissue (keratin, collagen, etc.). Simple proteins linked to nonprotein compounds are called "conjugated proteins"; these include (1) lipoproteins—proteins bound to fatty acids; (2) glycoproteins—proteins linked to carbohydrates; (3) phosphoproteins—proteins bound to phosphate; and (4) nucleoproteins—proteins bound to nucleic acids. Proteins are also linked to metals (metalloprotein) and complex organic heterocycles as in hemoglobin.

PROTEIN SOLUBILITY

Early classifications of proteins were based on solubility of proteins in aqueous salt solutions. For example, globulins are generally soluble in dilute salt solutions but are insoluble in pure water and 50% saturated salt solutions. Albumins are soluble in both pure water and 50% saturated salt solutions.

ELECTROPHORETIC MIGRATION

Classifications of proteins based on electrophoretic mobility (to be discussed subsequently in this chapter) include albumin, alpha-1-globulins, alpha-2-globulins, beta globulins, and gamma globulins.

Since many methods of studying proteins must first involve a separation step prior to quantitation, experimental techniques for classifying proteins differ mainly in the property difference used to separate the various proteins in plasma.

SELECTED EXAMPLES OF SEPARATION

Techniques for separating proteins include: (1) solubility—where variations in size and shape of protein molecules and the presence of ions or dehydrating solvents (polyethylene glycol) in the solution enable proteins to be separated; (2) electrophoresis—where different ionic charges on proteins in alkaline buffers allow separation and, to a lesser extent, size and shape of proteins influence electrophoretic separation (more acidic proteins tend to migrate faster due to the increased negative charges on the protein in an alkaline solution); (3) ultracentrifugation—protein separation based on variation in molecular mass; (4) chromatography—protein separation based on differences in size, shape, polarity, and differential affinity for fixed and mobile phases; and (5) immunoprecipitation or fixation—protein separation based on a unique antigenic determinant on specific proteins.

Ion exchange chromatography is also used to separate proteins often on the basis of protein acidity in cation exchange chromatography and basicity in anion exchange chromatography. Gel filtration uses the principle that smaller proteins get trapped in a molecular polymer (like dextran) while larger proteins do not.

Isoelectric focusing utilizes the concept that proteins have different PI values; hence, by varying the pH of the medium and applying an electric field, the protein will migrate until the pH of the agarose medium equals the PI of the protein. Counterimmunoelectrophoresis and immunoelectrophoresis use a combination of electrophoresis and immunochemicals for the identification and semiquantitation of specific proteins. Passive radial diffusion of various proteins in antibody-treated agar is determined by the size and shape of the protein along with the solubility of the protein–antibody complex. The rate of formation of protein–antibody complexes is important in rate nephelometry along with the size and shape of the complex sphere.

BIOLOGICAL FUNCTIONS

Proteins are often classified as to their biological functions, which include (1) enzymatic regulation of cellular function; (2) transport functions as in hemoglobin, transcortin, and lipoproteins; (3) infection defense as immunoglobulin and complement; (4) acid–base balance as protein buffers; (5) osmotic balance (albumin); (6) hemostasis as protein coagulation factors; and (7) body structure as bone osteoid.

PROTEIN PHYSIOLOGY

PROTEIN HYDROLYSIS

In the stomach and small intestine proteins are hydrolyzed to component amino acids through the action of the digestive enzymes. This process is required because macromolecular proteins are too large for absorption, while component amino acids are readily absorbed. Protein hydrolysis begins in the stomach with the proteolytic action of pepsin; further hydrolysis follows in the small intestines through action of the pancreatic proteolytic enzymes (i.e., trypsin, chymotrypsin, and carboxypeptidase). These enzymes are secreted as inactive precursors into the small intestine, whereupon they are activated by proteolytic cleavage. For example, trypsinogen is converted to trypsin through action of enterokinase from the small intestine. The resulting trypsin catalyzes the hydrolysis of chymotrypsinogen to chymotrypsin. Trypsin and chymotrypsin apparently hydrolyze proteins to polypeptides. Carboxypeptidase hydrolyzes these polypeptides to dipeptides, and finally dipeptides are absorbed and broken down to amino acids in the intestinal wall by action of dipeptidases.

NITROGEN BALANCE

Proteins are unique in metabolism in that they contain nitrogen. A positive "nitrogen balance" means that the net gain in protein in the diet is sufficient to meet the needs of growth and repair of the body. After surgery or during starvation, the amount of protein needed often exceeds the amounts of dietary replacement protein, hence tissue proteins are often depleted. This latter situation is referred to as "negative nitrogen balance."

MEASUREMENTS OF PROTEINS IN BODY FLUIDS

The routine measurement of total protein is generally carried out on serum using colorimetry. Since proteins make up approximately 7% of the mass of serum with the remaining 93% being solvent (or water), it is not surprising that the normal range for total protein is 6.0–7.8 g/dL in the recumbent patient. Ambulatory patients have slightly higher concentrations of serum protein because of shifts of vascular water to the extravascular space.

HYPERPROTEINEMIA

Patients usually have elevated protein levels either due to (1) dehydration, where the amount of protein is unchanged but the amount of plasma water is reduced, or (2) abnormal production of protein as seen in malignancies such as multiple myeloma, where levels of immunoglobulin are markedly increased.

HYPOPROTEINEMIA

Hypoproteinemia is usually due to renal protein loss as seen in nephrotic syndrome, where large amounts of low-molecular-weight proteins (especially albumin) are excreted in the urine. Also large quantities of protein can be lost in severe bleeding or burns.

Protein starvation or malabsorption of amino acids can restrict the amounts of amino acid needed for synthesis of proteins. Also diseases of the liver can cause significant reduction in the synthesis of protein especially albumin.

KJELDAHL TITRATION METHOD

All of the nitrogen in protein and other nitrogen-containing compounds is converted to ammonium ions (NH_4^+) by treating the sample with sulfuric acid, potassium sulfate, and copper sulfate. Alkalization liberates the ammonia vapors:

$$NH_4^+ + OH^- \rightarrow NH_3\uparrow + H_2O$$

which are distilled into a boric acid solution then titrated with standard acid to determine the percentage of nitrogen in the original sample. The amount of protein present in the sample is calculated by multiplying the Kjeldahl nitrogen by 6.25 (6.25 = 1.00/0.16), which assumes a protein nitrogen percentage of 16%. This method is not used for routine protein determinations in the chemical laboratory but is considered as the protein reference method.

COLORIMETRIC PROTEIN METHODS

Biuret Method

Two adjacent peptide bonds (tri- and higher polypeptides) form a blue–violet complex with cupric ions of the biuret reagent (copper sulfate in an alkaline medium). The complex is similar to that formed between Cu^{2+} and the "biuret" molecule (see Figure 6.7). Although the exact structure of the protein–Cu^{2+} complex is not known, one Cu^{2+} coordinates with up to six peptide bonds. The more peptide bonds that coordinate to the Cu^{2+}, the more intense the color, hence the reaction gives a good estimate of the amount of protein present. The biuret assay is most often performed on serum. Plasma may be used but fibrinogen present will increase the protein level. Hemoglobin, bilirubin, and triglycerides (or lipids) interfere with the reaction as does ammonium ions (NH_4^+). The biuret method is not recommended for urine protein because of NH_4^+ interference. However, if urine proteins are precipitated and redissolved, the biuret reaction can be performed on the isolated urinary protein.

Phenol Method

Phenolic side chains of tyrosine and tryptophane residues in proteins reduce phosphomolybdotungstic acid to form a blue color. This reaction is more sensitive than biuret; however, the accuracy of the method is generally poor in specimens with abnormal protein content. The Folin–Lowry method utilizes both biuret and phenol reduction in a very sensitive assay (100 times more sensitive than biuret). Protein levels as low as 30 μg can be measured using Folin–Lowry.

Ninhydrin Reaction

Ninhydrin reacts with free amino groups of proteins and amino acids to form a colored compound. This reaction can be used for the colorimetric measurement of proteins and amino acids. The imino acids proline and hydroxproline do not react with ninhydrin because they have no free amino group.

PROTEIN MEASUREMENTS IN OTHER BODY FLUIDS

Spinal Fluid Protein

Tricholoroacetic acid (TCA) is used in a turbidimetric procedure for the measurement of spinal fluid protein. TCA gives a higher turbidity with globulins than albumin, hence TCA is not recommended for urine protein measurements.

Normal levels of spinal fluid protein range from 15 to 45 mg/dL. Spinal fluid protein levels are increased in cerebral hemorrhage, acute and chronic infections, and tumors. In multiple sclerosis, the total spinal fluid protein is generally not

Figure 6.7. The biuret molecule.

increased; however, the gamma globulin level is increased indicating local im-munoglobulin production.

Urine Protein

Because sulfosalicylic acid (SSA, 3-carboxy-4-hydroxybenzenesulfonic acid) gives higher turbidity with albumin that trichloroacetic acid, SSA is preferred in the turbidimetric measurement of total urine proteins. Normal urine protein lev-els are less than 150 mg/24 hour specimen. Urine protein levels are increased significantly in renal disease due to either tubular or glomerular dysfunction or both.

ALBUMIN METHOD

Serum albumin is measured routinely using dye-binding techniques. These colori-metric methods take advantage of the fact that certain chromogenic dyes will form complexes with albumin. The requirements for a dye-binding method include the following: (1) the absorbance of the dye, when bound to albumin, must be at a different wavelength than that of the unbound dye; (2) the complex should absorb at wavelengths where hemoglobin and bilirubin will have minimal interference; and (3) the dye must bind specifically to albumin rather than other serum proteins or constituents. Since dye-binding procedures involve a dynamic equilibrium be-tween albumin-bound and unbound dye, temperature increases tend to shift the equilibrium to the left (or toward unbound reactants). Therefore, the temperature must remain constant when testing standards, controls, and patients' specimens. Bromocresol Green (BCG) is the recommended dye for albumin measurement. Other dye-binding methods employing methyl orange, bromocresol purple, and 2-(4'-hydroxyazobenzene) benzoic acid have been used.

Normal levels of albumin range from 3.5 to 6.0 mg/dL. Albumin levels are low in patients with liver disease, nephrosis, protein-losing enteropathy, and hemodi-lution.

ALBUMIN–GLOBULIN RATIO

By subtracting the albumin level from the total protein level, an indirect estimate of globulin can be obtained. By dividing the albumin level by the globulin level, a new parameter called the A/G ratio can be calculated. Ratios are useful and usually present a more sensitive parameter for detection of disease than either measurement used independently when (1) increases of one measurement (i.e., gamma globulin) is directly proportional to the presence (and severity) of the disease and (2) levels of the other parameter (i.e., albumin) are inversely propor-tional to the presence (and severity) of the disease.

Normal levels of the A/G range from 1.5 to 2.5. The A/G parameter is an especially sensitive indication of liver disease since the latter usually causes a decreased albumin with an increased gamma globulin level. In addition to liver disease, low A/G ratios are also seen in malnutrition, burns, diarrhea and nephrosis, lymphomas, myelomas, and granulomatous disease. Electrophoresis (to be discussed later in this chapter) has generally replaced the A/G ratio as a diagnostic tool.

PROTEIN ELECTROPHORESIS

ELECTROPHORESIS

This term is used to describe the movement of ions under the influence of an electric field. Cations and anions in aqueous solution will migrate toward the opposite charged electrodes. Anions migrate toward the positive electrode (anode); cations migrate toward the negative electrode (cathode). Large molecules like proteins can be separated by electrophoresis.

Proteins are amphoteric and lose protons in base solutions to form negative ions. The more acidic R-group residues on a protein (e.g., as in aspartate and glutamate residues), the greater the negative charge on the protein in alkaline buffers (pH = 8.6). Hence, the more rapidly the protein fraction will migrate on a solid medium toward the positive pole of an electrical circuit.

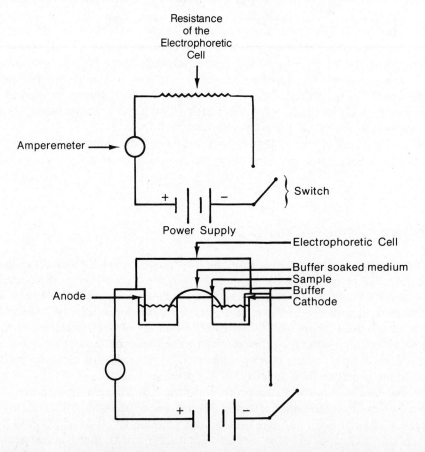

Figure 6.8. A comparison of an electrophoretic cell to a resistance in an electrical circuit.

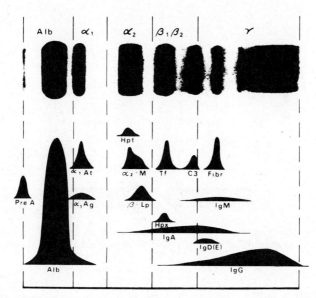

Figure 6.9. The electrophoretic separation of the proteins of plasma. [Adapted with permission from Laurell, C. B. *Clinical Chemistry,* **19,** 99 (1973) Copyright 1973 American Association for Clinical Chemistry.] Abbreviations: ALB, albumin; AT, antitrypsin; M, macroglobulin; Hp, haptoglobin; Tf, transferin; Lp, lipoprotein; C_3, complement; and Fibr, fibrinogen.

On comparing the electrophoresis to an electric circuit (see Figure 6.8), the electrical field is applied and proteins differentially migrate toward the anode owing to their negative charges in the alkaline buffer. After color development of the electropherogram, a densitometric tracing of the various protein bands on the medium is made (see Figure 6.9). Considerations other than charge on the protein influence migration; these include (1) size and shape of the protein molecule; (2) the medium used; (3) the ionic strength of the buffer; and (4) the effects of temperature.

ENDO-OSMOTIC EFFECT

The endo-osmotic effect of hydrate cations moving toward the cathode retard the migration of proteins toward the anode. Endo-osmosis can even cause weakly charged gamma globulins to migrate cathodically depending on the point of sample application (see Figure 6.10). The so called "wick effects" of the solvent interacting with the medium enhance the migration of proteins to the middle of the support medium, then retard further migration.

TEMPERATURE EFFECTS

Movement of electrons through a resistance such as electrophoretic cell (see Figure 6.8) generates heat. The increased heat lowers the resistance and accordingly will increase the current, if the voltage of the system is constant, in accor-

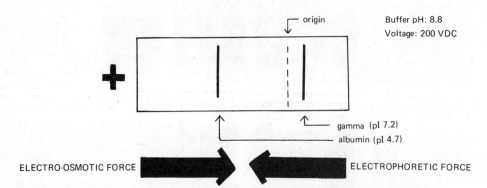

Figure 6.10. The electro-osmotic force versus the electrophoretic force and their combined influence on protein migration. (Reproduced with permission of Helena Laboratories, Beaumont, TX.)

dance with Ohm's law:

$$\text{current} = \frac{\text{voltage}}{\text{resistance}}$$

Subsequently, the increased current will further increase heat, decreasing resistance even more at constant voltage. The ever increasing temperature evaporates the solvent, concentrates the solute ions, and further increases current. The increased current flow accelerates the migration of proteins. However, the increasing heat can denature the proteins on the electrophoretic strip. Therefore, to avoid these problems, electrophoretic experiments are best performed at constant current rather than constant voltage.

SPECIFIC COMPONENTS OF ELECTROPHORESIS

SAMPLE

Serum samples are recommended for serum electrophoresis. Plasma or incompletely clotted blood give a separate electrophoretic band for fibrinogen. The sample may be stored prior to analysis as follows: (1) 5 days at room temperature; (2) 30 days at 4°C; or (3) up to 5 years frozen. However, resolution of protein bands decreases with time. Also, repeated freezing and thawing can denature proteins and is not recommended.

SUPPORT MEDIA

For five-zone protein electrophoresis (see Figure 6.9), cellulose acetate and agarose are often used with a Mylar backing to facilitate handling of the strip.

Agarose gel is the recommended medium for lipoprotein electrophoresis and for high-resolution serum electrophoresis, where as many as 12 different protein bands can be identified. Starch gel and polyacrylamide gel can also be used for protein electrophoresis. Polyacrylamide gels have small pore sizes and behave as a "molecular sieve," hence, proteins are separated both by size and charge with 20 or more separate protein bands being seen.

COLOR DEVELOPMENT

Protein bands are colorless and therefore must be stained prior to visualization and quantitation using densitometry. Ponceau S stains proteins red at pH less than 10. Other protein stains such as Amido Black 10 B, Lissamine Green, Nigrosin, and Coomassie Blue can be used. Coomassie Blue increases the sensitivity of the procedure and is therefore better for staining proteins present in trace quantities. Specific stains for lipoproteins include oil red O and Sudan Black. Periodic Acids Shiff's stain is used for carbohydrate containing proteins (glycoproteins and mucoproteins). Also, specific substrates are used to "stain" enzymes and generate either fluorescent bands (usually based on NADH or NADPH generation) or visible bands (based on reduction of formazan dyes with NADH).

QUANTITATION

After staining, the electrophoretic medium can be read visually or scanned spectrophotometrically on a densitometer. However, dye binding in the staining process is based on a combination of the dye with free amino groups; hence, globulins (which have fewer amino groups than albumin) bind less dye than albumin. Also, globulins differ in dye binding among themselves. The amount of dye bound also depends highly on the dye used.

GENERAL PROCEDURE FOR ELECTROPHORESIS

Initially, the cellulose acetate medium is soaked in a barbitol buffer (pH 8.6). After blotting the medium, the patients' samples and controls are innoculated. The cellulose acetate strip is then placed in a electrophoretic chamber and current is applied. After a fixed time period, the cellulose acetate film is removed and stained with Ponceau S. Quantitation is then performed by scanning protein bands on a densitometer. Figure 6.11 shows a densitometer tracing of the five major zones or bands of proteins observed: (1) albumin, (2) alpha-1-globulins; (3) alpha-2-globulins; (4) beta globulins; and (5) gamma globulins. Each of the globulin bands consist of several different proteins as illustrated in Figure 6.11.

ELECTROPHORETIC ARTIFACT

A frequent "application" artifact occurs giving an extra band appearing between the beta and gamma fractions. This artifact can be caused by (1) a dirty sample

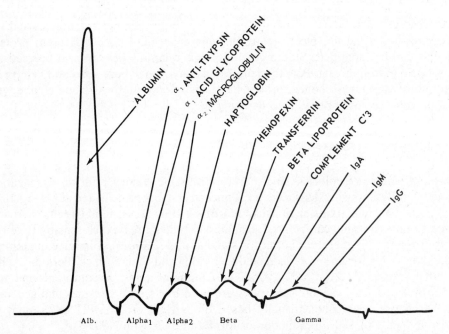

Figure 6.11. Five-zone electrophoretic densitometer pattern and component proteins of each fraction. (Reproduced with permission of Helena Laboratories, Beaumont, TX.)

applicator tip; (2) increased pressure on the sample applicator tip trapping slow moving proteins; or (3) nonmigratory debris in serum. Slight changes in methodology or buffer ionic strength can cause transferrin, beta lipoproteins, or complement proteins to migrate more slowly and hence cause the extra band.

CSF PROTEIN ELECTROPHORESIS

Owing to the low level of protein in spinal fluid (normal levels range from 15 to 45 mg/dL), CSF must be concentrated prior to electrophoretic studies. The electrophoretic pattern of CSF is similar to the five-zone pattern of protein electrophoresis with the following exceptions: (1) a fast migrating prealbumin bond anodic to albumin is observed and (2) the alpha-2 and gamma fractions are less intense. CSF electrophoresis is used primarily to identify an isolated rise in gamma globulin production in the CSF, an observation highly predictive of multiple sclerosis. Also, the appearance of oligoclonal bands in the gamma region that are not present in the serum pattern is suggestive of multiple sclerosis.

PLASMA PROTEINS

Of the major plasma proteins, the liver synthesizes prealbumin, albumin, and alpha and beta globulins. In general, the reticuloendothelial system of the liver, spleen, and bone marrow is responsible for the production of gamma globulins.

PREALBUMIN

Prealbumin, produced in the liver, migrates anodic to albumin and is normally not seen in serum protein patterns. Prealbumin functions to transport thyroxine and retinol-binding protein (which binds vitamin A). Prealbumin is decreased in patients with negative nitrogen balance as seen in malnutrition, malignancy, chronic infections, liver disease, etc. Prealbumin is routinely measured by radial immunodiffusion and nephelometry.

ALBUMIN

Albumin is a carbohydrate-free protein (molecular weight of 65,000), which normally makes up about 60% of total serum protein. Albumin is produced in the liver and has a half-life of about 4 weeks. Albumin migrates closest to the anode on electrophoresis and functions to transport substances, especially those that would otherwise be insoluble, to the organ of elimination or to body tissues where needed. For example, albumin-bound bilirubin is transported to the liver for excretion, while albumin-bound toxic metals are transported to the kidney for excretion. Also, albumin binds calcium, cortisol, sex hormones, drugs, and fatty acids; hence, these substances can be transported to tissue where needed. Albumin also maintains osmotic pressure and serves as a major reserve for amino acids.

Serum albumin elevations are almost exclusively due to dehydration, while decreases are seen in (1) excessive protein loss due to kidney damage or severe hemorrhage or burns; (2) impaired synthesis as seen in hepatic disease, toxemia of pregnancy or genetic disorders; (3) inadequate protein intake seen in malnutrition, and (4) GI disorders with malabsorption, vomiting, or diarrhea (prior to dehydration).

GLOBULINS

As described earlier, globulins are proteins other than albumin and make up the electrophoretic alpha, beta, and gamma fractions. Specific quantitative measurements for globulins usually involve radial immunodiffusion or nephelometry.

Alpha-1-Antitripsin (AAT)

This protein inhibits protein digesting enzymes (proteolytic enzymes) including trypsin, leukocyte proteases, elastases, and collagenases. Alpha-1-antitrypsin is especially important in inhibiting leukocyte proteases, whereby it protects the lung from proteolytic damage during inflammatory processes. An inherited gene determines whether patients will have low levels of AAT with both homo- and heterozygous patients having reduced levels. Individuals with low levels of AAT are predisposed to lung disease (emphysema) and liver disease (neonatal hepatitis). In the absence of AAT, neonatal hepatitis often progresses to cirrhosis.

There are many different genetic phenotypes for the production of AAT; these can be identified by counter immunoelectrophoresis. Elevations of AAT are seen in stress (e.g., acute infection); hence, it is classified as an "acute phase protein." Chronic infections, pregnancy, and estrogen administration can cause elevations of AAT.

Acid-1-Glycoprotein (Orsomucoid)

Acid-1-glycoprotein is elevated as an "acute phase protein" and is also elevated in malignancy. This poorly staining glycoprotein has no known physiological function.

Alpha-2-Macroglobulin

This glycoprotein is the major alpha-2-globulin and is elevated in the serum in nephrotic syndrome. Alpha-2-macroglobulin is very large (MW = 820,000); hence, it is not filtered as rapidly as smaller proteins in nephrosis.

Haptoglobin

This glycoprotein binds extracorpuscular hemoglobin and forms complexes that are catabolized by the reticuloendothelial system. This haptoglobin–hemoglobin complex limits the deposit of hemoglobin in the renal tubules during intravascular hemolysis. Haptoglobin is increased in stress (acute phase protein) and conditions such as chronic infections, other inflammatory diseases, and neoplasms. Haptoglobin is decreased in intravascular hemolysis along with erythrocyte counts. Also bilirubin measurement (direct and indirect), LDH isoenzymes, and direct Coombs tests are also useful for the diagnosis of intravascular hemolysis. Haptoglobin is decreased in liver disease due to reduced synthesis. Low or undetectable levels of haptoglobin are present in 4 to 6% of American Negros due to genetic predisposition. Haptoglobin is increased in a variety of diseases, especially malignancies.

Hemopexin

This protein binds the heme portion of hemoglobin and forms a complex too large to be filtered by the renal glomeruli. Albumin also binds "free" heme (to form methemalbumin). These two complexes function to conserve iron. Hemopexin has been reported to be both decreased and normal in pre-eclampsia as compared to normal pregnancies.

Transferrin (Siderophilin)

This glycoprotein is the major iron-transporting protein in plasma. Transferrin, normally 30% saturated with iron, transports iron from sites of absorption and erythrocyte breakdown to sites of erythropoiesis in the bone marrow. Transferrin is increased in iron-deficiency and in women taking oral contraceptives. Transferrin is decreased in patients with protein-losing conditions (e.g., nephrosis) infections, neoplastic disease, and atransferrinemia, a rare congenital deficiency of transferrin.

Measurements for transferrin are routinely made using radial immunodiffusion or nephelometry. Transferrin may also be measured indirectly by adding enough iron to the serum to saturate transferrin-binding sites, removing the excess iron with a resin, then measuring the serum iron. While this procedure measures the total iron-binding capacity (TIBC), it tends to overestimate transferrin-binding capacity by 10 to 20% because iron binds to proteins other than transferrin.

Beta-Lipoprotein

This lipoprotein, otherwise called low density lipoprotein (LDL), transports mainly cholesterol. It is increased in Fredrickson Type II lipoproteinemia and nephrotic syndrome.

Complement C_3 Component (Beta 1C-Globulin)

C_3 levels are often used as an indication of total complement levels. C_3 levels are low in autoimmune diseases (e.g., systemic lupus erythematosus, rheumatoid arthritis), septicemia, liver disease, and glomerulonephritis (poststreptococcol and membranoproliferative). Serum C_3 levels are elevated in malignancy.

GAMMA PROTEINS OR IMMUNOGLOBULINS

IMMUNOGLOBULINS

These protein molecules migrate primarily in the gamma region and represent the "humoral" component of the immune system. They carry antibody activity and, accordingly, bind specific foreign substances that elicit their formation. These foreign substances are called antigens. Immunoglobulins are produced by B lymphocytes (B cells) and their progeny, plasma cells.

CHEMICAL STRUCTURE OF IMMUNOGLOBULINS

The basic structural unit of immunoglobulins consists of four polypeptide chains held together by disulfide bonds. Two of these peptide chains are called "light chains," kappa (κ) and lambda (λ) type, present normally in a ratio of 2:1, respectively. The other two peptide chain components of immunoglobulin are called "heavy chains;" these are unique for each immunoglobulin class. Since there are five major immunoglobulin classes, specific heavy-chain designations are designated using Greek letters such as Ig-gamma (γ), Ig-alpha (α), Ig-mu (μ), Ig-delta (δ), and Ig-epsilon (ε) for immunoglobulin classes IgG, IgA, IgM, IgD, and IgE, respectively.

The structural arrangement of the four polypeptide chains is shown in Figure 6.12. Using papain as a proteolytic enzyme, the basic immunoglobulin structure is

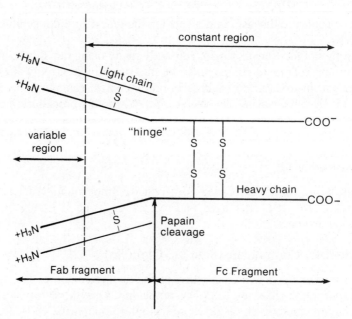

Figure 6.12. Basic structure of immunoglobulins.

broken into three subunits: (1) one Fc fragment of the heavy chain to the "hinge" (see Figure 6.12) and (2) two Fab subunits consisting of a light-chain portion and a heavy-chain portion of the "hinge." The Fd fragment consists of only the heavy-chain component of the Fab moiety. Fab fragments are variable in amino acid makeup and determine the ability of the immunoglobulin to bind specifically to antigens. The amino acid makeup of the Fc portion is fixed for each immunoglobulin class and binds complement.

DIFFERENT CLASSES OF IMMUNOGLOBULINS

Immunoglobulin G (IgG)

This protein has a molecular weight of 150,000 and a sedimentation coefficient of 7S. It exists as one immunoglobulin four-chain monomer (see Figure 6.8) and makes up about 75% of the total serum immunoglobulin concentration. IgG has four major subclasses: IgG1, IgG2, IgG3, and IgG4. IgG has both antibacterial and antiviral activity; however, it has less antibacterial activity than IgM and less antiviral activity than IgA. IgG is the only immunoglobulin that crosses the placenta; hence, at birth, humoral immunity persists for about three months. Gradually the infant synthesizes its own IgG, which reaches 60% of the adult level by 2 years. Normal serum levels of IgG range from 900 to 1500 mg/dL. IgG is elevated in infection, hepatocellular disease, "IgG-type" multiple myeloma, and autoimmune diseases. IgG is decreased in immune deficiency states such as Bruton's agammaglobulinemia.

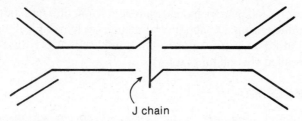

Figure 6.13.   Dimeric structure of secretory IgA depicting the "J chain" link.

Immunoglobulin A (IgA)

This protein is composed of one, two, and sometimes three immunoglobulin monomeric units, with approximate molecular weights ranging from 160,000 (monomer) to 400,000 (dimer) and sedimentation coefficients ranging from 7 to 15S (see Figure 6.13). Approximately 50% of IgA synthesis occurs in the lymphoreticular cells lying under the mucosa of the respiratory and alimentary tracts. In the mucosal cells, monomeric IgA is taken up and converted into "secretory" IgA (a dimer, see Figure 6.13) by the addition of a secretory peptide. These IgA dimers provide local defense against viral and bacterial infections. Normal adult levels of IgA in serum range from 140 to 220 mg/dL. IgA is elevated in infections of the respiratory and GI tracts and in "IgA-type" multiple myeloma; it is decreased in selective IgA deficiency and almost absent in newborns.

Immunoglobulin M (IgM)

IgM is composed of five immunoglobulin monomeric units arranged in a ring structure as shown in Figure 6.14. Because of the high molecular weight of IgM (850,000 g/mol), the sedimentation coefficient is high at 19S. Like IgG, IgM binds antigens (at Fab components) and fixes complement (at Fc components). However, upon antigenic stimulation (i.e., acute infection), IgM levels increase prior to IgG levels. This observation suggests IgM functions as the "first line" of humoral immunity. Later, in the infection, the IgG levels rise in response to antigen stimulation, especially to particulate antigens on the cell surfaces of bacteria. IgM levels in serum increase rapidly after birth and continue to rise the first year with the normal adult range for IgM being 80 to 120 mg/dL. Increases in IgM

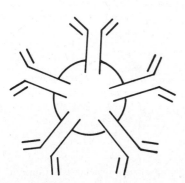

Figure 6.14.   Pentameric "ring" structure of IgM.

are seen in infections and usually indicate a recent (or acute) infection. IgM is also increased in Waldenström's macroglobulinemia, a lymphoproliferative malignancy. In rheumatoid arthritis, an apparent complex between IgM and IgG has been identified (called rheumatoid factor).

Immunoglobulin D (IgD)

This immunoglobulin monomer is normally present in serum only in trace amounts, 0.2% of total immunoglobulin, and has no apparent significant function.

Immunoglobulin E (IgE)

Reagin antibodies or antibodies associated with hypersensitivity of allergic responses are IgE antibodies. The Fc portion of monomeric, antigen-specific IgE binds to specific receptors on mast cells (particularly of the nasopharynx) in sensitized individuals (see Figure 6.15). When the specific antigen is present, the antibody binds the antigen through the Fab portion, an event that stimulates the associated mast cell to release its granules which are rich in vasoactive substances (i.e., histamine, bradykinin, and serotonin).

FORMULAS FOR IMMUNOGLOBULINS

Formulas for immunoglobulins are based on the Greek letters for heavy chains and light chains, with parenthesis used for polymeric structures. For example, the formulas for IgG are $\gamma_2\kappa_2$ and $\gamma_2\lambda_2$, while for pentameric IgM, the formulas are $(\mu_2\kappa_2)_5$ and $(\mu_2\lambda_2)_5$.

MONOCLONAL AND POLYCLONAL GAMMOPATHIES

As mentioned earlier, gamma globulins are produced by plasma cells and precursor B lymphocytes. A clone of plasma cells (or B lymphocytes) is a population of

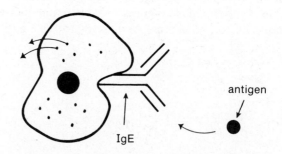

Mast cell with cytoplasmic
granules and allergen specific
IgE bound through Fc receptor.

Figure 6.15. Mast cell (with cytoplasmic granules) combined with allergen-specific IgE. Antigen binding to cell bound IgE stimulates the release of cytoplasmic granules.

cells derived from repeated mitosis from a single lymphocyte. This single clone of cells (or monoclone) produces only one type of antibody (i.e., IgG with kappa light chains; IgA with lamba light chains; etc.). A monoclonal gammopathy (MG) is a proliferative disease (often malignant) where a single clone of cells proliferates and produces a single type of immunoglobulin (often in very large amounts). Since all the immunoglobulin produced is identical molecularly, electrophoresis of the serum specimen reveals a single "spike" in the beta–gamma region. Other gamma globulin production is supressed in MG along with albumin. Albumin synthesis is modulated to maintain colloidal osmotic pressure; hence, when immunoglobulin production is increased, production of albumin is reduced. Monoclonal protein is also referred to as "paraprotein" or as an "M" electrophoretic band. Biclonal and triclonal gammopathies have been reported as well.

Polyclonal gamma globulin increases are a reflection of many different clones of cells producing immunoglobulin in response to antigenic stimulation; therefore, the immunoglobulins vary in molecular structure. A polyclonal gammopathy (e.g., caused by infection, liver disease, etc.) produces a broad, diffuse band on electrophoresis in the beta–gamma region.

Monoclonal spikes on electrophoresis are seen in the following: multiple myeloma, macroglobulinemia (and other malignant lymphoproliferative diseases), Franklin's disease (or heavy-chain disease), and idiopathic or benign "M" spike. To evaluate the significance of the appearance of a monoclonal "M" band in an electrophoresis study, other laboratory tests are needed; these may include: (1) serum immunoelectrophoresis to determine the exact type of immunoglobulin (or fragment thereof) being produced and ascertain the ratio of kappa to lambda light chains; (2) urine protein electrophoresis and immunoelectrophoresis to detect Bence Jones protein; (3) quantitation of IgG, IgA, and IgM to quantitate either increases or suppression in production; and (4) appropriate bone marrow examinations and examination of a lesion biopsy.

Bence Jones protein may escape detection in the urine since most urine protein assays are designed to measure albumin. In the Bradshaw tests, often used to screen for Bence Jones protein, urine is layered over several milliliters of concentrated hydrochloric acid. If Bence Jones protein is present (in significant quantities), a protein precipitate at the interface between the urine and hydrochloric acid layers is formed.

MULTIPLE MYELOMA

Most myelomas involve the excessive production of intact immunoglobulin molecules along with increased production of peptide fragments of immunoglobulins. In light-chain disease, however, more light-chain fragments of immunoglobulins are produced and released, which are readily filtered in the urine (owing to their low molecular weight). These light chains are detected in the urine as Bence Jones protein of Bence Jones myeloma or light-chain disease. Franklin's disease (or heavy-chain disease) is a myeloma involving the overproduction and release of heavy-chain fragments.

Most myelomas are of the IgG and IgA types, and Bence Jones protein is detected in the urine in 60% of all myelomas. Multiple myeloma incidence increases with age and is often associated with anemia, plasmacytosis with replacement of bone marrow with 10% (or more) plasma cells, lytic bone lesions and bone fractures, and a monoclonal protein concentration greater than 3 g/dL, increasing with time.

Kappa and Lambda Light Chains

Free and Bound Light Chains
Specific antisera have been developed to determine if the light chains are "bound" in the intact immunoglobulin or "free." The presence of free light chains indicates that these peptides were released from cells before being bound to heavy chains. An increase in free light chains often indicates a more malignant tumor.

Kappa/Lambda Ratio
The kappa to lambda ratio is normally between 2/1 and 3/2. In other words, two light chains of the same type are bound to a specific heavy-chain class. If the immunoglobulin molecule is formed by many different clones of cells (polyclonal), approximately 60% of the molecules will have kappa chains with the remaining 40% lambda type. However, if one clone of cells predominates (as in a malignant plasma cell tumor where 10^9 cells may develop from one precursor mother cell), the immunoglobulin class and light-chain type will be the same for all immunoglobulins produced. The kappa to lambda ratio will then be markedly elevated or markedly decreased in a monoclonal gammopathy.

Immunosuppression
Another feature of a malignant plasma cell tumor involves the suppression of immunoglobulin classes other than the class being produced in excess. Quantitation by radial immunodiffusion or rate nephelometry can be used to support either suppression or elevations of the various immunoglobulin classes.

WALDENSTRÖM'S MACROGLOBULINEMIA

Proliferating cells resembling B lymphocytes (rather than plasma cells) are seen in this disease, where complete IgM molecules as well as fragments thereof are found in the serum. The increased concentration of IgM in the plasma often causes a marked viscosity increase reducing circulation and encouraging thrombosis. Bence Jones protein is detected in only 10 to 20% of patients with Waldenström's macroglobulinemia. Bleeding symptoms or cerebral ischemia, cold intolerance, hyperviscosity of serum, and heart failure are seen in these patients; treatment often includes plasmapheresis.

AMYLOID DISEASE

Amyloid is a proteinaceous substance deposited throughout the blood vessels and organs. Organs particularly susceptible to amyloid deposits include the liver, kidney, spleen, and adrenal glands. Amyloid is identical to the variable portion of light chains and is possibly produced by plasma cells. Amyloidosis may exist as a primary disease or secondary to another disease involving prolonged stimulation of the immune system (i.e., rheumatoid arthritis, syphilis, etc.). Amyloid is identified on tissue slides by staining with Congo Red.

CRYOGLOBULINEMIA

Cryoglobulins are formed due to polymerization of immunoglobulins. These abnormal globulins precipitate when a serum (or plasma) sample is cooled, then redissolved when the specimen is warmed. Cryoglobulins are rather common in adults beyond 60 years of age. Cryoglobulins are often due to polymers of IgM (most common), IgG, or IgA and hence are frequently seen in myeloma and macroglobulinemia. Cryoglobulins may also be polyclonal and are seen in rheumatoid arthritis, systemic lupus erythematosis (SLE), and other autoimmune diseases. Mixed cryoglobulin complexes deposit in vessel walls, fix complement with the resulting inflammation causing a vasculitis. Renal damage and neurologic disease often follows. Monoclonal cryoglobulin is more commonly associated with Raynaud's phenomenon or vascular purpura.

ASSOCIATED RENAL DISEASE

Increased production of abnormal proteins greatly affects the kidneys due to either (1) the deposition of amyloid in amyloidosis; (2) the vasculitis of cryoglobulinemia; or (3) the formation of urinary casts (due to Bence Jones protein), which block renal tubules.

SERUM ELECTROPHORETIC PATTERNS IN DISEASE

Various and often distinct electrophoretic patterns are seen in diseases affecting protein production, catabolism, or excretion. Changes in globulin fractions (increases or decreases) may be masked by an increase or decrease in another protein migrating in the same fraction; therefore, care and experience are required for proper interpretation of electrophoretic patterns. Figure 6.16 depicts a normal electrophoretic pattern with the major proteins and their relative electrophoretic positions indicated.

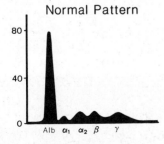

Figure 6.16.   Normal electrophoretic pattern of serum proteins.

STRESS PATTERN

This electrophoretic pattern (Figure 6.17) is seen in inflammation due to trauma (accident or surgery), infections, or acute cellular necrosis (e.g., heart attack) and is sometimes called an "immediate response pattern."

Decreased albumin is noted in the stress pattern with increases in acute phase "globulins" including alpha-1-antitrypsin, orosomucoid, haptoglobin, fibrinogen, and C-reactive protein. A similar pattern may be seen in malignancy.

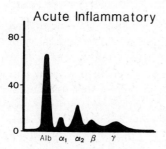

Figure 6.17. Stress protein pattern or immediate response pattern.

DELAYED RESPONSE PATTERN

A delayed response pattern is similar to the stress pattern (Figure 6.18); however, the gamma globulin fraction is increased. This pattern is especially seen in parasitic, fungal, and mycobacterial infections.

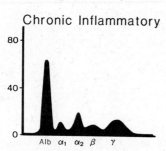

Figure 6.18. Delayed response pattern.

HEPATIC CIRRHOSIS PATTERN

This protein pattern is characterized by a decreased albumin, a polyclonal increase in gamma, and a "beta–gamma" bridge due to the increase in IgA (see Figure 6.19).

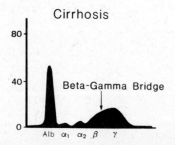

Figure 6.19. A cirrhosis pattern illustrating the "beta–gamma" bridge.

HYPOGAMMAGLOBULINEMIA PATTERN

This pattern shows a decreased gamma fraction due to immunosuppression (due possibly to radiation or chemotherapy), hereditary agammaglobulinemia, infants, and, rarely, light-chain disease (see Figure 6.20).

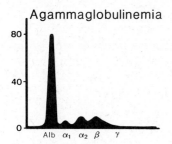

Figure 6.20.   Hypogammaglobulinemia pattern.

MONOCLONAL GAMMOPATHY

This pattern is unique in that a very intense protein "spike" is seen in the beta–gamma region. The exact position of the band depends on the type of immunoglobulin (or fragment thereof) (see Figure 6.21). Albumin is decreased in this pattern as well as other background immunoglobulins.

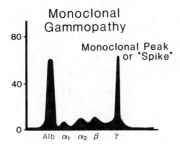

Figure 6.21.   Monoclonal "spike" pattern.

NEPHROTIC SYNDROME PATTERN

This pattern reflects the filtering of lower-molecular-weight proteins across the glomerular membranes, especially albumin. Higher-molecular-weight proteins such as alpha-2 macroglobulin are retained in the plasma (see Figure 6.22). Also,

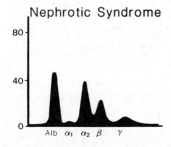

Figure 6.22.   Nephrotic syndrome pattern.

the gamma fraction is reduced. The beta fraction may be elevated owing to increases in low-density lipoprotein often seen in nephrotic syndrome (even though the latter protein stains poorly).

PROTEIN-LOSING ENTEROPATHY PATTERN

This pattern is similar to the nephrotic pattern and is seen in malabsorption. The pattern is characterized by decreased albumin and gamma fractions with an increased alpha-2 fraction (owing to a relative increased alpha-2 macroglobulin). The beta lipoprotein fraction is not increased in this pattern.

OTHER DISEASE PATTERNS

An alpha-1 antitrypsin deficiency pattern appears essentially normal; however, there is no apparent alpha-1 fraction (see Figure 6.23). In hemolytic anemia, the beta fraction will often be reduced because of the reduced haptoglobin. Also, haptoglobin migration relative to other beta globulin is determined by the patients phenotype.

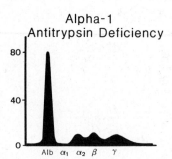

Figure 6.23.   Alpha-1 antitrypsin deficiency.

GLYCOPROTEINS AND MUCOPROTEINS

With the exception of albumin, essentially all proteins have some carbohydrate content. Glycoproteins contain trace amounts of carbohydrates (up to 15% total carbohydrate) with less than 4% as hexosamine; this carbohydrate is chemically bound and can only be removed chemically. On the other hand, mucoproteins contain 10 to 75% carbohydrate with over 4% as hexosamine with loose or labile bonding. Glycoproteins are found in plasma; mucoproteins are found in connective tissue. Mucoproteins are elevated in various diseases, particularly malignant conditions and collagen diseases. Specific immunochemicals are used for the analysis of carbohydrate-containing proteins. Periodic acid-Schiff's reagent can be used to stain protein–carbohydrate moieties on electrophoretic strips. Also, glycoproteins and mucoproteins do not precipitate upon treatment with trichloroacetic acid; hence, the reagent is used to extract glyco- and muco-proteins.

TUMOR MARKERS

Several glycoproteins are classified as tumor markers. For example, carcinoembryonic antigen (CEA) is normally found in fetal intestinal cells, disappears, and may be detected again in the plasma of adult patients with various malignancies, especially colon cancer. Alpha fetoprotein (AFP) may be detected in the plasma of patients with carcinoma of the liver and testicular carcinoma.

IMMUNODIFFUSION METHODS FOR PROTEIN MEASUREMENT

Routine protein electrophoresis is nonspecific in that the dye used to stain the electrophoretic strip stains all proteins. We know that when antigens and antibodies diffuse and react in an agar medium, a visible precipitate can form due to lower solubility of the large antigen–antibody complex. However, optimal concentrations of antigen and antibody are required for the precipitate to form; these concentrations fall into the "equivalence zone" for the antigen–antibody complex. If excess antigen is present (postzone), no precipitate will form. Similarly, if too much antibody is present (prozone), no visible precipitate will form.

Since proteins can serve as antigens in animals, specific antibodies can be recovered from animal serum and used in immunodiffusion experiments. More recently, antibodies have been made by hybridoma techniques and by genetic engineering. There are two types of diffusion studies.

SINGLE DIFFUSION (RADIAL IMMUNODIFFUSION) (RID)

The antibody is incorporated into the agar gel and antigen is allowed to diffuse radially, decreasing its concentration as it diffuses until the "equivalence zone" is reached. A circle forms with the radius proportional to the concentration of the diffusing protein. With standards, the analyst can determine the concentration of protein in the patient's samples. This procedure is called radial immunodiffusion.

DOUBLE DIFFUSION

Contrasting single diffusion where essentially only the antigen migrates, double diffusion allows migration of both antigen and antibody. In the Ouchterlony technique, antigen and antibody are placed in separate wells of an agar plate. Three types of precipitin lines are formed, as shown in Figure 6.24. Reacting antigen or antibody cannot cross the precipitin barrier; however, unrelated (or nonreacting) antigen and antibody can readily cross the precipitin band.

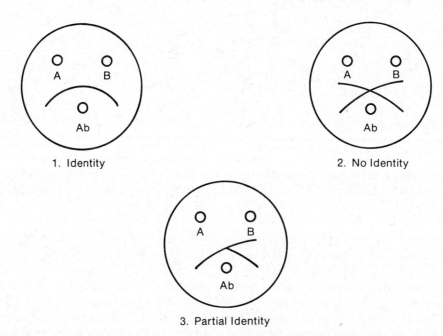

1. Identity   2. No Identity

3. Partial Identity

Figure 6.24. Ouchterlony technique employing double diffusion (1) A and B protein antigens in both wells are equal immunologically; (2) antigens are not equal; and (3) protein antigens A and B are not equal but do show partial identity. Precipitin acts as a barrier to diffusing antibody.

IMMUNOELECTROPHORESIS (IEP)

Another double-diffusion method of studying proteins involves immunoelectrophoresis that (1) employs electrophoresis to separate the various proteins and (2) immunoprecipitation for identification and semiquantitation. The procedure for IEP includes the following: (1) addition of the protein sample and control to wells on the agarose film; (2) electrophoretic separation of the proteins; (3) addition of specific antisera to antibody troughs between sample and control wells; and (4) incubation to allow double diffusion and precipitin arc formation for each protein (see Figure 6.25 for sample agarose film).

Precipitin arcs formed in IEP provide semiquantative information; increased protein concentration (compared to the control) is indicated by (1) a wider and darker precipitin arc; (2) a longer precipitin arc; and (3) a precipitin arc closer to the antibody trough.

INTERPRETATION OF LABORATORY DATA

The following examples will concentrate on the interpretation of electrophoresis data. (Several cases presented have been adapted from lectures by Richard B. Passey, Ph.D., University of Oklahoma, Department of Pathology.)

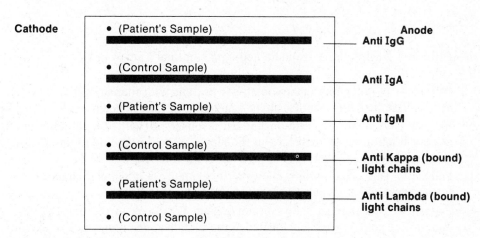

Figure 6.25. Agarose electrophoretic plate used for immunoelectrophoresis with alternating sample, controls and antibodies, for major immunoglobulins and light chains.

EXAMPLE 6.1.   Shown in Figure 6.26 are actual protein patterns observed in 10 different specimens. Match the protein pattern with the appropriate description. Serum samples are used unless otherwise specified.

_____	a.	Normal serum protein pattern
_____	b.	Chronic inflammation (or infection)
_____	c.	Nephrotic syndrome
_____	d.	Bence Jones protein (urine)
_____	e.	Waldenström's macroglobulinemia
_____	f.	Hypoproteinemia
_____	g.	Spinal fluid protein
_____	h.	Multiple myeloma (IgG type)
_____	i.	Glomerulonephritis (urine)
_____	j.	Alpha-1 antitrypsin deficiency

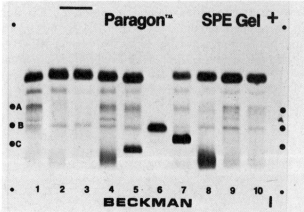

Figure 6.26

Answer:

a. Pattern 9. All fractions are identified with "normal" amounts of each being indicated.

b. Pattern 4. A polyclonal gammopathy is present with increases observed in acute phase proteins, that is, alpha-1 and alpha-2 globulins.

c. Pattern 1. A nephrotic protein pattern is observed with low albumin and gamma fractions with increases in the alpha-2 fraction (owing to an alpha-2 macroglobulin increase).

d. Pattern 6. This urine pattern shows a very intense monoclonal band due to the presence of kappa light chains. This is an example of Bence Jones protein.

e. Pattern 7. This pattern shows a monoclonal IgM peak in the beta–gamma region. The peak is precisely located at the sample application point. Waldenström's macroglobulinemia is a lymphoproliferative disorder associated with a IgM monoclonal peak. (Also note the suppression of other immunoglobulins, especially IgG. The albumin peak is also decreased.)

f. Pattern 10. All fractions are present, but in low concentrations.

g. Pattern 3. The presence of prealbumin, albumin, and beta globulins and the absence of other fractions suggests a spinal fluid pattern.

h. Pattern 5. This pattern shows a monoclonal band in the gamma region consistent with an IgG multiple myeloma. Suppression of other immunoglobulins is noted as well.

i. Pattern 2. This urine pattern shows no alpha-2 macroglobulin; however, the presence of immunoglobulins indicates glomerular loss of high-molecular-weight proteins. This observation is consistent with more extensive glomerulonephritis.

j. Pattern 8. No alpha-1 fraction is observed suggesting an alpha-1 antitrypsin deficiency.

EXAMPLE 6.2.   Case History: A 43-year-old white male was admitted to the hospital to be evaluated for persistent low-back pain. About five months previous to this admission, the patient had lifted a cabinet and a pain developed across his lower back. A few weeks later, the patient again lifted a heavy object and again experienced pain, this time in the upper back. Both of these pains have continued to the present time. No numbness was present. X-rays revealed severe osteoporosis with peculiar spotty demineralization in the pelvis and femur areas. The patient noted that the frequency of urination had increased over the past few months.

Laboratory data:

	BUN (7–21)[a] mg/dL	Creatinine (0.6–1.3) mg/dL	Uric Acid (2.6–7.0) mg/dL	Calcium (8.6–10.2) mg/dL	Phos. (2.6–4.5) mg/dL	Total Protein (6.0–7.4) g/dL	Alb. (3.6–4.9) g/dL	Glob. (1.6–3.5) g/dL
Admission	26		8.2	12.5	5.0	6.5	4.2	2.4
5 days later	39	2.9	9.2	13.8	5.6	6.7	4.4	2.3

[a] Reference ranges for components

Other laboratory data:

24-hour urine protein (normal, under 150 mg/24 hr) = 28,851 mg/24 hr
Creatinine clearance (normal, 75–140 mL/min) = 33 mL/min
CBC—revealed anemia with an elevated MCV.

Discussion: Myeloma was thought unlikely because the total protein and globulin were within normal limits. Vitamin B12, folate, and Fe/TIBC were ordered to evaluate the possible etiology of the anemia. The results of these tests were all within normal limits. Serum PTH was within normal limits. A bone marrow was performed and revealed 67% plasma cells compatible with the diagnosis of multiple myeloma.

A serum protein electrophoresis was performed and a hypogammaglobulinemia pattern was obtained. No monoclonal peak was observed. This pattern can be seen in light-chain myeloma, where the low-molecular-weight light-chain proteins are excreted and normal immunoglobulin production is suppressed.

A protein electrophoresis study was performed on urine along with an immunoelectrophoresis on urine. The resulting data are shown in Figure 6.27. Interpret the patterns.

Answer: The electrophoresis tracing shows a strong monoclonal peak in the urine consistent with Bence Jones protein. Bence Jones protein is often seen in myeloma (of all types) and is a bad prognostic sign. In this case, the patient has a multiple myeloma, or "Light-Chain Disease," kappa type.

EXAMPLE 6.3. The data presented in Figure 6.28 were obtained on a patient as part of a myeloma workup. Interpret the data.

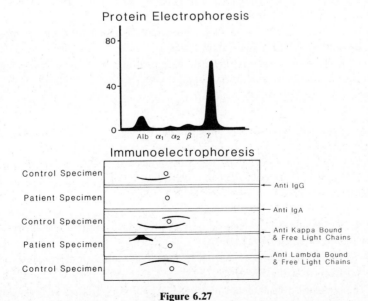

Figure 6.27

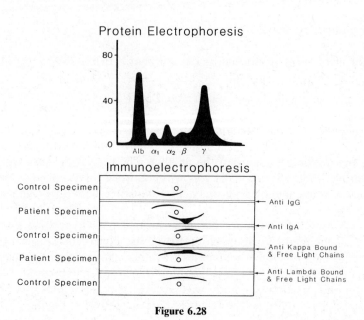

Figure 6.28

Answer: The serum protein electrophoresis shows a monoclonal band in the beta–gamma region. The immunoelectrophoresis pattern is consistent with an IgA monoclonal protein with kappa light chains. Also the kappa/lambda ratio is abnormally increased. All of these observations are suggestive of multiple myeloma. A bone marrow examination should be performed to check for plasmacytosis.

BIBLIOGRAPHY

Bakerman, S. *Lecture Notes from Clinical Chemistry Review at East Carolina University.* Greenville, NC: East Carolina University, 1977.

Electrophoresis Reference Chart, Serum Associated Protein Wall Chart, Product Literature. Beaumont, TX: Helena Laboratories.

Fudenberg, H. H., et al. *Basic and Clinical Immunology.* Los Altos, CA: Lange, 1978.

Golais, T. *Electrophoresis Manual.* Beaumont, TX: Helena Laboratories, 1974.

Henry, J. B. (ed.). *Clinical Diagnosis and Management,* 2nd ed. Philadelphia: Saunders, 1979.

Kyle, R. A. "The diagnosis of monoclonal gammopathies," *Clinical Immunology Newsletter* 1(6), 1 (1980).

Laurell, C. B. "Electrophoresis, specific protein assays, or both in the measurement of plasma proteins," *Clinical Chemistry* 19(1), 99 (1973).

Peck, J. M. "Laboratory tip: Application artifact in electrophoresis," *American Journal of Medical Technology* 44(2), 20 (1978).

Pribor, H. C. and Pugliese, X. "New laboratory procedures for multiple sclerosis," *Laboratory management* (May 27, 1980).

Ritzmann, S. E. and Daniels, J. C. *Serum Protein Abnormalities, Diagnostic and Clinical Aspects,* Boston: Little Brown, 1975.

Sun, T. "The present status of electrophoresis," *Laboratory Management* 43 (June, 1979).

Teitz, N. (ed.). *Fundamentals of Clinical Chemistry,* 2nd ed. Philadelphia: Saunders, 1976.

Zilva, J. F. and Pannall, R. P. *Clinical Chemistry in Diagnosis and Treatment.* Chicago: Year Book, 1979.

POST-TEST

1. The element present in all proteins that allows differentiation of proteins from carbohydrates and lipids is:
   a. sulfur
   b. nitrogen
   c. phosphorus
   d. hydrogen

2. Amino acids that cannot be synthesized in the body are known as _____ amino acids.
   a. dietary
   b. fundamental
   c. basic
   d. essential

3. Amino acids can act as buffers. The group on the α carbon that can act as a proton donor is the _____ group. The group on the α carbon that can act as a proton acceptor is the _____ group.
   a. carboxyl
   b. amino

4. Because of amphoteric properties, the charge on the protein molecule is dependent on pH. Indicate in the first blank, the charge on the protein molecule at the pH indicated; indicate in the second blank the direction of movement in an electrical field.
   1. _____ _____ pH above isoelectric point
   2. _____ _____ pH below isoelectric point
   3. _____ _____ pH equal to isoelectric point

5. The chemical linkage of amino acids that forms peptide bonds occurs between:
   a. amino and R groups
   b. two R groups
   c. amino and carboxylic groups
   d. carboxylic and R groups

6. Proteins acquire different shapes and forms due to bonding that occurs within the molecule or between polypeptide chains. Match the structure with the bonding mechanism.

   _____ 1. Primary structure
   _____ 2. Secondary structure
   _____ 3. Tertiary structure
   _____ 4. Quaternary structure

   a. interaction of R groups in a single polypeptide chain
   b. hydrogen bonding in a single polypeptide chain
   c. peptide bonds
   d. interaction of two or more polypeptide chains

7. When a protein is denatured, the disruption of bonding occurs in the
    _____ structure.
    a. primary
    b. secondary
    c. tertiary

8. The "Sia" test, a screening test to demonstrate elevated immunoglobulin
    levels, is based on the solubility properties of proteins. This test is based on
    the insolubility of immunoglobulins in:
    a. weak salt solutions
    b. 50% saturated salt solutions
    c. pure water
    d. saturated salt solutions

9. In the blank following each step in the protein digestion pathway, indicate
    the enzyme/s necessary for cleavage of the protein molecule.

    Food   a. trypsin
      ↓   b. dipeptidases
    Stomach   1. _____   c. carboxypeptidases
      ↓   d. pepsin
    Upper small   e. chymotrypsin
                   2. _____
    intestine
      ↓
    Small intestine 3. _____
      ↓
    Absorption

10. Negative nitrogen balance may be associated with all of the following
    except:
    a. starvation/malnutrition
    b. following surgery
    c. high protein diets
    d. following severe burns
    e. chronic blood loss

11. Using the Kjeldahl method, the nitrogen content of a sample was determined
    to be 1.2 g/dL. Assuming 16% nitrogen content, the total protein would be
    reported as _____ .

12. The smallest peptide chain required for the biuret reaction is:
    a. dipeptide
    b. tripeptide
    c. tetrapeptide
    d. hexapeptide

13. Match the procedure with test principle:
    _____ 1. Biuret   a. measures nitrogen present
    _____ 2. Phenol method   b. intensity of color depends on
                    (Folin–Ciocalteu)   the amount of tyrosine and
    _____ 3. TCA, SSA   tryptophane present

_____ 4. Kjeldahl   c. color based on the number of
_____ 5. Ninhydrin   peptide bonds present
_____ 6. Bromcresol Green   d. reacts with free amino groups
                                            e. colorimetric procedure based on albumin–dye complex formation
                                            f. denatures the protein and causes precipitation

14. Calculate the A/G ratio. Given: Total protein = 7.5 g/dL; Albumin = 4.0 g/dL. The A/G ratio is _____ .

15. The upper limit of normal for CSF protein is generally considered to be _____ mg/dL.
    a. 45
    b. 30
    c. 90
    d. 15

16. Match the disease with the expected serum protein findings. Two answers are required, one from each category. One has three answers.

_____ 1. Renal disease associated with glomerular damage   a. increased total protein
                                            b. decreased total protein
_____ 2. Hepatic disease
_____ 3. Multiple myeloma   c. decreased albumin
_____ 4. Malnutrition   d. increased globulin
_____ 5. Waldenström's

17. For protein electrophoresis in a clinical setting, a five-band separation is routinely performed. The electrophoretic support media most commonly used is:
    a. cellulose acetate
    b. agarose gel
    c. starch gel
    d. polyacrylamide gel

18. The site(s) of synthesis for albumin is _____ ; for prealbumin is _____ ; for alpha and beta globulins is _____ ; for immunoglobulins is _____ .
    a. liver
    b. reticuloendothelial system

19. Match the problem with the probable cause as it relates to a protein electrophoresis. There may be one or more than one answer.

_____ 1. Bands too tight, not well separated   a. wick effect
                                            b. endo-osmotic effect
_____ 2. Bands too diffuse   c. application technique
_____ 3. Proteins did not migrate   d. pH of buffer incorrect
                                            e. ionic strength of buffer too high
_____ 4. Gamma swept to cathode at pH 8.6
_____ 5. Bands on edges fuzzy   f. ionic strength of buffer weak

———————— 6. Underestimation of
protein by dye-binding

g. band too dense; amino
groups not exposed

———————— 7. Artifact in beta–gamma
region

20. The reference method for protein electrophoresis is ————————.
    a. biuret
    b. zone electrophoresis
    c. isoelectric focusing
    d. Tiselius moving boundary

21. Below is a five-band electrophoretic pattern as performed on cellulose ace-
    tate at pH 8.6. Each band is designated by number. Indicate the protein(s)
    that migrate in each area and whose serum concentration is of sufficient
    nature to alter the electrophoretic pattern in disease states. When more than
    one protein migrates in each band, designate each in sequence of migration
    from anode to cathode.

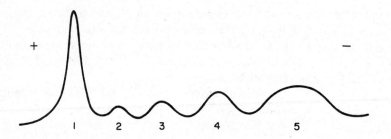

22. Proteins that are members of the "acute phase reactants" include all except:
    a. haptoglobin
    b. alpha-1-antitrypsin
    c. alpha-1-glycoprotein
    d. alpha-2-macroglobulin
    e. C-reactive protein
    f. fibrinogen
    g. complement C3

23. The functions of albumin include all except:
    a. transport functions
    b. regulates osmotic pressure
    c. reserve store for amino acids
    d. associated with immune response

24. Match the expected findings with the electrophoretic pattern.
    ———————— 1. Acute stress pattern
    ———————— 2. Alcoholic cirrhosis
    ———————— 3. Monoclonal gammopathy
    ———————— 4. Nephrotic syndrome
    ———————— 5. Normal
    ———————— 6. Agammaglobulinemia
    ———————— 7. α-1-Antitrypsin deficiency

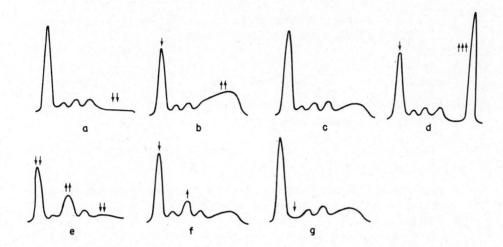

25. Using cellulose acetate as the support media, the migration of a protein during electrophoresis is most dependent on:
    a. shape
    b. size
    c. charge
    d. state of hydration

26. In reference to the immunoglobulin molecule, match the following.
    _____ 1. Number of heavy chains   a. one
    _____ 2. Types of heavy chains   b. two
    _____ 3. Number of light chains   c. three
    _____ 4. Types of light chains   d. four
    _____ 5. Following papain cleavage,   e. five
                 number of fragments   f. ten
    _____ 6. Number of different possible   g. twenty
                 immunoglobulins

27. Match the following. Each may have one or more answers.
    _____ 1. IgG   a. 19S
    _____ 2. IgA (secretory)   b. activates and fixes complement
    _____ 3. IgM   c. monomer
    _____ 4. IgE   d. dimer
    _____ 5. IgD   e. pentamer
                                    f. found on mucosal surfaces
                                    g. allergic or reagin
                                    h. offers first line of defense upon
                                       exposure to antigen
                                    i. crosses placenta
                                    j. present in normal plasma in
                                       greatest concentration
                                    k. delta heavy chain
                                    l. associated with Waldenström's

28. The portion of the immunoglobulin molecule that binds antigen is
    _____ ; the portion that binds complement is _____ .
    a. Fab
    b. Fc
    c. Fd

29. Bence Jones protein may be composed of:
    a. free kappa or lambda chains
    b. free, G, A, or M chains
    c. Fab fragments
    d. Fc fragments

30. A serum protein electrophoresis pattern may reveal a monoclonal spike in
    association with all of the following except:
    a. septicemia
    b. multiple myeloma
    c. lymphoma
    d. Franklin's heavy-chain disease
    e. Waldenström's

31. An individual was diagnosed as IgG myeloma. Indicate what would be the
    expected laboratory findings for the tests listed below. Use ↑ (increased); ↓
    (decreased); → (normal).
    _____   1. Total protein
    _____   2. Albumin
    _____   3. Globulin
    _____   4. Serum calcium
    _____   5. IgG
    _____   6. IgA
    _____   7. IgM

32. A technologist was evaluating results from a chemistry profile. It was noted
    that, among the abnormal results, both the total protein and albumin were
    increased. What can one deduce from these results?

33. In light-chain disease, the test that would be most diagnostic is:
    a. serum protein electrophoresis
    b. urine protein electrophoresis
    c. Bradshaw test
    d. serum IEP

34. A patient was diagnosed as light-chain disease. The results from a routine
    urinalysis revealed only 100 mg/dL protein present (dip-stix method). The
    quantitative protein using a SSA precipitation method revealed 10 g/24 hr.
    No error was made in either analysis. What is the reason for the discrepancy?

35. Match the test with test principle. Two answers are required, one from each
    category.
    _____   1. Double diffusion   a. both antigen and anti-
    _____   2. Single diffusion   body diffuse into a gel
    _____   3. Immunoelectrophoresis   from wells
                                               b. electrophoresis followed
                                                  by diffusion

    c.   one reactant (usually antigen) diffuses into a gel containing other reactant (usually antibody)

    d.   Ouchterlony technique
    e.   RID
    f.   IEP

36. A double-diffusion technique was used to identify antigens. The two upper wells contained antigens. The lower well contained antibody or antibodies. Beneath each pattern, interpret results. Use terms: pattern of (a) identity, (b) partial identity, or (c) nonidentity.

      1   2   3

37. The stain used to visualize carbohydrate is:
    a. Ponceau S
    b. Periodic-acid Schiff
    c. Oil-red-O
    d. Congo red

38. CEA as a tumor marker is used to monitor response to treatment in:
    a. breast cancer
    b. prostatic cancer
    c. lung cancer
    d. colorectal cancer

39. Alpha fetoprotein is a tumor marker for (two answers required):
    a. hepatocellular cancer
    b. testicular teratocarcinoma
    c. medullary carcinoma
    d. colorectal cancer

ANSWERS

1. b
2. d
3. a; b
4. (1) negative, anode; (2) positive, cathode; (3) doubly charged, will not migrate
5. c
6. (1) c; (2) b; (3) a; (4) d

7. c
8. c
9. (1) d; (2) a, c, e; (3) b
10. c
11. 1.2 g/dL × 6.25 = 7.5 g/dL
12. b
13. (1) c; (2) b; (3) f; (4) a; (5) d; (6) e
14. 1.1
15. a
16. (1) b, c; (2) b, c; (3) a, c, d; (4) b, c; (5) a, d
17. a
18. a; a; a; b
19. (1) e, d; (2) f; (3) d; (4) b; (5) a; (6) g; (7) c
20. d
21. (1) albumin; (2) alpha-1-antitrypsin, acid-1-glycoprotein; (3) α-2-macroglobu-lin, haptoglobin; (4) hemopexin, transferrin, beta lipoprotein, complement C3; (5) IgA, IgM, IgG
22. d
23. d
24. (1) f; (2) b; (3) d; (4) e; (5) c; (6) a; (7) g
25. c
26. (1) b; (2) e; (3) b; (4) b; (5) c; (6) f
27. (1) b, c, i, j; (2) d, f; (3) a, b, e, h, l; (4) c, g; (5) c, k
28. a, b
29. a
30. a
31. (1) increased; (2) decreased; (3) increased; (4) increased; (5) increased; (6) decreased; (7) decreased
32. The patient was probably dehydrated at the time of specimen collection. The profile should be repeated on a new specimen.
33. b
34. The dip-stix methodology is sensitive to albumin with little sensitivity to globulins. Dip-stix procedures will underestimate or give false negative results when the protein present is a globulin as in Bence Jones myeloma.
35. (1) a, d; (2) c, e; (3) b, f
36. (1) b; (2) c; (3) a
37. b
38. d
39. a, b

CHAPTER 7
LIPIDS

OBJECTIVES

On completion of this chapter, you will be able to:

1. Define, illustrate, or explain the following terms:
    a. high-density lipoprotein
    b. lipoprotein

    c. free fatty acid

    d. lactescent

    e. chylomicrons

    f. prebeta lipoprotein

    g. triglyceride

    h. sphingolipids

    i. L/S ratio

    j. sedimentation coefficients

2. Discuss the chemical and structural composition of lipid classes.

3. Describe the various methods used to measure lipids.

4. List and characterize the four major types of lipoproteins.

5. Discuss the six major classifications of hyperlipoproteinemias and describe experimental data used to make each classification.

6. Describe the methods used to measure lipoproteins and lipoprotein cholesterol content.

7. Discuss the use of the LDL-cholesterol/HDL-cholesterol ratio for the assessment of cardiovascular disease risk.

8. Discuss the major steps of lipid metabolism.

9. Describe the lipids and associated lipoproteins primarily associated with coronary atherosclerosis and possible mechanisms for the formation of atheromatous lesions in arteries.

10. List and describe the different types of lipids and diseases associated with abnormally increased levels of each.

11. List normal blood levels of major lipids.

GLOSSARY OF TERMS

Abetalipoproteinemia. A missing apolipoprotein B (apo-B) inhibits the formation of VLDL, LDL, and chylomicrons. Steatorrhea (fatty stools) is observed and malabsorption of fat-soluble vitamins occurs. Neurological symptoms and retinal degeneration follow later in the disease.

Apolipoproteins. The proteins found in the structure of the various lipoprotein micelles. The apolipoprotein groups are apolipoproteins A, B, C, D, and E. These apolipoproteins are presently quantitated by enzyme immunoassay (EIA) and radioimmunoassay (RIA).

Atherosclerosis. A disease associated with narrowing of the lumen of arteries due, in part, to lipid deposition (mostly cholesterol).

CAD. Coronary atherosclerosis or coronary artery disease.

CHD. Coronary heart disease.

Chylomicrons. A lipoprotein fraction containing mostly exogenous triglycerides. They are the least dense of the lipoproteins and float in serum upon storage at 4°C. They remain at the origin in most electrophoretic separations.

Friedewald's formula. The formula used to approximate the concentration of LDL-cholesterol:

$$\text{LDL-cholesterol} = \text{total cholesterol} - \left(\frac{1}{5}\,\text{triglyceride} + \text{HDL-cholesterol}\right)$$

The above relationship is based on the observation that the cholesterol content of the VLDL moiety can be estimated by dividing the total triglyceride by 5 (provided that the total triglyceride is less than 400 mg/dL and that no chylomicrons are present in the serum or plasma).

High-density lipoproteins (HDL). A lipoprotein fraction sometimes called alpha lipoprotein due to its migration in electrophoresis. It has the highest density of the four major lipoproteins. The lipid portion of HDL is mostly made up of cholesterol and phospholipids in a ratio of 1:2.

Hyperlipidemia. An elevation of one or more of the plasma lipids (usually triglyceride and/or cholesterol).

Hyperlipoproteinemia. An increase in one or more of the lipoproteins.

Ketosis. Formation of keto acids due to an accumulation of acetyl-coenzyme A from excessive fatty acid metabolism.

Lactescence. A milky appearance in serum or plasma due to the elevation of triglyceride.

Lipemic. A term used to describe the milky appearance of plasma due to the elevated triglyceride content.

Lipogenesis. Formation of fatty acids from intermediates of glucose metabolism (i.e., pyruvate and acetyl-coenzyme A).

Lipolysis. The breakdown of triglyceride to fatty acid and glycerol in the adipose tissue.

Lipoprotein electrophoresis. A generally convenient way of separating serum lipoproteins by their differential migration under the influence of an electrical field in carefully selected buffers and media.

Lipoproteins. A combination of lipid and protein in a spherical structure suitable for the transport of lipid in the plasma.

Lipoprotein ultracentrifugation (preparative). A lengthy and expensive technique for separating the various plasma lipoproteins. This method employs high-speed centrifugation and the addition of salt solutions for appropriate density gradients.

Low-density lipoproteins (LDL). A lipoprotein fraction often called beta lipoprotein because of its migration in electrophoresis. The lipid portion of this micelle contains mostly cholesterol and cholesterol esters along with some phospholipids and triglyceride. LDL and VLDL are both involved in the formation of the atheromatous lesions.

Polyanion precipitation. A convenient method of separating lipoproteins containing apolipoprotein B by selective precipitation using sulfated polyanions. This method is relatively simple, accurate, and inexpensive, and is routinely used in the clinical laboratory for lipoprotein fractionation.

Very-low-density lipoproteins (VLDL). A lipoprotein fraction commonly referred to as prebeta lipoprotein because of its characteristic migration in electrophoresis. This lipid fraction consists mostly of endogenous triglycerides.

Lipids play major roles in (1) the structure and composition of cell membranes; (2) intracellular storage depot of metabolic fuel; (3) transport form of metabolic fuel; and (4) the cushioning and support of vital body organs. Also, cholesterol, a lipid, serves as a precursor for some hormones. In clinical chemistry, an elevated level of lipids (especially cholesterol) in the blood is related to diseases such as atherosclerosis, where fatty lipids are deposited in the internal layers of arterial walls ultimately restricting the flow of blood. Lipid-storage diseases, especially in the newborn, and fetal lung maturity can be assessed by measurements of lipid levels.

ELEMENTAL COMPOSITION

Lipids are organic compounds that contain the elements carbon and hydrogen with a lesser amount of oxygen. Like most nonpolar organic compounds, lipids are generally insoluble in aqueous solvents and very soluble in organic solvents like ether and benzene. In human biochemistry, major compounds classified as lipids include sterols, fatty acids, triglycerides, and phosphatides. Fat-soluble vitamins (A, D, E, and K) and bile pigments are also classified as lipids.

FATTY ACIDS

FATTY ACID STRUCTURE

Fatty acids are organic acids and, as discussed in Chapter 1, most organic acids are carboxylic acids containing the (—COOH) functional group:

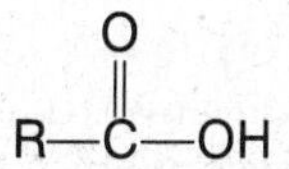

$$R-\overset{\overset{\displaystyle O}{\|}}{C}-OH$$

The R chain portion consists of repeating (—CH$_2$—) or (—CH—) units with a (CH$_3$—) grouping on the end as shown in Figure 7.1.

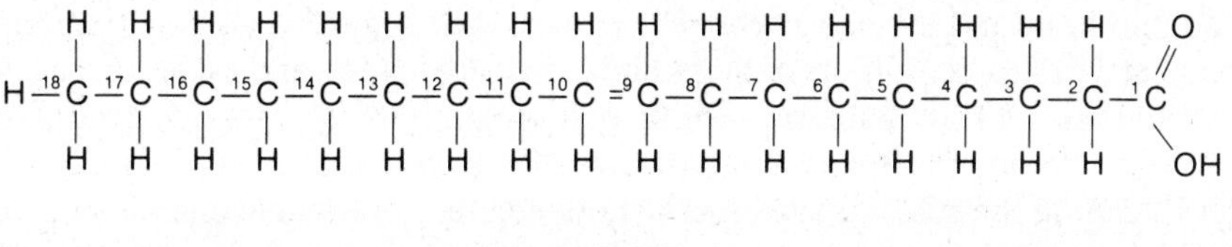

Oleic Acid

Figure 7.1. Chemical structure of a fatty acid showing one double bound (or unsaturated bond) at carbon number 9.

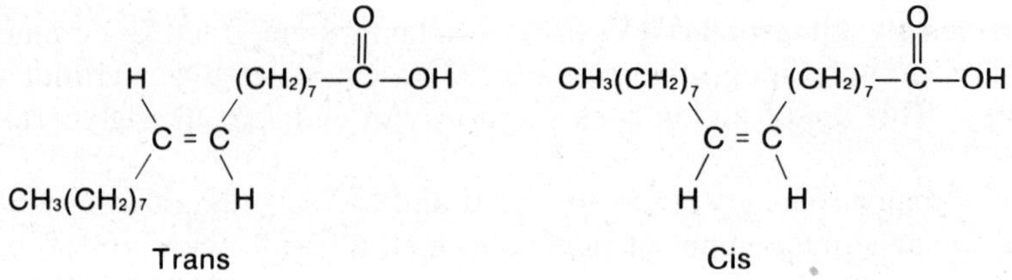

Figure 7.2. Cis and trans configuration of a fatty acid.

CIS AND TRANS ISOMERS

Recalling that compounds are isomers when they have the same formula and different structures, fatty acids have "cis" and "trans" isomers because of restricted rotation around carbon–carbon double bonds. In Figure 7.2, cis and trans isomers of oleic acid are illustrated. Double bonds of naturally occurring fatty acids are of the cis configuration.

CHARACTERISTICS OF FATTY ACIDS

Important fatty acids in humans include those that are synthesized by human cells and those that are not; this latter group of fatty acids must be included in the diet (e.g., linoleic and linolenic acids). In general, fatty acids in human metabolism and nutrition are monocarboxylic, straight-chain compounds (as illustrated in Figure 7.1) containing an even number of carbon atoms. This latter observation is not surprising since cellular production of fatty acids involves coupling reactions of two-carbon acetyl units.

CARBON CHAIN LENGTHS AND SATURATION

Human fatty acids have carbon chain lengths of 4–24, while dietary fatty acid chains generally have 16–18 carbons. Fatty acids may be saturated (no carbon–carbon double bonds), monounsaturated (one carbon–carbon double bond), or polyunsaturated (more than one carbon–carbon double bond). The presence of double bonds in fatty acids lowers the melting points of fats and, since most vegetable fats are polyunsaturated, they exist as oils.

Different fatty acid structures can be described by a structure code. For example, oleic acid (code $18:1^9$) indicates *18* carbons with *1* double bond; the *9* signifies the double bond is between carbons 9 and 10 (see Figure 7.1). Some important human fatty acids (common names) and structure codes are as follows: lauric, $12:0$; myristic, $14:0$; palmitic, $16:0$; palmitoleic, $16:1^9$; stearic, $18:0$; oleic, $18:1^9$: γ-lenolenic, $18:3^{6,9,12}$; arachidic, $18:3^{6,9,12}$; behenic, $22:0$; and lignoceric, $24:0$. Recalling the rules of systematic nomenclature of organic chemistry, names can be based on the number of carbon atoms with preceding numbers to denote double bond positions and an "oic" suffix (e.g., palmitoleic, $16:1^9$ is 9-hexadecanoic).

CLASSIFICATION OF LIPIDS

ESTERIFICATION

While fatty acids are, by themselves, classified as lipids, they serve as components of other lipids to form more complex molecular lipid compounds. Fatty acids are combined with other compounds in a chemical reaction called "esterification." Carboxylic acids and alcohols combine to form esters in this same reaction (see Figure 7.3). Note that H_2O is also a product of this reaction.

CLASSES OF LIPIDS

Four classes of lipids containing fatty acids as a component (bound by an ester linkage) are as follows: (1) cholesterol esters; (2) triglyceride; (3) glycerol phosphatides; and (4) sphingolipids. Cholesterol also can exist in an unesterified form (to be discussed below). These four groups of lipids are transported in the blood as components of lipoproteins.

LIPOPROTEINS

Lipoproteins are combinations of lipids and proteins that form a spherical micelle structure; this micelle structure facilitates transport of insoluble lipids in the plasma. (These will be discussed in greater detail in the chapter.) There are four major lipoproteins based on density (and size) differences: chylomicrons, very-low-density lipoprotein (VLDL), low-density lipoprotein (LDL), and high-density lipoprotein (HDL).

CHOLESTEROL

STRUCTURE

The chemical structure for cholesterol is shown in Figure 7.4. Cholesterol is a macromolecular alcohol (or sterol) that forms esters with fatty acids through the hydroxyl grouping located on carbon number 3. About two-thirds of the total cholesterol in plasma is "esterified," with the remaining fraction called unesterified cholesterol or "free" cholesterol. In cells, the unesterified cholesterol is a major component of cell membranes.

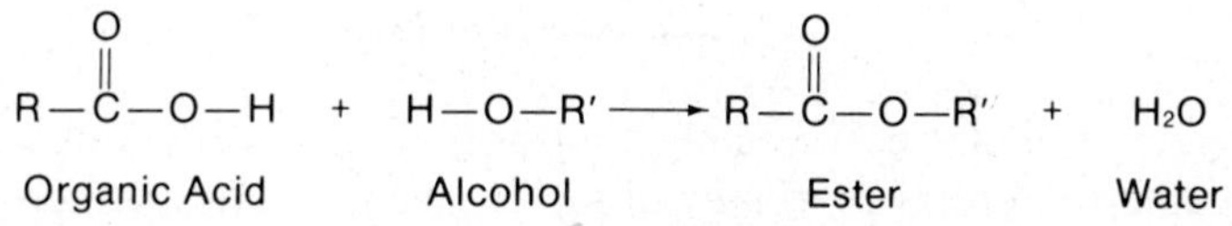

$$R-\overset{\overset{\displaystyle O}{\|}}{C}-O-H \;+\; H-O-R' \longrightarrow R-\overset{\overset{\displaystyle O}{\|}}{C}-O-R' \;+\; H_2O$$

Organic Acid   Alcohol   Ester   Water

Figure 7.3.   Formation of an ester from the reaction of a carboxylic acid and alcohol.

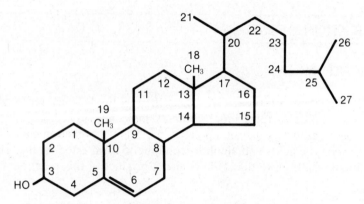

Figure 7.4. Structure and atom numbering of cholesterol.

CHOLESTEROL SYNTHESIS

Cholesterol can be synthesized by all cells in the body except erythrocytes, with cells in the liver and the mucosa of the small intestines producing significant quantities of cholesterol. Esterification of cholesterol with fatty acids also takes place in the liver and small intestines. Endogenous production of cholesterol requires acetyl coenzyme A. Therefore, amino acids, carbohydrates, and fatty acids, supplied in excess of metabolic needs, can contribute to the total cholesterol pool by furnishing acetyl coenzyme A.

CHOLESTEROL EXCRETION

Cholesterol is metabolized in the liver to bile salts by the conversion of the aliphatic cholesterol side chain to a carboxylic acid side chain (—COOH at carbon number 24) together with the addition of hydroxyl group(s) to the ring structure. Subsequently, amide linkage of the exocyclic carboxyl group with glycine or taurine results in the formation of conjugated bile salts. Because of similarity in structure, these bile salts associate with unmetabolized cholesterol and thus further facilitate the excretion of cholesterol in the bile. In the gall bladder, cholesterol often precipitates to form gall stones, a process enhanced by local infection. Unfortunately for people with elevated plasma cholesterol levels, excreted cholesterol in the form of bile is reabsorbed in the intestine and returned to the liver via the enterohepatic circulation. Also bile salts are essential for the normal intestinal absorption of fat-soluble vitamins and lipids.

CHOLESTEROL AND ATHEROSCLEROSIS

Elevated levels of cholesterol (mainly transported as LDL) are associated with accelerated atherosclerosis especially of the coronary arteries. Atherosclerosis is a disease where lipid containing deposits build up in arteries and eventually inhibit normal circulation. The disease occurs most often in adult men and postmeno-pausal females and leads to acute myocardial infarction (AMI).

Proposed Mechanism of Atherosclerosis

Cholesterol deposits, in the intima of arterial walls, are engulfed by phagocytic smooth muscle cells from the media layer of the arterial wall to form a proliferat-ing "foam cell," which appears grossly as lipid plaque. This plaque extends into and ultimately occludes the lumen of the artery. Symptoms of angina (heart pain) are seen when the coronary vessels become 70% blocked. Also, when the choles-terol plaques are disturbed and collagen is exposed, platelets aggregate and thrombi develop. The resultant blockage causes a myocardial infarction when occurring in coronary arteries. This disease process (atherosclerosis) can take decades before symptoms become apparent.

Chemotherapy

Drugs have been used to lower cholesterol, the most effective being the bile sequestriants such as cholestramine, which bind bile salts in the intestines and interfere with the enterohepatic recovery of bile salts. Several plant fibers act as bile sequestriants in a natural way and can lower cholesterol levels significantly while improving overall bowel function. Oat bran and beans are apparently most effective in lowering blood cholesterol levels. Strict adherence to a low fat diet is also encouraged in the patient with increased cholesterol levels.

MEASUREMENTS OF TOTAL CHOLESTEROL

A fasting or nonfasting serum specimen can be used for the measurement of cholesterol; however, lipemia in a nonfasting specimen can interfere with the assay. Hence, a fasting specimen is recommended. Hemolysis also interferes with cholesterol assays.

Nonenzymatic Colorimetric Methods

Chemical oxidation of cholesterol by oxidizing acids (H_2SO_4) leads to a green color (Liebermann–Burchard reaction or LB reaction) and, in the presence of iron and heat forms a red color with five times the absorbance observed in the absence of iron. Serum can be treated directly with H_2SO_4 in the LB reaction; however, (1) unequal chromogenicity between cholesterol and its esters is present; (2) bilirubin and hemoglobin may nonspecifically interfere; and (3) proteins may interfere due to charring. Extraction of cholesterol with organic solvents eliminates some of the interfering substances; however, the unequal chromogenicity problem of the ana-lyte and its ester remains. Therefore, preliminary hydrolysis of the esters followed by extraction prior to the LB reaction is recommended as a reference method (Abell–Kendall method). The hydrolysis is carried out by treating the specimen with KOH and heating; the resulting free cholesterol may be precipitated by addition of digitonin.

Enzymatic Cholesterol Measurements

As described, chemical oxidative measurements of cholesterol are nonspecific, often lack needed sensitivity, and are laborious. An enzymatic cholesterol method is based on (1) enzymatic liberation of free cholesterol using cholesterol esterase; (2) oxidation of liberated cholesterol with cholesterol oxidase to liberate peroxide (H_2O_2); and (3) oxidative coupling of phenol and 4-aminoantipyrine by H_2O_2 in the presence of peroxidase. This sequence of reactions results in the stoichiometric formation of a red quinoneimine product, which is measured at 500 nm. Reducing agents like ascorbate (vitamin C) are known to interfere with color formation.

OTHER DISEASES AND CHOLESTEROL LEVELS

Other than inherited primary defects of lipid metabolism, secondary causes of increased cholesterol are (1) diabetes mellitus (increases in LDL and VLDL); (2) hypothyroidism (increase in LDL); (3) nephrotic syndrome (increases in LDL and VLDL); (4) early hepatitis; and (5) obstructive liver disease (decreased cholesterol excretion). Pregnancy also tends to elevate cholesterol. Cholesterol varies with age, sex, diet, race, season, and geographical location. Normal ranges, depending on method, are in the neighborhood of 150–250 mg/dL.

GLYCEROL AND TRIGLYCERIDE

CHEMICAL STRUCTURE

Glycerol has a polyalcohol structure (see Figure 7.5). With glycerol serving as the alcohol, free fatty acids can form esters or triglycerides as illustrated in Figure 7.6. These triglycerides are also called triacylglycerols.

TRIGLYCERIDE SYNTHESIS

As described above, triglycerides are triesters of glycerol and fatty acids and are synthesized mainly in the small intestine, liver, and adipose tissue. In the small intestine, triglycerides are synthesized from monoglycerides (a glycerol ester with only one fatty acid) and fatty acids obtained from recently digested fat. Hence, these triglycerides are often called "exogenous triglycerides."

   In the liver and adipose tissue, triglycerides are synthesized from plasma fatty

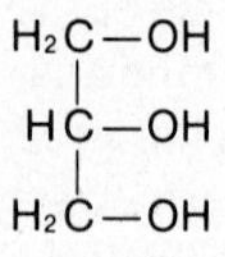

H_2C-OH
$\vert$
$HC-OH$
$\vert$
H_2C-OH

glycerol   **Figure 7.5.**   Chemical structure of glycerol.

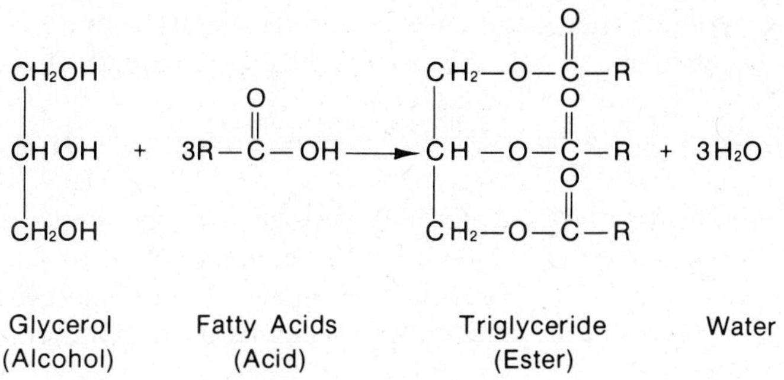

Figure 7.6. Formation of triglycerides from glycerol and fatty acids (R = alkyl groups).

acids; these triglycerides are sometimes referred to as "endogenous trigly-cerides." Triglycerides are stored in the adipose tissue and are readily broken down with the release of fatty acids in the plasma for energy. Triglycerides then readily contribute to the energy pool of the body; cholesterol does not.

The degree of saturation of the fatty acid chains on triglycerides can determine the physical state of the triglyceride fat. Unsaturated triglycerides (or triglycerides with unsaturated fatty acids) are often liquid at room temperature (as in vegetable oils), while saturated fatty acids are solid at room temperature (as in saturated animal fats).

TRIGLYCERIDE MEASUREMENTS

In recent years, enzymatic methods for the quantitation of triglycerides in serum have been developed and are generally preferred over the older colorimetric method of Hantzsch (to be discussed below). Kinetic (or UV) methods for the measurement of triglyceride are based on enzymatic hydrolysis of triglycerides to glycerol and fatty acids, with subsequent oxidation of liberated glycerol with NAD and glycerol dehydrogenase to form NADH and dihydroxyacetone. The rate of formation of NADH, which absorbs at 340 nm, is then used to calculate the triglyceride concentration. The reaction sequence is as follows:

$$\text{triglyceride} \underset{}{\overset{\text{lipase}}{\rightleftharpoons}} \text{glycerol + fatty acids}$$

$$\text{glycerol + NAD}^+ \underset{}{\overset{\text{glycerol dehydrogenase}}{\rightleftharpoons}} \underset{\text{absorbs at 340 nm}}{\text{NADH + H}^+} + \text{dihydroxyacetone}$$

Also, glycerol kinase can be substituted for glycerol dehydrogenase:

$$\text{glycerol + ATP} \underset{}{\overset{\text{glycerol kinase}}{\rightleftharpoons}} \text{glycerol-3-phosphate + ADP}$$

The resultant ADP product then can be used to form pyruvate:

$$\text{ADP + phosphoenolpyruvate} \underset{}{\overset{\text{pyruvate kinase}}{\rightleftharpoons}} \text{ATP + pyruvate}$$

The resulting pyruvate then can be reduced with NADH with a measurable decrease in absorbance at 340 nm, the basis for this kinetic methodology:

$$\text{pyruvate} + \text{NADH} + \text{H}^+ \xrightleftharpoons{\text{lactate dehydrogenase}} \text{lactate} + \text{NAD}^+$$

The kinetic method involving glycerol dehydrogenase (see above) can be converted to a colorimetric method by enzymatic reduction of tetrazolium salts with NADH and diaphorase. The tetrazolium bromide, 3-(4,5-dimethyl-thiazolyl-2)-2,5-diphenyl-2H-tetrazolium bromide (MTT), can be used as a chromophore:

$$\text{NADH} + \text{H}^+ + \text{MTT} \xrightleftharpoons{\text{diaphorase}} \text{NAD}^+ + \text{MTT} \cdot \text{H (colored product)}$$

Direct enzymatic measurement of triglyceride based on glycerol as an intermediate tend to overestimate triglyceride concentrations owing to the presence of free glycerol in the serum.

Hantzsch Condensation

An older, more laborious colorimetric method for the measurement of triglyceride involves the use of the Hantzsch condensation reaction. The method includes the (1) extraction of lipid with an organic solvent; (2) removal of phospholipids (and other interfering materials) by absorption with zeolite, alumina, or diatomaceous earth; (3) hydrolysis of isolated triglyceride using lipase or alcoholic KOH to liberate glycerol; (4) oxidation of glycerol to formaldehyde using periodic acid

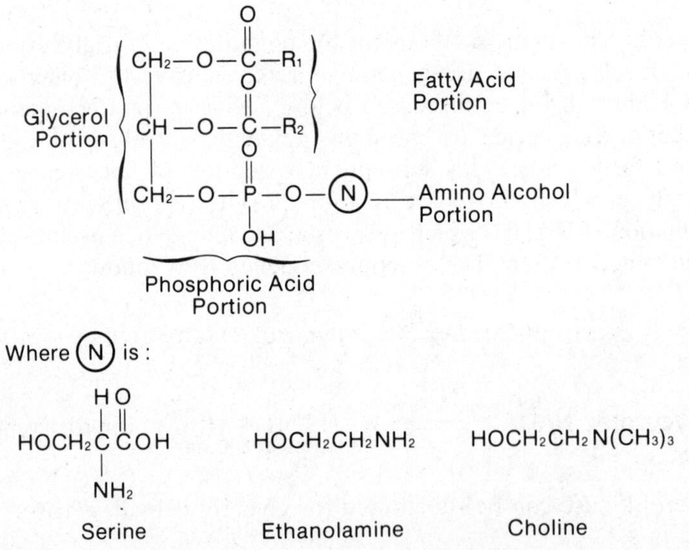

Figure 7.7. General structural formula of glycerolphosphatides. Structural components consist of glycerol, fatty acid, phosphoric acid, and aminoalcohol portions. The glycerphosphatides with the aminoalcohol choline is lecithin (or phosphytidylcholine). The glycerophosphatides where the amino alcohol is ethanolamine (phosphytidylethanolamine) or serine (phosphytidylserine) are both called cephalins.

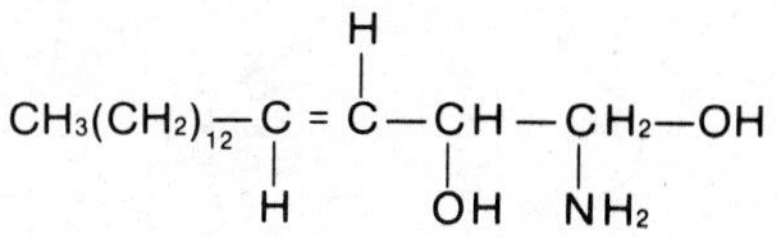

$$CH_3(CH_2)_{\overline{12}}C = \overset{\overset{\textstyle H}{|}}{C} - \overset{\overset{\textstyle }{|}}{C}H - \overset{\overset{\textstyle }{|}}{C}H_2 - OH$$

H   OH   NH₂

Figure 7.8. Chemical structure of sphingosine.

(HIO_4); and (5) condensation of formaldehyde with acetylacetone in the presence of NH_4^+ to form a colored product, which is measured at 412 nm. This latter product can also be measured using fluorescence with a 400 nm primary filter with a 485 nm secondary filter.

SPECIMEN REQUIREMENTS AND NORMAL LEVELS

Fasting (10–12 hours) serum or plasma can be used for triglyceride analysis; however, some triglycerides can be trapped during the clotting process; therefore, levels in plasma may be somewhat higher. Also, for enzymatic methods, anticoagulants may interfere. Normal values for triglyceride range from 10 to 190 mg/dL. Increased serum triglyceride levels are seen in oversyntheses and/or impaired catabolism. In most cases, elevated serum triglycerides are seen in obesity due to excessive ingestion of carbohydrates, fat, and alcohol.

PHOSPHOLIPIDS

GLYCEROPHOSPHATIDES

The general formula for glycerophosphatides is shown in Figure 7.7. Glycerophosphatides are synthesized mainly in the mucosa of the small intestines and in the liver. These phospholipids circulate in the plasma and play a significant role in cell structure, especially cells of the nervous system. It has been reported that phospholipids also increase the solubility of cholesterol and hence limit the uptake of cholesterol by the intimal lining of the arterial walls as seen in atherosclerosis.

SPHINGOLIPIDS

The structural formula for sphingosine is shown in Figure 7.8. With bonding of a phosphoryl choline group on the primary alcohol group of sphingosine and attachment of a fatty acid to the amine functional group, the resulting sphingolipid is sphingomyelin. Cerebrosides have a sphingosine backbone as well; however, a hexose group (commonly galactose) is attached to the primary alcohol group. Cerebrosides have no phosphate group. Cerebrosides may be sulfated on the hexose ring with the formation a derivative called cerebroside sulfate. Like glycerophosphatides, these lipids play a structural role in cell membranes, especially those of the nervous system.

L/S RATIO TEST

Measurements of Phospholipids

In premature infants, a common cause of death is due to collapsing of the fine air spaces of the lung, which interferes with the exchange of oxygen and carbon dioxide. This disease is called respiratory distress syndrome (RDS). As fetal lungs mature, the amount of lung surfactant in the form of glycerolphosphatides (specifically lecithin) enhances alveolar stability. Therefore, the measurement of lecithin in the amniotic fluid is an aid in the assessment of "fetal lung maturity." Another phospholipid, sphingomyelin, is also present in the amniotic fluid and is used as an internal control in the chromatographic measurement of lecithin owing to its somewhat constant level during embryonic development. The L/S ratio test is discussed in greater detail in Chapter 13.

Another phospholipid, phosphatidylglycerol (PG), is the second most abundant lung surfactant. PG normally appears around 36–37 weeks of gestation and increases thereafter. PG may signal the first biochemical stage of lung maturation; it also may act to stabilize lecithin.

The measurement of amniotic fluid lecithin levels [or the lecithin/sphingomyelin (L/S) ratio] is based on a thin-layer chromatographic separation of the extracted lipid with visualization of lipid fractions with iodine or molybdate. The chromatogram may be scanned with a densitometer for quantitation.

Normal Values of the L/S Ratio Test

In early gestation (26th week or 6.5 months), the L/S ratio is less than 1.0; however, after the 35th week, lecithin concentrations increase rapidly with L/S ratios of 2.0 or higher normally being observed.

Usually an L/S ratio of 2.0 or more indicates fetal lung maturity, a ratio of 1.5 or less indicates lung immaturity with the possibility of respiratory distress syndrome on delivery. Other laboratory procedures such as the amniotic fluid surfactant "shake test" and creatinine are also used to assess fetal maturity.

Phospholipid Phosphorus

Phospholipids (glycerolphosphatides and sphingomyelins) phosphorus measurements involve (1) the extraction of lipids; (2) the digestion of the lipid in an oxidizing acid to liberate free phosphate; and (3) the measure of free phosphate by the reactions

$$\text{free phosphate + ammonium molybdate} \rightarrow \text{phosphomolydic acid}$$

$$\text{phosphomolybdic acid} \xrightarrow{\text{reducing agent}} \text{molybdenum blue + phosphomolydous acid}$$

Normal levels of phospholipid phosphorus in serum are 3–11 mg/dL; phospholipid levels in serum range from 150 to 380 mg/dL.

LIPID-STORAGE DISEASES

The phospholipids (specifically sphingolipids) and their metabolites accumulate in certain inherited enzyme deficiencies causing significant and progressive central nervous system impairment. Gaucher disease involves an accumulation of ceramide glucose in the liver and spleen due to a deficiency of β-glucosidase. These diseases are usually confirmed by enzyme assay. Niemann–Pick disease is associated with accumulation of sphingomyelin due to a deficiency of sphingomyelinase. Most of the lipid-storage diseases have autosomal recessive inheritance. There are at least 10 hereditary disorders associated with lipid accumulation. Although symptoms vary, most are associated with mental retardation and/or CNS impairment.

FATTY ACIDS

LIPOPROTEIN LIPASE

Triglyceride hydrolysis in the blood is catalyzed by an enzyme called lipoprotein lipase. The liberated fatty acids are either taken up by tissue near the site of hydrolysis or remain in the blood. In adipose tissue, fatty acids are taken up and recombined with glycerol to form triglyceride, the storage form of fat in the adipose tissue. These stored fats can be readily broken down to fatty acids, which are released into the blood as part of the "energy pool." Enzymatic oxidation of fatty acids to carbon dioxide and water is accompanied by a large quantity of energy (exothermic reaction) with nearly 40% of the energy serving as useful energy to the body (or stored as energy-rich phosphate compounds such as ATP). Breakdown of fatty acids occurs through a process called beta oxidation with the formation of acetyl coenzyme A (acetyl CoA). Acetyl CoA is metabolized by reacting with oxaloacetate, a compound derived primarily from glucose metabolism.

DIABETIC KETOSIS

In diabetic ketosis, acetyl CoA levels increase, due to accelerated fatty acid metabolism, exceeding the amount of oxaloacetate required for normal metabolism to CO_2, water, and energy. Hence, an alternate metabolic route for acetyl coenzyme A is selected where it reacts with itself to form keto acids, acetoacetate, and beta-hydroxybutyrate. Acetoacetate further decomposes to form acetone, which is excreted mostly by the lungs. Unless carbohydrate metabolism is restored, a severe metabolic acidosis can cause diabetic coma. Figure 7.9 illustrates this alternate pathway of fatty acid metabolism in diabetic ketosis.

FREE FATTY ACIDS

As mentioned earlier, the general formula for fatty acids is RCOOH where the R group consists mainly of straight chains of carbon atoms with even numbers of

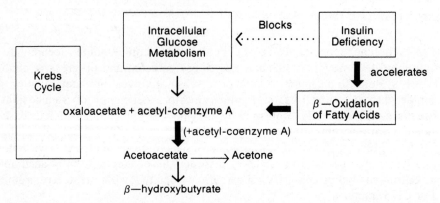

Figure 7.9. Formation of ketoacids caused by alternate metabolism of acetyl coenzyme A.

atoms. Fatty acids are present in the plasma in esterified form or as unesterified fatty acids, often referred to as "free fatty acids" (FFA). FFA are largely bound by albumin. Plasma levels of FFA rise due to increased hydrolysis of triglyceride in adipose tissue as influenced by lipase. Lipase activity is influenced by a number of hormones including epinephrine, norephinephrine, thyroid stimulating hormone (TSH), adrenocorticotropic hormone (ACTH), growth hormone, and glucagon.

Fatty acids can be measured by extraction followed by titration with dilute alkali using thymolphthalein as an indicator. Gas–liquid chromatography and thin-layer chromatographic methods can also be used for fatty acid determinations.

FATTY LIVER

Fat accumulation in the liver is often seen in alcoholism, malnutrition, obesity, and diabetes. High fat diets and chemical liver toxins have also been associated with "fatty liver." The disease is probably due to inadequate synthesis of phospholipids, which are required to transport lipids out of the liver.

LIPOPROTEINS

Since many lipids are essentially insoluble in aqueous solutions like plasma, the combination of lipid and protein into a lipoprotein is essential for the transportation of lipids in the vascular system. Lipoproteins have a spherical micellular structure with the hydrophobic lipid nonpolar groups (i.e., alkyl groups) directed toward the center of the sphere; the more polar hydrophilic lipid components are found near the surface of the lipoprotein sphere.

There are four different lipoprotein classes based on (1) the electrophoretic migration of lipoproteins and (2) the sedimentation coefficients (Sf) of each lipoprotein in the ultracentrifuge. The Sf values of lipoproteins are negative sedimentation coefficients (in Svedbergs) in sodium chloride solution (density of 1.063 g/mL). These four major classes of lipoproteins are depicted in Figure 7.10 with each containing different amounts of cholesterol, triglyceride, phospholipid, and protein. The properties and lipid composition of the lipoproteins are shown in Table 7.1.

APO-PROTEINS

Various apolipoproteins are found in the different lipoproteins. For example, chylomicrons contain apo A-I (33%), C (32%), A-IV (14%), E (10%), and B (5%).

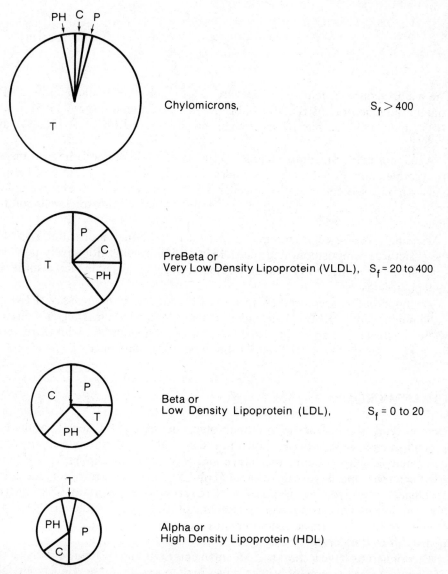

Figure 7.10. Four different classes of lipoproteins containing variable amounts of C (cholesterol), T (triglyceride), PH (phospholipid), and P (protein). Both the electrophoretic and ultracentrifuge names for the four lipid fractions are shown.

Table 7.1. Properties and Lipid Composition of Lipoproteins

Lipoprotein	Mobility on Electro- phoresis	Protein (%)	Triglyc- ceride (%)	Choles- terol (%)	Phos- pholipid (%)	Sf
Chylomicron	Origin	1	90	5	4	Greater 400
VLDL	Prebeta	10	60	15	15	20–400
LDL	Beta	25	10	45	20	0–20
HDL	Alpha	45	3	20	30	—

The major apolipoproteins of VLDL are C (55%), B (25%), and E (15%). LDL contains apolipoproteins B (95%) and E (3%). HDL_2 contains apo A-I (65%), A-II (10%), and C (13%), while HDL_3 contains mostly apo A-I (62%), A-II (23%), and C (5%).

Apolipoproteins functions include (1) transport—where apo B helps transport triglyceride and apo A-I facilitates cholesterol transport; (2) structure of lipoproteins—apolipoproteins stabilize micellar lipid structures; and (3) enzyme cofactors—C-I is a cofactor for acyltransferase and C-II is a cofactor for lipoprotein lipase.

The analysis of apolipoproteins has provided additional information regarding the structure, composition, and physiology of lipoproteins. Relatively new data suggest that many heretofore classified lipid disorders may be, in fact, disorders of apolipoproteins.

Apolipoproteins are currently measured by enzyme immunoassay (EIA) and radioimmunoassay (RIA). Deficiencies of apo A-I are seen in Tangier's disease, while deficiencies of apo B are called abetalipoproteinemia. Low levels of apo E are seen in familial type III Fredrickson hyperlipoproteinemia.

CHYLOMICRONS

Obviously, as the size of the four lipoproteins decreases, the density of the lipoprotein fraction increases (as shown in Figure 7.10). Also, the more triglyceride contained in a lipoprotein, the larger and less dense the lipoprotein fraction. Chylomicrons, the largest lipoprotein (110–120 nm in diameter) contain mostly "exogenous triglyceride" synthesized from recently digested fat. Chylomicrons are insoluble in plasma and, when a test tube of plasma is stored overnight at 4°C, chylomicrons will form a visible creamy layer on the surface of plasma. The density of chylomicrons is 0.95 g/mL.

Chylomicrons have a short half-life upon entry into the systemic circulation (5–15 minutes) and are rapidly broken down by lipoprotein lipase. Hence, chylomicrons are normally not present in fasting blood specimens.

VERY-LOW-DENSITY LIPOPROTEINS (VLDL)

Prebeta (or VLDL) lipoproteins contain "endogenous" triglyceride, or triglycerides produced mainly in the liver. Prebeta lipoproteins migrate with the alpha-2 globulins on electrophoresis. Increased levels of VLDL or prebeta lipoprotein cause a plasma specimen to be turbid or lactescent. The plasma half-life of VLDL is 6 to 12 hours.

LOW-DENSITY LIPOPROTEINS (LDL)

Beta lipoproteins (or LDL) contain mostly cholesterol (in the form of cholesterol esters) and migrate in the beta region on electrophoresis. Since LDL have densities from 1.006 to 1.063, they form the top layer after ultracentrifugation in a NaCl solution with a density of 1.063 g/mL.

HIGH-DENSITY LIPOPROTEINS (HDL)

Alpha lipoproteins (or HDL) contain about 50% lipid and 50% protein. HDL usually migrates with the alpha-1 globulins on electrophoresis. HDL, with densities of 1.063–1.21 g/mL, separate into the lower layer upon ultracentrifugation in NaBr (or KBr) with a density of 1.063 g/mL. Increases in either LDL or HDL impart no visible change to the plasma, even after overnight storage at 4°C. HDL enhances lecithin cholesterol acyltransferase (LCAT) activity, where a transfer of a fatty acid from phosphytidyl choline to cholesterol (forming cholesterol esters) is catalyzed. The apolipoprotein A-I of HDL activates LCAT. Precursors to HDL are produced in the liver and intestine with completion of development taking place in the plasma. HDL subfractions include HDL_1, HDL_2, and HDL_3 in order of increasing density; these HDL moieties all have different structure and function.

LIPOPROTEIN ELECTROPHORESIS

Electrophoresis studies on lipoproteins are carried out in a manner similar to routine protein electrophoresis; however, fat stains are used for color development (see Figure 7.11). Prebeta lipoproteins, although significantly larger than beta lipoproteins, migrate faster on agarose in alkaline buffers and do not appear as a separate band, but rather as a shoulder to the more intensely staining beta fraction. Chylomicrons are not shown in Figure 7.11 and are usually not seen in fasting specimens; these are identified as a fat-staining band at the origin. Alpha lipoproteins are the fastest migrating fraction but are less intensely staining than beta and prebeta fractions.

ELECTROPHORESIS

Lipoprotein electrophoresis studies are carried out on fresh EDTA plasma; EDTA is preferred because it delays deterioration of the lipoproteins (probably by inacti-

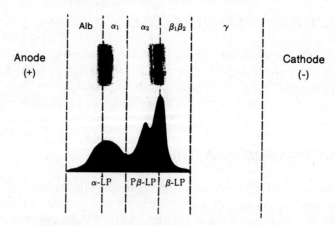

Figure 7.11. Electrophoretic separation of lipoproteins.

vating serum lipase). Fresh serum can also be used. If the electrophoresis is performed on an old specimen (after 24 hours), variable results may be obtained, especially in the VLDL fraction. Freezing irreversibly alters the electrophoretic pattern; hence, for long-term storage (7 days or less), storage at 4°C is recommended. The media for electrophoresis may be agarose gel, starch gel, paper, or polyacrylamide. Cellulose acetate is not recommended for lipoprotein electrophoresis. After electrophoretic separation, visualization of lipoproteins is achieved by staining with oil red O or Sudan Black. These stains react with triglyceride esters and cholesterol esters only; free cholesterol and phospholipids do not stain. The agarose gel is then scanned with a densitometer to produce a tracing as depicted in Figure 7.11.

HYPERLIPIDEMIAS

A hyperlipidemia is a disease where the blood level of one of the lipoproteins is abnormally increased. Hyperlipidemias can be inherited traits as primary hyperlipidemias. Hyperlipidemia can also be secondary to other diseases or disorders. For example, any disease that causes excessive release of "lipo-active" hormones (i.e., low insulin) can induce an elevation of fatty acids. The liver clears excess lipids and incorporates them into lipoproteins causing a hyperlipoproteinemia. Secondary hyperlipoproteinemia can be caused by diseases like diabetes mellitus, nephrotic syndrome, and hypothyroidism. As shown in Figure 7.8, low levels of insulin stimulate lipolysis; hence, patients with defective carbohydrate metabolism often have associated lipid disorders.

LIPOPROTEIN PHENOTYPES

It is often helpful to divide lipid disorders into six phenotypes or classes: Fredrickson Types I, IIA, IIB, III, IV, and V. Each lipid phenotype is associated with certain clinical and biochemical features, which are described in Table 7.2 along with recommendations for medical management. Phenotypes or Fredrickson types can be suggested by observing (1) serum appearance, after storage overnight at 4°C; (2) total cholesterol and triglyceride levels; (3) lipoprotein electrophoresis data; and (4) patient's signs and symptoms.

Type I

Type I is a rare lipid phenotype that manifests itself early in childhood, usually at infancy. The disorder is caused by a lipoprotein lipase deficiency; hence, chylomicrons, which contain exdogenous triglyceride, are not effectively broken down.

Table 7.2. Various Clinical and Biochemical Features of Lipid Phenotypes Along with Suggestions for Medical Management

Lipid Phenotype	Clinical Features	Genetic and/or Biochemical Features	Medical Management
I	Rare, lipemia retinalis, eruptive xanthomas, hepatosplenomegly, abdominal pain	Genetic recessive, lipoprotein lipase deficiency, chylomicron accumulation	Low triglyceride diet; medium chain fatty acid triglycerides; no drugs
IIA	Xanthelasma, tendon and tuberous xanthomas, accelerated atherosclerosis especially coronary	Genetics: monogenic, polygenic, or sporatic; LDL accumulation	Low cholesterol polyunsaturated fat diet, cholestyramine, colestipol, nicotinic acid, probucol; clofibrate has some minimal benefits; no dextrothyroxine due to increased cardiovascular risk
IIB	Common, accelerated atherosclerosis, coronary and peripheral	VLDL accumulation; LDL accumulation	
III	Palmar xanthomas, tuberous xanthomas, accelerated atherosclerosis especially peripheral	Production of abnormal lipoprotein with normal apo B and C in abnormal composition; increase in apo E-II, decreased in apo E-III and E-IV	Low cholesterol polyunsaturated fat diet; clofibrate
IV	Common, association with abnormal glucose tolerance, increased uric acid, accelerated atherosclerosis, especially peripheral	Increased VLDL; triglyceride markedly increased; total cholesterol may be increased due to VLDL elevation	Low carbohydrate diet; clofitrate
V	Rare, lipemia retinalis, eruptive xanthomas, hepatosplenomegly, abdominal pain	Commonly associated with insulin-requiring diabetic; rare form associated with lipoprotein lipase deficiency	Low triglyceride and low carbohydrate diet; insulin for even mild insulin-requiring diabetic; nicotinic acid, clofibrate

Recent studies have shown that these patients may lack apolipoprotein C, which is necessary for the activation of lipoprotein lipase; hence, the disorder may lie with a apolipoprotein disorder rather than a lack of lipoprotein lipase. These patients respond to low fat diets.

The serum appearance in these patients shows a heavy accumulation of chylomicrons on the surface of the serum test tube with a clear serum infranatant. The triglyceride will be markedly elevated with levels as high as a 1000 mg/dL or higher. The cholesterol total will be normal or slightly elevated. The electropherogram will show a heavy chylomicron band.

Type IIA

This lipoproteinemia may be either primary or secondary to hypothyroidism, obstructive liver disease, or nephrosis. Type IIA, although uncommon, has a homo- or heterozygous inheritance. Patients with IIA often die before adulthood from circulatory complications of accelerated atherosclerosis, especially coronary atherosclerosis. The measurement of cholesterol in cord blood may detect the primary IIA in infancy. The treatment of IIA patients includes diets low in fat and cholesterol together with aerobic exercise. Various drugs have been shown to reduce cholesterol, especially cholestyramine, a bile sequestriant, which binds bile salts (the main excretory form of cholesterol) in the intestines and facilitates their excretion. Diets high in fiber, especially oat and bean fiber, exhibit a bile-binding effect like cholestyramine and have been shown to lower serum cholesterol by 25% or more.

Studies have shown that IIA patients have defective receptors on their cell surfaces for LDL. LDL binds to cell receptors (on human fibroblasts, smooth muscle cells of arterial walls, and lymphocytes) and is internalized by the cell. Cellular metabolism of LDL produces free cholesterol, which decreases cellular production of cholesterol by inhibiting the activity of 3-hydroxy-3 methylglutaryl CoA reductase (HMG CoA reductase). Hence, patients with defective receptors for LDL have unregulated cellular production of cholesterol. For this reason, dietary management alone has not generally been effective in lowering serum cholesterol in these patients.

In IIA, the serum appearance is clear; cholesterol is markedly increased (three to four times normal), while serum triglyceride is normal. Electrophoresis reveals an increased beta fraction (or LDL fraction).

Type IIB

Type IIB is a common lipoproteinemia often associated with improper diet or may have secondary causes such as hypothyroidism, obstructive hepatic disease, or nephrotic syndrome. This disease is associated with accelerated atherosclerosis of both coronary and peripheral arteries. Patients often respond to diets low in cholesterol and saturated fat along with chemotherapy. Weight reduction to the "ideal weight" is encouraged along with moderate aerobic exercise.

The serum appearance in type IIB is clear or slightly turbid with no chylomicrons seen. Both cholesterol and triglyceride serum levels are increased. Electrophoresis in type IIB reveals increased beta (LDL) and prebeta (VLDL) bands.

Type III

Type III lipoproteinemia, an uncommon inherited disease, is characterized by an increase in an abnormal lipoprotein called IDL, intermediate density lipoprotein. The fraction is frequently called a "floating beta" because of its low density. The lipoprotein IDL contains normal apolipoproteins B and C, but they are present in abnormal amounts. Also, type III is associated with an increase in apo E-II with the absence of apo E-IV and E-V. Patients with this lipid disorder have accelerated atherosclerosis. A secondary cause of phenotype III is diabetes. Studies have shown that type III patients often have decreased lipoprotein lipase activity, which may be due to an inhibitor or a decrease in apo-protein C.

The serum appearance for type III is turbid; serum levels for both cholesterol and triglyceride are increased. The electropherogram will often show one band migrating in the beta and prebeta region. Patients with type III lipoproteinemia benefit from diets low in cholesterol and polyunsaturated fats. Some minimal benefit of chemotherapy with clofibrate has been reported.

Type IV

Type IV is a more common lipoproteinemia associated with increases in "endogenous" triglyceride. In these patients, excess carbohydrates influence the increased production of VLDL by the liver. Type IV is often associated with obesity, abnormal glucose tolerance, and accelerated atherosclerosis (especially of peripheral vessels). Type IV lipoproteinemia may be familial or secondary to diseases like diabetes, pancreatitis, or alcoholism. These patients respond best to low carbohydrate diets and reduction in total calories sufficient to lower body weight to the "ideal weight."

The serum appearance in type IV is turbid. Cholesterol may be normal to slightly increased owing to the presence of cholesterol in the VLDL moiety. The triglyceride level will be moderately to markedly increased; electrophoresis reveals an intense band in the prebeta region (VLDL) with no chylomicrons being observed.

Type V

Type V lipoproteinemia is uncommon and may be familial or secondary to diseases like diabetes, pancreatitis, or alcoholism. Signs and symptoms of type V are very similar to type I; however, onset is usually in adulthood.

The serum appearance is characterized by the heavy accumulation of chylomicrons floating on the surface of the serum. The serum infranatant is turbid due to an increase in VLDL. Cholesterol levels are normal to slightly elevated, while the total triglyceride is markedly elevated. Electrophoresis reveals a broad chylomicron band together with an intensely staining prebeta band.

SPECIAL PATIENT PREPARATION PRIOR TO LIPID STUDIES

The patient should be fasting for 12–14 hours with the last meal being low in fat. Nonfasting specimens will cause chylomicron appearance together with an increased triglyceride. The patient (1) should not be on a special diet at least two weeks prior to sampling; (2) should not be taking medications known to affect the lipid study; and (3) should avoid the consumption of alcohol at least 24 hours prior to the study. Sample electrophoretic tracings for selected Fredrickson types are shown in Figure 7.12.

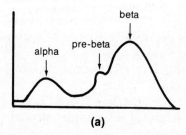

(a)

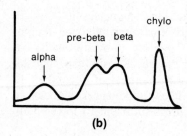

(b)

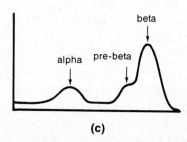

(c)

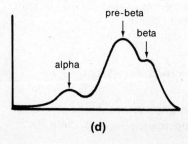

(d)

Figure 7.12. Selected lipoprotein electrophoretic patterns: (a) type IIA; (b) type I; (c) normal pattern; and (d) type IV.

CARDIOVASCULAR DISEASE RISK

HIGH-DENSITY LIPOPROTEIN CHOLESTEROL (HDL-C)

In recent years, studies have shown that the amount of cholesterol in the HDL fraction (alpha fraction) is inversely related to the incidence of coronary heart disease (CHD). The higher the level of HDL-C, the lower the incidence of CHD. Evidently, the HDL lipoprotein affords a protective effect against the development and progression of atherosclerosis. Patients with low levels of HDL-C have much higher risks for the development of atherosclerosis.

Tangier Disease

Tangier disease is associated with a familial HDL deficiency; hence, a very weakly staining electrophoretic band is observed in the alpha region. Apparently the disease is due to the ineffective synthesis of HDL. The plasma cholesterol is normal or reduced and triglyceride levels are normal or increased in Tangier disease.

HDL Measurements

High-density lipoprotein cholesterol is determined by (1) precipitating the LDL and VLDL from the serum by addition of a polyanion (heparin) and a divalent cation (e.g., Mg^{2+}); (2) centrifugation to separate the precipitated LDL and VLDL from the HDL in the supernatant; and (3) enzymatic cholesterol measurement in the HDL fraction in the supernatant. Normal levels of HDL-C are 35–65 mg/dL for males and 35–75 mg/dL for females.

LDL-C

Increased levels of cholesterol in the LDL fraction have been shown to be directly related to increased risk of coronary heart disease.

Calculating LDL-C Using Friedewald's Formula

Since cholesterol is found in HDL, VLDL, and LDL, by measuring the total serum cholesterol and subtracting HDL-C and VLDL-C, we get an estimate of LDL-C. VLDL-C is estimated by dividing the total triglyceride by five. This calculation for LDL-C is accurate only if (1) the total triglyceride is not greater than 400 mg/dL and (2) there are no chylomicrons present. The equation for calculating LDL-C cholesterol are:

$$\text{total cholesterol} = \text{LDL-C} + \text{VLDL-C} + \text{HDL-C}$$

$$\text{LDL-C} = \text{total cholesterol} - (\text{VLDL-C} + \text{HDL-C})$$

$$\text{LDL-C} = \text{total cholesterol} - \left(\frac{\text{triglyceride}}{5} + \text{HDL-C}\right)$$

Other Methods

LDL-C can be measured by separating the fraction using an ultracentrifuge and measuring the cholesterol level. Estimates of LDL-C can also be obtained electrophoretically using a colorimetric cholesterol oxidase substrate followed by integration of the electrophoregram tracing.

RISK FOR CORONARY HEART DISEASE

Although many factors must be considered in the assessment of coronary heart disease risk (e.g., diet, exercise, blood pressure, smoking), a rough assessment of coronary heart disease risk can be made on the basis of the LDL-C/HDL-C ratio (see Table 7.3).

By observation of the data in Table 7.3, it is apparent that an LDL-C/HDL-C ratio of 2 or lower is desirable and lowers the predicted risk of coronary heart disease below the average.

Lowering Coronary Heart Disease (CHD) Risk Factors

CHD is associated with certain risk factors including (1) high serum cholesterol [total cholesterol greater than 230 mg/dL, LDL-C/HDL-C ratio greater than 3.55 (males)]; (2) low HDL-C (less than 50 mg/dL); (3) hypertension (greater than 160/95); (4) cigarette smoking; (5) glucose intolerance; (6) physical inactivity; (7) obesity; (8) serum triglyceride elevation (greater than 210 mg/dL); and (9) stress.

Table 7.3. Coronary Heart Disease Risk Based on LDL-C/HDL-C[a]

	Risk	LDL-C/HDL-C
Men	One-half average	1.00
	Average	3.55
	2× average	6.25
	3× average	7.99
Women	One-half average	1.47
	Average	3.22
	2× average	5.03
	3× average	6.14

[a] Regression equations (Developed from data of the Framingham Heart Study):

average risk % (men) = 0.69 (age) − 17 (for men older than 30)

average risk % (women) = 0.94 (age) − 35 (for women older than 40)

(Note: Average risk for women between 30 and 40 years is less than 3%.)

Table 7.4. Apparent Phenotypes Based on Serum Appearance, Total Triglyceride, and LDL-C

Apparent Phenotype	Serum Appearance	Lipid Characteristics	Secondary Causes
I (Rare)	Chylomicrons, clear infranatant	Increased triglyceride as chylomicrons	Lupus, lymphomas
IIA	Clear	increased LDL-C	Hypothyroidism, obstructive liver disease, nephrosis
IIB or III	Turbid	Increased LDL-C and triglyceride	IIB-hypothyroidism, obstructive liver disease, nephrosis, III—diabetes
IV	Turbid	Increased triglyceride and normal LDL-C	Diabetes, estrogen, alcohol, nephrosis, pregnancy
V	Chylomicrons, turbid infranatant	Increased triglyceride as both Chylomicrons and VLDL	Alcoholism, diabetes, pancreatitis, estrogen

Reduction of CHD Risk

Reduction of CHD risk is often difficult and, in many cases, requires a change in "lifestyle." Blood lipids can be markedly lowered through a combination of dietary management, exercise, and, if necessary, chemotherapy. Both glucose intolerance and hypertension respond favorably to the weight loss that often accompanies control of lipid ingestion and exercise. HDL-C can be raised substantially by exercise.

APPARENT LIPOPROTEIN PHENOTYPES

As shown in Table 7.4, using serum appearance, total triglyceride, and LDL-C, an "apparent" phenotypic lipid classification can be made. For example, for specimens with heavy chylomicrons and markedly elevated triglyceride levels, if the subnatant is clear, the apparent phenotype is I; if the serum infranatant is turbid, the apparent phenotype is V.

For patients where the specimen is clear, the LDL-C is elevated, and the triglyceride is normal, an apparent phenotype of IIA is indicated. If, on the other hand, (1) the triglyceride is elevated, (2) the LDL-C is within normal limits, (3) the specimen is turbid, and (4) no chylomicrons are observed, an apparent phenotype of IV is suggested. When both the LDL-C and the triglyceride are elevated and the specimen is turbid with no chylomicrons, an apparent phenotype of IIB or III is indicated. Ultracentrifugation is usually recommended to identify a type III IDL fraction. IDL denotes "intermediate density lipoprotein."

EXAMPLE 7.1. (a) Calculate the LDL-C for a 40-year-old male patient with a total cholesterol of 350 mg/dL, total triglyceride of 245 mg/dL, and a HDL-C of 36 mg/dL. (b) What is the patient's risk? (c) What is the patient's "apparent phenotype" if the serum specimen was turbid after overnight storage at 4°C?

Answer: (a) Compute LDL-C:

$$\text{LDL-C} = \text{total cholesterol} - \left(\frac{\text{triglyceride}}{5} + \text{HDL-C} \right)$$

$$\text{LDL-C} = 350 \text{ mg/dL} - \left(\frac{245 \text{ mg/dL}}{5} + 36 \text{ mg/dL} \right)$$

$$\text{LDL-C} = 350 \text{ mg/dL} - 85 \text{ mg/dL} = 265 \text{ mg/dL}$$

(Note that this is an elevated LDL-C being greater than 170 mg/dL.)

(b) Compute LDL-C/HDL-C ratio:

$$\frac{265 \text{ mg/dL}}{36 \text{ mg/dL}} = 7.4$$

Using Table 7.3 for males, this ratio is slightly less than 7.99; hence, the patient is approaching three times normal risk for coronary heart disease.

(c) The specimen is turbid, the triglyceride level is increased, and the LDL-C is markedly elevated; hence, the apparent phenotype is Type IIB or III (see Table 7.4).

LIPID METABOLISM

Refer to Figure 7.13 for the metabolic steps. In the intestinal tract (1), ingested fats are broken down by pancreatic lipase (in the presence of bile salts) to glycerol, phospholipids, cholesterol, fatty acids, monoglycerides, and other products. In the absence of bile salts, fat and fat-soluble vitamins A, D, E, and K will not be absorbed; fatty stools (called steatorrhea) and vitamin deficiency may occur.

As shown in Figure 7.13, short chain fatty acids (12 or less carbons) are transferred from the intestine to the portal blood. These fatty acids are bound to albumin and are transported to the liver in portal blood (see step 2).

In the intestinal mucosa, longer chain fatty acids, glycerol, monoglycerides, and cholesterol are absorbed. In the intestinal mucosa, monoglycerides are converted to triglycerides. Triglycerides are then combined with apolipoprotein B to

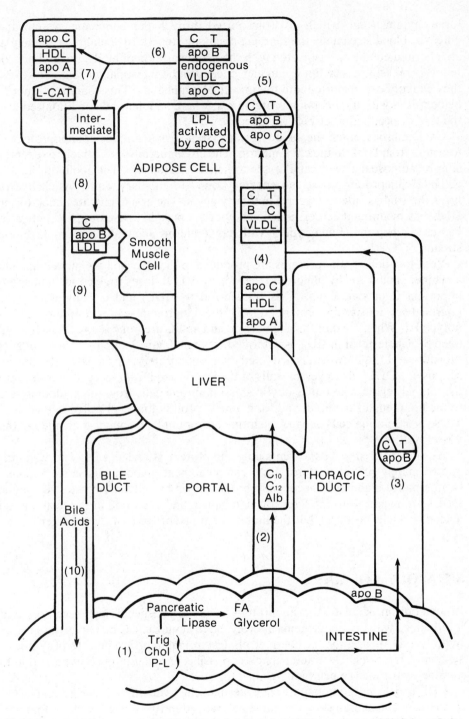

Figure 7.13. The metabolism of lipids. (Adapted and reprinted with permission of BioScience Laboratories, Van Nuys, CA.)

form chylomicrons, which are transported through the lymphatic system and enter the blood stream via the thoracic duct (see step 3). In abetalipoproteinemia, an inherited condition associated with the malabsorption of lipids, no apolipoprotein B is present, hence lipids can gain access to the intestinal mucosa; however, they cannot be transported into the intestinal lymphatics. This condition is not to be confused with hypobetalipoproteinemia, where an inherited condition reduces the rate of production of beta lipoprotein.

Chylomicrons enter the blood stream (step 4) whereupon apolipoprotein C transfers from HDL to the chylomicron. The chylomicrons then attach to receptor sites on adipose tissue or capillary endothelial cells (step 5). The apolipoprotein C activates lipoprotein lipase, an enzyme bound to capillary walls, which hydrolyzes the triglyceride to glycerol and fatty acids. The apo C migrates back to the HDL. As mentioned earlier, a deficiency of lipoprotein lipase or apo C activator causes a lipoproteinemia type I with chylomicron accumulation in the blood stream.

Very-low-density lipoproteins are produced in the liver and are metabolized in a manner similar to chylomicrons (see step 6). VLDL loses triglyceride and apolipoprotein C in stages (apo C is transferred to HDL) and is converted to an intermediate-density lipoprotein (step 7), the lipoprotein seen abnormally in phenotype III. When most of the triglyceride and apo-C are removed, LDL is formed (step 8). Cholesterol in LDL is esterified by the enzyme lecithin cholesterol acyl transferase (LCAT), which is activated by apolipoprotein A from HDL. In atherosclerosis, LDL binds to vessel walls of the arteries and from there it migrates into the intimal region where phagocytic smooth muscle cells from the media migrate to the lipid, engulf the lipid, and form a new, proliferating cell called a foam cell. These proliferating cells expand the intimal layer until it occludes the lumen of the vessel wall.

As shown in step 9, the remaining lipoprotein is called LDL and contains approximately 50% cholesterol. LDL can combine to smooth muscle membranes, be internalized, and be subsequently broken down by cellular enzymes. Unbound LDL is transported to the liver for metabolism and excretion in the form of bile acid. It is believed that HDL facilitates the transport of LDL to the liver.

METABOLISM OF LDL

In metabolism of LDL, the lipoprotein is bound by specific cell receptors and internalized with apolipoproteins hydrolyzed to amino acids and cholesterol esters hydrolyzed. Apparently, the liberated cholesterol inhibits the cholesterol synthesis system of the cells. This biofeedback is possibly missing in phenotype II due to missing or ineffective cell binding sites for LDL.

HDL is believed to transport cholesterol to the liver where it can be converted to bile salts and excreted via the intestinal tract (step 10). While the major function of other lipoproteins is the transport of lipid, HDL plays a more significant role in lipid metabolism by (1) activation of lipoprotein lipase (apo C) and (2) esterification of cholesterol in plasma (apo A).

BIBLIOGRAPHY

Assman, G. *Lipid Metabolism and Atherosclerosis*. Stuttgart, Germany: Schattauer-Verlag, 1982.

Bagade, J. D. and Albers, J. J. "Plasma HDL concentrations in chronic hemodialysis and renal transplant patients," *New England Journal of Medicine*, 1436 (June 23, 1977).

Bakerman, S. *Review of Clinical Chemistry*. Greenville, NC: East Carolina University, 1977.

Carew, T. E., et al. "A mechanism by which high-density lipoproteins may slow the atherogenic process," *Lancet* **L**, 315–317 (1976).

Castelli, W. P., et al. "HDL cholesterol and other lipids in coronary heart disease," *Circulation* **55**, 767–772 (1977).

Castelli, W. P. "High blood lipid levels," *Medical News, Journal of the American Medical Association* 1066 (Mar. 14, 1977).

Fisher and Truitt. "Mechanism of atherosclerosis," *Annals of Internal Medicine* **85**, 497–508 (1976).

Fredrickson, D. S. "It's time to be practical," *Circulation* **51**, 209 (1975).

Gambino, R. "Tips on technology: Do you really need lipoprotein electrophoresis?" *MLO*, 15 (Sept.–Oct 1973).

Glueck, C. J., et al. "Familial hyper-alpha-lipoproteinemia; studies of 18 kindreds," *Metabolism* **24**, 1234–1265 (1975).

Gordon, T., et al. "High density lipoprotein as a protective factor against coronary heart disease: The Framingham study," *American Journal of Medicine* **62**, 707–714 (1977).

Henck, C. C. and Schlierf. "Beta-lipoprotein cholesterol quantitation with polycations," *Clinical Chemistry* **23**(3), 536 (1977).

Ibbott, F. *Current Progress in Understanding the Lipoproteins and Their Metabolities*. Van Nuys, CA.: Bio-Science Laboratories.

Levy, R. I. and Fredrickson, D. S. "Diagnosis and management of hyperlipoprotein-emia," *American Journal of Cardiology* **22**, 576–583 (1968).

Levy, R, et al. In *Lipoprotein Metabolism*, Greten (ed.). Berlin, 1976, p. 56.

Levy, R. I. "Treatment of hyperlipidemia," *New England Journal of Medicine* **290**, 1295–1301 (1974).

Levy, R. I. "Drug therapy of hyperlipoproteinemia," *Journal of the American Medical Association* **235**(21), 2334 (1976).

Lewis, B. *In the Hyperlipidemias*, London: Blackwell, 1976, pp. 206–210.

Miller, G. J. and Miller, N. E. "Plasma-high-density lipoprotein concentration and development of ischemic heart disease," *Lancet* 16 (Jan. 4, 1975).

Orten, J. M. and Newhaus, O. W. *Human Biochemistry*, St. Louis: Mosby, 1975.

Pribor, H. C. and Morrell, C. "Are phospholipids/total lipids determinations obsolete," *Laboratory Management* (Sept. 1980).

Section on Lipoproteins, Laboratory for Molecular Diseases, National Heart and Lung Institute, NIH, Bethesda, Maryland, 1978.

Tietz, N. (ed.). *Fundamentals of Clinical Chemistry*, 2nd ed. Philadelphia: Saunders, 1976.

Wood, P. D. et al. "Plasma lipoprotein concentrations in middle aged runners," *Circulation* **50**, 111–115, (1974).

POST-TEST

1. Match the electrophoretic name with the name derived from Svedberg sedimentation constants.

   _____ 1. chylomicrons   a. extremely-low-density lipoproteins
   _____ 2. alpha   b. very-low-density lipoproteins
   _____ 3. prebeta   c. low-density lipoproteins
   _____ 4. beta   d. high-density lipoproteins

2. The lipoprotein that is the major carrier of cholesterol is:
   a. HDL
   b. LDL
   c. VLDL
   d. chylomicrons

3. The lipoprotein that is the major carrier of endogenous triglycerides is:
   a. HDL
   b. LDL
   c. VLDL
   d. chylomicrons

4. In lipoproteins, as the percentage of lipid increases, the Sf value will

   _____ .

   a. increase
   b. decrease

5. Define the following words by matching the word with the correct definition.

   _____ 1. hyperlipemia
   _____ 2. hyperlipidemia
   _____ 3. hyperlipoproteinemia
   _____ 4. lactesence
   _____ 5. lipogenesis
   _____ 6. lipolysis
   _____ 7. steatorrhea

   a. increase in plasma concentration of one or more lipoprotein families
   b. increase in concentration of any plasma lipid constituent
   c. increase in concentration of triglycerides in the form of VLDL or chylomicrons
   d. synthesis of fatty acids from glucose and intermediates of the glycolytic pathway
   e. hydrolysis of triglycerides to form fatty acids and glycerol
   f. presence of excessive lipid in the stool
   g. milky or turbid serum in appearance
   h. definition not given

6. A whole blood specimen was received in the laboratory. After centrifugation, the technologist noted that the serum was cloudy in appearance. The

cloudiness could be caused by (may be one or more answers):
   a. cholesterol
   b. endogenous triglycerides
   c. chylomicrons
   d. phospholipids

7. Match the basic formula with the lipid class:
   _____ 1. cholesterol
   _____ 2. free fatty acid
   _____ 3. triglyceride
   _____ 4. glycerolphosphatide

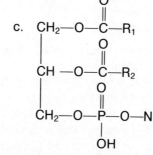

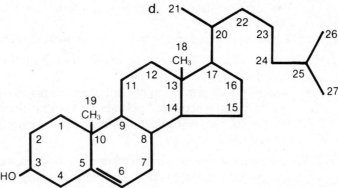

8. Two fasting lactescent serum samples were placed overnight in the refrigera-
    tor at 4°C. The next day the samples were removed. Sample #1 was still
    lactescent with a cream layer over a cloudy infranatant. Sample #2 was clear
    with only a cream layer. *From this observation,* you can conclude that (may
    be one or more answers):
    A. Sample #1 contains:
       a. elevated cholesterol
       b. elevated triglycerides

        c.   chylomicrons
        d.   elevated free fatty acids

   B. Sample #2 contains:
        a.   elevated cholesterol
        b.   elevated endogenous triglycerides
        c.   chylomicrons
        d.   elevated free fatty acids

9. Which of the complex lipids contain carbohydrates?
   a.   sphingomyelin
   b.   lecithin
   c.   cerebrosides
   d.   cholesterol
   e.   cephalins

10. From the list of vitamins below, circle all that are fat soluble:
   a.   vitamin A
   b.   vitamin B12
   c.   vitamin C
   d.   vitamin D
   e.   vitamin E
   f.   vitamin K

11. Below are statements describing the normal metabolism of lipids. Circle all true statements.
   a.   Bile is needed for proper fat digestion because it allows the lipids to form small micelles with greater surface area.
   b.   Pancreatic enzymes hydrolyze the lipids in the small intestines.
   c.   The longer chain fatty acids are incorporated into chylomicrons and then enter directly into the portal blood to be carried to the liver.
   d.   The liver synthesizes lipoproteins, which allow the lipids to be carried in the blood.
   e.   Lipids are stored in the liver, readily available for energy if needed.

12. It is believed the HDL fraction acts in all of the following ways except:
   a.   activates lipoprotein lipase
   b.   aids in the esterification of cholesterol in plasma
   c.   aids in the transport of lipids across the intestinal wall
   d.   transports excess cholesterol from extrahepatic cells to the liver for excretion

13. Circle all true statements below that apply to cholesterol.
   a.   The major portion of cholesterol in the plasma is in the free form.
   b.   Digitonin will precipitate cholesterol esters and allow the separation of free cholesterol from the esterified form.
   c.   The major route for cholesterol excretion is through the bile.
   d.   The cholesterol found in the blood is derived entirely from dietary cholesterol.
   e.   Cholesterol is known as depot or neutral fat and is the major constituent of adipose tissue.
   f.   Cholesterol contributes freely to the energy pool of the body.

14. A cholesterol ester is cholesterol combined with:
    a. glycerol
    b. fatty acids
    c. an amino alcohol
    d. phosphoric acid

15. The major site of cholesterol esterification is:
    a. intestinal mucosa
    b. liver
    c. heart
    d. vascular walls

16. The following are statements regarding ketone bodies. Circle all correct statements:
    a. Ketone bodies are products of triglyceride metabolism.
    b. An excess of ketone bodies produces acidosis.
    c. Ketone bodies consist of large amounts of beta-hydroxybutyric acid with lesser amounts of acetone and acetoacetic acid.
    d. Ketone bodies are produced when fat is metabolized.

17. In abetalipoproteinemia, apo B protein, the protein associated with lipid transport across the intestinal wall, is missing. Explain why the findings listed below may be associated with this disease.
    a. vitamin deficiency
    b. steatorrhea
    c. acanthocytosis
    d. electrophoresis shows absence of beta and prebeta bands

18. Below each Frederickson's type are statements that describe findings associated with each hyperlipidemia. Circle *all* correct answers or fill in the blanks.
    Type I
    a. The appearance of the serum is: clear/ slightly turbid/ turbid.
    b. After overnight refrigeration at 4°C, the appearance of this serum is: unchanged/ cream over clear/ cream over cloudy.
    c. The assay results will show a significant increase in: cholesterol/ triglycerides.
    d. The electrophoresis will show the: presence of chylomicrons/ absence of chylomicrons.
    e. The probable cause of this hyperlipidemia is _____
       _____

    Type IIA
    a. The appearance of the serum is: clear/ slightly turbid/ turbid.
    b. After overnight refrigeration at 4°C, the appearance of this serum is: unchanged/ cream over clear/ cream over cloudy.
    c. The assay results will show a significant increase in: cholesterol/ triglycerides.
    d. The electrophoresis will show the: presence of chylomicrons/ absence of chylomicrons.
    e. The probable cause of this hyperlipidemia is _____
       _____

Type IIB
a. The appearance of the serum is: clear/ slightly turbid/ turbid.
b. After overnight refrigeration at 4°C, the appearance of this serum is: unchanged/ cream over clear/ cream over cloudy.
c. The assay results will show a significant increase in: cholesterol/ triglycerides.
d. The electrophoresis will show the: presence of chylomicrons/ absence of chylomicrons.
e. The probable cause of this hyperlipidemia is _____
    _____

Type III
a. The appearance of the serum is: clear/ slightly turbid/ turbid.
b. After overnight refrigeration at 4°C, the appearance of this serum is: unchanged/ cream over clear/ cream over cloudy.
c. The assay results will show a significant increase in: cholesterol/ triglycerides.
d. The electrophoresis will show the: presence of chylomicrons/ absence of chylomicrons.
e. The demonstration of _____ is used to make a positive diagnosis of this Type.

Type IV
a. The appearance of the serum is: clear/ slightly turbid/ turbid.
b. After overnight refrigeration at 4°C, the appearance of this serum is: unchanged/ cream over clear/ cream over cloudy.
c. The assay results will show a significant increase in: cholesterol/ triglycerides.
d. The electrophoresis will show the: presence of chylomicrons/ absence of chylomicrons.
e. The probable cause of this hyperlipidemia is _____
    _____

Type V
a. The appearance of the serum is: clear/ slightly turbid/ turbid.
b. After overnight refrigeration at 4°C, the appearance of this serum is: unchanged/ cream over clear/ cream over cloudy.
c. The assay results will show a significant increase in: cholesterol/ triglycerides.
d. The electrophoresis will show the: presence of chylomicrons/ absence of chylomicrons.
e. Probable cause _____

19. Below is a list of items. Indicate how a nonfasting specimen might affect test results.
    a. serum appearance
    b. triglycerides
    c. lipoprotein electrophoresis
    d. cholesterol

20. Cholesterol may be elevated in all of the following (may be one or more answers):

    a.   obstructive disease of the liver
    b.   diabetes mellitus
    c.   nephrotic syndrome
    d.   after estrogen administration
    e.   hypothyroidism

21. The assay procedure for cholesterol using concentrated H_2SO_4 and acetic
     anhydride that results in a green colorimetric reaction is called:
    a.   Bloors
    b.   Zalkis, Zak, Boyle
    c.   Kessler and Lederer
    d.   Libermann–Burchard

22. In most triglyceride assay methods, the triglycerides are hydrolyzed either
     enzymatically or chemically. The portion of the molecule then used in quan-
     titation is _____ .
    a.   fatty acids
    b.   glycerol

23. Shown below are lipoprotein patterns obtained on agarose gel. Match each
     pattern with the lipoprotein phenotype.
     _____   1.   Normal
     _____   2.   Type IV
     _____   3.   Type IIA
     _____   4.   Type I

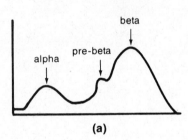

(a)

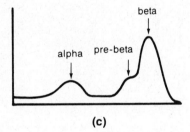

(c)

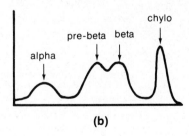

(b)

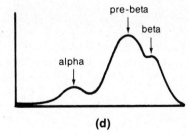

(d)

24. The following are statements regarding the lipid-storage diseases. Circle all
     true statements.
    a.   With the exception of Fabry's disease, the mode of transmission of lipid-
         storage disorders is autosomal dominant.

    b.   Sphingolipids are found primarily in cell membranes and CNS.

    c.   The disease results from inadequate synthesis of sphingolipids.

    d.   Most lipid-storage diseases cause mental retardation or CNS impairment.

25. Saponification refers to:
    a.   acid hydrolysis
    b.   alkali hydrolysis
    c.   enzymatic hydrolysis

26. The reference method for cholesterol is:
    a.   Kessler and Lederer
    b.   Shoenheimer–Sperry
    c.   Zalkis, Zak, Boyle
    d.   Abell–Kendall

ANSWERS

1. 1. a
    2. d
    3. b
    4. c

2. b

3. c

4. a

5. 1. c
    2. b
    3. a
    4. g
    5. d
    6. e
    7. f

6. b, c

7. 1. d
    2. b
    3. a
    4. c

8. A. b, c
    B. c

9. c

10. a, d, e, f

11. a, b, d

12. c

13. c

14. b

15. b

16. a, b, c, d
17. a. Fat-soluble vitamins are not absorbed
     b. When fat is not absorbed, it remains in the stool and produces fatty stools
     c. Essential fatty acids not absorbed—cell membrane damage of RBCs leads to formation of acanthrocytes
     d. Lipids are not absorbed and β-lipoprotein is needed for synthesis of chylomicron, VLDL, and LDL; therefore, no beta or prebeta bands are present on electrophoresis.
18. Type I
     a. turbid
     b. cream over clear
     c. triglycerides
     d. presence of chylomicrons
     e. deficiency of lipoprotein lipase
     Type IIA
     a. clear
     b. unchanged
     c. cholesterol
     d. absence of chylomicrons
     e. inherited disorder, probably defective receptor sites on cells for LDL uptake
     Type IIB
     a. clear or slightly turbid
     b. unchanged
     c. cholesterol and triglycerides
     d. absence of chylomicrons
     e. secondary lipidemia; may be associated with hypothyroidism, diabetes mellitus or nephrotic syndrome
     Type III
     a. turbid
     b. unchanged
     c. cholesterol and triglycerides
     d. absence of chylomicrons
     e. "the floating beta" (floating beta is an abnormal VLDL and migrates between beta and prebeta)
     Type IV
     a. turbid
     b. unchanged
     c. triglycerides
     d. absence of chylomicrons
     e. overproduction of VLDL by the liver or decreased catabolism of VLDL
     Type V
     a. turbid
     b. cream over cloudy
     c. cholesterol and triglycerides
     d. presence of chylomicrons
     e. deficiency of apo-C-II, activator of LPL

19. a. serum may be cloudy due to the presence of chylomicron or VLDL
    b. assay results will show an increase
    c. chylomicron may be present with an increased VLDL
    d. will not be affected
20. a, b, c, e
21. d
22. b
23. 1. a
    2. d
    3. c
    4. b
24. b, d
25. b
26. d

C. Magnesium Methodology
    1. Specimen Requirements
    2. Dye-Complexing Method
        a. Principle
        b. Test
    3. Atomic Absorption (Reference Method)
    4. Reference Range
D. Clinical Significance
    1. Decreased Magnesium (Hypomagnesemia)
    2. Increased Magnesium (Hypermagnesemia)

III. COPPER
    A. Metabolism
    B. Functions of Copper
    C. Copper-Storage Diseases
        1. Wilson's Disease (Hepatolenticular Degeneration)
            a. Laboratory Findings
        2. Menke's Kinky-Hair Syndrome
    D. Copper Methodology
        1. Specimen Requirements
        2. Test Methods
        3. Reference Range

IV. IRON
    A. Types of Iron-Containing Compounds
        1. Heme Compounds
            a. Hemoglobin
            b. Myoglobin
            c. Oxidation–Reduction Enzymes
        2. Nonheme compounds
            a. Ferritin
            b. Hemosiderin
            c. Transferrin
    B. Iron Metabolism
    C. Iron Methodology
        1. Specimen Requirements
        2. Test
            a. Total Iron Methodology With Protein Precipitation
            b. Total Iron Methodology With No Protein Precipitation
            c. Total Iron BInding Capacity (TIBC)
        3. Reference Range
    D. Clinical Interpretation
        1. Iron Deficiency Anemia
            a. Etiology
            b. Laboratory Findings
            c. Case History, Iron Deficiency Anemia

CHAPTER 8
INORGANIC IONS

OBJECTIVES

On completion of this chapter, you will be able to:

1. Define selected keywords associated with bone pathology.
2. List the functions of calcium and phosphorus in the body.
3. Describe the factors that influence mineral absorption variability.
4. List the percentage range for each of the three forms of calcium found in the body.
5. Discuss the effect of protein and pH changes in the interpretation of total calcium values.
6. Describe the three major hormonal mechanisms and feedback systems responsible for maintaining calcium homeostasis.
7. For the diseases listed below, discuss etiology and expected laboratory findings associated with each disease state:
   a. primary hyperparathyroidism
   b. secondary hyperparathyroidism
   c. primary hypoparathyroidism
   d. osteomalacia or rickets
   e. osteoporosis
   f. Paget's disease
8. Describe the specimen requirements for calcium and phosphorus.
9. Discuss the various types of assay methods that are used to determine calcium and phosphorus concentration.
10. List conditions that are associated with hypercalcemia.
11. Describe the distribution, functions, and regulation of magnesium in the body.
12. Describe the specimen requirements for magnesium assay.
13. Discuss the methodology used for magnesium assay.

14. Describe the etiology of hypermagnesemia and hypomagnesemia and correlate the clinical findings associated with each state.

15. Describe the distribution and trace the metabolism of copper in the body.

16. Discuss the functions of copper in its role as an oxidase enzyme.

17. List two types of copper-storage disorders.

18. Describe the etiology of Wilson's disease and correlate laboratory findings with this disease state.

19. Describe the specimen requirements for copper.

20. Discuss the methodology used for copper assay.

21. Describe the chemical composition of heme and hemoglobin.

22. Divide iron-containing compounds into heme and nonheme groups; describe the oxidation state of iron in each group and list three examples of compounds found in each group.

23. Trace the metabolism of iron in the body.

24. Describe specimen requirements for iron.

25. Discuss the methodology used for iron assay.

26. Describe the expected laboratory findings associated with:
    a. iron deficiency anemia
    b. anemia of chronic disorders
    c. iron overload disorders
    d. hepatitis
    e. nephrosis

27. Explain how ferritin assays may be used to differentiate between iron deficiency anemia and the anemia of chronic disorders.

GLOSSARY OF TERMS

Bone remodeling. The process whereby bone is destroyed and renewed. Remodeling involves both osteoblasts (bone formation) and osteoclasts (bone resorption).

cAMP. Cyclic adenosine monophosphate acts as the "second messenger" in the hormonal mechanism of action of peptide hormones (PTH, calcitonin).

Hydroxyapatite. Principal crystalline component of bone composed principally of calcium, phosphorus, and hydroxyl ions, $Ca_{10}(PO_4)_6(OH)_2$.

Hydroxyproline. Imino acid that is present in high concentration in collagen which forms the bone matrix (nonmineralized portion of the bone). In an imino acid, a $>NH$ group replaces the $—NH_2$ group.

Osteoblasts. Cells whose principal function is bone formation. They are located on the advancing surfaces of developing bone. Osteoblasts make collagen or the protein portion of the bone. It is postulated that osteoblasts, osteoclasts, and osteocytes are derived from the same group of osteoprogenitor cells. However, new evidence suggests that osteoblasts and osteocytes have a common precursor, but the osteoclast precursor may be a circulating monocyte.

Osteoclastic osteolysis. Osteolysis performed by osteoclasts.

Osteoclasts. Cells that participate in bone resorption. They are found in concavities in the surface of bone. A mechanism of absorption may involve secretion of proteolytic enzymes with collagenolytic activity, however the resorptive process has not been fully solved.

Osteocyte. Calcified bone cell found in fully formed bone. An osteoblast becomes an osteocyte when it becomes surrounded by newly synthesized unmineralized matrix (osteoid). It is believed osteocytes may also resorb bone.

Osteocytic osteolysis. Osteolysis performed by osteocytes. Osteocytes may release mineral from bone to blood.

Osteoid. Nonmineralized bone or the protein portion of the bone rich in the amino acid lysine, and the imino acids, proline and hydroxyproline.

Osteomalacia. Bone condition in which there is an increase in uncalcified bone (osteoid) due to impairment of mineralization. Bones are soft.

Osteoporosis. A bone condition associated with a decreased mass of bone with normal mineralization of the existing bone. The structure is fragile.

Rickets. Osteomalacia that occurs before the epiphyseal plates are closed.

Steatorrhea. The presence of fat in the feces. Lack of pancreatic enzymes, lack of bile, or intestinal transport deficiencies may lead to this condition.

Tetany. Increased neuromuscular irritability, which may be caused by hypocalcemia or hypomagnesemia. Symptoms may include muscle cramping, tingling sensations in fingers and toes, or contraction of muscle groups in severe tetany.

CALCIUM AND PHOSPHORUS

Calcium and phosphorus are distributed in the body in the following way: (1) over 99% of the body calcium and 80–85% of the body phosphorus are contained in bones and teeth and (2) approximately 1% of the body calcium and 15% of the body phosphorus are present in the extracellular fluids and soft tissue. Essentially all calcium in blood is in the plasma. Calcium is absent from red blood cells. The major portion of the phosphorus in the blood is bound to lipids (phospholipids) and proteins (phosphoproteins). Only a small amount is present as inorganic phosphorus. Serum phosphate decreases immediately after eating or following the administration of glucose or insulin (probably due to the phosphorylation of carbohydrate metabolites). Children have higher serum phosphate levels than adults.

FUNCTIONS OF CALCIUM AND PHOSPHORUS

In the body, calcium has the following functions:

1. *Structural Role.* Calcium in combination with phosphorus and hydroxyl ions form the crystalline (hydroxyapatite) portion of the skeleton.

2. *Activator.* Calcium may act as an effector or coupling factor in the cAMP second messenger mechanism of hormonal action, which involves phosphorylation of metabolic enzymes. Calcium is involved in activating selected enzymes.

3. *Blood Coagulation.* In blood coagulation, calcium is listed as factor IV. It is required for the conversion of prothrombin to thrombin.

4. *Skeletal and Cardiac Muscle Contraction.* In muscle, calcium activates a reaction between actinomysin and ATP that causes muscular contraction. The ability of the muscle to contract is lost in the absence of calcium. In the heart, calcium is responsible for contraction function, impulse formation, and impulse conduction.

5. *Nerve Impulse Transmission.* Calcium is believed to function as an important coupling factor in neurotransmitter release.

6. *Regulation of Membrane Ion Transport and Membrane Stability.* Calcium maintains rigidity of plasma membranes and influences membrane permeability to ions. A reduction in ionized calcium increases sodium permeability and enhances the excitability of tissues. Tetany may result. When ionized calcium is increased, the effect is opposite (muscle weakness).

7. *Milk Production.* Calcium is necessary for milk production in lactating females.

8. *Cellular Secretion.* Calcium induces the secretion of preformed cell products. In the presence of cytoplasmic calcium, amylase and insulin stored in secretory vesicles are secreted by exocytosis.

In the body, phosphorus has the following functions:

1. *Structural Role.* Phosphorus is a constituent of the crystalline portion of the bone. Phospholipids are major constituents of cell membranes. Nucleic acids are phosphate polymers.

2. *Carbohydrate Metabolism.* Phosphorus participates in the formation of hexose and triose phosphates in the intermediary metabolism of carbohydrates.

3. *Stored Energy.* Purine nucleotides provide the cell with stored energy in the form of adenosine triphosphate (ATP). Creatine phosphate is a stored-energy source in muscle.

4. *Regulatory Functions.* Cyclic adenosine monophosphate (cAMP) is a second messenger in a hormonal mechanism of action. The phosphorylation of proteins serves as a means of regulating their function.

5. *Blood Buffer.* The phosphate buffer system is an important urine and blood buffer.

ABSORPTION VARIABILITY

Milk and milk products are the major sources of dietary calcium and phosphorus. Many factors account for the variability in absorption of calcium and phosphate from the small intestine.

Concentration in the Diet

The higher the concentration of calcium and phosphorus in the intestine, the greater the absorption of both.

Calcium is absorbed by a combined process of active transport and passive diffusion. At normal calcium intake, the amount absorbed by the passive process is small (only about 15%). The active component is saturable, carrier-mediated, and represents the dominant process of calcium absorption at physiological concentrations. It is suggested that a protein, calcium-binding protein, which is induced by activated vitamin D, may serve as the actual carrier, but evidence indicates that the protein may not function in exactly this fashion. Calcium is actively secreted into the gastrointestinal tract in the digestive secretions (endogenous fecal calcium); therefore, the net calcium absorption is equal to the total calcium absorption minus the endogenous fecal calcium.

Like calcium, phosphate absorption proceeds by the combined processes of passive diffusion and active transport. The active phosphate component, believed to be the dominant process in phosphate absorption, cannot be saturated. Therefore, on a molar basis, the net absorption of phosphate is always higher than that of calcium (two to three times higher). The active transport system for phosphorus is responsive to activated vitamin D. Like calcium, phosphorus is also secreted into the gastrointestinal tract (endogenous fecal phosphorus).

A diet high in either calcium or phosphate tends to decrease the absorption of the other. For example, cow's milk, which is high in phosphate content, is believed to cause hypocalcemia in the newborn. A diet high in calcium decreases phosphate absorption because insoluble calcium phosphate is formed in the intestine.

pH

Calcium and phosphorus are both absorbed throughout the small intestinal tract. A low pH enhances absorption; therefore, the rate of absorption is highest in the duodenum, but owing to the greater length of the ileum, a greater quantity is absorbed in this portion of the small intestine. At an alkaline pH, both calcium and phosphorus form insoluble compounds.

Presence of Activated Vitamin D

Activated vitamin D is *essential* for calcium absorption and promotes phosphate absorption. However, vitamin D is not required for phosphate absorption. It is believed that the calcium-binding protein may be dependent on activated vitamin D or may be induced by the presence of activated vitamin D. The exact role of vitamin D in calcium transport has not been clarified. There is agreement, however, that the active transport component is under the influence of activated vitamin D.

Parathyroid Hormone

Parathyroid hormone (PTH) enhances calcium and phosphate absorption in an indirect way. PTH stimulates the hydroxylation of vitamin D in the kidney, thus increasing activated vitamin D. The activated vitamin D increases the absorption of both calcium and phosphorus.

High Protein Diet

The presence of certain amino acids (e.g., lysine and arginine) enhance calcium absorption.

Steatorrhea

Steatorrhea inhibits calcium absorption because fats form insoluble compounds with calcium, leading to calcium loss in the feces. Steatorrhea also decreases the intestinal absorption of vitamin D (a fat-soluble vitamin), and both calcium and phosphorus absorption are decreased in vitamin D deficiencies.

Diets

Both phytic acid (which occurs in cereal grains, especially oats) and oxalic acid (found in spinach) decrease calcium absorption due to the formation of insoluble compounds. Metal cations such as iron, lead, manganese, and aluminum form insoluble phosphates and hence decrease phosphate absorption.

Hormonal Influences

Growth hormone increases the intestinal absorption of phosphorus. In a normal child during puberty, the inorganic phosphorus concentration may exceed the adult reference range. Cortisol decreases the absorption of calcium by serving as an antagonist to the action of vitamin D.

Age

Calcium absorption decreases with age due to decreased intestinal absorption. Elderly persons may have decreased total calcium values.

SERUM FORMS OF CALCIUM

Calcium is present in the serum in three different forms: (1) ionized or free (45–50%); (2) complexed (5–10%); and (3) protein bound (40–50%).

The complexed form is bound to constituents present in serum, mostly citrate and phosphate. The protein bound form is bound mostly to albumin; however, a small amount may be bound to globulins. The ionized and complexed forms are diffusible and can cross cell membranes. The ionized fraction is considered to be the physiologically active component. The protein bound form is not diffusible, cannot cross cell membranes, and is not physiologically active.

PROTEIN AND pH EFFECTS ON CLINICAL INTERPRETATION OF TOTAL CALCIUM CONCENTRATION

In the clinical laboratory, calcium concentration is generally measured as total calcium, which includes calcium in all forms. Since only the ionized form is physiologically active, total calcium concentration may be misleading. Serum protein concentration and pH are factors that must be considered for clinical interpretation of total calcium values.

Protein Effects

In disease states that result in low albumin concentration (e.g., renal disease), the concentration of total calcium may be decreased since the protein-bound calcium is less than normal. However, the ionized calcium generally remains normal. Since the ionized calcium is the physiologically active form, tetany is rarely associated with hypoalbuminemia.

In clinical practice, total calcium concentration may be corrected for variations in serum albumin concentration by the following method. Determine the deviation of patient albumin from the mean normal albumin concentration that has been established for the laboratory. When the sample albumin is below the mean normal value, add 0.8 mg/dL to the total calcium value for every 1.0 g/dL deviation of albumin from normal; when the sample albumin is above the mean normal value, subtract 0.8 mg/dL from the total calcium value for every 1.0 g/dL deviation of albumin from normal. For example:

Laboratory mean normal albumin = 4.5 g/dL
Patient total calcium value = 8.0 mg/dL
Patient albumin concentration = 2.5 g/dL
Corrected total calcium = 8.0 + 1.6 = 9.6 mg/dL

Hyperproteinemia may cause increased total calcium. Some myeloma globulins can bind enough calcium to increase the total calcium without affecting the ionized fraction.

pH Effects

About 90% of the protein-bound calcium is bound to albumin (largely to carboxyl groups) and this binding is pH dependent. In acidotic conditions, symptoms of hypercalcemia may be present even though the total calcium is within normal limits. The acidosis causes an increase in the ionic calcium fraction. In alkalotic conditions, tetany may develop even though the total calcium is within normal limits. The alkalosis causes a decrease in the ionized fraction.

REGULATORY MECHANISMS FOR CALCIUM AND PHOSPHORUS

Blood calcium and phosphate levels are influenced directly or indirectly by: (1) parathyroid hormone (PTH); (2) calcitonin (CT, thyrocalcitonin, TCT); (3) vitamin D; and (4) other factors.

Parathyroid Hormone

Parathyroid hormone (MW 9500) is a single-chain polypeptide composed of 84 amino acids. There are no intrachain disulfide linkages. The hormonal mechanism of action is via adenylate cyclase and cyclic adenosine monophosphate (cAMP). Serum PTH exists in three forms: the intact molecule, an amino terminal fragment (N-terminal fragment), and a carboxy terminal fragment (C-terminal fragment). The PTH gland secretes predominately intact PTH and perhaps a few C-terminal fragments. The kidney (principally) and the liver cleave the intact molecule into two fragments: the amino terminal and the carboxy terminal fragments. Only the intact molecule and the N-terminal fragment are the biologically active components. The C-terminal fragment is not believed to be biologically active.

The biological half-life of the intact and the N-terminal portion is short (about 10 minutes); the half-life of the C-terminal portion is longer (1 hour or longer). PTH is eliminated via the urine by both glomerular filtration (principal route for C-terminal) and tubular secretion (N-terminal fragments only). The fragments are presumed to be degraded by the kidney, since urinary RIA measurements show minimal amounts present. The liver is also capable of molecular cleavage of PTH, but this process is not believed to be a factor in the elimination of the hormone from the circulation.

Usually, there are two pair or four parathyroid glands found in thyroid tissue. A few individuals may have as many as six to eight parathyroid glands. It is rare to find individuals with less than four glands.

PTH is released by parathyroid glands when extracellular ionized calcium is decreased. PTH secretion is regulated by serum calcium through a negative feedback mechanism. As serum ionized calcium increases, PTH secretion is inhibited. PTH acts on three organ or tissue sites to raise serum calcium levels; these are described below.

Bone
In bone, PTH causes the release of calcium from bone into blood. PTH stimulates both osteocytic osteolysis and the resorption of bone by preexisting osteoclasts. PTH also stimulates the formation of new osteoclasts. It causes the organic matrix of the bone to break down and the mineral component to dissolve, thereby raising blood calcium.

Bone is composed of two major components: (1) an organized organic matrix (osteoid) composed of collagen rich in the amino acid, lysine, and imino acids, proline and hydroxyproline; and (2) an inorganic crystalline portion composed largely of calcium and phosphorus (hydroxyapatite crystals) deposited in the organic matrix. Bone is a continuously metabolically active tissue with bone remodeling proceeding in the following manner:

1. Following activation (with PTH or local stimuli), precursor cells differentiate into osteoclasts that resorb a localized area of mineralized bone.
2. An intermediate phase follows in which mononuclear cells lie the surface of resorbed bone. These cells may represent preosteoblasts or postosteoclasts.
3. Osteoid is laid down followed by mineralization.

In humans, the complete remodeling process requires about four months.

Under abnormal conditions, such as prolonged elevated PTH activity, osteoblastic activity increases to rebuild destroyed bone. Osteoblasts make alkaline phosphatase hence the serum skeletal alkaline phosphatase concentration increases. Circulating skeletal alkaline phosphatase is regarded as a reflection of osteoblastic activity and bone turnover rate. As the organic matrix of the bone is destroyed, hydroxyproline is released. Hydroxyproline released from bone matrix is not reutilized for collagen biosynthesis rather it is excreted in the urine. Hydroxyproline excretion has been used clinically as an index of bone matrix resorption.

Kidney

In the kidney, PTH increases distal tubular reabsorption of calcium and decreases the reabsorption of phosphate in the proximal tubule (i.e., increases urinary loss of phosphates).

Under normal conditions, only the diffusible calcium and phosphate (ionized and complexed) are filtered by the renal glomeruli. Most of the diffusible calcium (97–99%) is reabsorbed by the tubules, and the reabsorption process occurs at multiple sites along the nephron. Approximately 65% of the filtered calcium load is reabsorbed in the proximal tubule, 25% in the ascending loop of Henle, and the remaining 10% in the distal portion of the tubule. The calcium reabsorption process in the proximal tubule and loop of Henle is not saturable and is insensitive to PTH. Calcium reabsorption in the distal nephron is PTH dependent and appears to be saturable. PTH is believed to regulate the set point for calcium reabsorption based on systemic requirements.

Phosphate reabsorption is largely limited to the proximal tubule and the reabsorptive process is saturable. Therefore, phosphorus may be regarded as a threshold substance. If the amount of filtered phosphorus exceeds the reabsorptive capacity, phosphorus will be lost in the urine.

Since the mechanism of hormonal action of PTH is via cAMP, urinary cAMP (nephrogenous cAMP) may be used as an indicator of PTH activity.

PTH also stimulates an increase in the 1-hydroxylase enzyme in the kidney leading to an increase in activated vitamin D.

Intestine

PTH is believed to increase calcium absorption from the intestinal tract in an indirect way. The increase in activated vitamin D from the kidney acts on the intestinal tract to increase calcium and phosphorus absorption from the gut.

The mechanisms of action of PTH are summarized in Figure 8.1.

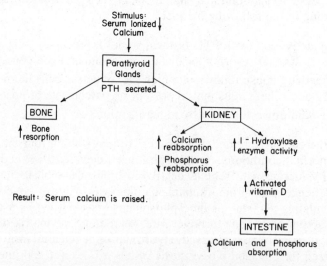

Figure 8.1. Mechanisms of action of parathyroid hormone.

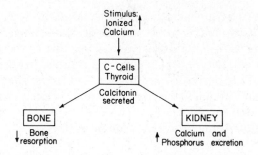

Figure 8.2. Mechanisms of action of calcitonin.

Calcitonin

Calcitonin (CT) or thyrocalcitonin (TCT) is a 32-amino-acid peptide with one disulfide bond. The molecular weight of the calcitonin monomer is 3000 g/mole. The circulation half-life for human calcitonin is approximately 5 minutes. Calcitonin is cleared from the blood and destroyed by liver and kidney. The hormonal mechanism of action is believed to be via cAMP.

Calcitonin is secreted by parafollicular or C cells of the thyroid in response to an increase in ionized calcium. It is believed that calcitonin protects against hypercalcemia. The bone and kidney are responsive to human calcitonin.

Bone

Calcitonin acts directly on the bone to inhibit bone resorption and the activity of bone resorbing cells (i.e., inhibits osteoclastic bone resorption and osteolytic osteolysis). Resorption of both the mineral and matrix phases of the bone are inhibited. Calcitonin also stimulates phosphorus uptake into bone.

Kidney

Calcitonin decreases the tubular reabsorption of calcium and phosphate as well as of sodium, potassium, and magnesium.

Calcitonin is also an important hormonal marker for the detection of medullary carcinoma of the thyroid. However, increased levels of calcitonin are found in a variety of other conditions (e.g., chronic hypercalcemia; thyroiditis; hypergastrinemia; renal failure; pheochromocytoma; and carcinoma of the lung, breast, and islet cell).

The mechanisms of action of calcitonin are summarized in Figure 8.2.

Vitamin D

Activated vitamin D refers to 1,25-dihydroxycholecalciferol [1,25-$(OH)_2D_2$ or 1,25-$(OH)_2D_3$]. Calciferol refers to two sterols: (1) cholecalciferol or vitamin D3, which is the form of the vitamin found in animal tissue, fish liver oils, and irradiated milk; and (2) ergocalciferol or vitamin D2, which is the form of the vitamin found in plants and irradiated yeast and bread. In humans, the hydroxylated forms of both calciferols are equivalent in potency and in biological activity. For the remainder of this discussion, D will be used to designate either the D2 or D3 form.

Most foodstuffs lack calciferols but contain inactive vitamin D precursors that require ultraviolet light for the conversion to calciferols. In the human, the natural supply of vitamin D is derived from the diet and from 7-dehydrocholesterol, a precursor of vitamin D found in the skin. Vitamin D in excess of needs is stored or metabolized to inactive products and excreted in the bile. Adipose tissue and muscle are the major storage sites for vitamin D.

Activation of Vitamin D

The activation of vitamin D is shown in Figure 8.3. The key below is used to interpret each numbered step of Figure 8.3.

1. The provitamins are activated in subcutaneous tissue by UV light (sunlight). Photolytic cleavage results in the formation of calciferol as shown in Figure 8.4.
2. A transport protein, vitamin-D-binding protein (alpha globulin), transports the vitamin D sterols to the liver where hydroxylation occurs at C-25 under the influence of the 25-hydroxylase enzyme to produce 25-hydroxyvitamin D (25-OHD).
3. The 25-OHD lacks metabolic activity and must undergo a second hydroxylation in the kidney. Two hydroxylase enzymes are present in the kidney: the 25-OHD-1-hydroxylase enzyme, which produces the physiologically active 1,25-(OH)$_2$D metabolite, and the 25-OHD-24-hydroxylase enzyme, which produces the inactive 24,25-(OH)$_2$D metabolite.

Functions of Activated Vitamin D

There are three target tissues for activated vitamin D: the intestine, bone, and kidney.

Intestine.   Activated vitamin D increases the intestinal absorption of both calcium and phosphorus. It is absolutely required for calcium absorption. Vitamin D increases the synthesis of high affinity calcium-binding protein. Phosphorus will be absorbed in vitamin D deficiency states. The process of phosphate absorption is independent of active calcium transport.

Bone.   Activated vitamin D is involved in bone resorption; PTH and 1,25-(OH)$_2$D potentiate each other in bone resorption. Present evidence suggests that

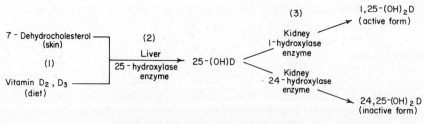

Figure 8.3.   Activation of vitamin D.

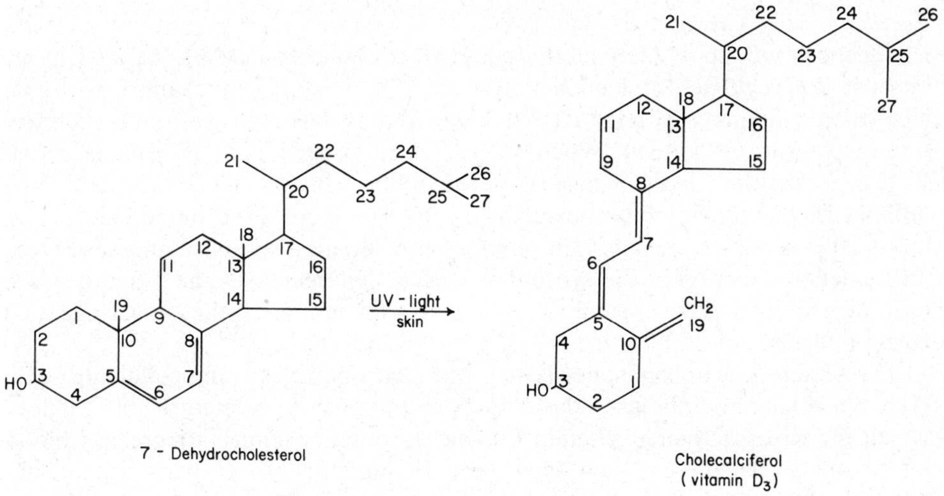

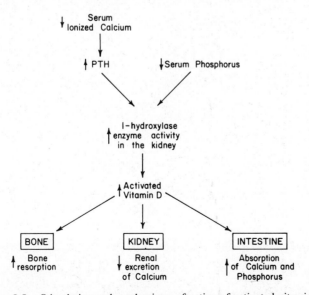

Figure 8.4. Formation of calciferol by photolytic cleavage.

activated vitamin D may stimulate osteolytic osteocytes and preformed osteo-clasts, but the formation of new osteoclasts is the domain of PTH.

Activated vitamin D does not appear to have any direct effect on bone forma-tion or mineralization. There is evidence that another vitamin D metabolite may influence the process of bone formation.

Kidney. Activated vitamin D enhances calcium retention in the serum (i.e., decreases renal excretion). The effect of the vitamin on renal handling of phos-phate is not totally clear at present.

The mechanisms of action of activated vitamin D are summarized in Figure 8.5.

Figure 8.5. Stimulation and mechanisms of action of activated vitamin D.

Feedback Mechanisms

Endogenous vitamin D (activated vitamin D) is now considered to be a hormone because it is regulated by feedback systems. The feedback mechanism with ionized calcium operates through PTH. PTH increases the activity of the 1-hydroxylase enzyme in the kidney. When ionized calcium decreases, PTH is secreted. PTH increases the 1-hydroxylase enzyme activity, which then increases activated vitamin D, thus raising blood calcium levels. The increase in blood calcium inhibits PTH secretion, which then removes the initiating stimulus. Increased ionized calcium decreases PTH secretion, which then decreases the 1-hydroxylase enzyme activity and the concentration of activated vitamin D, thus lowering blood calcium levels.

The feedback with phosphorus is direct and operates in the following way. Hypophosphatemia increases the 1-hydroxylase enzyme activity in the kidney, which increases activated vitamin D. The activated vitamin D increases blood phosphate levels, thereby removing the initiating stimulus. Hyperphosphatemia decreases the 1-hydroxylase enzyme activity and the concentration of activated vitamin D, thus lowering blood phosphate levels. These feedback mechanisms are illustrated in Figure 8.6. The result of these feedback systems is the regulation of calcium and phosphorus homeostasis within narrow limits. The proposed hormonal mechanism of action of activated vitamin D is the same as that for steroid hormones (see Chapter 12). The exogenous or dietary vitamin D still retains the vitamin status.

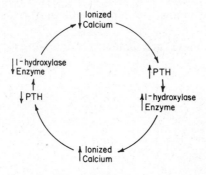

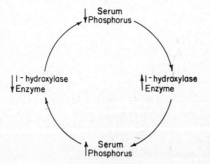

Figure 8.6. The 1-hydroxylase enzyme feedback mechanisms with ionized calcium and PTH (upper) and with phosphorus (lower).

Other Influencing Factors

Besides PTH, calcitonin, and activated vitamin D, several other hormones influence the blood concentration of calcium and phosphorus.

Glucocorticoids decrease bone formation by inhibiting osteoblast differentiation and function and increasing bone resorption. Glucocorticoids lower the serum concentration of calcium and 1,25-$(OH)_2$D leading to secondary hyperparathyroidism. Glucocorticoids also interfere with intestinal absorption of calcium.

Both androgens and estrogens are believed to produce a general anabolic effect on bone formation. Recent research shows evidence that estrogen stimulates the activity of the 1-hydroxylase enzyme. The development of osteoporosis following menopause is thought to be due to the decrease in estrogen levels. Estrogens inhibit the activity of osteoclasts.

Growth hormone increases absorption of phosphate. Children, at puberty, may have serum phosphate levels that exceed the normal adult reference range. Growth hormone also increases renal excretion of calcium.

LABORATORY FINDINGS IN DISEASES ASSOCIATED WITH ABNORMALITIES IN CALCIUM AND PHOSPHORUS

Primary Hyperparathyroidism

Primary hyperparathyroidism initially was regarded as rare, but the disorder is now recognized as being more common. The disease is an adult disease with a higher incidence in females than males (2 : 1).

Etiology
Primary hyperparathyroidism is caused by a single parathyroid adenoma (greater than 80%) or hyperplasia of parathyroid tissue. Parathyroid carcinoma is rare.

Clinical Presentations
Some patients may have no symptoms. The type of symptoms vary with the extent and the severity of the disease.

Renal stone formation is a complication most often associated with chronic mild hypercalcemia and modestly abnormal parathyroid function tests. The development of renal stones is probably due to the presence of increased calcium and phosphorus in the urine plus the alteration in urinary pH caused by the PTH effect on the proximal renal tubule. PTH inhibits renal bicarbonate resorption, thereby increasing urinary pH. Calcium phosphate is less soluble at an alkaline pH, thus promoting the formation of stones.

Bone disease is usually associated with moderate to severe hypercalcemia and markedly abnormal parathyroid function tests. Probably all patients with hyperparathyroid disease have an increased rate of bone turnover, but bone disease develops when the rate of bone resorption exceeds the rate of bone formation. The bone disease associated with hyperparathyroidism is called *osteitis fibrosa cystica*, which results in so-called brown tumors of the bone. These tumors are composed largely of osteoclasts. Radiographic findings of osteitis fibrosa cystica are considered to be diagnostic of hyperparathyroidism.

Gastrointestinal symptoms may be present. Ill defined abdominal pain, anorexia, nausea, vomiting, and constipation may be associated with hyperparathyroidism. There appears to be an increased incidence of peptic ulcer disease and pancreatitis in individuals with this disease.

The most common neuromyopathic symptoms include weakness and fatigability. Central nervous system symptoms include lethargy and confusion. Psychiatric symptoms include depression, memory loss, and personality changes.

Vascular abnormalities include increased incidence of hypertension. Hypercalcemia may also cause changes in the electrocardiogram.

Also in primary hyperparathyroidism, the renal clearance of uric acid is reduced leading to hyperuricemia. The elevated uric acid may lead to deposition of uric acid crystals in joints provoking acute gouty attacks. Calcium crystal deposition in joints also occurs as demonstrated by the presence of calcium pyrophosphate dihydrate crystals in joint fluid.

Laboratory Findings
The laboratory findings associated with hyperparathyroid disease include: (1) serum calcium, increased; (2) serum phosphate, decreased; (3) serum alkaline phosphatase, increased in about 25% of cases; (4) serum magnesium, decreased in 10–15% of cases; (5) serum chloride, increased; (6) serum uric acid may show modest increase; (7) serum PTH level inappropriately high when considered with the calcium level; (8) tendency toward metabolic acidosis due to decreased bicarbonate resorption caused by the PTH effects on the proximal renal tubule; and (9) increased renal phosphate clearance.

Case History, Primary Hyperparathyroidism
A 45-year-old female, who was employed as a medical technologist, was assigned to a reference range study project. She used her own blood sample as one of the normal donor subjects. The calcium result on her chemistry profile was elevated. She had no symptoms of illness; therefore, she did not visit her physician at that time. For the next four years, she followed the disease with periodic chemistry profiles and additional laboratory tests. Renal function was normal. Test results are listed in Table 8.1.

Discussion.   The intermittent pattern of serum calcium fluctuation may be a common finding associated with mild hypercalcemia. In mild hypercalcemia, the serum calcium value may periodically fall within the upper portion of the reference range; therefore, more than one sampling may be necessary to demonstrate the elevated calcium. The ionized calcium was increased, a finding associated with increased PTH. The 24-hour urinary excretion of calcium was increased. There were no significant changes in alkaline phosphatase, inorganic phosphorus, or uric acid. The serum magnesium was within normal limits. Thyroid hormone was assayed to rule out hyperthyroidism, a disease that can elevate serum calcium. The chloride:phosphate ratio was borderline. Although the PTH assay was within normal limits, the level was inappropriate when the serum calcium value was considered. Under normal circumstances, an elevated serum calcium would result in undetectable PTH levels. Only an autonomous, abnormal parathyroid secretes significant PTH in the presence of inappropriate hypercalcemia. The

Table 8.1. Hyperparathyroid Case History Laboratory Results

Year	Total Ca (8.5–10.2)[a] mg/dL	Ionized Ca (1.14–1.29)[a] mmol/L	PO4 (2.6–4.5)[a] mg/dL	ALP (17–60)[a] IU/L	Cl (98–106)[a] mmol/L	Uric Acid (2.6–7.0)[a] mg/dL
1978	10.9		3.5	21	105	3.8
1979	10.8		3.5	26		3.9
1980	10.3		3.4	24		3.7
1981	10.6		3.9	24	104	3.6
1982	10.4	1.35	3.2	23	106	4.0

Year	Urine Ca (100–250)[a] mg/24 hr	C-Terminal PTH (20–70)[a] μIEq/mL	Mg (1.8–2.4)[a] mg/dL	Thyroxine (5.5–11.5)[a] μg/dL
1982	337	37	2.1	8.0

[a] Reference range for components is indicated in parentheses.

above findings, that is, hypercalcemia with normal C-terminal PTH, are also findings in malignancy. This diagnosis was ruled out due to the length of time the disease persisted. Assay of the N-terminal PTH has been suggested as a test which would be helpful in differentiating malignancy from primary hyperparathyroidism if the diagnosis is questionable.

Secondary Hyperparathyroidism of Chronic Renal Disease

Etiology

Chronic renal disease may lead to secondary hyperparathyroid disease with resulting bone disease. Excessive PTH secretion results from hypocalcemia. The sequence of events believed to occur is illustrated in Figure 8.7. In renal disease, serum phosphates increase because the damaged nephrons are unable to excrete phosphates. There is a feedback mechanism with serum phosphate and the 1-hydroxylase enzyme in the kidney. When serum phosphates are increased, the 1-hydroxylase enzyme activity decreases, which reduces the synthesis of activated

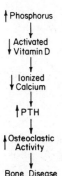

Figure 8.7. Proposed etiology of secondary hyperparathyroidism of chronic renal disease.

vitamin D. The decreased activated vitamin D reduces calcium absorption from the intestinal tract. The resulting hypocalcemia stimulates the release of PTH. PTH stimulates osteoclastic activity leading to bone disease (osteomalacia, osteitis fibrosa cystica).

Laboratory Findings
The laboratory findings associated with secondary hyperparathyroidism of renal disease include: (1) serum BUN, creatinine, and uric acid, increased; (2) serum albumin, decreased; (3) serum phosphate, increased; (4) serum calcium, decreased; (5) serum alkaline phosphatase, increased; and (6) serum PTH, increased.

Primary Hypoparathyroidism

Etiology
Hypoparathyroid disease is usually the result of postoperative complications following hyperparathyroid or thyroid surgery. PTH cannot be secreted in response to hypocalcemia. Rarely is hypoparathyroid disease spontaneous.

Clinical Presentations
In hypoparathyroid disease, the hypocalcemia causes an increase in neuromuscular irritability, which may result in the development of tetany. The most typical symptoms of tetany are muscle cramps and tingling sensations in fingers and toes. In severe tetany, cramps give way to contraction of muscle groups.

Mental symptoms may include irritability, impaired memory, and psychotic behavior. The skin may be dry and flaky, the nails and hair coarse and brittle. Electrocardiogram abnormalities may be present. Cataracts may develop in prolonged untreated hypocalcemia.

Laboratory Findings
The laboratory findings associated with hypoparathyroidism include: (1) serum calcium, decreased; (2) serum phosphate, normal or increased; (3) serum alkaline phosphatase, normal; and (4) urinary calcium and phosphorus, low.

Osteomalacia or Rickets

Osteomalacia may be defined as a bone condition in which there is a decrease in calcified bone due to an impairment of mineralization.

Etiology
Osteomalacia or rickets may be divided into two groups: (1) disease associated with the reduction of circulating vitamin D, which may result from inadequate UV exposure, inadequate dietary intake, vitamin D malabsorption, or abnormal vitamin D metabolism; and (2) disease associated with hypophosphatemia, which may result from inadequate dietary phosphorus intake, reduced intestinal absorption of phosphorus, or decreased renal phosphate reabsorption (renal wasting of phosphate).

Laboratory Findings

The laboratory findings associated with the lack of vitamin D are: (1) serum calcium and phosphorus, usually decreased; (2) serum alkaline phosphatase, increased; (3) serum PTH, increased due to the hypocalcemia; and (4) urinary hydroxyproline and cAMP may be increased due to the PTH effects on bone and renal tissue.

The laboratory findings associated with hypophosphatemia are: (1) serum phosphate, decreased; (2) serum calcium, normal; (3) serum alkaline phosphatase, increased; and (4) serum PTH, seldom increased.

If the etiology of the osteomalacia is questionable, 25-OHD may be measured. In osteomalacia associated with lack of vitamin D, the 25-OHD will be decreased.

Osteoporosis

Osteoporosis is a bone condition associated with a decrease in bone mass but with normal mineralization of the existing bone. Osteoporosis is the most common metabolic bone disease.

Etiology

Osteoporosis is associated with aging, hormonal deficiencies and excesses, nutritional deficiencies, genetic disorders, hematological malignancies, immobilization, and other disorders. It is found commonly in postmenopausal women and in those individuals being treated with long-term high doses of glucocorticoids.

Laboratory Findings

Most patients have no significant abnormalities of calcium, phosphorus, or alkaline phosphatase. PTH has been reported to be normal, low, or high. Radiologic studies may be helpful in diagnosis.

Paget's Disease or Osteitis Deformans

Paget's disease is a disorder of the bone associated with increased bone remodeling. The disease may involve only one bone or be more or less generalized.

Etiology

Once thought to be a rare disorder, X-ray studies have revealed Paget's disease to be a more common, often asymptomatic, affliction. It is most common in persons of Anglo-Saxon origin. An increased incidence is seen in both males and females with a positive family history of Paget's disease. The causative agent may be due to a slow virus infectious agent; however, evidence of a viral etiology has not been documented. Other hypotheses include an autoimmune process, an inborn error of connective tissue metabolism, neoplastic transformation of bone cells, an abnormality of vascular supply, or an abnormality of hormone secretion.

Laboratory Findings

The laboratory findings associated with Paget's disease include: (1) serum calcium and phosphorus, usually normal; (2) serum alkaline phosphatase, increased, the degree of increase may be very high depending on the extent of skeletal involvement; and (3) urine hydroxyproline, increased.

CALCIUM METHODOLOGY

Specimen Requirements

For calcium assay, the specimen should be fresh, fasting serum or heparinized plasma. Other anticoagulants remove calcium (EDTA chelates calcium, oxalates precipitate calcium as calcium oxalate), hence would be unsatisfactory. Hemoconcentration should be avoided upon collection. The determination should be performed within 24 hours. As with protein-bound substances, total calcium values change with posture. In the recumbent position, the values will be lower.

Total Calcium Methodology

Clark–Collip or Kramer–Tisdall
  Principle. Calcium is precipitated by oxalate to form calcium oxalate:

$$Ca^{2+} + C_2O_4^{2-} \rightarrow CaC_2O_4 \downarrow$$

The precipitate is washed and redissolved in sulfuric acid, which forms oxalic acid:

$$CaC_2O_4 + H_2SO_4 \rightarrow H_2C_2O_4 + CaSO_4$$

Redox titration of the oxalic acid with potassium permanganate is performed. Potassium permanganate acts as its own indicator, an excess is signaled by a pink color:

$$2KMn^{(VII)}O_4 + 5H_2C_2O_4 + 3H_2SO_4 \xrightarrow{70°C} K_2SO_4 + 2Mn^{(II)}SO_4 + 10CO_2 + 8H_2O$$

  Sources of Error. Some organic substances and magnesium (as magnesium oxalate) may coprecipitate with the calcium oxalate and these are capable of reducing the potassium permanganate, thus producing false high values.

Chelation with EDTA
  Principle. Dissociated metal ions (e.g., Ca^{2+}) and certain indicators form colored compounds. When all the metal ions are removed from the indicator, the indicator changes color. Calcium forms chelated complexes with ethylenediaminetetraacetic acid (EDTA). EDTA is used as a titrating agent to remove the calcium from the indicator complex. As the EDTA removes calcium, the indicator changes color. Indicators that are used include calcein, Cal-Red, Eriochrome Black T, and ammonium purpurate.

  Sources of Error. Calcium is removed slowly from the Ca-indicator complex resulting in a gradual color change making the end point hard to detect. Magnesium and other divalent ions (Fe, Cu, Zn) may also form complexes with EDTA depending on the pH; therefore, the pH is critical in this type of analysis. An alkaline pH is used to avoid interferences by other ions.

Cresolphthalein Complexone

Principle. Calcium, in the presence of HCl, is dissociated from the protein. Calcium then reacts with cresolphthalein complexone to form a color complex at a high pH (10–12). The pH is maintained by a buffer (diethylamine buffer). The intensity of the color is proportional to the calcium concentration. Another reagent (8-hydroxyquinoline) binds magnesium to prevent interference with this ion.

Chloranilic Acid

Principle. Calcium is precipitated by chloranilic acid to form Ca-chloranilate:

$$Ca^{2+} + chloranilate \rightarrow Ca\text{-}chloranilate \downarrow$$

The precipitate is washed and redissolved in EDTA at a high pH. EDTA chelates calcium and releases chloranilic acid. The liberated chloranilic acid is colored and can be measured photometrically:

$$Ca\text{-}chloranilate + EDTA \rightarrow Ca\text{-}EDTA + chloranilic\ acid\ (purple\ color)$$

Flame Photometry

The determination of calcium by flame emission photometry has met with little success. Sodium and potassium interfere and phosphates and sulfates inhibit calcium emission. Calcium is not easily excited, even in a hot flame.

Atomic Absorption (Reference Method)

As in flame emission methods, phosphates bind calcium and form compounds that do not dissociate in the flame. To prevent this interference, lanthanum (preferred) or strontium is added which binds preferentially to phosphates and prevents the formation of calcium phosphate (see Chapter 4). In some methods, proteins are precipitated (TCA) and removed before introduction of the sample into the flame. This increases the accuracy and sensitivity of the procedure.

Ionized Calcium Methodology

Since the physiologically active calcium is the ionized calcium, measurement of the ionized fraction is desirable. In the past, measurement of the ionized fraction was impractical because anaerobic samples were required. New instruments, now available, do not require anaerobic samples. The Radiometer Company offers an instrument that measures pH and ionized calcium in samples exposed to air. The instrument determines the sample pH, the actual ionized calcium and the ionized calcium corrected to a normal pH (7.4). On aerobic samples, the corrected ionized calcium value is valid as long as the actual blood pH of the patient is within normal limits.

In ion-selective electrode methodology, both a measuring and a reference electrode are necessary to measure the potential of a solution. The reference electrode maintains a constant voltage and the potential of the measuring electrode is made to respond proportionally to the concentration of the ion of interest. A membrane (i.e., for calcium and potassium electrodes) or glass (i.e., for sodium electrode) separates the electrode from the sample. When the electrode is in contact with the sample, selective ion transport, ion exchange, or ion complexing takes place in the membrane or glass material producing a voltage change across the membrane surface. This change in potential is measured by comparing the new potential across the measuring electrode to the reference electrode, which is in contact with the measuring electrode by a conductive pathway. This change in potential is recorded on a voltmeter and converted to a concentration value.

It has been suggested that the calcium level in CSF may be used as an indicator of ionized calcium. Cerebrospinal fluid is an ultrafiltrate of plasma; protein-bound calcium does not cross the blood brain barrier.

Various nomograms have been constructed to calculate ionized calcium based on serum total calcium and serum total protein. However, these calculated results are valid only when the patient's pH, protein species, and cephalin concentration (phospholipid) are normal. The extent of calcium binding varies with different protein and cephalin concentration (which bind calcium similar to protein). Therefore, any change in serum pH, in proportion of serum proteins, or in the concentration of phospholipids may alter the values obtained by these nomograms. A formula based on the nomogram by McLean and Hastings is:

ionized Ca^{2+} (mg/dL)

$$= \frac{(\text{total serum calcium in mg/dL} \times 6) - (\text{serum protein in g/dL} \times \frac{1}{3})}{\text{serum protein in g/dL} + 6}$$

URINE CALCIUM

The amount of calcium excreted per day is dependent on dietary factors; therefore, the patient should be placed on a controlled calcium diet three days prior to the urine collection. Twenty-four hour collections are usually obtained and the amount of calcium excreted per 24 hours is determined. Urinary calcium will precipitate as calcium oxalate and calcium phosphate crystals; therefore, the urine should be acidified with HCl before analysis is performed (if alkaline) to redissolve any crystals that may have precipitated. The calcium may then be determined by the same methodology used for serum.

Reference Range

Serum calcium (total)	8.5–10.2 mg/dL or 2.12–2.55 mmol/L
Serum calcium (ionized)	1.14–1.29 mmol/L (Radiometer Instrument)
Urine calcium	100–250 mg/24 hours

INORGANIC PHOSPHORUS METHODOLOGY

Specimen Requirements

For inorganic phosphorus assay, a whole blood specimen is obtained and allowed to clot, and the serum is then separated from the cells as soon as possible. The specimen should be free from hemolysis, since red blood cells contain large amounts of organic phosphates. A fasting specimen is recommended to eliminate the effects of carbohydrate metabolism (inorganic phosphorus decreases immediately following meals).

Acid Molybdate or Fiske and Subbarow Method

A protein-free TCA filtrate may be used. Tungstate filtrates cannot be used because phosphotungstic acid is formed from the tungstic acid and inorganic phosphate. When protein-free filtrates are used, most interfering substances and colors are removed. Newer methods, using smaller sample sizes, do not use protein-free filtrates.

Principle
The phosphate in the serum or filtrate is reacted with molybdate under controlled pH conditions to form a phosphomolybdate complex:

$$phosphate + ammonium\ molybdate \rightarrow ammonium\ phosphomolybdate$$

A reducing agent is added that reacts with the phosphomolybdate complex to form molybdeum blue, which is proportional to the amount of phosphate present:

$$phosphomolybdate\ complex + ascorbic\ acid \rightarrow molybdeum\ blue$$

Different reducing agents are used including ascorbic acid, stannous chloride, and aminonaphtholsulfonic acid. An acid pH favors the reduction of the phosphomolybdate complex. However, if the pH is too acid, phosphate esters undergo hydrolysis leading to erroneous values for inorganic phosphorus.

Urine Phosphorus

As with calcium, phosphorus values vary by dietary factors. Usually a 24-hour urine collection is obtained, and the amount of phosphorus excreted per 24 hours is determined. The methodology for urine phosphorus is the same as that used for serum phosphorus.

Reference Range

Serum inorganic phosphorus (adult)	3.0–4.5 mg/dL or 1.0–1.5 mmol/L
Serum inorganic phosphorus (infants in first year)	up to 6.0 mg/dL
Urine phosphorus	under 1.3 g/24 hours

CLINICAL SIGNIFICANCE

Many diseases are associated with alterations in the blood level of calcium and phosphorus. A partial list follows.

Increased Serum Calcium

Parathyroid-Mediated Conditions

Excessive PTH may be produced (1) in hyperparathyroid disease; (2) by tumors, especially those of the lung, kidney, and pancreas, which secrete PTH hormone (ectopic PTH production); and (3) in an inherited disease called familial hypocalcinuric hypercalcemia (FHH).

FHH is an uncommon autosomal dominant disease formerly called familial benign hypercalcemia. These individuals do not develop bone or renal disease. Parathyroidectomy is not useful in treatment. The PTH concentration may be elevated but not as elevated as in primary hyperparathyroid disease. Urinary calcium excretion is used to differentiate FHH from primary hyperparathyroidism. In FHH, the renal calcium excretion is low or normal, while in primary hyperparathyroidism, most individuals have increased urinary calcium excretion. When FHH is suspected, laboratory testing of family members may be helpful.

Malignancies

Some tumors produce PTH-like substances that cause osteolytic activity in bone or stimulate the activity of osteoclasts. Certain tumors are commonly associated with hypercalcemia. These include breast carcinoma; multiple myeloma; renal cell carcinoma; hematological malignancies; and squamous cell tumors of the lung, head, neck, and other sites.

Drugs

The ingestion of certain drugs and vitamin preparations may elevate blood calcium levels.

The mechanism for thiazide-induced hypercalcemia is not clear. It has been suggested that thiazides may increase renal tubular reabsorption of calcium or that the kidney glomerular filtration rate is reduced resulting in decreased clearance of calcium from blood.

In hypervitaminosis D, both serum calcium and phosphorus may be elevated due to increased gut absorption. Both vitamins A and D in excessive doses stimulate osteoclastic bone resorption.

Lithium carbonate, used in treating depression, can elevate PTH by reducing the responsiveness of parathyroid glands to serum calcium. Lithium may also reduce calcium excretion by the kidney.

Chronic excessive ingestion of absorbable antacids (milk-alkali syndrome) may also elevate blood calcium.

Other Causes

Blood calcium elevation may occur during prolonged immobilization of part or all of the body. The mechanisms believed responsible include increased bone resorption and a decrease in osteoblastic activity.

Several diseases may be associated with elevated blood calcium levels. In hyperthyroidism, increased bone turnover elevates calcium. In sarcoidosis, tuberculosis, histoplasmosis, and coccidioidomycosis, the intestinal absorption of calcium is increased.

Decreased Serum Calcium

Serum calcium levels may be decreased in (1) hypoparathyroid disease; (2) vitamin D deficiency; (3) chronic renal disease; (4) fat malabsorption; and (5) pancreatitis. In fat malabsorption, dietary calcium is lost in the feces. In pancreatitis, it is thought that calcium is removed from the serum because it forms insoluble complexes with fatty acids that are released from tissues destroyed by pancreatic enzymes.

Increased Serum Phosphates

Serum phosphate may be increased in (1) hypoparathyroid disease; (2) hypervitaminosis D; (3) renal insufficiency; and (4) acromegaly. In renal insufficiency, phosphates are not excreted and accumulate in the blood. In acromegaly, the increase in growth hormone increases intestinal absorption of phosphates.

Decreased Serum Phosphates

Serum phosphates may be decreased in (1) rickets; (2) osteomalacia; (3) hyperparathyroid disease; (4) hypovitaminosis D; (5) fat malabsorption; (6) Fanconi syndrome; and (7) use of antacids such as aluminum hydroxide. In the Fanconi syndrome, several transport systems in the proximal renal tubule are defective. Amino acids and phosphates are not reabsorbed and are lost in the urine. Antacids, such as aluminum hydroxide, form insoluble compounds with phosphorus leading to loss of phospates in the feces.

MAGNESIUM

Magnesium is distributed in the body as follows: (1) approximately 65% is found in bones and teeth; (2) approximately 35% is present in intracellular fluid (cells of muscle, liver and other soft tissue); and (3) a small amount is found in extracellular fluid. Magnesium is the second most abundant cation in intracellular fluid; potassium is first.

FUNCTIONS OF MAGNESIUM IN THE BODY

Magnesium is an important cofactor of enzymes. All or almost all phosphate-transfer reactions (that is, $ATP \rightleftharpoons ADP$) require magnesium as a cofactor. Phosphate-transfer enzymes are called kinase enzymes; hexokinase, fructokinase, and creatine kinase are all examples of phosphate-transfer enzymes. Magnesium also acts as a cofactor for the hydrolase enzymes, alkaline phosphatase and prostatic acid phosphatase, which catalyze the cleavage of organic phosphate ester bonds.

The magnesium concentration together with calcium influences nerve conduction and neuromuscular contractions. A tetany may be associated with hypomagnesemia. This tetany will respond only to magnesium therapy and will not be corrected by the administration of calcium.

REGULATION

Magnesium is present in many foods (whole grain cereals, legumes, milk, nuts, bananas, and most leafy vegetables), and a dietary deficiency is rarely seen. Magnesium does not require vitamin D for absorption. The magnesium in serum is present as follows: (1) approximately two-thirds in a diffusible form, which is the physiologically active form, and (2) approximately one-third bound to proteins (three-quarters to albumin and one-quarter to globulins). Like calcium, the binding to protein is influenced by pH.

Magnesium is filtered by the glomerulus and tubular reabsorption occurs. Like calcium, PTH enhances tubular reabsorption of magnesium. Factors that decrease renal tubular reabsorption include hypercalcemia, alcohol ingestion, thyroid hormone, calcitonin, growth hormone, high sodium intake, chronic mineralocorticoid effect, and certain drugs.

The body has great ability to conserve magnesium by tubular reabsorption. In magnesium deficiency, urine excretion is extremely low.

MAGNESIUM METHODOLOGY

Specimen Requirements

The sample for magnesium assay may be serum or heparinized plasma drawn with no venous stasis. Since the concentration of magnesium is higher in red blood cells (three times higher) than plasma, the presence of hemolysis will cause false high values. The serum or plasma should be separated from the cells as soon as possible to avoid passage of magnesium from erythrocytes. Any sample contaminated with tissue fluid will be falsely elevated since the intracellular magnesium level is much higher than the extracellular fluid level. Magnesium is stable for several days at refrigerator temperatures; for longer storage, the sample should be frozen.

If the test is performed on urine, the pH should be adjusted to 1.0 and the urine should be mixed well prior to analysis.

Dye-Complexing Method

Principle
Colorimetric methods are based on the formation of a magnesium complex with a reagent. The complex is colored and the intensity of the color is proportional to the magnesium concentration. Since calcium interferes, an agent is added which chelates calcium. The calcium chelate does not absorb at the wavelength chosen to read the test.

Test
In the ACA methodology, the dye used is methylthymol blue (MTB). Since calcium will also bind MTB, the barium salt of ethylenebis (oxyethylenenitrilo) tetraacetic acid (Ba-EGTA), is used to bind calcium to reduce this interference.

$$Mg^{2+} + MTB \rightarrow Mg\text{-}MTB \text{ complex (absorbs 600 nm)}$$

$$Ca^{2+} + Ba\text{-}EGTA \rightarrow \text{complex (does not absorb at 600 nm)}$$

Other dyes used for magnesium assay include Calmagite, Magon, and Eriochrome Black T.

Atomic Absorption (Reference Method)

The reference method for magnesium assay is atomic absorption spectrophotometry. Phosphates may interfere; therefore, lanthanum or strontium is used to bind phosphates.

Reference Range

Serum magnesium   1.8–2.4 mg/dL or 1.5–2.0 meq/L or 0.75–1.0 mmol/L

CLINICAL SIGNIFICANCE

Decreased Magnesium (Hypomagnesemia)

Serum magnesium may be decreased in (1) conditions that are associated with loss of magnesium through the GI tract (diarrhea, malabsorption); (2) conditions associated with renal loss of magnesium, that is, where glomerular filtration predominates over reabsorption (diuretic agents, osmotic diuresis); (3) chronic alcoholism; (4) deficient dietary intake; (5) hyperthyroidism; (6) hyperaldosteronism; (7) newborns or infants on high phosphate diets; (8) hypercalcemia; and (9) hyperparathyroidism.

Low magnesium levels may cause weakness, irritability, delirium, convulsions, cardiac arrhythmias, and tetany. Both hypo- and hypermagnesemia may be associated with an impairment of PTH secretion. This PTH impairment is most sensitive to low magnesium levels; the inhibition produced at high levels is only partial. At normal physiological levels, the effect of magnesium on PTH secretion is not thought to be significant. Hypocalcemia may accompany hypomagnesemia.

Increased Magnesium (Hypermagnesemia)

Serum magnesium may be increased in (1) acute renal failure (glomeruli are unable to excrete magnesium and the blood level rises); (2) hypothyroidism; (3) ingestion of gastroenteric medications (Milk of Magnesia, Gelusil, Maalox, Epsom salts), especially in those patients with impaired renal function; and (4) in dehydration.

Excess magnesium acts as a sedative. Magnesium increases the atrioventricular conduction time. Extreme excesses may cause cardiac arrest.

COPPER

Copper is an essential trace element. It is found (1) in cells in combination with proteins and as metalloenzymes (enzyme that contains an inorganic ion as part of the enzyme molecule) and (2) in serum. The copper in serum is present in three forms. Most serum copper (greater than 90%) is firmly bound to the transport protein, ceruloplasmin. Less than 10% is loosely bound to albumin and a very small amount is present as a free dialyzable form.

METABOLISM

Copper is ingested in food (copper-containing food includes liver, mushrooms, nuts, and oysters) and only a small amount is absorbed from the stomach and the upper part of the small intestine. Most of the ingested copper is not absorbed and is lost in the stool. The absorbed copper enters the portal blood and is bound to albumin and certain amino acids then transported to the liver. A hepatocyte membrane receptor for the copper–albumin complex allows transport of the copper into liver cells.

In the hepatocyte, the copper is bound to hepatic metallothionine, used for synthesis of copper-containing enzymes, or excreted in the bile. Most of the copper retained in the hepatocyte is used for the synthesis of ceruloplasmin. Ceruloplasmin, the copper-transport protein, then enters the systemic circulation.

The major pathway for copper excretion is through the bile; a minor pathway is urinary excretion of the free dialyzable form. A copper deficiency has never been observed in human beings.

FUNCTIONS OF COPPER

Copper-containing enzymes include ascorbic acid oxidase, delta aminolevulinic acid dehydrase, dopamine beta-hydroxylase, galactose oxidase, monoamine oxidase, superoxide dimutase, ferroxidase II, tyrosinase, cytochrome oxidase, and ceruloplasmin (ferroxidase I). The function of the last three enzymes is discussed below.

Tyrosinase is the enzyme that controls the pigmentation of the skin. Tyrosinase catalyzes the oxidation of the amino acid, tyrosine, and other phenolic compounds, to produce the black pigment melanin. In animals, albinism is believed to be caused by a lack of the tyrosinase enzyme. In humans, albinism is not believed to be caused by a lack of the enzyme. The failure of cells to produce melanin may be due to a deficiency of the amino acid, tyrosine in cells, or more likely, due to the presence of some substance or defect that inhibits the activity of the tyrosinase enzyme.

All energy in animals is derived from oxidative metabolism. Cytochrome oxidase is the principle "terminal" oxidase enzyme in all animals. The terminal oxidase is the last enzyme involved in the oxidation of a substrate; when the final reaction cannot proceed, all intermediate carriers remain in a reduced state and cannot be oxidized in the usual way.

Ceruloplasmin is also called ferroxidase I. Ceruloplasmin catalyzes the oxidation of ferrous iron to ferric iron. The ferric iron can then be incorporated into apotransferrin to form transferrin, the iron-transport protein as illustrated in Figure 8.8.

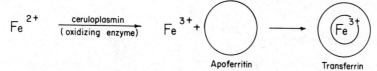

Figure 8.8. The oxidative role of ceruloplasmin in transferrin synthesis.

Transferrin can donate its iron directly to the reticuloendothelial cells of the bone marrow for hemoglobin synthesis. Without ceruloplasmin, transferrin production would probably be inadequate. Sufficient iron, therefore, would not be available for the formation of hemoglobin and other iron-containing proteins.

COPPER-STORAGE DISEASES

Wilson's Disease (Hepatolenticular Degeneration)

This disease was named after the British physician Kinnier Wilson. Wilson's disease is a rare inherited disease (autosomal recessive) in which copper is deposited in various organs. It is believed that the disease is caused by a defect associated with metallothionine, a sulfur-containing protein, which binds copper and other trace metals. In Wilson's disease, it has been shown that the liver metallothionine has greater avidity for copper than the metallothionine in normal persons. Normally, copper is released from metallothionine to be excreted in bile or to be used in the synthesis of ceruloplasmin. When metallothionine cannot release its copper, copper accumulates in cells and less ceruloplasmin is synthesized. The copper accumulates, especially in liver and brain cells, and in the cornea of the eye.

The symptoms are usually present in childhood and present as liver disease (chronic hepatitis), neurological disease (due to degenerative changes in the basal ganglia of the brain), or hemolytic disease. Corneal rings called Kayser–Fleischer rings may be present. A family history may be helpful in making a diagnosis.

Laboratory Findings
The laboratory findings associated with Wilson's disease include: (1) serum copper, decreased; (2) serum ceruloplasmin, decreased in most cases; (3) urine copper, increased in most cases probably due to an increase in free copper in the serum; and (4) liver biopsy, increased amount of copper per dry weight of liver.

Menke's Kinky-Hair Syndrome

Menke's kinky-hair syndrome is another inherited (sex-linked recessive) copper-storage disease. As in Wilson's disease, this disease reveals an increase in cytosol copper in cells. In the kinky-hair disease, the defect is also believed associated with the metallothionine protein, but the defect is believed to be different from that in Wilson's disease. There appears to be a defect in the efflux of copper from cells.

COPPER METHODOLOGY

Specimen Requirements

The type of sample for copper assay should be serum. The collection vessel and syringe should be metal free; the needle used in collection should contain no copper.

Test Methods

Atomic absorption is the assay method of choice. A colorimetric method is available. The copper is split from the protein with acid and the proteins are precipitated. A color reagent (cuprizone) forms a blue compound with cupric ions.

Reference Range

Serum copper   0.75–1.45 μg/mL.

IRON

The total amount of iron in an adult is approximately 4–5 g. About three-fourths of this iron has an active vital physiological role (hemoglobin, myoglobin, oxidation–reduction enzymes), the remaining portion is present in various storage forms (ferritin, hemosiderin) that can be mobilized if necessary. In the body, little or no iron exists unbound. Free iron is toxic to the body. The approximate distribution of iron in the normal adult male is shown in Table 8.2.

Table 8.2. Approximate Distribution of Iron in a Normal Adult Male

Component	Percent
Hemoglobin	68.3
Ferritin	12.7
Hemosiderin	11.7
Myoglobin	3.3
Enzyme iron	0.19
Transferrin	0.17

Reproduced with permission from Henry, J. B. *Clinical Diagnosis and Management by Laboratory Methods,* 16th ed. Philadelphia: Saunders, 1979, Vol. I.

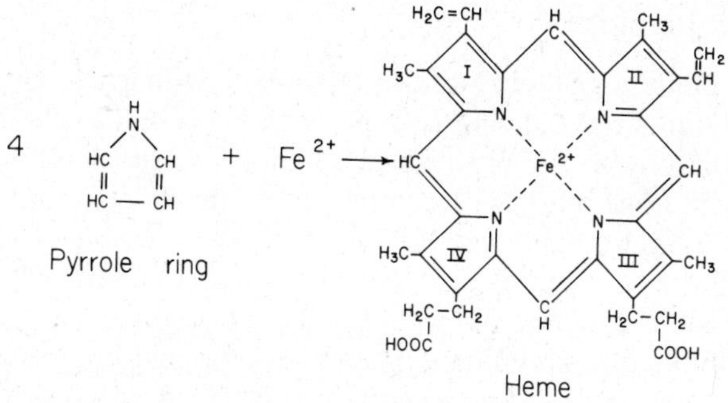

Figure 8.9. Synthesis of heme.

TYPES OF IRON-CONTAINING COMPOUNDS

There are two general types of iron-containing compounds: those that contain heme and those that do not contain heme.

Heme Compounds

Heme is composed of four pyrrole rings (i.e. porphyrin) in combination with ferrous (2+) iron. The molecular structure of heme is shown in Figure 8.9.

In heme-containing compounds, iron is in the lower oxidation state (2+). Heme-containing compounds are most often associated in some way with the oxygenation of tissues. Examples of heme-containing compounds include hemoglobin, myoglobin, and oxidation–reduction enzymes.

Hemoglobin
Hemoglobin carries oxygen from the lungs to metabolizing tissue. Hemoglobin is composed of four heme groups and four polypeptide (globin) chains. The molecular weight of hemoglobin is 64,456 g/mole. Hemoglobin gives blood its deep red color. Hemoglobin with iron in the ferric (3+) oxidation state is called methemoglobin and is unable to bind and transport oxygen.

Myoglobin
Myoglobin, the protein pigment of cardiac and skeletal muscle, is believed to provide a storage form of oxygen for muscles. This oxygen is available when tissue oxygen is low. Myoglobin, unlike hemoglobin, is able to release oxygen only when oxygen tension is very low. Myoglobin is composed of one heme moiety and one polypeptide chain; hence, its molecular weight is about one-fourth that of hemoglobin.

Oxidation–Reduction Enzymes
Heme-containing enzymes include the cytochromes, a group of enzymes located inside mitochondria that transfer electrons during glucose metabolism, and catalase and peroxidase, which are enzymes that destroy hydrogen peroxide, a toxic compound formed by the cell.

346   Inorganic Ions

Nonheme Compounds

Nonheme iron-containing compounds are involved with the absorption, transport, and storage of iron. In the nonheme type, iron is present as the ferric ($3+$) form and is bound directly to the protein. The major storage sites for iron are in the liver, spleen, and bone marrow. Examples of nonheme compounds include ferritin, hemosiderin, and transferrin.

Ferritin

Ferritin is a soluble iron-containing protein and the major iron-storage protein in the body. It is composed of an apoprotein shell filled with iron in the form of a ferric hydroxide, ferric phosphate complex. Each molecule of ferritin can accommodate up to 4,000 atoms of iron, and when fully saturated, consists of over 20% iron by weight. The molecular weight of ferritin is approximately 450,000 g/mole.

Ferritin may be found in many tissues, but is especially abundant in the cytoplasm of reticuloendothelial cells located in the liver, spleen, and bone marrow. Ferritin acts as a store from which the iron portion of the molecule can be mobilized as needed for the synthesis of heme compounds. The measurement of serum ferritin directly reflects the available iron stores in the body.

Hemosiderin

Hemosiderin is a complex mixture, of which ferritin probably forms a part; it may represent denatured ferritin. Hemosiderin, an insoluble compound, may be present in the cell as large granules (iron oxide granules) visible by light microscopy. When iron body stores are high, the greater portion of the stored iron may be in the form of hemosiderin.

Transferrin

Transferrin, a beta-1-globulin synthesized in the liver, is the iron transporting protein. Transferrin carries the iron needed for hemoglobin synthesis directly to the developing red blood cells in the bone marrow. When erythropoietic activity is increased, iron may be drawn from both storage forms. The molecular weight of transferrin is approximately 90,000 g/mole. Transferrin can bind two Fe^{3+} ions per molecule.

IRON METABOLISM

The iron needs of the body are usually met by dietary intake and from recycled iron derived from destroyed red blood cells. Iron is lost through the skin, nails, hair, urine, and feces, and through menstrual flow. During pregnancy, as much as 400 mg of iron may be lost to the fetus and as a result of blood loss during delivery.

A small amount of iron is absorbed from the stomach, but most is absorbed in the upper small intestine (duodenum and jejunum). Only iron in the ferrous form can be absorbed. Ferric iron is the type commonly found in foodstuff, therefore, the presence of reducing substances such as ascorbic acid enhance absorption. Phosphates and phytate (found in cereal) reduce absorption because of the formation of insoluble iron complexes.

Once iron enters the intestinal mucosal cells, it is oxidized to the ferric (3+) form, and most becomes attached to the transferrin protein for transport in the blood to various body-storage sites, such as the liver and bone marrow. Any remaining iron in the mucosal cell attaches to apoferritin to form ferritin. At the storage site, the iron is released from the transferrin protein and stored inside cells as either ferritin or hemosiderin. The transferrin may then be used again for further iron transport. In normal persons, the saturation of transferrin with iron is only 25–30%.

The absorption of iron from the intestinal tract must be controlled since the body's capability to eliminate iron is limited. Tissue accumulations of iron may cause toxic reactions. Iron stored in the heart causes myocardial damage and arrhythmias. The least amount of iron required to produce cardiac damage is not known but injury may occur at low levels. According to the "mucosal block theory," intestinal iron absorption continues only until the apoferritin in the mucosal cells is fully saturated with iron. Further absorption is then blocked. This prevents the uptake of excessive iron from the intestinal tract. When the amount of iron in the cell decreases, by passing into the plasma, additional iron can enter the cell. It is known, however, that high oral doses of iron increase the amount of iron absorbed; therefore, absorption must not be governed by apoferritin alone. It is possible that a transport system may regulate intestinal absorption of iron, but the mechanism of transport into intestinal mucosal cells has not been fully resolved.

The life span of a normal red blood cell is approximately 120 days. Erythrocytes are removed daily by the mononuclear phagocyte system. The elimination of aged normal erythrocytes is believed to occur by a physical process. As a cell ages, lipids are lost from the cell membrane, and the red cells lose their ability to change shape easily. They then rupture in splenic sinuses, and the debris is phagocytosed by splenic macrophages. However, recent electron microscopy pictures suggest that an immune mechanism of elimination may occur. An IgG antibody binds to a glycoprotein molecule that appears on the surface of old and damaged cells. This antibody is recognized by macrophages that reside in the spleen, and the macrophage attaches to the red blood cell. (Macrophages have IgG-Fc receptors.) The erythrocyte is then engulfed by the macrophage, a vacuole is formed, and the red blood cell is digested by enzymes. Regardless of the process of destruction, the iron that is contained in the red blood cell is returned to the bone marrow for use in the production of new erythrocytes.

IRON METHODOLOGY

Specimen Requirements

For iron assay, a fasting plasma or serum sample may be used. Serum iron shows marked diurnal variation, with higher levels occurring in the morning and decreased levels in the evening (decreased as much as 30%). In newer procedures, slight hemolysis can be tolerated, because iron in porphyrin rings (as present in hemoglobin) will not react with the color reagent. Extensive hemolysis causes false high values.

Test

Colorimetric procedures are the most common methodology for iron determination. Atomic absorption spectrophotometry has been used to measure iron, but this methodology is not recommended due to matrix interferences and limited sensitivity.

In colorimetric methods, the first step in most iron procedures is the disruption of the iron protein complex, since most of the iron in the serum or plasma is bound to protein. Some procedures precipitate the protein after iron is disrupted, others do not. If proteins are precipitated, there may be some loss of iron with the protein, resulting in lower iron values.

Total Iron Methodology With Protein Precipitation

1. Protein is precipitated by hot TCA and the iron is freed simultaneously.

2. An Aliquot of the protein-free filtrate is reacted with a chromogen, forming a Fe-chromogen complex and producing a color in proportion to the iron present. Since most color reagents react only with ferrous iron (2+), the ferric iron (type in transferrin) must be treated with a suitable reducing agent. Reagents used for color development include sulfonated bathophenanthroline, ferrozine, terosite, and 2,4,6-tripyridyl-triazine. Thiocyanate may also be used as the chromogen. It does not require a reducing agent, since it will react with ferric iron.

Total Iron Methodology With No Protein Precipitation

1. Iron is released from the protein complex by adjusting the serum pH to approximately 6.0. Under these conditions, serum proteins remain in solution.

2. The ferric ion is then reduced to the ferrous state and a chromogen (same color reagents as method above) is added for color development.

3. Correction is made for interference by nonspecific colors or opacity of the serum by the use of individual serum blanks.

Total Iron Binding Capacity (TIBC)

The TIBC correlates with the concentration of transferrin. Transferrin is measured indirectly by TIBC since the amount of iron transferrin can bind is determined.

1. Excess iron is added to the serum. Transferrin will then bind enough iron to become completely saturated.

2. The excess iron, or the iron that did not combine with transferrin, is removed either by ion-exchange resins or by magnesium carbonate.

3. The iron remaining in the supernatant is determined by the iron procedure described above. The value obtained represents the TIBC.

4. The percentage saturation of transferrin with iron is determined by dividing the total serum iron by the capacity to bind times 100:

$$\% \text{ saturation} = \frac{\text{serum iron}}{\text{TIBC}} \times 100$$

Reference Range

Iron levels show sex variation; men normally have higher levels than women. In women, iron levels vary with menstrual cycles. Serum iron levels tend to decrease with age.

Serum iron	42–135 μg/dL
TIBC	250–350 μg/dL
% saturation	30–44%

CLINICAL INTERPRETATION

Iron Deficiency Anemia

Etiology
When the amount of iron available is insufficient to meet the needs of erythropoiesis, iron deficiency anemia results. Deficient iron may result from (1) blood loss; (2) dietary insufficiency (especially during pregnancy or rapid growth); and (3) malabsorption (especially after gastrectomy and in sprue).

The stages in the development of iron deficiency anemia occur in the following order. Iron stores in the liver, spleen, and bone marrow become depleted. As the iron stores are depleted, the serum ferritin decreases. Serum transferrin then increases, elevating the serum TIBC. Serum iron falls, indicated by the presence of a low serum total iron, followed by clinical anemia. The blood picture in iron deficiency anemia includes a fall in hemoglobin, hematocrit, red cell count, mean corpuscular volume (MCV), and mean corpuscular hemoglobin (MCH) and if severe enough, a fall in the mean corpuscular hemoglobin concentration (MCHC).

The red cell morphology shows microcytosis, hypochromia, poikilocytosis, and anisocytosis.

Laboratory Findings
The laboratory findings in iron deficiency anemia include (1) serum iron, decreased; (2) TIBC, increased; (3) percentage saturation, decreased; (4) serum ferritin, decreased; (5) bone marrow, absent iron stores: and (6) peripheral blood smear, microcytic, hypochromic red blood cells.

Case History, Iron Deficiency Anemia
A 55-year-old white female was admitted to the hospital for evaluation of persistent back pain. Her admission laboratory work revealed a low hemoglobin. The results of her laboratory tests are given in Table 8.3. The blood smear revealed microcytes, with moderate hypochromia, a few elliptocytes, and an occasional teardrop cell.

Discussion. The laboratory studies revealed the findings associated with iron deficiency anemia, namely, low serum iron and elevated TIBC with decreased iron saturation of transferrin. The bone marrow was negative for stainable iron. The serum ferritin was low. The hemoglobin, hematocrit, red cell count, MCV, MCH, and MCHC were all decreased. The blood smear revealed hypochromic, microcytic cells, the type of cells seen in iron deficiency anemia and thalassemia.

Table 8.3. Iron Deficiency Anemia Case History Laboratory Results

Iron (42–135)[a] μg/dL	TIBC (250–350)[a] μg/dL	Ferritin (20–120)[a] ng/mL	Bone Marrow
less than 20	423	18	No stainable iron present

Hemoglobin (12–16)[a] g/dL	RBC (4.2–5.4)[a] × 10^6	HCT (37–47)[a] percent	MCV (81–99)[a] μm³fL	MCH (27–31) pg	MCHC (33–37) g/dL
9.5	3.9	31.1	79.7	24.5	30.7

[a] Reference range for components is indicated in parentheses.

The etiology of the iron deficiency could not be determined. In an adult, iron deficiency anemia is usually secondary to blood loss, commonly GI bleeding or menstrual loss. Stool occult blood studies were all negative, and there was no past history of bloody stools. The patient was free of any gastrointestinal disturbances. The patient was postmenopausal.

To rule out thalassemia, hemoglobin electrophoresis was performed and hemoglobin A_2 was quantitated; the results of these tests were normal. The Coombs direct and indirect test and the sucrose hemolysis test were all negative, ruling out hemolysis. It was concluded that a defect in iron transport might be present. The patient was placed on supplemental iron and was to be reevaluated based on the results of post therapy blood tests.

Anemia of Chronic Disease

Etiology
Chronic inflammatory disorders such as infections, inflammation (rheumatoid arthritis), and neoplastic disease may result in an anemia with a blood picture similar to that seen in iron deficiency anemia. In the early stages, the anemia is usually normochromic and normocytic, but with long-standing disease, a microcytic, hypochromic picture may develop. In these disorders, the release of iron from the reticuloendothelial cells is blocked. Transferrin does not increase as in a true iron deficiency anemia. The patient may have low serum iron levels, but iron deficiency is not present, and iron replacement therapy is not indicated. Treatment with iron could result in iron overload and heart failure.

Laboratory Findings
The laboratory findings in anemia of chronic disease include (1) serum iron, usually decreased; (2) TIBC, usually decreased; (3) percentage saturation, usually decreased; (4) serum ferritin, normal to increased, and (5) bone marrow, normal to increased iron stores.

Iron Overload Disorders

Etiology
Iron overload may result from (1) hematological disorders such as thalassemia, aplastic anemia, or sideroblastic anemia; (2) hemosiderosis caused by repeated

blood transfusions or excessive iron therapy; and (3) idiopathic hemochromatosis.

Idiopathic hemochromatosis is a hereditary disorder associated with a body iron overload due to a defect in the mechanism controlling iron absorption from the intestine. Iron absorption is abnormally high over a period of years, and clinical manifestations usually appear between the ages of 40 and 60. Iron, as hemosiderin, accumulates in the liver, pancreas, heart, and skin of the affected individual. The skin becomes discolored (bronzed); there is degeneration and fibrosis of the liver and pancreas ending in hepatic cirrhosis and diabetes mellitus. Cardiac failure may develop as well. The disease is treated with therapeutic phlebotomy plus deferroxamine, an iron chelating agent.

In iron overload disorders, the total iron is increased. Transferrin does not increase but becomes more saturated with iron than would be the situation in a normal individual, resulting in increased percent saturation.

Laboratory Findings

The laboratory findings in iron overload disorders include (1) serum iron, increased; (2) TIBC, decreased; (3) percentage saturation, increased; (4) serum ferritin, increased; and (5) bone marrow, increased iron stores.

Other Diseases

Serum iron and TIBC are also altered in disorders indirectly related to iron metabolism, such as liver and kidney conditions. In liver disease, iron as ferritin may be released from necrotic liver cells, increasing serum iron and ferritin levels. In kidney disease (nephrosis), transferrin may be lost in the urine, resulting in loss of iron and decreased TIBC.

SERUM FERRITIN

The measurement of serum ferritin has been proposed as a test to measure iron stores. In healthy adults, the ferritin concentration in serum is directly related to the storage iron that can be mobilized in the body. The levels do not show the diurnal variation characteristic of serum iron levels.

Ferritin is a protein manufactured by cells that store iron. Ferritin synthesis is induced by iron in almost all tissues. Ferritin was once thought to be an intracellular protein only. With the development of sensitive RIA procedures, it has been shown that ferritin is a normal component of the serum. It is believed that the ferritin in serum is derived from the breakdown of tissue that contains iron, especially macrophages of the reticuloendothelial cells of the liver, spleen, and bone marrow.

Low Ferritin Levels

Ferritin levels are thought to reflect the reticuloendothelial iron stores accurately. Low serum ferritin levels indicate depletion of iron stores and are thought to provide an early warning of iron deficiency or to give confirmation of an iron deficiency state. In the development of iron deficiency anemia, serum ferritin decreases as iron stores are depleted. This occurs before the increase in transfer-

Table 8.4. Laboratory Findings in Iron Disorders

	Serum Fe	TIBC	Percentage Saturation	Serum Ferritin	Bone Marrow (Iron Stores)
Fe deficiency anemia	↓ᵃ	↑	↓	↓	Absent iron
Anemia of chronic diseases	↓	↓→	↓	→↑	→↑
Fe overload	↑	↓	↑	↑	↑
Viral hepatitis	↑	↑	↑→	↑	
Nephrosis	↓	↓	↑		

ᵃ ↑ = increased; ↓ = decreased; → = normal.

rin and the decrease in serum iron, thereby alerting the physician to an iron deficiency state early in the disease process. Treatment can be initiated at this stage, and the anemia associated with iron deficiency can be averted.

In addition to the detection of iron deficiency anemia, serum ferritin levels are especially useful in distinguishing between anemia of chronic disease and iron deficiency anemia. In both disorders, the serum iron may be decreased. The measurement of serum ferritin will allow differentiation. In iron deficiency anemia, the serum ferritin is decreased; in anemia of chronic disease, the serum ferritin is normal or increased.

High Ferritin Levels

Serum ferritin levels may be elevated irrespective of iron stores in certain liver diseases (e.g., hepatitis, alcoholic cirrhosis, and malignant disorders of the liver), malignancies (e.g., leukemias, Hodgkin's disease), and chronic inflammatory diseases. In these disorders, the rise in serum ferritin is thought to be derived from the release of ferritin from inflamed tissue or in malignancies, possibly from the production of ferritin by proliferating tumor cells. Serum ferritin is also increased in iron overload disorders.

A summary of laboratory findings in iron disorders is given in Table 8.4.

Reference Range

Serum ferritin, adult females	20–120 ng/mL
Serum ferritin, adult males	20–300 ng/mL
Serum ferritin, newborn	25–200 ng/mL
Serum ferritin, 6 months–15 years	7–142 ng/mL

In either men or women, values between 10 and 20 ng/mL are borderline.

BIBLIOGRAPHY

Bakerman, S. and Khazanie, P. "Calcium metabolism and hypercalcemia," *Laboratory Management*, 17–25 (July 1982).

Bates, H. M. "How to detect iron deficiency before anemia develops," *Laboratory Management* 17–22 (Jan. 1980).

Brown, D. M. "Vitamin D and bone metabolism," *Diagnostic Medicine* **1**(1), 82 (1978).

Engel, R. H. "The dynamic role of vitamin D in calcium and phosphate metabolism, part I," *Laboratory Management,* 51–56 (Mar. 1979).

Engel, R. H. "The dynamic role of vitamin D in calcium and phosphate metabolism, part II," *Laboratory Management,* 53–56 (Apr. 1979).

Felig, P., Baxter, J., Broadus, A., and Frohman, L. (eds.). *Endocrinology and Metabolism.* New York: McGraw-Hill, 1981.

Frieden, E. "The Biochemistry of Copper," *Scientific America,* **218**(5), 103 (1968).

Gagel, R. F. "Calcitonin in thyroid carcinoma and other disease." *Laboratory Management,* 35–48 (June 1982).

Ham, A. W. *Histology,* 7th ed. Philadelphia; Lippincott, 1974.

Harper, H. H. *Review of Physiological Chemistry,* 15 ed. Los Altos, CA: Lange, 1973.

Henry, J. D.(ed.). *Clinical Diagnosis and Management by Laboratory Methods,* 16th ed. Philadelphia; Saunders, 1979.

Hunt, J. S., Chapman, A. L., and Wood, G. W. "Mononuclear phagocyte interaction with complement-coated, antibody coated, and aged erythrocytes," *Laboratory Medicine* **13**(3), 170 (1982).

Kay, M. "A pictorial: Removing 360 billion old red cells," *Lab World,* 41–43 (Dec. 1981).

Krause, J. R. "Serum ferritin test and its relationship to iron deficiency," *Laboratory Medicine* **12**(9), 536 (1981).

LaRue, L. "Stored iron and coronaries," *Lab World,* 1 (Aug. 1981).

Lente, F. V. and Galen, R. S. "Finding the cause of asymptomatic hypercalcemia," *Diagnostic Medicine* **4**(2), 81 (1981).

Marx, S. J., Spiegel, A. M. et al. "Familial hypocalciuria hypercalcemia," *New England Journal of Medicine* **307**(7), 416 (1982).

Simon, M. and Cuan, J. "Diagnostic utility of C-terminal parathyrin measurement as compared with measurements of N-terminal parathyrin and calcium in serum," *Clinical Chem.* **26**(12), 1672 (1980).

Skikne, B. S. and Cook, J. D. "Serum ferritin in the evaluation of iron status," *Laboratory Management,* 31–35 (May 1981).

Speicher, C. E. *Magnesium. Check Sample Program CC-119,* ASCP, 1979.

Stearns, F. M. "The importance of serum ionized calcium," *Laboratory Management,* 11–12 (Oct. 1978).

Tietz, N. (ed.). *Fundamentals of Clinical Chemistry,* 2nd ed. Philadelphia; Saunders, 1976.

Williams, R. H. (ed.). *Textbook of Endocrinology,* 6th ed. Philadelphia; Saunders, 1981.

POST-TEST

---

1. Indicate increased (↑) or decreased (↓) absorption of calcium from the GI tract:
    1. _____ Acid pH
    2. _____ Presence of activated vitamin D
    3. _____ Diet high in dairy products
    4. _____ Alkaline pH
    5. _____ Diet high in phosphates
    6. _____ Presence of PTH
    7. _____ Steatorrhea
    8. _____ Old age

2. Match the following:
   1. _____ Deficiency in bone mass (thin
               bones)   a. Osteomalacia
   2. _____ Bone mass normal, increased or   b. Osteoporosis
               decreased but proportion of calci-
               fied bone deficient
   3. _____ Type of bone disease that occurs
               after menopause in women
   4. _____ Type of bone disease associated
               with vitamin D deficiency in chil-
               dren (rickets)
   5. _____ Type of bone disease associated
               with steatorrhea and malabsorption

3. Osteoblasts are associated with bone _____, while osteoclasts are associated with bone _____ .
   a. construction
   b. destruction

4. The mineralized portion of the bone is called _____ and the nonmineralized (protein) portion is called _____ .
   a. osteoid
   b. hydroxyapatite

5. Alkaline phosphatase is made by:
   a. osteoclasts
   b. osteoblasts
   c. osteocytes
   d. none of the above

6. The physiologically active form(s) of calcium is (are):
   a. complexed calcium
   b. ionized calcium
   c. protein bound calcium
   d. total calcium

7. Match the following:
   1. _____ Acidosis   a. Increased ionized calcium
   2. _____ Alkalosis   b. Decreased ionized calcium
   3. _____ Hypoproteinemia   c. Increased total calcium
   4. _____ Hyperproteinemia   d. Decreased total calcium
                                    e. No match

8. Match the following (one has two answers):
   1. _____ Stimulated by a decrease in   a. PTH
               ionized calcium   b. Calcitonin
   2. _____ Stimulated by an increase in
               ionized calcium
   3. _____ Increases phosphate excretion
               via the urine by decreasing
               resorption
   4. _____ Acts on bone to cause bone resorption

5. _____ Inhibits bone resorption
6. _____ Raises blood calcium levels
7. _____ Protect against hypercalcemia
8. _____ Produced by the parathyroid glands located in the thyroid
9. _____ Produced in C-cells in the thyroid

9. Discuss the following as each relates to vitamin D:
   a. source
   b. role of UV light
   c. name of activated vitamin D
   d. organ sites of hydroxylation
   e. hormonal status of activated vitamin D, that is, feedback mechanisms
   f. functions of vitamin D:
   g. effect of hypervitaminosis D on calcium and phosphorus blood levels

10. Match the following:
    1. _____ Magnesium   a. Constituent higher in RBC
    2. _____ Iron   than in plasma
    3. _____ Phosphate   b. Constituent higher in
    4. _____ Calcium   plasma than in RBC

11. The reference method for calcium is:
    a. flame photometry
    b. atomic absorption
    c. cresolphthalein complexone
    d. Clark–Collip titration method
    e. chloranilic acid

12. In the atomic absorption method for calcium, interference by phosphate is avoided by:
    a. dilution of the specimen prior to analysis
    b. EDTA anticoagulation collection tube
    c. careful control of pH
    d. addition of lanthanum

13. The procedure for phosphorus determination using molybdate is called:
    a. Clark–Collip method
    b. Sulkowitch test
    c. Fiske–Subbarow method
    d. phosphorus complexone method

14. Match the following. Each has three answers. Choose one answer for each category:
    1. ___ ___ ___ Rickets   a. increased serum cal-
                                              cium
    2. ___ ___ ___ Primary hyper-   b. decreased serum cal-
                   parathyroidism   cium

3. ___ ___ ___   Primary hypo-   c. normal serum calcium
                 parathyroidism
                                      d. increased serum phos-
                                          phorus
4. ___ ___ ___   Normal growing   e. decreased serum phos-
                 child at puberty   phorus
                                      f. normal serum phos-
                                          phorus

                                      g. increased alkaline
                                          phosphatase
                                      h. decreased alkaline
                                          phosphatase
                                      i. normal alkaline phos-
                                          phatase

15. Secondary hyperparathyroidism may be characteristically associated with all the following *except:*
    a. osteomalacia
    b. malabsorption
    c. steatorrhea
    d. renal disease
    e. hypervitaminosis D

16. Tetany may be caused by decreased:
    a. total calcium
    b. ionized calcium
    c. ionized magnesium
    d. phosphorus
    e. b and c

17. A 3-year-old poorly nourished child was seen in a pediatric clinic. The mother worked at night and slept during the day. The child's sleeping pattern was altered to fit the mother's work schedule. Play time was largely limited to indoor activities. The child was diagnosed as osteomalacia associated with vitamin D deficiency (rickets). The expected laboratory findings associated with this disease would be [use ($\uparrow$) increased, ($\downarrow$) decreased, ($\rightarrow$) normal; justify each answer given]:
    Serum calcium _____
    Serum phosphorus _____
    Serum alkaline phosphatase _____
    Serum PTH _____

18. An elevated serum calcium was noted on a routine chemistry profile. The patient had no history of illness, was not on any medication, and did not take vitamin preparations. The final diagnosis of primary hyperparathyroid disease was made after ruling out other causes of hypercalcemia. Based on the above diagnosis, the expected laboratory test values associated with this disease would be [use ($\uparrow$) increased, ($\downarrow$) decreased, ($\rightarrow$) normal; justify each answer given]:
    Serum calcium _____
    Serum phosphorus _____

Serum alkaline phosphatase _____
Renal phosphate clearance _____
Serum PTH hormone _____
Urinary hydroxyproline _____

19. Paget's disease of the bone is associated with all the following *except:*
    a. increased osteoclastic activity
    b. increased osteoblastic activity
    c. very high serum alkaline phosphatase levels
    d. increased excretion of hydroxyproline
    e. increased PTH

20. Match the following:
    1. _____ Transferrin   a. Contains heme
    2. _____ Myoglobin   b. Does not contain heme
    3. _____ Hemosiderin
    4. _____ Cytochrome enzymes
    5. _____ Hemoglobin
    6. _____ Ferritin

21. Heme compounds contain iron in _____ oxidation state while nonheme iron compounds contain iron in _____ oxidation state.
    a. 2+
    b. 3+

22. Indicate the oxidation state of iron:
    1. _____ iron in ingested food   a. ferric (3+)
    2. _____ usual form in which iron is b. ferrous (2+)
                absorbed
    3. _____ form in ferritin
    4. _____ form in transferrin
    5. _____ form in oxyhemoglobin

23. The transport protein for iron is:
    a. transcortin
    b. ceruloplasmin
    c. transferrin
    d. ferritin

24. Serum iron levels are altered by
    a. age
    b. sex
    c. diurnal variation
    d. day to day variation
    e. a, b, and c
    f. all of the above

25. The *main* storage form(s) for iron in *normal* persons is (are):
    a. hemosiderin
    b. ferritin
    c. transferrin
    d. a and b
    e. all of the above

26. Complete the blanks below to describe the molecular configuration of the compounds listed:
    a. Heme
       1. _____ number of polypeptide chains
       2. _____ number of porphyrin rings
       3. _____ number of iron atoms
    b. Hemoglobin
       1. _____ number of polypeptide chains
       2. _____ number of porphyrin rings
       3. _____ number of iron atoms

27. Major sites for iron storage include:
    a. spleen, bone marrow, liver
    b. bone marrow, liver, intestinal cells
    c. spleen, bone marrow, intestinal cells
    d. bone marrow, liver, pancreas

28. Total iron binding capacity indirectly measures:
    a. transferrin
    b. unsaturated iron binding sites on the transferrin molecule
    c. serum iron
    d. the iron saturation state of transferrin

29. Match the following:

    _____ 1. The concentration of iron when one-   a. 100 μg/dL
                   third of binding sites on transferrin are   b. 100 mg/dL
                   saturated with iron.   c. 300 μg/dL
    _____ 2. The concentration of iron when all   d. 300 mg/dL
                   binding sites on the transferrin mole-   e. 200 μg/dL
                   cule are saturated with iron.   f. 200 mg/dL
    _____ 3. The unsaturated iron binding capacity
                   would be _____ .

30. In the routine spectrophotometric analysis of serum iron, the iron is usually first:
    a. split from the protein by exposure to acid
    b. reduced with ascorbic acid
    c. adsorbed on cation-exchange resin
    d. precipitated with oxalate
    e. reacted with bathophenanthroline

31. Name two conditions that would result in an iron deficiency anemia.
    In this type of anemia, the serum would be _____ and the TIBC would be _____ . (Use increased or decreased.)

32. Name two conditions that would result in an iron overload.
    In this type of condition, the serum iron would be _____ and the TIBC would be _____ . (Use increased or decreased.)

33. The findings in iron deficiency anemia versus anemia of chronic disorders would be [use (↑) increased, (↓) decreased, (→) normal]:

	Fe Deficiency	Anemia of Chronic Disease
Serum iron	_____	_____
Serum ferritin	_____	_____

34. In the body, the trace element copper is important because of its role as:
    a. reductase enzyme
    b. an activator for enzyme reactions
    c. oxidase enzyme
    d. transferase enzyme

35. The transport protein for copper is:
    a. transferrin
    b. ceruloplasmin
    c. hepatocuprein
    d. transcortin

36. Wilson's disease (hepatolenticular degeneration) is associated with all the following *except:*
    a. increased serum copper
    b. increased urine copper
    c. decreased ceruloplasmin
    d. deposition of copper in tissues

ANSWERS

1. (1) increased; (2) increased; (3) increased; (4) decreased; (5) decreased; (6) increased; (7) decreased; (8) decreased
2. (1) b; (2) a; (3) b; (4) a; (5) a
3. a, b
4. b, a
5. b
6. b
7. (1) a; (2) b; (3) d; (4) c
8. (1) a; (2) b; (3) a, b; (4) a; (5) b; (6) a; (7) b; (8) a; (9) b
9. *Source:* diet, skin precursors; *role of UV light:* photolytic cleavage to yield calciferol; *name of activated vitamin D:* $1,25(OH)_2D$ or 1,25-dihydroxycalciferol; *organ sites of hydroxylation:* liver at C-25, kidney at C-1 and C-24; *feedback mechanisms:* see Figure 8.6; *Functions of vitamin D:* (a) increases calcium and phosphorus absorption in the gut, essential for calcium absorption; (b) increases bone resorption; (c) decreases renal excretion of calcium; *effect:* increases both.

10. (1) a; (2) a; (3) a; (4) b
11. b
12. d
13. c
14. (1) b, e, g; (2) a, e, g; (3) b, d or f, i; (4) c, d or f, g
15. e
16. e
17. *Serum calcium:* decreased because vitamin D absolutely necessary for calcium absorption from gut; *Serum Phosphorus:* decreased because vitamin D enhances phosphate absorption from gut and because PTH, which will be increased, enhances renal phosphate clearance; *serum alkaline phosphatase:* increased since osteoblastic activity is increased to build bone; *serum PTH:* increased because the decreased serum calcium stimulates PTH secretion.
18. *Serum calcium:* Increased PTH increases serum calcium; *serum phosphorus:* decreased to normal . . . increased PTH causes loss of phosphate in the urine. Phosphate concentration can reflect the extent of PTH secretion; *serum alkaline phosphatase:* increased to normal . . . Increased destruction of bone (by PTH) leads to osteoblastic activity to build bone. The alkaline phosphatase will be increased when bone involvement is large. Alkaline phosphatase is made by osteoblasts; *renal phosphate clearance:* increased PTH increases loss of phosphates in the urine; *serum PTH hormone:* increased to normal . . . PTH will always be increased, however, may be within normal range. PTH serum levels have to be interpreted by considering the serum calcium; *urinary hydroxyproline:* increased . . . hydroxyproline is found in the osteoid portion of the bone. With increased bone turnover due to increased PTH, the hydroxyproline excretion in the urine will be increased.
19. e
20. (1) b; (2) a; (3) b; (4) a; (5) a; (6) b
21. a, b
22. (1) a; (2) b; (3) a; (4) a; (5) b
23. c
24. f
25. b
26. Heme: (1) zero; (2) 1; (3) 1; Hemoglobin: (1) 4; (2) 4; (3) 4
27. a
28. a
29. (1) a; (2) c; (3) e
30. a
31. decreased dietary intake, blood loss or malabsorption; decreased, increased
32. hemochromatosis, hemosiderosis, hematological disorders; increased, decreased

33. *Iron deficiency anemia:* serum iron, decreased; serum ferritin, decreased; *anemia of chronic disorders:* serum iron, decreased; serum ferritin, increased to normal

34. c

35. b

36. a

CHAPTER 9
ACID–BASE AND ELECTROLYTES

OBJECTIVES

On completion of this chapter, you will be able to:

1. Define the following terms:
    a. acid
    b. base
    c. pK_a
    d. Dalton's Law
    e. Henry's Law
    f. partial pressure
    g. anion gap
    h. buffer anions (buffer base)
    i. hypocapnia
    j. hypercapnia
    k. hypoxemia
2. List normal values for:
    a. blood pH
    b. pH range considered compatible with life
    c. total CO_2 content of normal blood
    d. concentration of H_2CO_3 in normal blood
    e. concentration of HCO_3^- in normal blood
    f. ratio of the concentration of HCO_3^-/H_2CO_3 in normal blood
    g. dissociation constant of carbonic acid as found in the body
    h. solubility factor of CO_2
    i. partial pressure of CO_2 in normal arterial blood
    j. partial pressure of O_2 in normal arterial blood
    k. the percentage of hemoglobin saturated with oxygen in normal arterial blood
3. Discuss how the Henderson–Hasselbalch equation is used to determine the acid–base status of an individual. Label the metabolic or basic component; label the respiratory or acidic component.
4. Calculate the third unknown factor of the Henderson–Hasselbalch equation when given any two of the three variables.

5. Show by diagram and discuss the compensatory mechanisms of the kidneys and lungs.

6. List the four types of acid–base disturbances. Under each category, discuss pathophysiology and list situations that may cause each type of acid–base disturbance.

7. Calculate the anion gap when given electrolyte concentration values.

8. Discuss other buffer systems of the blood, both those in the plasma and those inside the red blood cell.

9. In regard to the oxygen dissociation curve, list conditions that may cause the curve to shift to the right or to the left. Describe the physiological effect associated with each type of shift.

10. Describe specimen requirements for blood gases. Discuss the effect on blood gas results in the following situations:
    a. patient hyperventilates during collection
    b. the sample is exposed to room air
    c. venous blood is used rather than arterial blood
    d. the sample is allowed to stand at room temperature before analysis
    e. plastic syringe is used instead of glass
    f. actual patient temperature varies from normal body temperature

11. Discuss the principle of each electrode (pH, pO_2, and pCO_2) in blood gas instruments.

12. Calculate the theoretical partial pressure of calibrating gases based on Dalton's Law.

13. Discuss the role of electrolytes in the body.

14. Trace the absorption and excretion pathway for each electrolyte.

15. Discuss assay methodology for each electrolyte.

16. List conditions that may be associated with abnormal electrolyte values.

17. List reference ranges for each electrolyte.

18. Determine the validity of electrolyte measurements.

19. Discuss the theory and technique for the measurement of osmolality by freezing-point depression.

20. Calculate the expected osmolality when given Na^+, glucose, and BUN assay values.

GLOSSARY OF TERMS

Acid. Substances that dissociate to produce protons (hydrogen ions) in solution.

Anoxia. Anoxia means without oxygen, but accepted terminology means a deficiency of oxygen in the blood.

Base. Substances that accept or remove protons or hydrogen ions from solution.

Base, negative. Negative deviation from normal values of buffer base in arterial blood. Normal buffer base is considered to be represented by zero because normal individuals do not have excess or deficit base. Negative base is usually interpreted as negative deviation from zero greater than 2.5 mmol/L. Negative base is also referred to as base deficient.

Base, positive.   Positive base is usually interpreted to be positive deviation from zero to greater than 2.5 mmol/L. Positive base is also referred to as base excess.

Bicarbonate (HCO_3^-).   Bicarbonate is the conjugate base of carbonic acid. When the Henderson–Hasselbalch equation is expressed using the carbonic acid–bicarbonate buffer system, bicarbonate represents the basic or metabolic component. Normal HCO_3^- in arterial blood is 20–26 mmol/L.

Buffer base (buffer anions).   Buffer base represents the sum of the concentrations (in mmol/L or meq/L) of all buffer anions present in the blood. Buffer anions include bicarbonate, phosphate, hemoglobin, and plasma proteins. The concentration of buffer anions will vary depending on hydrogen ion concentration. Buffer base measures the ability of the buffer systems of the blood to buffer acids. Normal buffer base of arterial whole blood at pH 7.4 is 46–50 mmol/L.

Dalton's law.   Dalton's law states that total pressure is equal to the sum of the partial pressures of individual gases.

Henderson–Hasselbalch equation.   The Henderson–Hasselbalch equation is used to explain the action of buffers and their role in maintaining the pH of a solution.

Henry's law.   Henry's law states that the solubility of gas in a liquid is directly related to the partial pressure of the gas above the liquid.

Hypercapnia.   Increased amount of carbon dioxide in the blood or elevated arterial blood pCO_2.

Hypocapnia.   Lack of carbon dioxide in the blood.

Hypoxemia.   Refers to a low oxygen content in arterial blood.

Hypoxia.   Refers to the lack of oxygen at the tissue and cellular level.

Oxygen content (ctO_2) (total oxygen).   Oxygen content is the sum of free oxygen (dissolved) and bound oxygen (to hemoglobin) in the blood. At the pO_2 of ambient air, very little oxygen is present in blood as the free gas.

Oxygen saturation.   Oxygen saturation measures the amount of oxygen combined with hemoglobin compared to the amount of oxygen that can be combined with hemoglobin, expressed as a percent. Normal percent oxygen saturation of arterial blood is 95–98%.

pCO_2.   The partial pressure or tension exerted by carbon dioxide in the blood usually expressed in mm Hg. pCO_2 of normal arterial blood is 32–45 mm Hg.

pO_2.   The partial pressure or tension exerted by oxygen in the blood usually expressed in mm Hg. pO_2 of normal arterial blood is 75–100 mm Hg.

pH.   pH expresses the concentration of free hydrogen ions in solution. Hydrogen ion concentration is expressed in log units ($pH = -\log H^+$). The pH of a solution is inversely proportional to the log of the concentration of hydrogen ions in solution (i.e., as the hydrogen ion concentration increases, pH decreases).

pK or pK_a.   pK equals the negative log of the dissociation constant of a buffer acid. The lower the value of pK, the stronger is the acid. In a solution having a pH equal to pK, the concentrations of the acid and its conjugate base are equal. The pK of the carbonic acid–bicarbonate buffer system is 6.1 at normal body temperature. At this pH, this buffer pair has peak efficiency.

P_{50} or $pO_2(0.5)$.   Refers to the arterial pO_2 required to saturate 50% of the hemoglobin with oxygen. The P_{50} at pH 7.4 (adults) $= 25$–29 mm Hg.

Total CO_2 (CO_2 content) (ctCO_2).   This represents the sum of the concentrations of dissolved carbon dioxide, carbonic acid, bicarbonate, and carbamino CO_2 (CO_2 bound to protein). If the sample is whole blood, total CO_2 would also include the CO_2 bound to hemoglobin. Normal total CO_2 arterial blood is 21–27 mmol/L.

The pH of the body must be maintained within very narrow limits. Many reactions in the body are dependent on enzymes, and each enzyme functions maximally within a narrow pH range. If the degree of deviation from normal pH is great enough, enzyme activity is depressed and life ceases.

    The end products of metabolism are primarily acid components. Sulfuric acid is produced from amino acid residues that contain sulfur. Phosphoric acid is produced from phospholipid catabolism. The end product of carbohydrate metabolism, carbon dioxide, when hydrated, forms carbonic acid. In fact, 15,000 mmol of CO_2 are produced each day in metabolism. Hence, the body must constantly guard against attempts to lower blood pH. In clinical practice, the most common disorders of acid–base disturbance are those associated with acidosis.

DEFINITIONS AND NORMAL VALUES

Acids may be defined as proton donors or substances that release hydrogen ions in solution. Bases may be defined as proton acceptors or substances that remove hydrogen ions from solution. The processes of dissociation (release of hydrogen ions) and association (removal of hydrogen ions) occur simultaneously in acid, neutral, or basic solutions. pH is the term used to describe hydrogen ion concentration over a range of $1 \cdot 10^{-1}$ to $1 \cdot 10^{-14}$ moles/L.

    Blood pH and plasma pH are essentially the same; however, owing to the effects of red blood cells on the liquid junction potential of the pH glass electrode, whole blood values will be approximately 0.01 pH unit lower than plasma values. Blood or plasma pH usually accepted as normal is 7.4 ± 0.05 pH unit or 7.35–7.45. The pH range usually accepted as *compatible* with life is 7.0–7.8. However, pH values as low as 6.8 have been recorded with patient survival.

    The concentration of ions may be expressed as meq/L or mmol/L. The concentration of molecules is expressed as mmol/L. When the concentration consists of ions and molecules, the concentration should be expressed only in mmol/L. The International System of Units expresses concentration (ions and molecules) in mmol/L only.

    In normal arterial blood, total CO_2 content (ctCO_2) is made up of (1) HCO_3^- (bicarbonate); (2) cdCO_2 (dissolved CO_2); (3) H_2CO_3 (carbonic acid); (4) CO_3^{2-} (carbonate ion); and (5) CO_2 bound to protein called carbamino CO_2. Most of the CO_2 in arterial blood is in the form of HCO_3^- with a mean concentration of 24 meq/L or mmol/L. The mean concentration of cdCO_2 + H_2CO_3 is 1.2 mmol/L, the dissolved form being present in the greatest amount. The carbonate ion and carbamino CO_2 add only insignificant amounts to total CO_2 content. Therefore, the mean value for total CO_2 content of normal arterial whole blood is 25.2 mmol/L. Total CO_2 content in normal arterial blood is summarized in Table 9.1.

Table 9.1. Total CO_2 Content in Normal
Arterial Blood

Component	Mean Concentration
HCO_3^-	24.0 mmol/L
$cdCO_2$ and $H_2CO_3^a$	1.2 mmol/L
CO_3^{2-}	insignificant
Carbamino CO_2	insignificant
Total CO_2	25.2 mmol/L

^{a} In the plasma, dissolved CO_2 reacts with water in an equilibrium reaction to form H_2CO_3 as: $cdCO_2 + H_2O \rightleftharpoons H_2CO_3$.

The reference range for total CO_2 content in arterial whole blood is 21–27 mmol/L. Venous whole blood contains more CO_2 (usually not more than 2 mmol/L) because it contains the CO_2 from metabolizing cells that is being carried to the lungs for excretion. The reference range for total CO_2 content in venous whole blood is 23–29 mmol/L.

BUFFER SYSTEMS AND HENDERSON–HASSELBALCH EQUATION

When an acid molecule dissociates, liberating a hydrogen ion or proton, the residue molecule (or ion) is defined as the conjugate base of that acid. The conjugate base of a weak acid (e.g., acetic acid) has strong affinity for the hydrogen ion and binds the hydrogen ion quite firmly allowing little dissociation to occur. The conjugate base of a strong acid (e.g., hydrochloric acid) has little affinity for the hydrogen ion, allowing greater dissociation to occur resulting in more hydrogen ions in solution. Therefore, weak acids have strong conjugate bases; strong acids have weak conjugate bases. Each acid and its conjugate base form a conjugate pair. A buffer is a weak acid in solution with its conjugate base. Buffers guard against pH shifts.

The carbonic acid–bicarbonate conjugate pair normally constitutes approximately 60% of the total buffering capacity of whole blood. This buffer pair may be illustrated as:

$$\frac{H_2CO_3 \text{ carbonic acid (weak acid)}}{HCO_3^- \text{ bicarbonate (conjugate base)}}$$

The carbonic acid or proton donor buffers against changes in pH due to the addition of a base. The buffering mechanism of carbonic acid is illustrated in the example below:

OH^- added
$\downarrow$
OH^- (strong base) + H_2CO_3 (proton donor) $\rightarrow$ H_2O + HCO_3^- (weaker base)

The bicarbonate or proton acceptor buffers against changes in pH due to the addition of an acid. The buffering mechanism of bicarbonate is illustrated in the example below:

H$^+$ added

↓

H$^+$ (strong acid) + HCO$_3^-$ (proton acceptor) → H$_2$CO$_3$ (weaker acid)

In normal arterial blood, the ratio of the concentration of HCO$_3^-$ and the concentration of H$_2$CO$_3$ is 20:1. This is illustrated in the example below. Brackets [ ] are used to designate concentration of:

$$\frac{[\text{HCO}_3^-]}{[\text{H}_2\text{CO}_3]} = \frac{24 \text{ mmol/L}}{1.2 \text{ mmol/L}} = \frac{20}{1}$$

If the concentration of any component is changed, the ratio may change; however, if the ratio still remains 20:1, the pH will be 7.4. These concepts are illustrated in Table 9.2. In Table 9.2, you will note that the total CO$_2$ value may be very abnormal, but as long as the ratio remains 20:1, the pH will be 7.4.

If the ratio of the concentration of the two components is plotted on linear graph paper against pH, a nonlinear curve results. However, the log of the ratio versus pH on linear graph paper will show a straight line relationship as illustrated in Figure 9.1. The straight line plot in Figure 9.1 shows that for each log ratio plotted on the graph, the correct pH can be obtained by adding 6.1 to the log ratio value. These data are illustrated in Table 9.3. This number (6.1) is the negative log of the dissociation constant of carbonic acid and is written as pK or pK_a. The value for pK changes with variations in ionic strength, temperature, and pH. For blood plasma at 37°C, 6.1 represents the normal mean value and is the value used in the Henderson–Hasselbalch equation.

pK refers to the pH at which both buffer components are present in equal amounts. Buffers are most effective in their buffering capacity at a pH close to their pK value. A 6.1 pH is over 1 pH unit away from the normal blood pH of 7.4. However, the carbonic acid–bicarbonate buffer pair is present in high concentra-

Table 9.2. The Ratio of [HCO$_3^-$]/[H$_2$CO$_3$] determines blood pH

Component Concentration [HCO$_3^-$]/[H$_2$CO$_3$]	Ratio[a]	Total CO$_2$ (mmol/L)	pH[a]	Component Change
24/1.2	20:1	25.2	7.4	HCO$_3^-$ normal H$_2$CO$_3$ normal
24/2.4	10:1	26.4	7.1	HCO$_3^-$ normal H$_2$CO$_3$ increased
48/2.4	20:1	50.4	7.4	HCO$_3^-$ increased H$_2$CO$_3$ increased
48/1.2	40:1	49.2	7.7	HCO$_3^-$ increased H$_2$CO$_3$ normal

[a] If the ratio remains 20:1, the blood pH will always be 7.4.

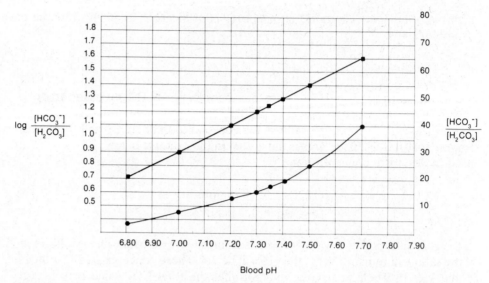

Figure 9.1. The log of the ratio of [HCO$_3^-$]/[H$_2$CO$_3$] (■) versus pH will show a linear relationship. The ratio of [HCO$_3^-$]/[H$_2$CO$_3$] versus pH (●) is nonlinear.

tion, and the concentration of the components can be regulated by either the kidneys or lungs; therefore, this buffer pair represents the most important buffer system in the body.

The Henderson–Hasselbalch equation is used to express the mathematical relationship between buffers and pH and their role in maintaining the pH of a solution. The pH value of whole blood or plasma may be calculated by the following formula known as the Henderson–Hasselbalch (H-H) equation.

$$pH = pK + \log \frac{[\text{conjugate base}]}{[\text{acid}]}$$

Using the carbonic acid–bicarbonate buffer pair, the H-H equation may be written

Table 9.3. Derivation of the pK of Carbonic Acid at 37°C

Log Ratio	Measured pH	Number Added to Log of Ratio to Derive the Correct pH	Calculation
0.7	6.8	6.1	0.7 + 6.1 = 6.8
0.9	7.0	6.1	0.9 + 6.1 = 7.0
1.1	7.2	6.1	1.1 + 6.1 = 7.2
1.2	7.3	6.1	1.2 + 6.1 = 7.3
1.3	7.4	6.1	1.3 + 6.1 = 7.4
1.4	7.5	6.1	1.4 + 6.1 = 7.5
1.6	7.7	6.1	1.6 + 6.1 = 7.7

$$pH = 6.1 + \log \frac{[\text{HCO}_3^-] \text{ basic or metabolic component}}{[\text{H}_2\text{CO}_3] \text{ acidic or respiratory component}}$$

Although the acidic or respiratory component is composed of the dissolved CO_2 and hydrated CO_2, it is generally written as the hydrated form (H_2CO_3) only. This terminology will be used throughout this chapter.

The application of the Henderson–Hasselbalch equation to determine blood pH is shown in the examples below. The first example shows the substitution of normal values; the second example shows the substitution of an abnormal bicarbonate content:

Example 1 $HCO_3^- = 24$ mmol/L
             $H_2CO_3 = 1.2$ mmol/L
             log of ratio (20) = 1.3
             pH = 6.1 + 1.3 = 7.4

Example 2 $HCO_3^- = 36$ mmol/L
             $H_2CO_3 = 1.2$ mmol/L
             log of ratio (30) = 1.48
             pH = 6.1 + 1.48 = 7.58

CHARACTERISTICS OF GASES

In the lungs, the air (gas) is separated from the blood (liquid) by a membrane that is permeable to gas molecules. The gas molecules escape into the blood or exit from the blood depending on the concentration of gas molecules on either side of the gas permeable membrane. The force that causes gas molecules to enter the liquid is called the pressure of the gas. The force that causes gas molecules to escape from the liquid is called the tension of the gas.

DALTON'S GAS LAW

Dalton's law states that the total gaseous pressure is equal to the sum of the partial pressures of individual gases. The individual pressures of all the gases in the atmosphere (including water vapor) is 760 mm Hg at sea level. Dalton's law may then be written:

total atmospheric pressure (760 mm Hg) = $pO_2 + pCO_2 + pN_2 + p(H_2O)$

The partial pressure of a gas is the pressure that each individual gas exerts and is independent of the other gases of a mixture. Partial pressure is expressed as "p." "p" may stand for either the pressure or tension of a gas and is expressed in mm Hg, torr, or kilopascals (kPa); 1 mm Hg is equal to 1 torr or 0.133 kPa. Most laboratories in the United States express partial pressure in mm Hg. The International System of Units expresses "p" in pascal units. The normal arterial pCO_2 is 40 mm Hg; the normal alveolar pCO_2 is 40 mm Hg.

HENRY'S GAS LAW

Henry's law states that the solubility of a gas in a liquid is directly related to the partial pressure of the gas above the liquid. The concentration of H_2CO_3 in the blood is directly related to pCO_2. The solubility factor for carbon dioxide in the blood is 0.03. The solubility factor is derived as follows:

$$\text{solubility factor} = \frac{\text{mmol/L of } H_2CO_3 \text{ in normal blood}}{pCO_2 \text{ (mm Hg) in normal alveolar air}} = \frac{1.2}{40}$$

$$= 0.03 \text{ mmol/L/mm Hg}$$

The Henderson–Hasselbalch equation may then be written:

$$pH = 6.1 + \log \frac{HCO_3^-}{pCO_2 \times 0.03} \quad \text{or} \quad 6.1 + \log \frac{ctCO_2 - (pCO_2 \times 0.03)}{pCO_2 \times 0.03}$$

COMPENSATORY MECHANISMS

When the concentration of the carbonic acid–bicarbonate buffer system deviates from normal, physiological mechanisms are initiated that tend to restore the ratio of the two components to normal. As long as the ratio of the concentration of HCO_3^-/H_2CO_3 remains $20:1$, the blood pH will be 7.4. When the normal $20:1$ ratio is disturbed, the kidneys and lungs tend to change the concentration of either HCO_3^- or H_2CO_3 so that a ratio close to $20:1$ is maintained. As a general rule, compensation will restore the pH toward normal but will never totally normalize the pH, because to do so would remove the stimulus to compensate. For a primary or single acid–base disturbance, the degree of compensation can be predicted. If the degree of compensation is inappropriate, the assumption is made that more than one type of acid–base disorder (mixed) is present.

KIDNEYS

Na⁺–H⁺ Exchange

The kidneys tend to correct for primary abnormalities in the basic or metabolic (HCO_3^-) component. Bicarbonate is filtered by the kidney glomerulus and can be handled in several ways depending on the acid–base status of the individual. The bicarbonate may be (1) excreted in the urine; (2) reabsorbed, either partially or almost totally, depending on need; or (3) synthesized by renal tubular cells. Changes in bicarbonate affect the numerator or metabolic component of the Henderson–Hasselbalch equation. The kidneys require three to four days to reach maximal compensatory capacity. The mechanism for bicarbonate synthesis and reabsorption (Na⁺–H⁺ exchange) is illustrated in Figure 9.2. The key below is used to interpret each numbered step of Figure 9.2.

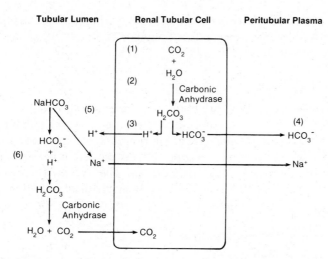

Figure 9.2. The kidney mechanism for bicarbonate synthesis and reabsorption (Na^+–H^+ exchange). The numbers in parentheses refer to reactions described in the text. (Adapted with permission from Valtin, H., *Renal Function: Mechanisms Preserving Fluid and Solute Balance,* Boston: Little, Brown, 1973.)

1. Carbon dioxide in renal tubular cells is derived from within the cell by metabolic processes and from diffusion of CO_2 into the cell from blood and tubular filtrate.
2. The CO_2 reacts with water in the presence of the enzyme, carbonic anhydrase, to form carbonic acid.
3. The carbonic acid dissociates to form $H^+ + HCO_3^-$.
4. The HCO_3^- enters the peritubular plasma.
5. The H^+ enters the kidney tubule lumen and is exchanged for Na^+ present in the tubular fluid. Electroneutrality is therefore maintained.
6. Inside the renal tubular lumen, the H^+ combines with filtered HCO_3^- to form carbonic acid. The carbonic acid dissociates to form CO_2 and H_2O. The reaction is accelerated in the proximal tubule by carbonic anhydrase found on epithelial cell brush borders. The CO_2 diffuses into the renal tubular cell and enters the reaction as in step 1. In this way, bicarbonate is indirectly returned to the blood.

The reaction with carbonic anhydrase is as follows:

$$CO_2 + H_2O \overset{CA}{\rightleftharpoons} H_2CO_3 \rightleftharpoons H^+ + HCO_3^-$$

The dehydration of H_2CO_3 forms CO_2 and H_2O; the hydration of CO_2 forms H_2CO_3. The law of mass action determines the direction the reaction will take.

The net effect of all the above reactions is the indirect reabsorption of urinary bicarbonate and generated tubular bicarbonate being reabsorbed into the circulation. Bicarbonate will be lost in the urine when blood levels are greater than normal (renal threshold approximately 28 mmol/L).

When the pH of the tubular fluid reaches 4.5, the Na^+–H^+ exchange stops. Therefore, the hydrogen ions have to be bound or buffered so that bicarbonate can continue to be produced.

Urinary Buffer Systems

In the urine, the hydrogen ions are buffered or bound by (1) the phosphate buffer system; (2) the ammonia buffer system; and (3) the carbonic acid buffer system.

The phosphate buffer system (pK 6.8) represents an important urinary buffer system. When the conjugate base accepts a hydrogen ion, it becomes lipid insoluble; therefore, it cannot cross cell membranes. The accepted hydrogen ion will remain in the urine. This buffer system and its mechanism of buffering hydrogen ions is

$$\frac{HPO_4^{2-} \text{ conjugate base}}{H_2PO_4^- \text{ acid}} \qquad HPO_4^{2-} + H^+ \rightarrow H_2PO_4^-$$

Ammonia is formed in the distal renal tubular epithelial cells from deamination of the amino acid glutamine. Ammonia diffuses across cell membranes and enters the nephron tubular lumen. Ammonia, the conjugate base, accepts hydrogen ions to form ammonium. The acid form (NH_4^+) is lipid insoluble and does not cross cell membranes and will be excreted in the urine. The pK of the ammonia–ammonium buffer system is 9.0. This buffer system and its mechanism of buffering hydrogen ions is illustrated below.

$$\frac{NH_3 \text{ conjugate base}}{NH_4^+ \text{ acid}} \qquad NH_3 + H^+ \rightarrow NH_4^+$$

Filtered bicarbonate, the conjugate base of the carbonic acid–bicarbonate buffer pair (pK-6.1), accepts hydrogen ions in the urine to form carbonic acid. The carbonic acid dissociates to form CO_2 and H_2O. The CO_2 diffuses into renal tubular epithelial cells to form H_2CO_3, which then dissociates to form HCO_3^- and H^+ (see Figure 9.2). This mechanism allows for reabsorption of filtered bicarbonate. The buffer system and its mechanism of buffering hydrogen ions is

$$\frac{HCO_3^- \text{ conjugate base}}{H_2CO_3 \text{ acid}} \qquad HCO_3^- + H^+ \rightarrow H_2CO_3$$

LUNGS

The lungs may compensate by altering the acid or respiratory component. During hypoventilation, CO_2 is retained, increasing the pCO_2 in the alveolar space. This increases the concentration of H_2CO_3 in the blood (Henry's law). During hyperventilation, CO_2 is "blown off," decreasing the pCO_2 in the alveolar space. This decreases the H_2CO_3 in the blood (Henry's law). Therefore, an inverse relationship exists between the rate of alveolar ventilation and the concentration of H_2CO_3 in the blood. These concepts are illustrated in Table 9.4.

Table 9.4. Relationship between Alveolar pCO_2 and the Concentration of H_2CO_3 in the Blood

Alveolar pCO_2	Concentration of H_2CO_3 in the Blood
10 mm Hg	$10 \times 0.03 = 0.3$ mmol/L
20 mm Hg	$20 \times 0.3 = 0.6$ mmol/L
40 mm Hg	$40 \times 0.03 = 1.2$ mmol/L (normal)
50 mm Hg	$50 \times 0.03 = 1.5$ mmol/L
80 mm Hg	$80 \times 0.03 = 2.4$ mmol/L

Changes in ventilation affect the denominator or respiratory (acidic) component of the Henderson–Hasselbalch equation. The lungs require one day to reach maximal compensatory capacity.

TYPES OF ACID–BASE DISTURBANCES

Acid–base disturbances are named according to the component of the carbonic acid–bicarbonate buffer system that is primarily affected and according to the pH deviation that results. The HCO_3^- is the metabolic or basic component; the H_2CO_3 is the respiratory or acidic component. There are four types of acid–base disturbances: (1) metabolic acidosis; (2) metabolic alkalosis; (3) respiratory acidosis; and (4) respiratory alkalosis.

METABOLIC ACIDOSIS

When the primary defect *decreases* the HCO_3^-, the component that is primarily affected is the metabolic or basic component. Since the basic component is lowered, the pH will be lowered. This type of acid–base disturbance is called metabolic (component primarily affected) acidosis (pH deviation that results).

METABOLIC ALKALOSIS

When the primary defect *increases* the HCO_3^-, the component that is primarily affected is again the metabolic or basic component; however, the basic component is increased resulting in a pH that increases. This type of acid–base disturbance is called metabolic (component primarily affected) alkalosis (pH deviation that results).

RESPIRATORY ACIDOSIS

When the primary defect *increases* the H_2CO_3, the component that is primarily affected is the respiratory or acidic component. Since the acidic component is

Table 9.5. The Etiology of the Four Types of Acid–Base Disturbances

Component Primarily Affected	Resulting pH Deviation	Type of Acid–Base Disturbance
HCO_3^- Metabolic or basic component decreased	Decreased	Metabolic acidosis
HCO_3^- Metabolic or basic component increased	Increased	Metabolic alkalosis
H_2CO_3 Respiratory or acid component increased	Decreased	Respiratory acidosis
H_2CO_3 Respiratory or acid component decreased	Increased	Respiratory alkalosis

increased, the pH will decrease. This type of acid–base disturbance is called respiratory (component primarily affected) acidosis (pH deviation that results).

RESPIRATORY ALKALOSIS

When the primary defect *decreases* the H_2CO_3, the component that is primarily affected is again the respiratory or acidic component; however, the acidic component is decreased resulting in a pH that increases. This type of acid–base disturbance is called respiratory (component primarily affected) alkalosis (pH deviation that results).

The etiology of each type of acid–base disturbance is summarized in Table 9.5.

ANION GAP (UNDETERMINED ANIONS)

The ionized or charged particles present in plasma are illustrated in Figure 9.3. The law of electroneutrality states that although the types of positive and negative charges may be variable, the total number of positively charged ions and the total number of negatively charged ions must be equal. The concentration of total cations in plasma must equal the concentration of total anions in plasma.

In the clinical laboratory, sodium, bicarbonate, and chloride are routinely measured. Note in Figure 9.3 that the sum of the remaining anions (undetermined anions), which includes the organic acids, phosphates, sulfates, and protein, exceed the sum of the remaining cations (undetermined cations), which includes calcium, magnesium, and potassium. This difference $(24 - 12 = 12)$ is referred to as the anion gap (AG). The term, anion gap, is a misnomer because it implies that there is a gap between anion and cation concentration. The term undetermined anions or unmeasured anions is a more accurate description. An estimation of the undetermined anions (anion gap) may be calculated by using the following formula:

$$AG = Na^+ - (Cl^- + HCO_3^-)$$

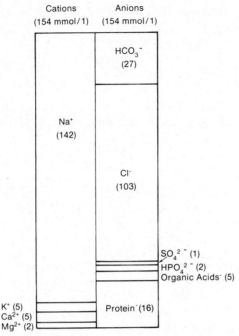

Cations
(154 mmol/1)

Anions
(154 mmol/1)

HCO_3^-
(27)

Na^+
(142)

Cl^-
(103)

SO_4^{2-} (1)
HPO_4^{2-} (2)
Organic Acids⁻ (5)

K^+ (5)
Ca^{2+} (5)
Mg^{2+} (2)

Protein⁻ (16)

Figure 9.3. Distribution of cations and anions in normal plasma. The concentration of each ion in mmol/L is indicated in parentheses. [Adapted from *The Chicago Medical School Quarterly*, **22**, 156 (1962).]

In using the formula, HCO_3^- and $ctCO_2$ may be interchanged, since the difference in total carbon dioxide content (normal arterial, approximately 25.2 mmol/L) and bicarbonate (normal arterial, approximately 24.0 mmol/L) is small. The anion gap reference range is 8–16 mmol/L; the mean value is 12 mmol/L.

The application of the anion gap formula is illustrated in the examples below:

Example 1 Na^+ = 140 mmol/L
Cl^- = 103 mmol/L
$ctCO_2$ = 27 mmol/L
AG = 140 − (103 + 27) = 10 mmol/L

Example 2 Na^+ = 130 mmol/L
Cl^- = 95 mmol/L
HCO_3^- = 10 mmol/L
AG = 130 − (95 + 10) = 25 mmol/L

ELEVATED ANION GAP

The addition of an acid to plasma alters the AG. Acids are buffered (or neutralized) by HCO_3^-, which is consumed in the process, therefore decreasing the concentration of HCO_3^-. However, the concentration of Na^+ and Cl^- are not altered. In substituting values into the AG formula, the difference between the concentration of Na^+ and the concentration of (Cl^- + HCO_3^-) increases, resulting in an elevation of the anion gap.

REDUCED ANION GAP

A reduced or low anion gap may be due to (1) laboratory error, which results in erroneous electrolyte values; (2) hypoalbuminemia; (3) bromide intoxication; and (4) multiple myeloma.

Laboratory Error

Falsely low sodium values or falsely elevated bicarbonate or chloride values reduce the anion gap.

Hypoalbuminemia

At normal pH 7.4, almost all serum proteins are negatively charged and counter-balance a portion of the sodium charge. Albumin, the protein present in the highest concentration in serum, is primarily responsible for the charge balancing effect of proteins. In diseases associated with hypoalbuminemia (i.e., renal disease and liver disease), part of the anion charge associated with albumin is replaced by the measured anions chloride and bicarbonate, resulting in a reduced anion gap. Hypoalbuminemia is the most common cause of a decreased anion gap.

Bromide Intoxication

In bromide intoxication, bromide accumulates in the blood and replaces some of the chloride in the body. Colorimetric chloride methods are generally not specific and measure both halides. The intensity of the color produced with bromide is greater than the color produced with chloride. The "high chloride value" with unchanged values for Na^+ and HCO_3^- results in a reduced anion gap or even a negative value.

Multiple Myeloma

Some of the myeloma proteins carry a positive charge and cause the retention of Cl^- to off-set their positive charge. If Na^+ remains constant but (Cl^- + HCO_3^-) increases, the anion gap is reduced. It has also been suggested that the cationic paraproteins displace sodium-containing water from serum, leading to decreased Na^+ values, thus reducing the anion gap.

USE OF THE ANION GAP

Clinically, the determination of the anion gap is most useful in diagnosing *metabolic acidosis*. With few exceptions, the anion gap will be normal in other acid–base disturbances. The anion gap may also be used as a tool to monitor response to therapy in those individuals with an increased anion gap, especially in diabetes ketoacidosis.

The anion gap is also useful as a quality control tool. With the use of multichannel instruments, the AG can be monitored daily on all electrolyte profiles with trends noted. If all AG values fall within a very narrow range or if all values are either too high or too low, a systematic error in electrolyte measurements may be indicated.

ACID–BASE DISTURBANCES

METABOLIC ACIDOSIS

In metabolic acidosis, the metabolic or basic component (HCO_3^-) is decreased. Bicarbonate may be lost because it is excreted unchanged from the body or because it is consumed in buffering acids (keto acids, lactic acid, sulfuric acid, phosphoric acid, hydrochloric acid, and others). The consumption of bicarbonate by acids may be illustrated as

$$H^+ + HCO_3^- \rightarrow H_2CO_3 \rightarrow H_2O + CO_2 \uparrow \quad \text{(blown off)}$$

As carbonic acid is formed from bicarbonate and hydrogen, it dissociates to form carbon dioxide and water. A decrease in pH (with consumption of HCO_3^-) and an increase in pCO_2 stimulate the respiratory center in the brain, and the CO_2 is blown off.

Metabolic acidosis may be divided into two types: (1) acidemia with a normal anion gap (hyperchloremic metabolic acidosis) and (2) acidemia with an increased anion gap (normochloremic metabolic acidosis).

Metabolic Acidosis With a Normal Anion Gap

In metabolic acidosis with a normal anion gap, HCO_3^- is lost directly, either in the gastrointestinal (GI) tract or urine, or an acid is added whose associated anion is measured and is used in the AG formula (i.e., chloride).

Administration of Chloride-Containing Acids (HCl, NH₄Cl, Arginine · HCl)
Although HCO_3^- is used to buffer the H^+, the anion gap will be normal because the anion, Cl^-, is measured with the anions commonly determined in the laboratory (see anion gap formula). With the addition of other acids, the anion gap is increased because the resulting released anion is not included in the measurement of the anions used in the AG formula.

Gastrointestinal Bicarbonate Loss
Direct loss of bicarbonate can occur when bicarbonate-containing body fluids (i.e., pancreatic secretions) are lost from the body. In severe diarrhea or in the presence of a pancreatic fistula, loss of HCO_3^- occurs through the gastrointestinal tract.

Inherited or Acquired Renal Tubular Defects
There are several variants of renal tubular acidosis. The defect is failure to acidify the urine despite an acidemia. The defect may involve the proximal tubule (defec-

tive reabsorption of HCO_3^-) or distal tubule (defect in H^+ secretion). Certain drugs inhibit the carbonic anhydrase enzyme. In all of these situations, HCO_3^- is lost in the urine.

Metabolic Acidosis With An Increased Anion Gap

In metabolic acidosis with an increased anion gap, the HCO_3^- is decreased due to the consumption of HCO_3^- by acid.

Uremia

Acids of sulfur and phosphorus are normal end products of protein and phospholipid catabolism. In renal disease, these acids accumulate in the blood because of defective excretion. This is the most common cause of acidosis with an increased anion gap.

Ketoacidosis

Ketoacids are produced when fats are broken down. Any situation associated with increased fat catabolism (diabetes mellitus or starvation) may result in ketosis. Alcoholism is also associated with ketoacidosis.

Of the three ketone bodies, acetoacetic acid (AcAc), beta-hydroxybutyric acid (BHB), and acetone, only AcAc and BHB are acids. Acetone is not an acid and does not alter the HCO_3^- concentration. BHB does not react in the nitroprusside test for ketone bodies.

Isopropyl alcohol (rubbing alcohol) is metabolized to acetone and may give a positive nitroprusside test. The acetone will produce a fruity odor in the breath. In isopropyl intoxication, the AG, blood sugar, and bicarbonate will be normal.

Lactic Acidosis

Increased blood levels of lactic acid may occur in those situations associated with inadequate oxygen delivery to tissues (i.e., shock, congestive heart failure, anemia) (see Chapter 5). In normal persons, small amounts of lactic acid are produced by muscle cells, but this is readily removed and processed by the liver. In certain types of liver disease, blood levels of lactic acid will increase because the liver is unable to catabolize the lactic acid normally produced. The ingestion of phenformin, an oral hypoglycemic agent, may also cause lactic acidosis.

Methanol Toxicity

Methanol is metabolized to formaldehyde then formic acid, which are distributed throughout total body water and concentrated in the aqueous humor of the eye and in cerebrospinal fluid. Blindness may result and the patient may appear to be intoxicated. Toxic metabolites also stimulate overproduction of organic acids.

Ethylene Glycol (Antifreeze) Toxicity

Ethylene glycol is metabolized to toxic products including oxalic acid. The patient appears to be intoxicated, but the smell of ethanol in the breath is absent. The urinary sediment may show calcium oxalate crystals either in the envelope-shaped dihydrate form or in the needle-shaped monohydrate form.

Salicylate Toxicity (Especially in Children)

Toxic doses of salicylates may stimulate the central nervous system resulting in hyperventilation with respiratory alkalosis. Salicylates also stimulate organic acid production; the respiratory alkalosis is followed by metabolic acidosis. The increased AG is due to the replacement of HCO_3^- with unmeasured organic anions and salicylates.

Paraldehyde Ingestion

Paraldehyde causes neurological depression and metabolic acidosis. Paraldehyde spontaneously decomposes to acetic acid; however, other anions are also present. Paraldehyde gives a distinct odor to the breath.

Laboratory Findings in Metabolic Acidosis

The laboratory findings in metabolic acidosis include (1) pH, decreased; (2) $ctCO_2$, decreased; (3) HCO_3^-, decreased; (4) pCO_2, decreased; (5) base, negative; and (6) AG, may be increased.

The compensatory mechanisms try to restore the normal 20:1 ratio of HCO_3^-/H_2CO_3. Since the primary disturbance decreased the HCO_3^-, the kidneys try to restore the HCO_3^- concentration toward normal. The kidney (1) increases the reabsorption of HCO_3^-, (2) increases the Na^+–H^+ exchange, thus generating more HCO_3^-; and (3) increases the ammonia formation to buffer the H^+ that is secreted. The pH of the urine will be decreased. The lungs compensate through hyperventilation (Kussmaul respiration) to try to lower the H_2CO_3, thereby maintaining a ratio as close to 20:1 as is possible.

Case History, Metabolic Acidosis

A 37-year-old white comatose male was seen in the emergency room. Past history revealed that the patient had developed diabetes at the age of eight. For approximately two years, he had experienced recurrent episodes of nausea and vomiting. Prior to this admission, he had been vomiting for a period of four days. At the time of admission, his blood pressure was 60/40 with pressure trousers in place. Admission laboratory data are given in Table 9.6.

Table 9.6. Metabolic Acidosis Case History Laboratory Data

Serum Glucose (74–112)[a] mg/dL		Na (135–145)[a] mmol/L		K (3.5–5.0)[a] mmol/L		Cl (98–106)[a] mmol/L
1520		119		8.6		79

pH (7.34–7.45)[a]	pCO₂ (32–45)[a] mm Hg	pO₂ (75–100)[a] mm Hg	HCO₃⁻ (20–26)[a] mmol/L	ctCO₂ (21–27)[a] mmol/L	Base (0)[a]	Saturation (95–98)[a] percent
6.96	19	125	4.1	4.6	−25.7	98.7

[a] Reference range for components.

Discussion

The patient was in severe metabolic acidosis at the time of admission. The laboratory data show the characteristic findings of metabolic acidosis, that is, decreased pH, pCO_2, $ctCO_2$, and HCO_3^-. The pO_2 was elevated because the patient was breathing rapidly (Kussmaul respiration) to ''blow off'' the CO_2 produced as the bicarbonate was consumed to buffer keto acids. The pO_2 of ambient air is much higher than arterial pO_2. The rapid exchange of air increased the alveolar pO_2, thus increasing the arterial pO_2 (Henry's law). The anion gap was elevated (36 mmol/L) due to the presence of keto acids. The patient was given intravenous insulin, saline, and bicarbonate. The electrolytes returned to normal as rehydration occurred. The patient improved over a period of two days and was discharged to be followed as an outpatient.

Numerous complications associated with diabetes were present in this patient, that is, retinopathy, nephropathy, neuropathy, and hypertension. The recurrent episodes of nausea and vomiting were thought to be a manifestation of diabetic gastroparesis.

METABOLIC ALKALOSIS

In metabolic alkalosis, the metabolic or basic component (HCO_3^-) is increased. Bicarbonate may be gained as a result of the ingestion of drugs whose end products of metabolism include bicarbonate, or from the loss of nonvolatile acid from extracellular fluid via the GI tract or kidneys.

Exogenous Alkali Ingestion

The administration or ingestion of drugs whose end products of metabolism include bicarbonate will result in a direct gain of bicarbonate. These drugs include sodium bicarbonate, sodium lactate, sodium citrate, and sodium acetate.

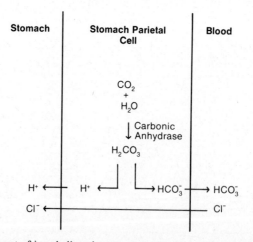

Figure 9.4. Movement of ions believed to occur as gastric acid is secreted into the stomach.

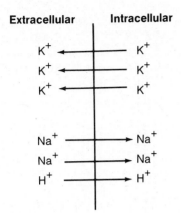

Figure 9.5. Reciprocal shift of ions in potassium deficiency.

Prolonged Vomiting or Nasogastric Suctioning

When gastric acid is formed, the movement of ions believed to occur is illustrated in Figure 9.4. Immediately after eating, the blood becomes slightly alkaline (alkaline tide) due to the addition of HCO_3^- to the blood as gastric acid is secreted into the stomach. As the gastric contents enter the intestine, the hydrogen ions of gastric juice are reabsorbed into the blood and bicarbonate is secreted into the intestinal lumen, thereby restoring the pH to normal.

The alkalosis that develops in prolonged vomiting or from prolonged nasogastric suctioning is believed to result from several mechanisms: (1) direct loss of HCl and other ions in gastric fluid and (2) through the hypokalemia that develops. In prolonged vomiting or suctioning, H^+, Cl^-, Na^+, and K^+ are lost. However, there is a proportionately greater loss of chloride than sodium. Chloride plays a vital role in the renal handling of sodium. In the proximal renal tubule, sodium is reabsorbed as sodium chloride or it is exchanged for hydrogen (Na^+–H^+ exchange). When sodium is reabsorbed as sodium chloride, water reabsorption follows passively due to the osmotic gradient produced by the sodium chloride. With the proportionately greater loss of chloride, less sodium is reabsorbed as sodium chloride and, therefore, less water is reabsorbed. This leads to extracellular volume contraction. The volume contraction stimulates the renin–angiotensin–aldosterone system. The loss of potassium in vomitus and in urine (due to the effect of aldosterone) produces hypokalemia. In hypokalemia, the shift of ions believed to occur is illustrated in Figure 9.5.

The reciprocal shift of ions (as illustrated in Figure 9.5) results in extracellular alkalosis and intracellular acidosis. The random urine chloride is low (less than 10 mmol/L). Once the vomiting stops, fluid and electrolyte (NaCl) replacement will reverse the alkalosis.

Severe Potassium Depletion

Any disorder associated with excess aldosterone (i.e., primary aldosteronism, Cushings, Bartter's syndrome, ACTH-secreting tumors) may lead to metabolic alkalosis. In the kidney, aldosterone enhances sodium reabsorption from the glomerular filtrate and enhances potassium secretion into the urine. The hypokalemia that develops causes K^+ to move from cells to plasma. Plasma H^+ and Na^+ move into the cell in order to maintain intracellular and extracellular electroneutrality (see Figure 9.5). Since H^+ is transferred into the cell, there is an extracellular alkalosis and an intracellular acidosis. In this group, chloride is not depleted, and the random urine chloride is usually greater than 20 mmol/L. This type of disturbance requires potassium replacement to correct the acid–base imbalance.

Laboratory Findings in Metabolic Alkalosis

The laboratory findings in metabolic alkalosis include (1) pH, increased; (2) ctCO_2, increased; (3) HCO_3^-, increased; (4) pCO_2, increased; and (5) base, positive.

The compensatory mechanisms try to restore the normal $20:1$ ratio of HCO_3^-/H_2CO_3. Since the primary disturbance increased the HCO_3^-, the kidneys try to lower the HCO_3^- towards normal. The kidney (1) decreases the reabsorption of HCO_3^-; (2) decreases the Na^+–H^+ exchange, thus conserving H^+ ions in the blood and generating less HCO_3^-; and (3) decreases ammonia formation, since less H^+ ions are secreted. The urinary pH is alkaline (usually).

The lungs compensate by hypoventilation in order to conserve CO_2, hence increasing the acidic or H_2CO_3 component. However, the pulmonary compensation for metabolic alkalosis is limited because hypoventilation also reduces arterial pO_2. A fall in arterial pO_2 stimulates the carotid and aortic bodies (chemoreceptors), which then stimulate the respiratory center to increase alveolar ventilation.

RESPIRATORY ACIDOSIS

Etiology

In respiratory acidosis, the respiratory or acidic component (H_2CO_3) is increased. Any disease or disorder that causes retention of CO_2 and, therefore, increased pCO_2, raises the concentration of H_2CO_3 in the blood with resultant acidosis. Decreased alveolar ventilation results from (1) generalized disease of the lungs such as in asthma, emphysema, and bronchopneumonia, (2) malfunction of the thoracic cage as would occur in respiratory muscle paralysis; and (3) depression of the respiratory center, which occurs in narcotic drug overdose.

Laboratory Findings in Respiratory Acidosis

The laboratory findings in respiratory acidosis include: (1) pH, decreased; (2) pCO_2, increased; (3) HCO_3^-, increased; and (4) ctCO_2, increased.

The compensatory mechanisms try to restore the normal $20:1$ ratio of HCO_3^-/H_2CO_3. Since the primary disturbance increased the H_2CO_3, the kidneys try to increase the HCO_3^- component. The kidney conserves bicarbonate by (1) increasing reabsorption of HCO_3^-; (2) increases the Na^+–H^+ exchange, which generates HCO_3^- and allows H^+ ions to be secreted into the urine; and (3) increases ammonia formation, which is needed to buffer the H^+ ions that are secreted. The urinary pH will be decreased.

Case History, Respiratory Acidosis

A nine-year-old white male was brought to the emergency room. He was unconscious when found on the playground at school. Resuscitation had been attempted and 100% oxygen given prior to admission to the hospital. There was no previous history of illness. Blood gases were performed and a toxicology screen was ordered. Admission electrolytes and blood gas laboratory results are listed in Table 9.7.

Discussion

At the time of admission, the patient was in severe respiratory acidosis. The pH was acidotic with increased pCO_2. The HCO_3^- was within the reference range because the kidneys require time to compensate (3–4 days for maximal compensation) and, therefore, due to the time element, HCO_3^- was not altered to reflect the sudden increase in the H_2CO_3 component. The pO_2 and the hemoglobin oxygen saturation were low because the child was not exchanging air.

Multiple drugs were administered in the emergency room with no clinical improvement. All attempts to revive the child were futile. The electroencephalogram was grossly abnormal and showed no electrical activity. The child became increasingly edematous with no urine output. A drug screen was performed on gastric contents, but the results were inconclusive. At autopsy, gross and microscopic finding were all within normal limits. Blood and gastric samples (no urine could be obtained) were sent to a reference laboratory, but no toxic products were identified.

Table 9.7. Respiratory Acidosis Case History Laboratory Data

	Na (135–145)[a] mmol/L	K (3.5–5.0)[a] mmol/L	Cl (98–106)[a] mmol/L
	137	5.0	87

	pH (7.34–7.45)[a]	pCO₂ (32–45)[a] mm Hg	pO₂ (75–100)[a] mm Hg	HCO₃⁻ (20–26)[a] mmol/L	ctCO₂ (21–27)[a] mmol/L	Base (0)[a]	O₂ Saturation (95–98)[a] percent
Admission	7.06	84	22	22	25	−8.4	18
1 hour later	7.05	90	21	24	26	−7.7	16
3 hours later	7.01	79	47	19	21	−12.3	59

[a] Reference range for components.

RESPIRATORY ALKALOSIS

Etiology

In respiratory alkalosis, the respiratory or acidic component (H_2CO_3) is decreased. Any disease or disorder that causes hyperventilation with loss of CO_2 and, therefore, decreased pCO_2, lowers the concentration of H_2CO_3 in the blood with resultant alkalosis. Increased alveolar ventilation may result from (1) stimulation of the respiratory center in the medulla of the brain and (2) stimulation of peripheral chemoreceptors located in artery walls.

The respiratory center in the medulla of the brain responds to an increase in pCO_2 or to a fall in pH of the cerebrospinal fluid. Even though the brain stem chemosensitive area primarily responds to hydrogen ions, the pCO_2 is the major chemical factor regulating alveolar ventilation. Carbon dioxide combines with water in the cerebrospinal fluid to form carbonic acid that dissociates, thus increasing the hydrogen ion concentration.

The peripheral chemoreceptors are located primarily in the bifurcation of the common carotid arteries (carotid bodies) and along the arch of the aorta (aortic bodies). These chemoreceptors are stimulated by a fall in pO_2 or a fall in pH of the fluid surrounding the cells. Signals are transmitted to the respiratory center to help regulate respiratory activity.

Laboratory Findings in Respiratory Alkalosis

The laboratory findings in respiratory alkalosis include (1) pH, increased; (2) pCO_2, decreased; (3) HCO_3^-, decreased; and (4) $ctCO_2$, decreased.

The compensatory mechanisms try to restore the normal $20:1$ ratio of HCO_3^-/H_2CO_3. Since the primary disturbance decreased the H_2CO_3, the kidneys try to decrease the HCO_3^-. The kidney excretes bicarbonate by (1) decreasing the reabsorption of HCO_3^-; (2) decreases the Na^+-H^+ exchange, which generates less HCO_3^- and allows H^+ ions to be conserved in the blood; and (3) decreases ammonia formation. The urinary pH will increase.

DIAGNOSING SINGLE ACID–BASE DISTURBANCES

A summary of acid–base disturbances and associated laboratory findings is presented in Table 9.8. In the table, note that in the two types of acidosis and in the two types of alkalosis, the concentration of pCO_2 and $ctCO_2$ are opposite. In one type of acidosis (or alkalosis), the pCO_2 and $ctCO_2$ are both increased; in the other type of acidosis (or alkalosis), the pCO_2 and $ctCO_2$ are both decreased. By noting the direction of pH change and the direction of change in pCO_2 or $ctCO_2$, the type of acid–base disturbance may be anticipated, providing it is a single type of disturbance.

BUFFER SYSTEMS OF THE BODY

Both intracellular and extracellular buffer systems are important in regulating the narrow limits of pH compatible with life (about 7.0–7.8). Buffers are found in

Table 9.8. Summary Table of Primary Acid–Base Disturbances and Associated Laboratory Findings

Acid-Base Disturbance	Primary Component Affected	Compensatory Mechanism	pH	ctCO₂	pCO₂
Metabolic acidosis	HCO_3^- ↓ [a]	H_2CO_3 ↓	↓	↓	↓
Metabolic alkalosis	HCO_3^- ↑	H_2CO_3 ↑	↑	↑	↑
Respiratory acidosis	H_2CO_3 ↑	HCO_3^- ↑	↓	↑	↑
Respiratory alkalosis	H_2CO_3 ↓	HCO_3^- ↓	↑	↓	↓

[a] ↑ = increased; ↓ = decreased.

plasma, interstitial fluid, intracellular fluid, and bone. End products of metabolism consist primarily of acids. Volatile acid is in the form of CO_2, which is derived from oxidative metabolism. Nonvolatile or fixed acid, produced from noncarbonic acid sources, is derived from protein catabolism (sulfates), phospholipid catabolism (phosphates), lipid metabolism (keto acids), and incomplete carbohydrate metabolism (lactate). The buffer systems of the body are especially effective in buffering acids.

BICARBONATE BUFFER SYSTEM

The pK of the bicarbonate–carbonic acid buffer pair is 6.1, and over 1 pH unit away from normal blood pH. However, this buffer pair represents the most important buffer found in plasma and interstitial fluid because it is present in high concentration, and the concentration of the buffer components can be regulated by the lungs and kidneys.

PROTEIN BUFFERS

Plasma proteins, especially albumin with a pK of 7.3, represent the second most important plasma buffer. At pH 7.4, most proteins assume a negative charge and act as proton acceptors in buffering acids:

$$H_2CO_3 + protein^- \rightarrow H \cdot protein + HCO_3^-$$

Hemoglobin (Hb), a protein in combination with heme, represents an important intracellular buffer. The pK_a of oxygenated hemoglobin is lower than that of deoxygenated hemoglobin. In other words, deoxygenated hemoglobin is less acidic than oxygenated hemoglobin. Therefore, deoxygenated hemoglobin is able to accept protons (H^+) and serves as an intracellular buffer for the acids produced in metabolism. When oxygen is released from hemoglobin in capillary blood, the deoxygenated hemoglobin can then buffer hydrogen ions generated from metabolic CO_2 that has diffused into the red blood cell. The reactions that take place in the red blood cell in CO_2 transport are referred to as the isohydric and chloride shift and are illustrated in Figure 9.6. The key below is used to interpret each numbered step of Figure 9.6:

1. Carbon dioxide from metabolizing tissue diffuses into plasma.
2. In the plasma, the hydration of CO_2 is very slow because the plasma contains no carbonic anhydrase. The little H^+ that is formed is buffered by HCO_3^-, proteins, and phosphate. Most of the CO_2 diffuses into red blood cells (RBC).
3. Within RBCs, a small portion of the CO_2 exists as dissolved CO_2 and in combination with hemoglobin as carbamino CO_2. Since erythrocytes contain abundant carbonic anhydrase, most of the CO_2 is rapidly hydrated to form H_2CO_3, which then dissociates to form H^+ and HCO_3^-.
4. The increase in H^+ causes the pH to fall, which enhances the release of oxygen from oxyhemoglobin. The oxygen then diffuses from RBCs to plasma. The deoxygenated hemoglobin is then able to buffer or accept the hydrogen ions.
5. As the bicarbonate concentration increases within RBCs, HCO_3^- diffuses from cells into plasma. In order to maintain electroneutrality, Cl^- shifts into erythrocytes from plasma. As a result of these reactions, most of the CO_2 generated by metabolic processes is carried to the lungs as HCO_3^- in plasma. These reactions are reversed when the oxygen concentration is elevated (lung alveoli) and oxygen diffuses into cells. Because of the chloride shift, venous blood will have less chloride than arterial blood (about 1 mmol/L lower).

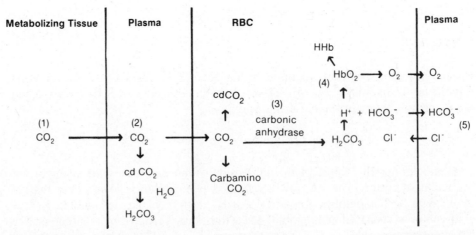

Figure 9.6. The hemoglobin buffer system. The numbers in parentheses refer to reactions described in the text. (Adapted with permission from Tietz, N., (ed.), *Fundamentals of Clinical Chemistry*, 2nd ed. Philadelphia: Saunders, 1976.)

PHOSPHATE BUFFERS

The inorganic phosphate buffer system is found in the plasma and in interstitial fluid, but, because of its low concentration, contributes very little to the buffering activity in these fluids:

$$HPO_4^{2-} + H^+ \rightleftharpoons H_2PO_4^- \text{ (proton acceptor)}$$

$$H_2PO_4^- + OH^- \rightleftharpoons HPO_4^{2-} + H_2O \text{ (proton donor)}$$

The intracellular organic phosphate, 2,3-diphosphoglycerate, represents an important intracellular buffer.

OXYGEN

TRANSPORT OF OXYGEN IN THE BLOOD

Oxygen is carried in the blood both as free gas and in a combined form with hemoglobin (oxyhemoglobin). At the pO_2 of ambient air, very little is carried as free gas. More oxygen will be carried in the free form in hyperbaric oxygenation or when breathing oxygen enriched air:

oxygen content (ctO_2) = oxygen bound to Hb + free dissolved oxygen (cdO_2)

The amount of oxygen that is bound by hemoglobin is dependent on (1) pO_2; (2) hemoglobin concentration; (3) temperature; (4) pH; (5) pCO_2; (6) type of hemoglobin present; and (7) the concentration of 2,3-diphosphoglycerate (DPG). (The affinity of oxygen for hemoglobin is regulated by DPG.) The saturation of hemoglobin with oxygen is expressed as a percent:

$$\% \text{ saturation} = \frac{\text{oxygen content}}{\text{oxygen capacity}} \times 100$$

Percent oxygen saturation is a measure of the amount of oxygen that is combined with hemoglobin compared to the total amount of oxygen that can combine with hemoglobin. The percent saturation may be directly measured in the following way. A blood sample is collected anaerobically and divided into two parts. The oxygen content is determined using a Van Slyke gasometer apparatus. First all gases are liberated from the sample and the pressure (P-1) is measured. P-1 is equal to the pressure exerted by all gases present. The oxygen is resorbed and the pressure is measured again (P-2). P-2 is equal to the pressure exerted by all gases except oxygen. The oxygen content is calculated by finding the difference, (P-1) − (P-2) = O_2 content. The second part of the sample is equilibrated with room air. This will allow for full saturation of hemoglobin with oxygen. The oxygen content is measured in the second sample as before. This will measure the capacity of the

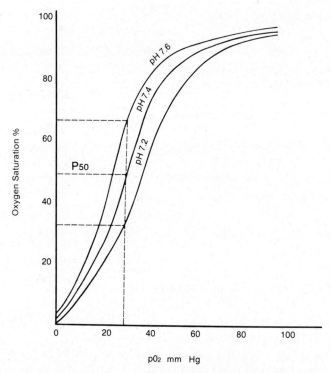

Figure 9.7. The oxygen dissociation curve. The P_{50} (dotted line to the center curve) represents the pO_2 required to saturate 50% of the hemoglobin with oxygen at normal pH and pCO_2.

hemoglobin to bind oxygen and the percentage saturation may then be calculated using the preceding formula.

At normal alveolar conditions ($pO_2 = 100$ mm Hg, $pCO_2 = 40$ mm Hg, temperature $= 37°C$), the saturation of hemoglobin with oxygen is approximately 95–98%. Blood gas instruments do not measure oxygen saturation directly. The oxygen saturation is calculated from sample pO_2 and pH measurements using the oxygen dissociation curve (see Figure 9.7).

In individuals with hemoglobinopathies, abberrations in 2,3-DPG, or significant amounts of carboxyhemoglobin or methemoglobin, the determination of oxygen saturation is more accurately done by using an oximeter. The oximeter is an instrument that uses spectrophotometric measurements to determine total hemoglobin, percent oxyhemoglobin, carboxyhemoglobin, and methemoglobin. The oxygen content is calculated based on the measured total hemoglobin and the percent oxyhemoglobin present. The oximeter methodology correlates well with the Van Slyke gasometric method, which is still considered the reference method for oxygen saturation.

OXYGEN DISSOCIATION CURVE

The oxygen dissociation curve shows the relationship between pO_2 and the saturation of hemoglobin with oxygen. The P_{50} or $pO_2(0.5)$ refers to the arterial pO_2

required to saturate 50% of the hemoglobin with oxygen. For adults, the P_{50} at pH 7.4 is 25–29 mm Hg (mean is 27 mm Hg). The oxygen dissociation curve is illustrated in Figure 9.7.

In Figure 9.7, the center curve is considered the standard curve, representing a pH of 7.4 and a pCO_2 of 40 mm Hg. The curve to the right represents a "shift" to the right, and the curve to the left represents a "shift" to the left. Note that at the same pO_2 (dotted line) a shift of the curve to the right will cause a fall in hemoglobin oxygen saturation, and a shift of the curve to the left will increase the hemoglobin oxygen saturation.

Shift to the Right (High P_{50})

A high P_{50} means that a higher than normal pO_2 (greater than 29 mm Hg) is required to saturate 50% of the hemoglobin with oxygen (see oxygen dissociation curve). Displacement of the curve to the right occurs under the following conditions: (1) acid pH; (2) increased pCO_2 (hypercapnia); (3) hyperthermia; (4) increased, 2,3-DPG; and (5) the presence of an abnormal hemoglobin with decreased affinity for oxygen.

Physiological Effects of a Right Shift

When the curve is shifted to the right, the affinity of oxygen for hemoglobin is decreased. Less oxygen is bound by hemoglobin in the lungs, but the physiological effect is small because the affinity is still high enough to bind adequate oxygen.

In metabolizing tissue, the lower pH (from increased CO_2) facilitates the release of oxygen from hemoglobin (shifts the curve to the right) thus ensuring adequate oxygen delivery to the tissues. Under normal conditions, the effect of pH on oxygen dissociation allows oxygen to be delivered to metabolizing tissues.

Shift to the Left (Low P_{50})

A low P_{50} means that a lower than normal pO_2 (less than 25 mm Hg) is required to saturate 50% of the hemoglobin with oxygen (see oxygen dissociation curve). Displacement of the curve to the left occurs under the following conditions: (1) alkaline pH; (2) decreased pCO_2 (hypocapnia); (3) decreased 2,3-DPG; (4) hypothermia; and (5) the presence of an abnormal hemoglobin with increased affinity for oxygen.

Physiological Effects of a Left Shift

When the curve is shifted to the left, the affinity of oxygen for hemoglobin increases. Less oxygen will be delivered to the tissues because less oxygen is dissociated from the hemoglobin.

CLINICAL SIGNIFICANCE

A portion of the oxygen dissociation curve has a flat area and a portion drops precipitously (see Figure 9.7). The saturation of hemoglobin with oxygen is minimally affected by changes in pO_2 when the pO_2 remains above 50–55 mm Hg. At pO_2 levels less than 50–55 mm Hg, small decreases in pO_2 drastically reduce the oxygen saturation, and, hence, the oxygen content of hemoglobin. In the treatment of individuals with lung disorders, the pO_2 is maintained to at least 50–55 mm Hg, or higher (if possible), to provide a greater margin of safety. The lowest pO_2 value that can be tolerated for long periods of time is approximately 40 mm Hg.

The physiological effects of moderately decreased pO_2 are (1) stimulation of the peripheral chemoreceptors (aortic and carotid bodies) leading to hyperventilation and (2) stimulation of erythropoietin production (produced in the kidney), which causes the bone marrow to increase the production of red blood cells (i.e., increases the hematocrit).

COLLECTION OF BLOOD FOR BLOOD GASES

When a sample is exposed to air, the gases in the sample will escape from the sample or gases in the air will enter the sample, depending on the parital pressure of each individual gas in each of the two phases (air or blood). The partial pressure of carbon dioxide in blood is greater than the partial pressure of carbon dioxide in ambient air. Therefore, when blood is exposed to air, the CO_2 will escape from the blood, decreasing the blood pCO_2 with a resulting increase in pH. The partial pressure of oxygen in ambient air is higher than that of blood; therefore, exposing blood to air will increase the sample pO_2. In order to avoid gas exchange, samples for blood gas analysis are collected in a closed system (anaerobic collection).

When blood is allowed to stand before analysis, the cells continue to metabolize glucose (glycolysis continues). Glycolysis consumes oxygen and increases lactic acid formation with a resultant decrease in pH. In order to avoid the changes associated with glycolysis, the blood gas determination should be performed immediately (within 10 minutes). If the determination is delayed, the specimen should be placed in ice water or placed on crushed ice (then covered with ice) but should be assayed within 1 hour.

Whole blood samples (using sodium heparin as the anticoagulant) are used for blood gas analysis. The blood may be obtained from arteries, capillaries, or veins. Arterial blood is preferred because it is more uniform in composition. If respiratory problems are to be evaluated, only arterial blood may be used. Arterial samples may be obtained from the radial, brachial, or femoral artery. The radial artery is the preferred site because there is less complication risk to the patient following collection.

If arterial blood cannot be obtained, capillary blood may be used as a substitute as long as moist heat is used before collection. Arterialized blood is obtained by warming a limb to 45°C. The heat dilates the capillaries, accelerates blood flow, and decreases the change in blood composition caused by tissue respiration. This method of collection is routinely used when samples are obtained from infants. Heparinized capillary tubes are used for collecting arterialized blood. After collection, a small piece of wire is inserted into the capillary tube before sealing. The wire may be moved with a magnet to allow mixing of the blood prior to analysis.

If oxygen data are not needed, venous blood may be an acceptable sample to evaluate metabolic problems. If a tourniquet is used in collection, the patient should not be allowed to pump his or her fist and the tourniquet should not be removed until after the specimen is obtained. If the tourniquet is removed while the blood is being drawn, accumulated metabolic products (largely lactic acid and CO_2) will contaminate the specimen resulting in decreased HCO_3^- and pH and increased pCO_2.

Patient hyperventilation before or during sample collection may introduce errors in blood gas values. During hyperventilation, the pCO_2 will decrease, which will increase the pH, and the pO_2 will increase. If the patient is anxious during sample collection, he should be reassured before drawing the sample. After entry into the artery, the phlebotomist should wait a few seconds for the hyperventilation to subside before drawing the sample.

The gastight glass syringe is considered the "reference" receptacle. Plastic syringes can be used; however, certain types of plastic may allow leakage of gases through walls. With a glass syringe, the plunger will move by arterial pressure when an artery is punctured. Plastic syringes, which require manual movement of the plunger, may introduce bubbles into the specimen as the plunger mechanism is pulled. Air bubbles introduce significant errors in pO_2 measurements (increase pO_2) and small errors in pCO_2 measurement (decrease pCO_2).

Vacutainer tubes may be used for routine acid–base measurements, but are not recommended for use if oxygen tension is to be evaluated. The air space left above the sample after the tube is filled contains enough oxygen to alter results.

Before analysis, all blood should be mixed thoroughly and warmed to body temperature (37°C). Temperature affects pH, pCO_2, and pO_2. The blood gas analyzer measures only the physically dissolved gases. As temperature increases, solubility of gases decrease. In a closed system, as the temperature increases, the pCO_2 and pO_2 increase and pH falls. At cold temperatures, the pCO_2 and pO_2 fall and the pH increases.

When the temperature of the patient deviates significantly from normal body temperature, the blood gas values need to be corrected to reflect actual patient temperature. In hypothermia, the pCO_2 and pO_2 will be lower, and the pH higher than actual measured values. For example, blood gas values measured at 37°C are pH, 7.20; pCO_2, 40 mm Hg; and pO_2, 80 mm Hg. The true values for this patient whose body temperature is 30°C are pH, 7.31; pCO_2, 29 mm Hg; and pO_2, 44 mm Hg.

BLOOD GAS INSTRUMENT

The blood gas instrument measures pH, pCO_2, and pO_2 (using three electrodes) and calculates HCO_3^-, $ctCO_2$, oxygen saturation, and deviation of base from normal buffer base.

pH ELECTRODE

Principle

The pH electrode consists of a glass electrode and a reference (calomel) electrode and is illustrated in Figure 9.8.

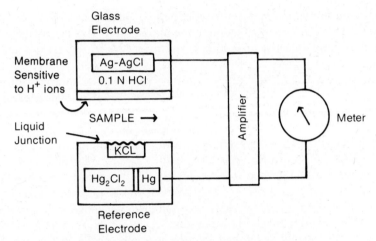

Figure 9.8. Schematic representation of a pH electrode.

The thin glass membrane of the glass electrode is composed of a lattice of metallic ions that can be exchanged for hydrogen ions. One surface of the thin glass membrane is in contact with a solution of a constant pH (0.1 N HCl). The other surface of the glass membrane is in contact with the sample. Hydrogen ions of the sample are exchanged for metallic ions of the glass membrane. A potential difference develops across the glass membrane that is proportional to the hydrogen ion activity in the test solution. To measure this potential difference, a metal connection (Ag–AgCl wire) is inserted in the 0.1 N HCl solution.

The reference electrode produces a constant reference potential to which the glass electrode may be compared.

Calibration

Two aqueous buffers of known pH are used for calibration (pH 7.383 and 6.841 are commonly used). The first buffer is aspirated and the balance control is adjusted so that the readout displays the value of the buffer. The second buffer solution is then aspirated and the slope control is adjusted until the readout displays the value of the buffer. (The slope determines the millivolt per pH unit response.)

Validation

Controls are used to validate the slope adjustment. Commercial controls are available to check three levels—acidic, normal and alkalotic pH.

Maintenance and Sources of Error

The temperature of the electrode chamber should be checked daily. The potassium chloride solution in the liquid junction should be changed periodically. Contamination (blood contamination or air bubble entrapment) of the liquid junction solution causes drift and erratic results.

Protein contamination of the electrodes produces slow response times and low recovery values in controls and samples. To prevent protein contamination, a proteolytic wash solution is flushed through the electrode chamber periodically. Distilled water should not be used to rinse the electrode chamber. Globulins may be precipitated and contaminate the electrode surfaces.

Glass electrodes deteriorate over time and should be checked periodically for small cracks. The cracks produce leaks leading to drift and erratic sample results.

When the blood gas instrument is not in use, the electrode chamber should be filled with a buffer solution near the pH of normal blood in order to prevent dehydration of the glass membrane.

Reference Range

Arterial pH (whole blood)	7.35–7.45
Venous pH (whole blood)	7.31–7.42

pCO_2 ELECTRODE (SEVERINGHAUSE ELECTRODE)

Principle

The pCO_2 or Severinghause electrode (see Figure 9.9) is a glass electrode, like the pH electrode, surrounded by a bicarbonate buffer solution and separated from the blood by a membrane permeable to CO_2 but impermeable to hydrogen ions. When the sample comes in contact with the membrane, the dissolved CO_2 gas diffuses through the membrane and the hydrogen ion concentration of the buffer solution changes according to the following equation:

$$CO_2 + H_2O \rightleftharpoons H_2CO_3 \rightleftharpoons H^+ + HCO_3^-$$

The difference in hydrogen ion concentration on each side of the hydrogen responsive glass membrane of the pH electrode creates a potential that is measured.

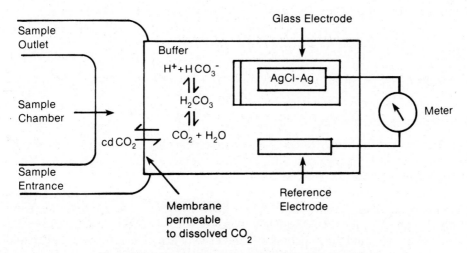

Figure 9.9. Schematic representation of a pCO_2 electrode.

Calibration

Humidified calibrating gases or liquids with known pCO_2 values are used for calibration. The pCO_2 for each calibrating gas is calculated from the ambient barometric pressure, the water vapor pressure (which is equal to 47 mm Hg at 37°C), and the fractional concentration of CO_2 in the gas. These calculations are performed as follows:

1. The ambient barometric pressure (which is in inches of mercury) is read.
2. The inches of mercury are converted to mm Hg. For example,

$$\text{Barometric pressure} = 29 \text{ inches of Hg}$$

$$1 \text{ inch} = 2.54 \text{ cm or } 25.4 \text{ mm}$$

$$29 \text{ inches} \times \frac{25.4 \text{ mm Hg}}{\text{inch}} = 736.6 \text{ mm Hg}$$

3. The partial pressure of each calibrating gas is calculated using Dalton's law. Dalton's law states that $p(\text{atm}) = pO_2 + pCO_2 + pN_2 + p(H_2O)$. The partial pressure of water at 37°C (the temperature in the blood gas instrument) is equal to 47 mm Hg. The corrected barometric pressure is equal to, $p(\text{atm}) - p(H_2O) = pO_2 + pCO_2 + pN_2$. The corrected barometric pressure, times the percent gas, is equal to the theoretical pressure of the gas. For example:

Ambient barometric pressure = 737 mm Hg
Temperature = 37°C
Corrected barometric pressure = 737 mm Hg − 47 mm Hg = 690 mm Hg

If 5% CO_2 gas is used, $pCO_2 = 690$ mm Hg × 0.05 = 34.5 mm Hg
If 10% CO_2 gas is used, $pCO_2 = 690$ mm Hg × 0.10 = 69.0 mm Hg
If 12% O_2 gas is used, $pO_2 = 690$ mm Hg × 0.12 = 82.8 mm Hg
If room air is used (21% O_2), $pO_2 = 690$ mm Hg × 0.21 = 144.9 mm Hg

Validation

The slope is validated by using controls with known pCO_2 values.

Maintenance and Sources of Error

The temperature of the electrode chamber should be checked daily. The electrode membrane must be changed periodically. As the membrane ages, stretching occurs and small holes develop that allow hydrogen ions to contaminate the bicarbonate buffer solution causing drift and erratic results. Protein buildup on the membrane can also slow response time, producing low values. When the membrane is changed, care should be taken to make sure no bubbles are trapped in the buffer solution behind the membrane. Bubbles may lead to erratic results with low values.

Reference Range

Arterial pCO_2 (whole blood)	32–45 mm Hg
Venous pCO_2 (whole blood)	39–55 mm Hg

pO_2 ELECTRODE (POLARIGRAPHIC OR CLARK ELECTRODE)

Principle

The pO_2 electrode or Clark oxygen electrode, which is illustrated in Figure 9.10, contains a platinum cathode and a silver–silver chloride anode, which are polarized at 0.6 volt. The electrode is separated from the sample by a semipermeable membrane that is permeable to oxygen but not to ions, proteins, or erythrocytes. An electrolyte solution surrounds the electrodes to provide an electrical connection between the anode and cathode and to buffer any peroxide or hydroxide that may accumulate. The anode supplies electrons necessary for the electroreduction that takes place at the cathode. As dissolved oxygen (cdO_2) diffuses through the membrane and is reduced at the cathode, a current flow is generated in proportion to the amount of oxygen reduced. The reactions that take place are as follows:

$$\text{Anode: } 2\,Ag \rightarrow Ag^+ + 2e^-$$

$$\text{Cathode: } \tfrac{1}{2}O_2 + 2e^- + H_2O \rightarrow 2OH^-$$

Calibration

To calibrate the pO_2 electrode, two humidified gases or liquids are used. The instrument is electronically adjusted to zero with a humidified gas or liquid that contains no oxygen. A second calibrating humidified gas or liquid with a known

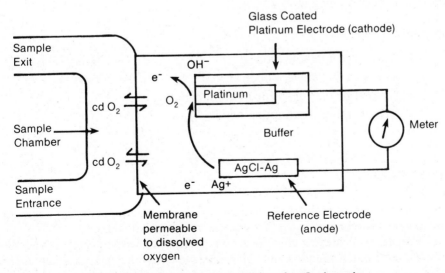

Figure 9.10. Schematic representation of a pO$_2$ electrode.

oxygen concentration is used to set the slope (the common gas used contains 12% oxygen).

The type of calibrator material (i.e., liquid or gas) influences the pO_2 value. Most pO_2 electrodes will yield lower pO_2 values for liquids than for gases. The bias is typically 1–6% depending on the type of membrane used. Since patient samples are liquids, the most accurate results would be obtained using tonometered liquid calibrators. Studies have shown that either type of calibrator material produces accurate results for most clinical samples. A significant bias exists when pO_2 values are high (above 500 mm Hg). If needed, a correction factor, called the blood gas factor, can be used to correct the pO_2 values when a gas calibrator is used. The blood gas factor may vary in different instruments because of differences in membrane construction (type of membrane, membrane thickness, diameter of the electrode, flow changes).

Validation

Controls are used to validate slope adjustment.

Maintenance and Source of Error

The temperature of the electrode chamber should be checked daily. The electrode membrane must be changed periodically. As the membrane ages, stretching occurs and small holes develop that lead to contamination of the electrode, producing erratic results. Protein buildup on the membrane can slow response time, producing low values. When the membrane is changed, care should be taken to make sure sufficient electrolyte solution is present so that no bubbles are trapped behind the membrane. Bubbles will lead to erratic results. Deposition of Ag^+ on the platinum cathode can be avoided by cleaning the platinum electrode when membranes are changed.

The sample compartment should be periodically sterilized with antiseptics to eliminate microorganisms that grow in blood retained in crevices. Microorganisms consume oxygen leading to low values.

Reference Range

Arterial pO_2 (whole blood)	75–100 mm Hg
Venous pO_2 (whole blood)	30–50 mm Hg
Arterial oxygen saturation (whole blood)	95–98%
Venous oxygen saturation (whole blood)	60–85%

NOMOGRAMS

A nomogram may be defined as a graphic display of an equation. The nomogram contains no information that is not present in the equation itself.

The Siggaard–Andersen alignment nomogram is a graphic display of the Henderson–Hasselbalch equation and can be used to determine unknown values from two known variables. Evaluation of the acid–base status of an individual requires the estimation of three interdependent factors—pH, pCO_2, and HCO_3^-. When two of these three variables are known, estimation of the third is possible, based on the Henderson–Hasselbalch equation. The Siggaard–Andersen nomogram can also be used to determine base excess or deficit when hemoglobin content is known. The Siggard–Andersen alignment nomogram is illustrated in Figure 9.11.

To use the Siggaard–Andersen alignment nomogram, two known variables are plotted and the plotted points are connected with a straight edge extending the line so all scales of the nomogram are dissected by the line. Unknown values are read from each scale by noting the point of dissection by the plotted line. To determine base excess, the "angel wing" portion of the nomogram is used. The hemoglobin concentration in g/dL is noted on the lower part of the wing. The vertical line of the wing corresponding to the hemoglobin concentration is followed upward to the point of dissection by the plotted line. The horizontal line of the wing at the point of dissection is followed to the side of the wing to determine base excess or deficit. For example,

Given:	$ctCO_2$ = 20 mmol/L
	pH = 7.3
	Hb concentration = 15 g/dL
Derived from nomogram:	pCO_2 = 39.5 mm Hg
	HCO_3^- = 19.0 mmol/L
	Base excess = -7 mmol/L

ELECTROLYTES

Electrolytes are classified as anions or cations depending on their direction of movement in an electrical field. If the electrolyte bears a positive charge, it will migrate to the cathode or negative pole and is called a cation (opposite charges attract). Sodium (Na^+) and potassium (K^+) are examples of cations. If the electrolyte bears a negative charge, it will migrate to the anode or positive pole and is called an anion. Chloride (Cl^-) and bicarbonate (HCO_3^-) are examples of anions.

The body electrolytes include Na^+, K^+, Cl^-, HCO_3^-, HPO_4^{2+}, Ca^{2+}, Mg^{2+}, $Fe^{2+,3+}$, $Mn^{2+,4+}$, Co^{2+}, $Cr^{3+,6+}$, Cd^{2+}, Zn^{2+}, Br^-, and I^-.

In describing electrolyte concentration, the proper expression of concentration is mmol/L. The concentration of electrolytes was formerly expressed as meq/L. For monovalent ions (i.e., Na^+, K^+, Cl^-, HCO_3^-), 1 meq/L is the same as 1 mmol/L.

SODIUM (Na^+)

Regulation

Sodium is the major cation in extracellular fluid. It is especially important for the maintenance of normal hydration and osmotic pressure. The diet contains sodium (as NaCl or table salt), which is nearly completely absorbed from the gastrointestinal tract. The excess is excreted by the kidneys. In most sections of the nephron,

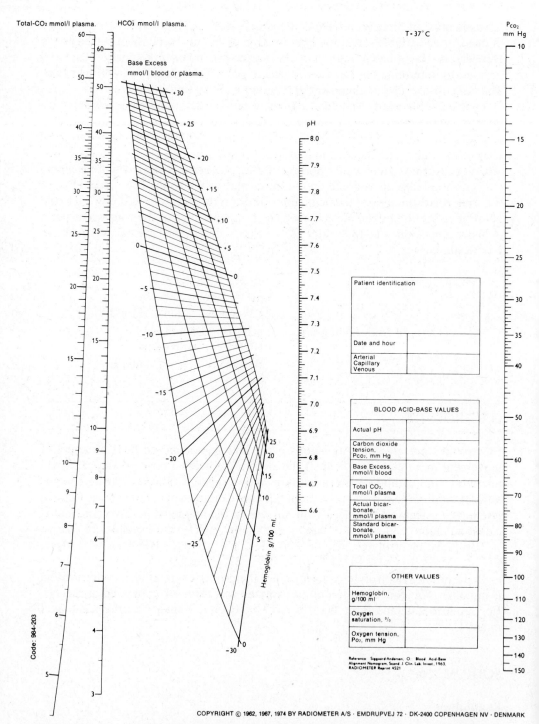

Figure 9.11. The Siggaard–Andersen alignment nomogram. (Copyright © 1962, 1967, 1974 by Radiometer A/S. Emdrupvej 72. DK-2400 Copenhagen NV. Denmark. This alignment nomogram is not to be used in actual clinical practice.)

sodium reabsorption is active; its active reabsorption is followed by the passive reabsorption of water. Under normal conditions, about 67% of the filtered sodium is reabsorbed in the proximal tubules, 25% in the loops of Henle, 5% in the distal tubules, and about 3% in the collecting ducts. In the distal tubule, the reabsorption of sodium is greatly affected by aldosterone (mineralocorticoid). Aldosterone enhances sodium reabsorption and potassium secretion. Sodium is a threshold substance (threshold 110–130 mmol/L), so that, normally, the amount excreted in the urine varies with sodium and water intake.

Hyponatremia (Low Serum Sodium)

A decrease in serum sodium may occur due to either solute (Na^+) loss or solvent (H_2O) excess. Decreased serum sodium is found in (1) diarrhea (loss of dietary sodium and sodium contained in pancreatic fluid); (2) Addison's disease (lack of aldosterone leads to sodium loss in the urine); (3) metabolic acidosis (cations are excreted with the increased anions); (4) diuretic therapy; (5) renal tubular disease (loss of sodium in the urine due to a kidney defect in sodium reabsorption); and (6) syndrome of inappropriate antidiuretic hormone (ADH) secretion (SIADH). In SIADH, water is retained in the plasma secondary to a continuous secretion of ADH with no regulatory feedback mechanism with plasma osmolality. When ADH is increased, free water or water without electrolytes, is returned to the plasma from the glomerular filtrate. Sodium stays in the urine. Urinary sodium will be high and serum sodium will be decreased.

Hypernatremia (Increased Serum Sodium)

Increased serum sodium is found in (1) any situation where water is lost in excess of salt as in profuse sweating, polyuric states, diarrhea, and renal disease; (2) certain types of brain injury in which the hypothalamus is injured leading to a lack of an adequate thirst mechanism; and (3) excessive treatment with sodium salts.

Sodium Methodology

Specimen
Either serum or heparinized (lithium heparin) plasma that is free from hemolysis may be used for sodium assay. A 24-hour urine or a random collection may be used for urine sodium determination.

Assay
The most common methodology used to determine sodium concentration is flame emission photometry using an internal standard (lithum or cesium). (See Chapter 4.) Ion-selective electrode methodology is also available, which measures ion activity. (Ion-selective electrodes are discussed in Chapter 8.)

False Low Sodium
Sodium is present only in the water portion of the serum. Anything that displaces water from the sample can cause false low sodium values in those methodologies that require dilution of the sample before analysis. The flame photometer normally dilutes the sample 1:200 before analysis. If some substance was present in the

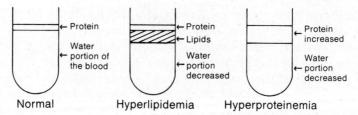

Figure 9.12. Effect of hyperproteinemia and hyperlipidemia on water volume displacement.

sample that displaced water, a volume displacement error can occur. Two substances that can displace water are protein (hyperproteinemia) and lipids (hyperlipidemia). The error introduced can be visualized in Figure 9.12.

Ion-selective electrode methodology requires no sample dilution prior to analysis. Lipids and proteins will not, therefore, introduce error. The ion-selective electrode measures sodium in plasma water.

Elevated glucose levels may also cause false low serum sodium values. When glucose is high, the osmotic effect of glucose draws water from the intracellular fluid into the vascular space causing a dilutional error, resulting in false low sodium values.

Reference Range

Serum sodium	135–148 mmol/L
Panic values	low—equal to or less than 110 mmol/L
	high—equal to or more than 170 mmol/L

Sodium values below 110 mmol/L are usually life-threatening and below 105 mmol/L cause irreversible brain damage. A rapid rise to greater than 170 mmol/L may yield permanent brain damage.

Urine sodium excretion depends on diet and state of hydration. The concentration of sodium in the cerebrospinal fluid is almost identical to that found in serum.

POTASSIUM (K+)

Regulation

Potassium is the major intracellular cation. Potassium, once absorbed by the intestinal tract, is partially removed by glomerular filtration. Regardless of the blood potassium level, at least 80% of the filtered load is reabsorbed in the proximal tubules and loops of Henle. In the distal tubules, potassium is secreted into the tubular fluid; however, in certain conditions (such as potassium deprivation), reabsorption of this ion may occur in this section of the nephron. In the collecting ducts, there is usually net reabsorption. Normally, about 10–20% of the filtered potassium load is excreted in the urine. When potassium intake is low, less will be excreted; and when the potassium intake is high, more than the filtered amount may be excreted. These changes are possible because the kidney can change the rate of potassium secretion in the distal tubule.

Potassium secretion is influenced by blood potassium levels, aldosterone, blood sodium levels, and hydrogen ion balance. There is no threshold level for potassium, and even in a deficiency state, potassium will be lost in the urine.

Hypokalemia (Low Serum Potassium)

Decreased serum potassium is found in (1) gastrointestinal fluid loss (potassium becomes depleted due to K^+ loss in vomitus or in feces as in diarrhea); (2) adrenal mineralocorticoid excess (aldosterone increases potassium secretion in the distal tubule resulting in urinary loss); (3) alkalosis (in alkalosis, the shift of H^+ out of the cell is followed by a shift of K^+ into the cell; the increased K^+ inside distal renal tubular cells leads to enhanced distal tubule secretion of K^+ with loss in the urine); and (4) diuretic-induced loss. The diuretics—chlorothiazide, ethacrynic acid, and furosemide—promote loss of fluid from the body by increasing the urine flow. Increased urine flow can lead to increased distal tubular secretion of K^+ with loss in the urine.

Hyperkalemia (Increased Serum Potassium)

Increased serum potassium is found in (1) renal disease (potassium may accumulate in the blood because the kidney is unable to regulate the blood potassium level as it normally does); (2) renal tubular acidosis [during potassium secretion, tubular fluid Na^+ may be exchanged for either H^+ or K^+; when the Na^+–H^+ (or K^+) mechanism does not function properly, potassium is retained in the serum]; and (3) acute acidosis. In acute acidosis, H^+ moves into the cells, K^+ moves out. The concentration of K^+ within cells (including the distal renal tubular cells) is decreased. Thus, less potassium is secreted resulting in elevated blood levels.

Abnormal potassium levels (increased or decreased) can cause cardiac arrhythmias.

Potassium Methodology

Specimen
Either serum or heparinized plasma that is free from hemolysis may be used for potassium assay. The potassium concentration within cells is much higher than the potassium concentration in serum; therefore, even minimal hemolysis may result in false high potassium levels.

A fasting specimen is preferred, because, immediately after eating, potassium will move into cells as glucose is phosphorylated.

The serum or plasma should be separated from cells shortly after collection to prevent any leakage of intracellular potassium into extracellular fluid. Potassium is released from thrombocytes in clotting; thus, plasma levels will be slightly lower than serum levels. In patients with thrombocythemia, the difference will be greater.

Prolonged use of the tourniquot or clenching of the fist during collection may increase potassium as a result of leakage of potassium from damaged cells.

A 24-hour urine or a random collection may be used for urine potassium determination.

Assay

The most common methodology used to determine potassium concentration is flame emission photometry using an internal standard (lithium or cesium). (See Chapter 4.) Ion-specific electrode methodology is also available, which measures ion activity. (Ion-selective electrodes are discussed in Chapter 8.)

Reference Range

Serum potassium	3.5–5.3 mmol/L
Panic values	low—equal to or less than 2.5 mmol/L
	high—equal to or more than 6.5 mmol/L

Potassium levels in the newborn are higher than those of adults. The 24-hour urine potassium excretion will vary with the diet and is influenced by other factors affecting potassium secretion.

CHLORIDE (Cl⁻)

Regulation

Chloride is the major extracellular anion. Chloride ions, ingested in food, are absorbed from the intestinal tract, removed from the blood by glomerular filtration, and then are passively reabsorbed by both the proximal and distal tubular cells. Under certain circumstances, chloride reabsorption in the distal tubules may be an active process. The sequence of events believed to occur is active transport of sodium followed by passive transport of chloride, with passive diffusion of water in response to the osmotic gradient set up by the sodium chloride.

Chloride may be lost in sweat, but during periods of profuse sweating, aldosterone is secreted, which causes the sweat glands to secrete sweat of lower sodium and chloride concentration. Blood levels of chloride are slightly lower immediately after eating due to the synthesis of gastric juice HCl (alkaline tide).

There is a reciprocal relationship between the plasma concentration of chloride and bicarbonate. In most situations, as bicarbonate rises, chloride falls, and vice versa.

Hypochloremia (Low Serum Chloride)

Decreased serum chloride may be found in (1) prolonged vomiting (due to loss of HCl, producing hypochloremic metabolic alkalosis); (2) metabolic acidosis due to renal failure or diabetic ketoacidosis (accumulation of phosphates and keto acids replace some of the chloride ions); (3) salt-losing renal disease (probably due to the inability of the kidney to reabsorb chloride ions); and (4) Addisonian crisis (serum chloride concentration is usually close to normal in Addison's disease except in crisis).

Hyperchloremia (Increased Serum Chloride)

Increased serum chloride may be found in (1) metabolic acidosis associated with prolonged diarrhea (due to loss of pancreatic fluid, which contains bicarbonate; a reciprocal relationship exists that increases chloride due to the loss of bicarbonate); (2) severe renal tubular pathology (hydrogen ion excretion is decreased resulting in decreased bicarbonate reabsorption with a reciprocal increase in chloride); (3) excessive ingestion of chloride; and (4) some cases of hyperparathyroidism. In hyperparathyroid disease, the elevated serum chloride is probably due to the effect of PTH on renal tubular reabsorption of bicarbonate. PTH inhibits renal reabsorption of bicarbonate resulting in a reciprocal increase in serum chloride.

Chloride Methodology

Specimen
The recommended specimen for the determination of chloride is a fresh non-hemolyzed serum sample. The serum should be separated from cells as soon as possible.

Schales and Schales (Titrametric Procedure)

1. A trichloracetic acid (TCA) filtrate is prepared from the sample.
2. A portion of the TCA filtrate is titrated with mercuric nitrate solution. Diphenylcarbazone is used as an indicator. The mercuric ions combine with chloride ions to form an undissociated but soluble mercuric chloride.

$$2 \, Cl^- + Hg^{2+} \rightarrow HgCl_2$$

3. After all chloride has reacted, the excess Hg^{2+} will combine with the indicator to form a blue–violet complex, thus indicating the endpoint.

A protein-free filtrate is desirable because the endpoint is very difficult to detect in icteric or hemolyzed samples. The endpoint is sensitive to pH; it is essential that the standard and samples are titrated at the same pH. Other halogens may react with mercury to give a positive error (bromide, iodide) as well as CN^-, CNS^-, and sulfhydryl groups of proteins (—SH groups).

Mercuric Thiocyanate (Colorimetric Procedure)

1. The specimen is mixed with a solution of mercuric thiocyanate.
2. Mercury has a high affinity for chloride and an undissociated mercuric chloride is formed and free thiocyanate is liberated.

$$Hg(SCN)_2 + 2Cl^- \rightarrow HgCl_2 + 2(SCN)^-$$

3. The thiocyanate reacts with ferric ions (present in a ferric nitrate reagent) to form a colored complex, ferric thiocyanate:

$$3(SCN)^- + Fe^{3+} \rightarrow Fe(SCN)_3$$

Other halogens that may be present may interfere.

Coulometric–Amperometric (Reference Method)

1. The sample is diluted in a nitric acid and acetic acid solution containing a small amount of gelatin. The nitric acid provides for good conductivity. The acetic acid makes the solution less polar; therefore, the AgCl formed is less soluble, creating a sharper endpoint. The gelatin allows the reaction to occur over the entire electrode surface making a more reproducible titration curve.

2. The titration is performed with a Buchler–Cotlove Chloridometer. Silver is generated at the anode (coulometric titration) by a silver generator electrode, which is fed a constant current so that the quantity of silver ions generated is constant. The silver ions combine with chloride to form a precipitate of silver chloride.

$$Ag^0 \rightarrow Ag^+ + e^-$$

$$Ag^+ + Cl^- \rightarrow AgCl \downarrow$$

3. After sufficient silver has been generated to react with all chloride present, the generation of additional silver ions will cause an increase in electroactivity of the titrant. The electroactivity is measured amperometrically by a set of silver indicator electrodes. The increase in current stops the generation of additional silver ions. Since the current is constant, the amount of silver generated is constant, the *time* necessary to reach the titration end-point is a measure of the chloride concentration.

4. Standards are employed and titration times are determined in order to calculate unknown values:

$$\frac{\text{titration time of unknown} - \text{titration time of blank}}{\text{titration time of standard} - \text{titration time of blank}}$$

$$\times \text{ standard concentration} = \text{chloride concentration}$$

Other halogens, as well as CN^-, CNS^-, and —SH groups can interfere.

Reference Range

Serum chloride   98–106 mmol/L

BICARBONATE (HCO$_3^-$) AND TOTAL CARBON DIOXIDE (ctCO$_2$)

Bicarbonate composes the second largest fraction of anions in the plasma. Bicarbonate is most frequently determined from nomograms constructed from the Henderson–Hasselbalch equation.

Definition of Total CO_2

Total CO_2 measures all CO_2 present, that is, $ctCO_2 = HCO_3^- + cdCO_2 + H_2CO_3 +$ carbamino $CO_2 + CO_3^{2-}$. Carbamino CO_2 is carbon dioxide bound to amino groups of hemoglobin and plasma proteins.

Total CO_2 Methodology

Natelson Gasometer (Manometric Procedure)
For manometric procedures, the specimen must be collected anaerobically. Manometric procedures measure the pressure of a gas at a fixed volume. The principle of measurement is based on Dalton's gas law, $p(\text{atm}) - p(H_2O) = pO_2 + pCO_2 + pN_2$:

1. An anaerobic sample (serum, plasma, or whole blood) is injected into the microgasometer.
2. Lactic acid is introduced, which releases all forms of CO_2 and other gases. A vacuum, applied to the reaction chamber, allows the released gases to escape from the sample into the vacuum above the liquid.
3. The pressure of the released gas is measured manometrically. This pressure is equal to P-1.
4. Alkali, such as NaOH, is introduced, which causes a total reabsorption of the CO_2 gas as Na_2CO_3.
5. The pressure required to confine the rest of the gases (N_2, O_2) in a fixed volume is measured. This pressure is equal to P-2.
6. P-1 minus P-2 is equal to the CO_2 present.

Autoanalyzer Method
The specimen for the autoanalyzer method should be collected under anaerobic conditions, and the collection tube should remain sealed until analyzed. In the autoanalyzer, sample cups are exposed to atmospheric air, which allows CO_2 gas to escape from the sample. To prevent this loss, all cups are completely filled and covered with parafilm. The probe pierces the parafilm when the sample is siphoned:

1. Sulfuric acid is added to the sample and CO_2 gas is released.
2. The released gas is absorbed into a weak alkaline carbonate–bicarbonate buffer solution containing phenophthalein.
3. The CO_2 absorbed into the buffer solution causes the pH of the buffer to decrease with a resulting decrease in the intensity of the red color.
4. The change in color of the phenophthalein (inverse colorimetry) is proportional to the concentration of the sample CO_2.

Blood Gas Instrument
The total CO_2 is a calculated value based on the Henderson–Hasselbalch equation. The pH and pCO_2 are directly measured. The HCO_3^- and $ctCO_2$ are calculated.

pCO₂ Electrode Methodology

Total CO_2 is measured by a pCO_2 electrode. This method does not require an anaerobic sample.

The pCO_2 electrode is a combination of a glass pH electrode and a reference electrode in contact with a weakly buffered electrolyte solution behind a membrane permeable to CO_2 gas. (See pCO_2 electrode picture in blood gas instrument discussion.) The CO_2 gas is released from the sample, diffuses through the membrane, and reacts with water in the buffer solution causing a change in the hydrogen ion activity of the buffer:

$$CO_2 + H_2O \rightarrow H_2CO_3 \rightarrow H^+ + HCO_3^-$$

The pH electrode senses the change in hydrogen ion activity, which is an indication of the concentration of CO_2 present.

Reference Range

Arterial ctCO₂ (whole blood)	21–27 mmol/L
Venous ctCO₂ (whole blood)	23–29 mmol/L
Venous ctCO₂ (serum)	24–32 mmol/L
Arterial HCO₃⁻ (whole blood)	20–26 mmol/L
Venous HCO₃⁻ (whole blood)	22–28 mmol/L

FORMULA FOR CHECKING THE ACCURACY OF ELECTROLYTE DETERMINATIONS

The number of cations in the body must equal the number of anions in the body. There are 154 mmol/L cation charges balanced by 154 mmol/L anion charges. When the concentration of cations (or anions) changes, a corresponding change in the anions (or cations) must occur in order to maintain electroneutrality. The following formula may be used to check the accuracy of electrolyte determinations:

$$Na^+ + K^+ + 7 \text{ (undetermined cations)} :: Cl^- + HCO_3^-$$
$$+ 24 \text{ (undetermined anions)}$$

(:: means the same as.) In the formula, ctCO₂ may be substituted for HCO_3^- since the majority of CO_2 is in the form of HCO_3^-. The undetermined cations consist of calcium and magnesium. The undetermined anions consist of protein, organic acids, phosphates, and sulfates (see Figure 9.3). The charge range should agree within 10 mmol/L. With new methodology and more accurate measurements, the difference should be less. These limits may be exceeded in diseases associated with the presence of abnormal proteins (i.e., monoclonal gammopathies) or in diseases associated with increases in keto acids (i.e., diabetes mellitus) or metabolites (i.e., renal disease). Use of the formula is illustrated below.

The following electrolyte values were obtained:

$$
\begin{array}{ll}
\text{Na}^+ & 120 \text{ mmol/L} \\
\text{K}^+ & 5.5 \text{ mmol/L} \\
\text{Cl}^- & 90 \text{ mmol/L} \\
\text{ctCO}_2 & 15 \text{ mmol/L}
\end{array}
$$

$$120 + 5.5 + 7 :: 90 + 15 + 24$$

$$132.5 :: 129.0$$

The charges agree within the limits; therefore, the electrolyte values are probably correct.

OSMOLALITY

DEFINITION OF OSMOLALITY AND OSMOLARITY

Osmolality measures the total number of dissolved solute particles in solution. The nature of the particles, that is, their shape, weight, or charge, does not influence osmolality. According to Avogadro, 1 gram molecular weight of any substance that does not dissociate will yield 6.023×10^{23} individual molecules or species when dissolved in a solvent. A substance that dissociates to form two species should therefore yield twice as many particles when dissolved in a solvent.

The term osmol refers to the weight of any element or compound that will exert the same osmotic pressure. One osmol of a substance which does not dissociate will be equal to 1 gram molecular weight (mole) of the substance. One osmol of any substance that does dissociate will be less than 1 mole of that substance. One osmol of any substance is equal to

$$
1 \text{ osmol} = \frac{\text{gram molecular weight}}{\text{number of particles or ions formed in solution}}
$$

One osmol of glucose is equal to its gram molecular weight (180 g) since it does not dissociate in solution. One osmol of NaCl is equal to one-half of its gram molecular weight (29 g) because it will dissociate into one sodium ion and one chloride ion in solution.

Osmolality is an expression of concentration, which measures the number of osmols contained in 1 kg of solvent; in the body, the solvent is water. Osmolality, a weight to weight relationship, is expressed as the number of osmols/kilogram (Osm/kg) or milliosmols/kilogram (mOsm/kg). One osmol is equal to 1000 milliosmols. The term osmolarity is a weight to volume relationship, and is expressed as the number of osmols or milliosmols/liter. Osmolality, the term used in the laboratory, is a more accurate expression of concentration than is osmolarity, since volume can fluctuate with changes in temperature. Since body fluids are low in solute, osmolality is expressed in milliosmols rather than in osmols.

In the preparation of standard solutions with known osmolalities, dissociation of an electrolyte into individual atoms may not be complete or the individual ions may form chemical bonds with solvent molecules; therefore, a correction factor, called the osmotic coefficient, must be used to determine the true osmolality. The true osmolality may be calculated by the following formula:

$$osmolality = \phi nc$$

where ϕ = osmotic coefficient (this coefficient is derived by dividing the measured osmolality by the theoretical osmolality of the electrolyte solution)

n = number of atoms formed upon dissociation

c = concentration of the electrolyte in mol/kg water

Problem: Calculate the true osmolality of a solution of NaCl prepared by dissolving 58 g of NaCl (1 mol) in 1 kg of water.

ϕ = 0.93 for NaCl

n = 2 since NaCl will dissociate into two particles

c = 1 mol/kg

True osmolality = $0.93 \times 2 \times 1$ = 1.86 Osm/kg or 1860 mOsm/kg

MEASUREMENT OF OSMOLALITY

One osmol of a substance when dissolved in 1 Kg of water will (1) lower the freezing point 1.86°C; (2) lower the vapor pressure 0.3 mm Hg; (3) increase the osmotic pressure 17,000 mm Hg; and (4) increase the boiling point 0.52°C. These properties, called colligative properties, may be used to measure osmolality. Instruments, which are used to measure osmolality (osmometers), measure either freezing-point depression or vapor-pressure (dew-point) depression. For concentrations in the biological range, measurements are more reproducible when performed by freezing-point depression than when performed by vapor-pressure depression. At low concentrations, freezing point is directly porportional to the number of dissolved particles, that is, twice as many particles will depress the freezing point twice as much, three times as many, three times as much, and so forth.

OSMOLALITY BY FREEZING-POINT DEPRESSION

The freezing-point osmometer consists of (1) a refrigerated water bath to hold, supercool, stir, and freeze the body fluid; (2) a thermistor or probe that changes its resistance with temperature changes; and (3) a readout device that translates the temperature-related resistance change into mOsm/kg body water.

The sample is placed into a special cuvette. The sample is automatically lowered into the cooling bath as the head of the instrument is lowered. The head contains the thermister and stir wire that enters the sample. The cooling bath rapidly supercools the sample (cools the sample below the freezing state, but the sample remains in a liquid state). The stir wire, suspended in the supercooled solution, is then vibrated, which initiates the process of crystallization. As crystallization occurs, the heat of fusion is released and the temperature rises to an equilibrium plateau, which is the freezing point of the solution. (The freezing point is the temperature at which the solvent or solution will turn from a liquid to a solid.) The osmolality is then displayed directly in mOsm/kg water. Standards are used to verify that the instrument is correctly indicating the known osmolality. Instrument adjustments may be made if the readings are incorrect. Controls are used to verify that the instrument has been adjusted correctly.

CLINICAL USE

The osmolality of normal serum generally falls within the range of 280–300 mOsm/kg. The serum electrolytes, particularly sodium, chloride, and bicarbonate, contribute about 275 mOsm, and glucose and nonprotein nitrogen substances contribute less than 10 mOsm. The effect of protein on osmolality by freezing-point depression is negligible.

The osmolality of normal urine is dependent on the state of hydration and diet as well as normal kidney function. The kidneys have the ability to alter the urine osmolality from 50 to 1400 mOsm/kg depending on diet and state of hydration. On a medium protein diet with average salt content, an adult excretes approximately 1200 mOsm of solute per day. Of this amount, urea comprises about 500 mOsm and Na^+, K^+, NH_4^+, and their attendant anions comprise the remainder. A normal adult on a normal diet with normal fluid intake will produce urine in the 500–850 mOsm/kg range.

Urine Osmolality/Serum Osmolality

The ratio of urine osmolality to serum osmolality is used as a tool to determine the ability of the kidneys to concentrate or as a tool that is used to evaluate water balance. Normally, 24-hour urine osmolality/serum osmolality (U/S) should produce a ratio greater than 1.0. After an overnight fast, the U/S ratio should be greater than 3.0. If the ratio is consistently only slightly greater than 1.0 or equal to 1.0, the kidneys have lost their ability to concentrate or dilute the urine, a finding in chronic renal disease. In individuals with diabetes insipidus, the ratio will be less than 1.0. In diabetes insipidus, with water deprivation, the ratio will remain less than 1.0. In psychogenic water ingestion, the ratio will increase to 2.0 or 3.0 under conditions of water deprivation.

Osmolal Gap

There are approximately 14 formulas that have been used to calculate serum osmolality from electrolyte measurements. According to Dorwart and Chalmers, the equation found to yield the most accurate calculated osmolality is Osmolality = 1.86 (Na^+) + (glucose/18) + (BUN/3) + 9. For ease in mental calculation, the formula frequently used is, Osmolality = 2 (Na^+) + (glucose/20) + (BUN/3). The difference between measured osmolality and calculated osmolality is referred to as the osmolal gap, also called delta osmolality. In normal individuals, measured and calculated osmolality are almost equal; however, the difference may vary slightly due to the formula used to perform the calculations.

If the calculated osmolality exceeds the measured osmolality, either a mathematical error or laboratory error in the measurement of Na^+, BUN, or glucose is present. If the measured osmolality exceeds the calculated osmolality significantly (usually greater than 10 mOsm/kg water is used), the assumption is made that some other toxin or drug is present that will add to the measured osmolality but would not affect the calculated osmolality. The most common cause of an osmolal gap is ethanol overdose. Other alcohols—isopropanol and methanol—also increase the osmolal gap. The measurement of serum osmolality has been suggested as a test that would be useful in screening for alcohol and other volatiles, especially in the assessment of comatose patients for whom no history is available.

In alcohol intoxication, the delta osmolality will only be increased when the measurements are performed by freezing-point depression. Instruments that measure osmolality by vapor-pressure depression are not affected because the volatility of the alcohol raises the vapor pressure of the solution and negates the solute effect.

The osmolal gap will also be increased in renal and liver disease and due to the ingestion of certain drugs.

REFERENCE RANGE

Serum osmolality	280–300 mOsm/kg
Urine osmolality	250–1200 mOsm/kg
24 hour urine osmolality	400–600 mOsm/kg

BIBLIOGRAPHY

Advanced Instruments, Inc. *The Physical Chemistry, Theory and Technique of Freezing Point Determinations*. Booklet from Advanced Instruments, Inc., Needham Heights, MA, 1971.

Aguanno, J. M., Ritzmann, S. E. and McCoy, M. *The Anion Gap*. Check Sample Program CC-140, ASCP, 1983.

Bakerman, S. *Acid-Base Lecture Notes*. Seminar, Greenville, N.C., Mar. 1983.

Beck, L. H., Hermany, P. L., Pellegrino, E. D., and Johnson, M. E. *Hydrogen Ion Regulation: Respiratory and Metabolic Acidosis*. Network for Continuing Medical Education No. 324 Telecourse Workbook, 1979.

Boyd, D. R. and Baker, R. J. "Osmometry: A new bedside laboratory aid for the management of surgical patients," *Surgical Clinics of North America* **51**(1), 241 (1971).

Broughton, J. O. *Understanding Blood Gasses*. Ohio Medical Products Article Reprint Library, Reprint No. 456, Aug. 1971.

Coe, F. L. "Metabolic alkalosis," JAMA **238**(21), 2288 (1977).

Dharan, M. "Blood gas analysis: What you should know," *Laboratory Management,* 18–21 (July 1977).

Dharan, M. "Increase your awareness of clinical osmometry," *Laboratory Management,* 38–42 (Jan. 1978).

Dorwart, W. V. and Chalmers, L. "Comparison of methods for calculating serum osmolality from chemical concentrations, and the prognostic value of such calculations," *Clin. Chem.* **21**(2), 190 (1975).

Emmett, M. and Narins, R. G. "Clinical use of the anion gap," *Medicine* **56**(1), 38 (1977).

Fleischer, W. R. and Hartmann, A. E. *Blood Gasses and Their Measurement Part I and Part II.* Booklet for audiovisual program, ASCP Education Center, 1977.

Garg, A. K. and Nanji, A. A. "The anion gap and other critical calculations," *Diagnostic Medicine* **5**(2), 32 (1982).

Hartmann, A. E. *Acid-Base Balance.* Check Sample Program CC-121, ASCP, 1980.

Henry, J. B. (ed). *Clinical Diagnosis and Management by Laboratory Methods,* 16th ed. Philadelphia: Saunders, 1979.

Huehns, E. R. "Oxygen delivery to the tissues," *Lab-Lore* **8**(9), 575 (1979).

Labelle, K. J. "Ethylene glycol poisoning," *Journal of the Indiana State Medical Association,* 249–252 (May 1977).

Lobdell, D. H. *Osmometry Revisited.* Booklet from Advanced Instruments, Inc., Needham Heights, MA.

McLendon, W. W. *Serum and Urine Osmolality.* Booklet from Advanced Instruments, Inc., Needham Heights, MA.

Moran, R. F. and Bradley, F. "Blood gas systems—Major determinants of performance," *Laboratory Medicine* **12**(6), 353 (1981).

Narins, R. G. and Emmett, M. "Simple and mixed acid–base disorders: A practical approach," *Medicine* **59**(3), 161 (1980).

North, J. W. "New electrolyte assays: A technologist's primer," *Laboratory Management,* 52–61 (July 1977).

Oh, M. S. and Carroll, H. J. "The anion gap," *New England Journal of Medicine* **297**(15), 814 (1977).

Schnur, M. J., Appel G. B., Karp, G., and Osserman, E. P. "The anion gap in asymptomatic plasma cell dyscrasias," *Annals of Internal Medicine* **86,** 304 (1977).

Smithline, N. and Gardner, K. D. "Gaps—Anionic and osmolal," *JAMA* **236**(14), 1594 (1976).

Soloway, H. B. "Things your mother never told you about acid-base," *Diagnostic Medicine,* **1**(4), 33 (1978).

Stolar, v. *Human Acid-Base Chemistry.* New York: Medical Programs Incorporated, 1973.

Tietz, N. (ed.). *Fundamentals of Clinical Chemistry,* 2nd ed. Philadelphia: Saunders, 1976.

Tomec, R. J. *Serum Osmolality.* Check Sample Program CC-134, ASCP, 1982.

Valtin, H. *Renal Function: Mechanisms Preserving Fluid and Solute Balance in Health.* Boston: Little, Brown and Co., 1973.

Valtin, H. *Renal Dysfunction: Mechanisms Involved in Fluid and Solute Imbalance.* Boston: Little, Brown and Co., 1979.

POST-TEST

1. Match the following:

   1. _____ acids
   2. _____ bases
   3. _____ pH

   a. term used to describe the pH at which acid and conjugate base are present in equal amounts

4. _____ pK b. proton donors
                             c. term used to describe hydrogen ion concentration
                             d. proton acceptors

2. The pH range usually accepted as normal is:
   a. 7.30–7.40
   b. 7.35–7.45
   c. 7.00–7.80
   d. 6.80–7.80

3. The concentration of bicarbonate in normal blood, pH 7.4, is _____.
   The concentration of carbonic acid in normal blood, pH 7.4, is _____.

4. At a pH of 7.4, the ratio of bicarbonate/carbonic acid will be:
   a. 1 : 20
   b. 20 : 1
   c. 10 : 1
   d. 1 : 10

5. The pK for carbonic acid in serum is:
   a. 1.6
   b. 6.0
   c. 6.1
   d. none of these

6. Partial pressure of a gas may be expressed in all of the following ways *except:*
   a. mm Hg
   b. torr
   c. kilopascals
   d. volumes percent

7. The pCO_2 in normal alveolar space is _____ mm Hg.

8. The solubility factor for CO_2 is:
   a. 0.003
   b. 0.30
   c. 0.03
   d. 3.0

9. Using the Henderson–Hasselbalch equation, write the formula used to determine blood or plasma pH using the bicarbonate/carbonic acid buffer system as found in the human body. Label the basic or metabolic component; label the acidic or respiratory component.

10. Through a diagram, show the buffering mechanism of bicarbonate.

11. Through a diagram, show the buffering mechanism of carbonic acid.

12. Given: $HCO_3^- = 36$ mmol/L; $pCO_2 = 40$ mm Hg. The pH is _____. Is this compatible with life? _____

13. Given: $ctCO_2 = 18.9$ mmol/L; $pCO_2 = 30$ mm Hg. The pH is _____. Is this compatible with life? _____

14. The kidney compensates for changes in:
    a. metabolic component
    b. respiratory component
    c. both
    d. neither

15. The lungs compensate for changes in:
    a. metabolic component
    b. respiratory component
    c. both
    d. neither

16. There is a direct/inverse relationship between the rate of respiration and the concentration of carbonic acid in the arterial blood.

17. Show by illustration the reactions that occur in the kidney in:
    a. bicarbonate synthesis
    b. bicarbonate reabsorption
    c. buffering hydrogen ions

18. Fill in the following chart, Use arrows to show increased ($\uparrow$) or decreased ($\downarrow$).

	Primary component affected. How?	Compensatory mechanism. How?	pH	ctCO$_2$	pCO$_2$
Metabolic acidosis					
Metabolic alkalosis					
Respiratory acidosis					
Respiratory alkalosis					

19. Fill in the following chart for kidney compensatory mechanisms. Use arrows to show increased ($\uparrow$) or decreased ($\downarrow$).

	Na$^+$–H$^+$ Exchange	Ammonia Formation	HCO$_3^-$ Reabsorption	Urinary pH
Acidosis				
Alkalosis				

20. Match the following:
    Respiratory compensatory mechanisms:
    1. _____ acidosis   a. hyperventilation
    2. _____ alkalosis   b. hypoventilation

21. What are undetermined anions?

22. An elevated anion gap is associated with:
    a. respiratory alkalosis
    b. respiratory acidosis
    c. metabolic alkalosis
    d. metabolic acidosis

23. A patient had the following laboratory results: Na = 123 mmol/L; Cl = 85 mmol/L; $ctCO_2$ = 16 mmol/L. The anion gap is _____ . What acid–base disturbance does this patient probably have? _____ .

24. Given the following laboratory data: blood pH = 7.25; pCO_2 = 25 mm Hg; Na = 123 mmol/L; K = 4.5 mmol/L; Cl = 85 mmol/L; $ctCO_2$ = 15 mmol/L; BUN = 50 mg/dL; Glucose = 100 mg/dL. The anion gap is _____ . In view to the laboratory data, what acid–base disturbance do you think is likely? _____ . The primary cause of this acid–base disturbance is _____ .

25. Match the following:
    1. _____ diabetes mellitus   a. metabolic acidosis
    2. _____ vomiting gastric juice   b. metabolic alkalosis
    3. _____ renal tubular acidosis   c. respiratory acidosis
    4. _____ potassium deficiency   d. respiratory alkalosis
    5. _____ renal disease
    6. _____ narcotizing drug overdose
    7. _____ hysteria
    8. _____ emphysema
    9. _____ overdose salicylate (several hours after ingestion)

26. Match the acid–base disturbance with the blood gas results:

	pH	pCO_2	$ctCO_2$	HCO_3^-	pO_2	% O_2 Saturation
a.	7.40	40	25	24	100	98
b.	7.24	62	28	26	49	77
c.	7.64	36	39	38	49	92
d.	7.06	10	3	3	125	97
e.	7.55	27	26	25	104	100

    1. _____ metabolic acidosis
    2. _____ normal
    3. _____ metabolic alkalosis
    4. _____ respiratory acidosis
    5. _____ respiratory alkalosis

27. The most important blood buffer is:
    a. hemoglobin buffer system
    b. bicarbonate–carbonic acid buffer system
    c. disodium phosphate–monosodium phosphate buffer system
    d. the ability of the plasma proteins to buffer

28. The isohydric chloride shift refers to:
    a. The shift of chloride into the stomach during HCl acid secretion
    b. The loss of chloride during severe vomiting
    c. The "alkaline tide" that occurs following meals
    d. The reactions that take place in the RBC in CO_2 transport

29. Buffer base refers to the buffer capacity of:
    a. bicarbonate–carbonic acid buffer system
    b. hemoglobin buffering system
    c. protein buffering system
    d. disodium phosphate–monosodium phosphate buffer system
    e. all of the above

30. On a blood gas report, oxygen saturation refers to:
    a. the amount of free oxygen present in the blood
    b. the amount of oxygen present when the blood is fully saturated with oxygen
    c. the amount of oxygen present when the blood is equilibrated with room air
    d. the amount of oxygen in the blood that is combined with hemoglobin

31. Oxygen saturation is measured by the following formula:
    a. $\dfrac{\text{oxygen content}}{\text{oxygen capacity}} \times 100$

    b. $\dfrac{\text{oxygen capacity}}{\text{oxygen content}} \times 100$

    c. $\dfrac{\text{oxygen capacity} \times \text{oxygen content}}{100}$

    d. $\dfrac{100}{\text{oxygen content}} \times \text{oxygen capacity}$

32. The percent of oxygen in room air is approximately:
    a. 21%
    b. 50%
    c. 10%
    d. 15%

33. Indicate how the following situations would affect blood gas results. Use increased ($\uparrow$), decreased ($\downarrow$), no change ($\rightarrow$).
    1. Sample exposed to air before analysis.
       a. _____ pCO_2
       b. _____ pO_2
       c. _____ pH
    2. Patient hyperventilates during collection.
       a. _____ pCO_2
       b. _____ pO_2
       c. _____ pH
    3. Venous blood obtained rather than arterial.
       a. _____ pCO_2
       b. _____ pO_2
       c. _____ pH

4. Blood gases measured at 37°C. Patient temperature 30°C. True patient values after correction will show the following changes.
    a. _____ pCO_2
    b. _____ pO_2
    c. _____ pH

34. The oxygen dissociation curve is illustrated in Figure 9.13. Match the correct curve with the situations listed below. Use: (R) = right shift; (L) = left shift; (M) = middle.
    1. _____ Patient pH = 7.2
    2. _____ Patient temperature = 30°C
    3. _____ Normal oxygen dissociation curve
    4. _____ High P_{50}
    5. _____ Low P_{50}
    6. _____ Patient has an abnormal Hb which has greater affinity for oxygen than normal
    7. _____ Decreased 2,3-DPG
    8. _____ Less oxygen delivered to tissues

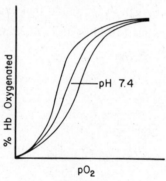

Figure 9.13

35. The preferred collection vessel for blood gasses is:
    a. capillary tubes with arterialized blood
    b. vacutainer tubes completely filled
    c. plastic syringe
    d. glass syringe

36. The correct anticoagulant to use for blood gas collection is:
    a. EDTA
    b. heparin
    c. oxalate
    d. no anticoagulant is used, serum is preferred

37. Blood Gas Instrument Principle. Match the following:
    1. _____ pH electrode
    2. _____ pCO_2 electrode
    3. _____ pO_2 electrode

    a. a membrane selectively permeable to the dissolved gas allows only the gas of interest to pass; the gas reacts with a weak buffer solution; the change in hydrogen activity in the buffer solution is measured

b. reduction takes place at the cathode; the change in potential is proportional to the rate of reduction

c. ions of the glass membrane are exchanged for hydrogen ions; the change in potential across the glass membrane is measured

38. Given: barometric reading = 29 inches of Hg; gas to be used to calibrate Blood Gas Instrument = 21% O_2; 10% CO_2.
The theoretical partial pressure of the oxygen is _____ .
The theoretical partial pressure of the carbon dioxide is _____ .

39. Match the following:
1. _____ anions   a. move to the anode in an electric field
2. _____ cations   b. move to the cathode in an electric field

40. Match the following:
1. _____ major intracellular cation   a. sodium
2. _____ major intracellular anion   b. potassium
3. _____ major.extracellular cation   c. no match
4. _____ major extracellular anion
5. _____ threshold substance
6. _____ not a threshold substance; loss still occurs when body levels are low
7. _____ increased reabsorbance from the glomerular filtrate in the presence of aldosterone
8. _____ increased loss in the urine in the presence of aldosterone
9. _____ cardiotoxic at high levels
10. _____ helps to maintain osmotic pressure and proper hydration

41. Total carbon dioxide refers to:
a. dissolved CO_2
b. bicarbonate
c. carbamino CO_2
d. carbonic acid
e. all of the above

42. All of the following can lead to increased potassium in the blood *except:*
a. serum sample from a patient with increased platelets
b. hemoconcentration caused by prolonged use of the tourniquet in drawing blood
c. clinching of fist during drawing of blood
d. hemolyzed sample
e. heparinized plasma sample

43. False hyponatremia as determined by flame photometry may result from:
a. lipemic sample
b. sample with elevated proteins

   c. sample with high blood glucose levels
   d. a and b
   e. all of the above

44. Hyponatremia is found in all of the following *except:*
   a. metabolic acidosis
   b. severe diarrhea
   c. renal tubular disease
   d. dehydration due to water loss
   e. Addison's disease

45. A technologist assayed two samples for electrolytes. Her results were as follows:

Sample #1	Sample #2
Na = 120 mmol/L	Na = 160 mmol/L
K = 5.5 mmol/L	K = 4.5 mmol/L
Cl = 100 mmol/L	Cl = 118 mmol/L
$ctCO_2$ = 23 mmol/L	HCO_3^- = 26 mmol/L

Are these valid reports? Circle yes or no below.
Sample #1   Yes No Why? _____
Sample #2   Yes No Why? _____

46. Of the substances listed below, the component that would cause the greatest elevation in osmolality is:
   a. glucose
   b. NaCl
   c. protein
   d. $CaCl_2$

47. If a solution depressed the freezing point 3.72°C below that of water, the osmolality of this solution is _____ Osm/kg. In mOsm/kg _____.

48. If a solution depressed the freezing point 0.55°C below that of water, the osmolality of this solution is _____ Osm/kg. In mOsm/kg _____.

49. The following values were obtained by laboratory measurement: Na = 140 mmol/L; glucose = 100 mg/dL; BUN = 12 mg/dL. Using the formula of convenience, the calculated osmolality is _____ mOsm/kg.

ANSWERS

---

1. (1) b; (2) d; (3) c; (4) a
2. b
3. 24 mmol/L; 1.2 mmol/L
4. b
5. c
6. d

7. 40

8. c

9. $pH = 6.1 + \log \dfrac{HCO_3^- \text{ basic or metabolic component}}{H_2CO_3 \text{ acidic or respiratory component}}$

10. $H^+ + HCO_3^- \rightarrow H_2CO_3$

11. $OH^- + H_2CO_3 \rightarrow H_2O + HCO_3^-$

12. $pH = 7.58$; yes

13. $pH = 7.40$; yes

14. a

15. b

16. inverse

17. See Figure 9.2

18. *Metabolic Acidosis:* HCO_3^-, decreased; H_2CO_3, decreased; pH, decreased; $ctCO_2$, decreased; pCO_2, decreased. *Metabolic Alkalosis:* HCO_3^-, increased; H_2CO_3, increased; pH, increased; $ctCO_2$, increased; pCO_2, increased. *Respiratory Acidosis:* H_2CO_3, increased; HCO_3^-, increased; pH, decreased; $ctCO_2$, increased; pCO_2, increased. *Respiratory Alkalosis:* H_2CO_3, decreased; HCO_3^- decreased; pH, increased; $ctCO_2$, decreased; pCO_2, decreased.

19. Acidosis: increased, increased, increased, acid
    Alkalosis: decreased, decreased, decreased, alkaline

20. (1) a; (2) b

21. Anions not normally determined in the clinical laboratory. They include keto acids, phosphoric sulfuric, hydrochloric acids, and protein.

22. d

23. 22 mmol/L; metabolic acidosis

24. 23 mmol/L; metabolic acidosis; renal disease (increased BUN)

25. (1) a; (2) b; (3) a; (4) b; (5) a; (6) c; (7) d; (8) c; (9) a

26. (1) d; (2) a; (3) c; (4) b; (5) e

27. b

28. d

29. e

30. d

31. a

32. a

33. 1. a. decreased, b. increased, c. increased; 2. a. decreased, b. increased, c. increased; 3. a. increased, b. decreased, c. decreased; 4. a. decreased, b. decreased, c. increased

34. (1) R; (2) L; (3) M; (4) R; (5) L; (6) L; (7) L; (8) L

35. d

36. b

37. (1) c; (2) a; (3) b

38. 144.8 mm Hg; 68.9 mm Hg

39. (1) a; (2) b
40. (1) b; (2) c; (3) a; (4) c; (5) a; (6) b; (7) a; (8) b; (9) b; (10) a
41. e
42. e
43. e
44. d
45. Sample #1: no; charge difference too large; chloride value too high for sodium, therefore, recheck electrolyte values. Sample #2: yes; charge difference within limits, results probably correct
46. d
47. 2; 2000
48. 0.296; 296
49. 289

CHAPTER 10
ENZYMES

I. ENZYME CLASSIFICATION AND NOMENCLATURE
  A. International Union of Biochemistry (IUB) Classification System
    a. Reactions
    b. Name
  B. Reaction Classes
    1. Oxidoreductases
    2. Transferases
    3. Hydrolases
    4. Lyases
    5. Isomerases
    6. Ligases

II. FACTORS GOVERNING ENZYME REACTIONS
  A. Substrate Concentration
  B. Hydrogen Ion Concentration or pH
  C. Temperature
  D. Enzyme Cofactors
    1. Coenzymes
    2. Activators

III. SUBSTRATE SPECIFICITY OF ENZYMES

IV. ENZYME INHIBITION
  A. Competitive
  B. Noncompetitive
  C. Other Inhibitors

V. ISOENZYMES

VI. ETIOLOGY OF ENZYMES IN THE BLOOD

VII. ENZYME ASSAY METHODS
  A. Specimen Requirements
  B. Assay Methodology
    1. Fixed Time or End Point
    2. Kinetic or Continuous Monitoring
    3. Immunoassay
  C. Units of Measure

OBJECTIVES

On completion of this chapter, you will be able to:

1. Define the function of an enzyme in a reaction.
2. List the six classes of enzyme-catalyzed reactions. Discuss the role that the enzyme serves in each reaction class.
3. Discuss the Michaelis–Menten theory regarding substrate concentration and reaction velocity. Define the term, Michaelis constant (K_m), and describe practical application of K_m.
4. Describe the effect of pH on reaction velocity and discuss why enzymes are usually assayed at the pH of maximal activity.
5. Discuss the effect of temperature on reaction velocity and define the assay temperature proposed by the Commission on Enzymes of the International Union of Biochemistry.
6. Define the terms coenzyme and activator. Discuss the role that each serves and the effect that each has in the regulation of reaction velocity. List two hydrogen transfer coenzymes that are used in the assay of many of the enzymes in the clinical laboratory. List two nonhydrogen transfer coenzymes.
7. Discuss the specificity of enzyme for substrate. Define what is meant by the active site of the enzyme and the effect of denaturation on enzyme–substrate complex.

8. Discuss competitive and noncompetitive inhibition and the effect of each type of inhibition on K_m. Discuss how the presence of metal ions may affect enzyme reaction rates.

9. Define the term isoenzyme and list methods that are used to separate isoenzymes.

10. Discuss the meaning of plasma-specific and non-plasma-specific enzymes. Discuss the etiology of non-plasma-specific enzymes in the blood.

11. Discuss general specimen requirements for all enzyme assays, noting exceptions.

12. List the three methods that may be used to determine enzyme concentration and discuss the theory of each method. Designate the method that is preferred when enzymes are measured by activity rates.

13. Define what is meant by an International Unit of enzyme activity.

14. For each enzyme listed below, discuss: (1) function, (2) source, (3) specimen requirements, (4) assay methods, (5) isoenzyme separation (if applicable), and (6) clinical significance:
    a. lactate dehydrogenase
    b. creatine kinase
    c. aspartate aminotransferase
    d. alanine aminotransferase
    e. alkaline phosphatase
    f. acid phosphatase
    g. pseudocholinesterase

15. For each enzyme listed below, discuss: (1) function, (2) source, and (3) clinical significance:
    a. isocitrate dehydrogenase
    b. glucose-6-phosphate dehydrogenase
    c. ceruloplasmin
    d. gamma-glutamyl transferase

GLOSSARY OF TERMS

Activator. An inorganic ion that increases the catalytic activity of an enzyme.

Coenzyme. An organic nonprotein molecule that is necessary for enzymatic activity.

Denature. Disruption of the tertiary bonds of a protein molecule that alters the folded structure. Denaturation results in loss of enzymatic, physical, and chemical properties.

End-point methods. End-point methods are also called fixed-time methods. In fixed-time methodology, the reaction is terminated after a period of time and the concentration of an end product or drop in substrate concentration is measured as an indication of enzyme activity.

Enzyme.   A protein that acts as a catalyst by affecting the rate at which a reaction proceeds.

First-order kinetics.   In first-order kinetics (for a given amount of enzyme), enzyme activity rate becomes dependent on substrate concentration when other reaction conditions are held constant (i.e., time, temperature, presence or absence of catalysts).

Holoenzyme.   The combination of an enzyme and its appropriate coenzyme that forms the catalytically active enzyme.

International Unit (U).   The unit of enzyme activity established by the International Union of Biochemistry defined as that amount of enzyme that catalyzes 1 μmol of substrate or coenzyme per minute under specified conditions (i.e., substrate concentration, temperature, pH).

Isoenzyme.   Enzymes that catalyze the same reaction but have different molecular structures and vary in their immunological, biochemical, and physical properties.

Kinetic methods.   Kinetic refers to motion. Kinetic methods, also known as continuous monitoring methods, measure the reaction rate (of a chemical reaction) as the reaction is proceeding. In enzyme kinetics, a change in absorbance per unit of time is measured.

Kinetics.   In chemistry, kinetics refers to the study of the reaction mechanism, the rate at which a chemical reaction proceeds, and the intermediate steps involved in the overall reaction.

Lineweaver–Burk equation.   The linearized form of the Michaelis–Menten equation. The reciprocals of reaction velocity ($1/V$) and substrate concentration ($1/[S]$) when plotted as $1/V$ on y-coordinate and $1/[S]$ on x-coordinate will form a straight line from which the maximal velocity can be determined at the y-intercept and K_m can be determined at the x-intercept. The Lineweaver–Burke equation allows calculation of K_m values from experimental data and is also used to study enzyme inhibition.

Metalloenzyme.   An enzyme that contains tightly bound metal atoms as part of its structure. The presence of the metal is essential for enzyme activity.

Michaelis constant (K_m). K_m refers to the substrate concentration at which the enzyme activity rate is proceeding at one-half maximal velocity. It is used to measure enzyme–substrate binding.

Michaelis–Menten equation.   An equation that describes the rate of a reaction catalyzed by an enzyme involving a single substrate and product under steady-state conditions:

$$E + S \underset{k_2}{\overset{k_1}{\rightleftharpoons}} ES \overset{k_3}{\rightarrow} E + P$$

Monoclonal antibody.   Antibodies, produced by a clone of plasma cells, that are immunochemically identical, thus all antigen-binding sites are identical.

Zero-order kinetics.   In zero-order kinetics, enzyme activity rate is dependent only on enzyme concentration when substrate concentration is in excess and other reaction conditions are held constant (i.e., time, temperature, presence or absence of catalysts).

Enzymes may be defined as protein catalysts for chemical reactions in biological systems. In order for a chemical reaction to begin, a certain amount of kinetic energy is needed; this energy is called the energy of activation. When enzymes are present, the energy of activation needed to initiate a reaction is decreased. Enzymes make spontaneous reactions proceed rapidly under conditions prevailing in the cell. Enzymes change only the rate at which the equilibrium is established between reactants and products. The enzyme is not consumed or permanently altered in the reaction. Only a small quantity of the enzyme is needed, and it can be used over and over again. Enzymes catalyze only a small number of reactions, frequently only one (reaction-specific catalysts). Most enzyme-catalyzed reactions are reversible. The directions of the reactions are usually determined by the reaction conditions.

ENZYME CLASSIFICATION AND NOMENCLATURE

Before the turn of the century, enzymes were given trivial names such as trypsin, diatase (amylase), ptyalin, and pepsin, and some of these names are still in use. By the turn of the century, enzymes were named for the substrate or group on which they acted and the suffix, -ase, added. For example, the enzyme that hydrolyzes urea was called urease, enzymes that hydrolyze phosphate ester bonds were called phosphatases, and those hydrolyzing lipids were called lipases. In some instances, the type of reaction involved was also identified. For example, the enzyme that catalyzes the removal of water from carbonic acid was called carbonic anhydrase.

INTERNATIONAL UNION OF BIOCHEMISTRY (IUB) CLASSIFICATION SYSTEM

For purposes of consistency and clarity, the International Union of Biochemistry (IUB) established a classification system for enzymes based on reaction types and reaction mechanisms. These proposals were published in 1964 and have been accepted by all workers in the field.

The major features of the IUB system for classification of enzymes follow.

Reactions

Enzyme-catalyzed reactions are divided into six major classes. There are subclasses within these classes. Based on reaction type, enzymes are classified as (1) oxidoreductases, (2) transferases, (3) hydrolases, (4) lyases, (5) isomerases, or (6) ligases.

Name

The enzyme name has two parts. The first portion of the name is derived from the name of the substrate or substrates acted upon by the enzyme. The second portion of the name indicates the type of reaction catalyzed, which is one of the six major reactions listed above, with a suffix "-ase". The -ase is no longer attached to the substrate. If additional information is needed to clarify the nature of the reaction, this information follows in parentheses. These two parts, the substrate(s) and reaction type, form the systematic name:

substrate name(s) + (reaction type)-ase = systematic name

In association with each systematic name, a code number name is also assigned designated as EC or Enzyme Commission number. The numbering system for each enzyme consists of four numbers separated by periods. The first number refers to one of the six major reaction classes as cited above; the second and third numbers give additional information, such as the designation of integral substrates and coenzymes; and the fourth number is the serial number for a specific enzyme.

For example, the enzyme, ATP : D-hexose-6-phosphotransferase, which catalyzes the transfer of a phosphate group from ATP to the hydroxyl group on carbon six of D-glucose, is designated by the number name EC 2.7.1.1. The first number denotes reaction class 2 (a transferase enzyme), subclass 7 (transfer of phosphate), subsubclass 1 (the phosphate acceptor is an alcohol), and the final number denotes the enzyme. Thus, by using this system, each enzyme can be identified by a system of four figures.

In addition to the IUB systematic and number name, a trivial name is also given to each enzyme. The trivial name is short and uncomplicated and is the name commonly used in the clinical laboratory.

REACTION CLASSES

Oxidoreductases

Enzymes in this class catalyze reactions involving the removal of hydrogen or the addition of oxygen. These enzymes were formerly known as dehydrogenases (remove H) or oxidases (add O_2). They catalyze oxidoreductions between two substrates, S and S':

$$S \text{ (reduced)} + S' \text{ (oxidized)} \xrightleftharpoons{\text{enzyme}} S \text{ (oxidized)} + S' \text{ (reduced)}$$

Groups oxidized or reduced by these enzymes included CH—OH, CH—CH, C=O, CH—NH$_2$, and CH=NH. Many of the oxidoreductase enzymes need an acceptor (or donor) substrate called a coenzyme in order to function.

A representative enzyme of this class is lactate dehydrogenase or LD (trivial name). This enzyme catalyzes the final step in the glycolytic pathway and is found in glucose-metabolizing tissue. In the presence of NAD^+ (coenzyme), LD catalyzes the removal of two hydrogen atoms from lactate to form pyruvate. One hydrogen is added to the coenzyme, NAD^+, which becomes NADH, and the other is liberated into cytoplasmic fluid as H^+ (see reaction in LD discussion). The reaction is reversible, and pyruvate can be reduced back to lactate. The systematic name for this enzyme is L-lactate:NAD oxidoreductase or EC 1.1.1.27.

Transferases

Enzymes in this class catalyze the transfer of a specific group, G (other than hydrogen), from one substrate to another:

$$S—G + S' \underset{}{\overset{enzyme}{\rightleftharpoons}} S'—G + S$$

For example, in transamination, an $—NH_2$ group is transferred; in transmethylation, a $—CH_3$ group is transferred; in phosphorylation, a $—PO_4$ group is transferred.

A representative enzyme of this class is aspartate aminotransferase or AST (formerly called SGOT). This enzyme catalyzes the reversible transfer of an amino group from glutamic acid (amino acid) to oxalacetic acid (keto acid). The amino acid that donates the amino group becomes a keto acid (α-ketoglutaric acid), and the keto acid or acceptor molecule becomes an amino acid (aspartic acid) (see reaction in AST discussion). The α-ketoglutaric acid, an intermediate in the Krebs cycle, provides a direct link between amino acid and carbohydrate metabolism. After entering the Krebs cycle, this compound can be utilized for energy (energy derived from protein), or amino acids may be derived from corresponding α-keto acids via specific transaminase enzymes (protein derived from carbohydrate). The systematic name for this enzyme is L-aspartate:2-oxoglutarate aminotransferase or EC 2.6.1.1.

Hydrolases

Enzymes in this class catalyze the cleavage by water of a large variety of bonds (ester, ether, peptide, glycosyl, acid–anhydride, C—C, C–halide or P—N). This group includes the digestive enzymes, namely, the proteases, lipases, and amylases. These enzymes usually do not require coenzymes but often need activators.

Representative enzymes of this class include acid and alkaline phosphatase. Both of these enzymes act on the same substrate and cleave phosphate ester bonds by hydrolysis; however, they are distinguished by their differing pH optima.

Lyases

The lyases include nonhydrolytic enzymes that catalyze the removal of specific groups from substrates, leaving double bonds:

$$\underset{}{\overset{X \quad Z}{-C-C-}} \underset{}{\overset{enzyme}{\rightleftharpoons}} X—Z + —C≡C—$$

The enzymes included in this group are those that split C—C bonds (decarboxylases), C—O, C—N, C—S, and C—halide bonds.

Carbonic anhydrase (CA) is an example of an enzyme in this group. Carbonic acid is converted to carbon dioxide and water (C—O bond split) in the presence of CA.

Isomerases

Isomers are compounds with the same atomic composition but with different structures. Thus, isomers of a given compound possess different chemical properties. Isomerase enzymes catalyze an intermolecular rearrangement of one isomeric compound to another.

Ligases

Ligase enzymes catalyze the linking together of two molecules, coupled with the breakdown of ATP or a similar energy-yielding triphosphate. These enzymes are involved in building new compounds and are sometimes known as synthetase enzymes. The formation of C—O, C—S, C—N, and C—C bonds are catalyzed by ligase enzymes.

Most of the enzymes commonly assayed in clinical laboratories belong to the first three reaction classes. A list of enzymes of diagnostic importance which are commonly assayed in clinical laboratories is given in Table 10.1.

Table 10.1. Enzymes of Clinical Importance Commonly Assayed in Clinical Laboratories

Trivial Name	Common Abbreviation	Enzyme Class	Principal Diagnostic Use
Acid phosphatase	ACP	Hydrolase	Prostatic cancer
Alkaline phosphatase	ALP	Hydrolase	Bone disease; cholestatic liver disease
Alanine transferase (glutamate pyruvate transaminase)	ALT (GPT)	Transferase	Parenchymal liver disease
Amylase	AMS	Hydrolase	Acute pancreatitis
Aspartate transferase (glutamate oxalacetate transaminase)	AST (GOT)	Transferase	Parenchymal liver disease
Ceruloplasmin		Oxidoreductase	Wilson's disease
Cholinesterase (pseudo)	CHS	Hydrolase	Organophosphorus exposure; sensitivity to succinylcholine
Creatine kinase (creatine phosphokinase)	CK (CPK)	Transferase	Acute myocardial infarction; skeletal muscle disease
Gamma-glutamyl transferase	GGT	Transferase	Cholestatic liver disease; chronic alcohol ingestion
Glucose-6-phosphate dehydrogenase	GPD	Oxidoreductase	Hemolytic disease
Isocitrate dehydrogenase	ICD	Oxidoreductase	Parenchymal liver disease
Lactate dehydrogenase	LD or LDH	Oxidoreductase	Acute myocardial infarction; skeletal muscle disease
Lipase	LPS	Hydrolase	Acute pancreatitis

FACTORS GOVERNING ENZYME REACTIONS

SUBSTRATE CONCENTRATION

If the concentration of a substrate in an enzyme reaction system is gradually increased while all other conditions are kept constant, the rate of reaction will increase as a function of substrate concentration until a maximum level is reached. The rate of the reaction depends on the concentration of the enzyme–substrate complex, rather than on the concentration of the enzyme itself. After all enzyme present is "bound", there is no further increase in reaction rate when additional substrate is added. In 1913, Michaelis and Menten proposed that an enzyme–substrate complex is the basis for enzymatic reactions. According to this theory, the enzyme (E), reacts with the substrate (S), to form an enzyme–substrate complex (ES), which then dissociates to yield free enzyme (E) and product (P):

$$E + S \underset{k_2}{\overset{k_1}{\rightleftharpoons}} ES \overset{k_3}{\rightarrow} E + P$$

In the formula, k_1 represents the rate of combination of E and S; k_2 is the rate of dissociation back to E and S; and k_3 is the rate of dissociation of ES to E and P. The breakdown of ES into E and P is determined by the concentration of ES. Once the enzyme is free, it again attaches to more substrate. When the substrate concentration is high, the enzyme binds immediately to more substrate, forming more ES and, therefore, more product. If the substrate concentration is low, the probability of binding is decreased. Thus, the rate of ES and subsequent P formation is dependent on substrate concentration as well as free enzyme concentration.

The relationship between substrate concentration and reaction velocity (Michaelis–Menten plot) is seen in Figure 10.1. In the plot, any point on the slope of the curve (as represented by A) is said to follow first-order kinetics; that is, the rate of the reaction is not proceeding at maximum velocity and is directly proportional to the substrate concentration. Point C on the graph represents the maximum rate of enzyme activity (V_{max}), and this rate of activity does not change as the concentration of the substrate is increased. This is referred to as zero-order kinetics. Zero-order kinetics refers to the situation in which all available sites on the enzyme molecule are saturated with substrate and the reaction is proceeding at its maximum velocity. Increasing the substrate concentration will not increase the velocity. Whenever possible, enzymes are assayed at zero-order kinetics. This means that the rate of the reaction is constant with time and is dependent only on the concentration of the enzyme. As long as sufficient substrate is present and the temperature and pH are held constant, the concentration of the enzyme is the only factor that affects the reaction rate. Under zero-order kinetics, the same amount of substrate is converted to product during each unit of time. Under these circumstances, the plot of the amount of product (measured by absorbance, y-axis), versus time (x-axis) on linear graph paper will produce a straight line with direct proportionality. As the substrate is depleted, zero-order kinetics changes to first-order kinetics.

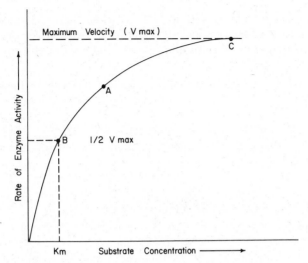

Figure 10.1. Substrate rate curve or Michaelis–Menten plot.

Some enzymes may have to be assayed under conditions other than zero-order kinetics. For example, in enzyme assays using coupled reactions, the substrate for the enzyme under assay may act as an inhibitor of another enzyme, which is used as a reagent in a coupled reaction. Also, substrate may be limited in solubility properties, thus sufficient substrate cannot be dissolved to reach the concentration required for zero-order kinetics.

After enzyme and substrate have reacted for a period of time, products are formed and substrate is consumed. In some reactions, these products, unless removed, inhibit reactions and cause the equilibrium constant to change. Products formed in the reaction may also alter pH. In actual practice, determined values are considered accurate if no more than 20% of the substrate has been used during the course of reaction. If more is consumed, the reaction should be repeated with a smaller quantity of enzyme. If the diluent employed to dilute the enzyme is water or saline, the activity change may become disproportionate; therefore, some chemists prefer a protein-based diluent such as heat inactivated serum or a pure albumin solution.

Point B on the substrate rate curve (Figure 10.1) represents the substrate concentration at which the reaction proceeds at one-half the maximum velocity ($\frac{1}{2}V_{max}$). This substrate concentration is called K_m or Michaelis constant. K_m measures the affinity of an enzyme for a particular substrate. The K_m number is characteristic for each enzyme and substrate pair. When the K_m is small, the affinity of the enzyme for the substrate is high, and half-maximal velocity is attained at low substrate concentration. When the K_m is large, the affinity of the enzyme for the substrate is low, and the enzyme may not be completely saturated even at high substrate concentration.

K_m values may be used to determine the substrate concentration to be used to measure enzyme activity at optimal substrate concentration. To ensure that zero-order kinetics is achieved, the substrate concentration should be at least 20 times, and preferably 100 times, the K_m value. Since few enzymes will readily permit the determination of K_m when velocity (V) is plotted against substrate concentration (S), a linearized form of the Michaelis–Menten equation is used. Using the double-reciprocal or Lineweaver–Burk plot of $1/V$ versus $1/[S]$ (S expressed in molarity), K_m may be estimated.

With relatively nonspecific enzymes, K_m values may also be used to determine which compound will make the most suitable substrate for those enzymes that use several substrates. K_m has also been used as an aid in isoenzyme studies where it is employed to identify the organ source for a particular enzyme. For example, using the same substrate, it was found that K_m values for the isoenzymes of alkaline phosphatase varied.

HYDROGEN ION CONCENTRATION OR pH

The activity of an enzyme varies with hydrogen ion concentration. Since enzymes are charged molecules, moderate pH changes can affect the ionic state of the enzyme and sometimes the ionic state of the substrate. The alteration in ionic charge may affect the conformation of the polypeptide chain (alters the tertiary or quaternary structure) of the enzyme protein or may alter the amino acid residues that function in the binding of the enzyme to the substrate. As the pH changes, ionized substrates may vary in ionic properties. Such changes may alter binding properties.

Most of the enzymes in plasma show maximal activity in the pH range of 7–8. If an enzyme-catalyzed reaction were performed and all factors known to alter activity rate were kept constant except pH, the rate of reaction observed at different hydrogen ion concentrations would vary from no activity to maximal activity. The plot of activity against pH produces a bell-shaped curve, as seen in Figure 10.2. Most enzyme assays are carried out at the pH of optimal activity for the following reasons: (1) the sensitivity of the measurement is optimal at this pH and (2) the error introduced by a small change in pH is minimal since the pH–activity curve has minimal slope at the pH of maximal activity. Buffers are incorporated into substrates because products formed or substrate loss can cause alterations in pH. The pK of the buffer system should be within one pH unit of the optimal pH for a particular enzyme system.

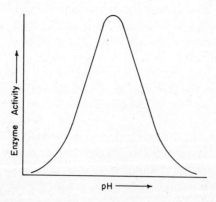

Figure 10.2.   Enzyme activity-versus-pH curve.

TEMPERATURE

Enzyme activity increases as temperature increases. For many biological reactions, the velocity of the reaction roughly doubles for each 10°C rise in temperature, and for each 10°C drop in temperature, the reaction velocity is halved. However, if the temperature increases much above the optimal temperature, the protein enzyme may denature. Denaturation may be defined as loss of biological activity and does not involve covalent bond cleavage, but results from a rearrangement of secondary and tertiary structure. Marked denaturation of enzymes occurs above 40–45°C. Therefore, the increase in rate of activity is counterbalanced by a loss of active enzyme.

The optimal temperature for most enzymes approximates the temperature of the cell (37°C). The preferred temperature for enzyme analysis is controversial. The Commission on Enzymes of the International Union of Biochemistry recommended that enzyme analysis be performed at 25°C; later this recommendation was amended to 30°C. In the clinical laboratory, the temperature preferred by many analysts is 37°C, since this approximates body temperature, the optimal temperature for most of the enzymes assayed.

For accurate enzyme assays, the temperature must be maintained to within ±0.1°C, irrespective of the temperature chosen for analysis.

ENZYME COFACTORS

Certain enzymes require cofactors before they are able to perform their catalytic activity. Other enzymes require cofactors for maximal catalytic activity. Cofactors are nonprotein materials. If the cofactor is an organic compound, it is called a *coenzyme*. If the cofactor is an inorganic ion, it is called an *activator*. If the inorganic ion is a metal cation and part of the enzyme molecule, the enzyme is referred to as a metalloenzyme. Not all enzymes need cofactors.

Coenzymes

The combination of an enzyme with a coenzyme is called a *holoenzyme*. The protein portion of the holoenzyme is called the apoenzyme; the nonprotein portion or coenzyme is bound to the apoenzyme protein. Many coenzymes are loosely bound to the enzyme and act as cosubstrates. As cosubstrates, coenzymes must be present in excess concentration for the reaction to proceed at the maximum rate. As the reaction proceeds, the coenzyme is consumed or its structure is altered. The chemical changes in the coenzyme exactly counterbalance the changes taking place in the substrate. For example, as one molecule of substrate is dehydrogenated (oxidized), one molecule of the coenzyme is hydrogenated (reduced). This concurrent change is used as the basis for the measurement of enzyme activity rates using continuous monitoring methods. Enzymes that belong to reaction classes 1 (oxidoreductases), 2 (transferases), and 5 (isomerases) frequently require the participation of coenzymes. Coenzymes function in group transfer reactions and frequently contain water-soluble vitamins, especially B vitamins, as part of their structure.

Coenzymes may be divided into two groups: (1) coenzymes that transfer hydrogen and (2) coenzymes that transfer groups other than hydrogen. Hydrogen-transfer coenzymes include $NAD^+/NADH$ (nicotinamide adenine dinucleotide or coenzyme-I) and $NADP^+/NADPH$ (nicotinamide adenine dinucleotide phosphate or coenzyme-II). Coenzymes that transfer groups other than hydrogen include ATP/ADP (adenosine triphosphate), a nonvitamin coenzyme that transfers a phosphate group, and pyridoxal phosphate, a vitamin B coenzyme that transfers an amino group in transamination reactions.

Activators

Many enzymes require the presence of metal ions to function. Those enzymes that bind the metal loosely are called metal-activated enzymes. The metal can be removed from metal-activated enzymes during purification processes. Metalloenzymes bind the metal more firmly, and the metal remains attached to the enzyme throughout the steps of normal purification procedures. Common cation activators include Mg^{2+}, Mn^{2+}, Fe^{2+}, Ca^{2+}, Zn^{2+}, and K^+. Anions may also function as activators. Magnesium is an obligate activator for all kinase enzymes (phosphate-transfer enzymes). Amylase is a calcium metalloenzyme that displays full activity in the presence of a variety of inorganic ions (Cl^-, Br^-, NO_3^-, and HPO_4^{2-}). In activator-dependent enzyme reactions, the activator as well as the substrate should be present in excess concentration. Excess activator may also function to inhibit activity; therefore, in some cases, optimal activator should be used.

SUBSTRATE SPECIFICITY OF ENZYMES

The concept of a template or "lock and key" model for the binding of enzyme and substrate has largely been abandoned for a model that emphasizes flexibility. This modern conception of enzyme–substrate interaction is called the induced-fit model. In the induced-fit model, the presence of the substrate is believed to induce conformational changes in the enzyme, causing the correct alignment of the amino acid residues or other groups on the enzyme for substrate binding, catalysis, or both. Only a small portion of the enzyme polypeptide chain is actually employed in the attachment of the enzyme to the substrate. This area is referred to as the catalytic site or active site and is formed by only three or four amino acids separated considerably from each other in the polypeptide chain but spatially near each other in the tertiary or folded structure. Side chains of these amino acids serve to anchor the enzyme to the substrate.

Enzymes generally show high specificity for at least a part of the substrate molecule. An enzyme may catalyze only a single unique reaction and no others. For example, glucose oxidase acts only on β-D-glucose; it cannot oxidize α-D-glucose. Human amylase attacks only $\alpha(1 \rightarrow 4)$ linkages; it cannot cleave cellulose in which the glucose units are attached in $\beta(1 \rightarrow 4)$ linkages.

Other enzymes exhibit group specificity and act on particular chemical groupings. For example, the phosphatase enzymes split phosphate from a large variety of organic phosphate esters, although at different rates. Pepsin and trypsin act on peptide bonds, thus allowing a large number of substrates to be catalyzed. This property reduces the number of digestive enzymes that would otherwise be required.

ENZYME INHIBITION

Inhibitors fall into two broad classes: (1) competitive and (2) noncompetitive.

COMPETITIVE

In competitive inhibition, the inhibiting agent binds to the enzyme at the same site as the substrate. Thus, the substrate and inhibitor compete for the same binding sites on the enzyme molecule. Competitive inhibitors have chemical structures very similar to the substrate; however, the inhibitor–enzyme complex forms no product. If enough inhibitor is present, it may saturate the enzyme so that no substrate may attach, thereby eliminating enzyme activity. Increasing substrate concentration overcomes competitive inhibition. As more substrate is added, the probability that the enzyme will combine with substrate rather than with inhibitor increases. At sufficiently high concentrations of substrate, the rate of activity is the same as the rate observed in the absence of the inhibitor. Competitive inhibition is a form of reversal inhibition. In competitive inhibition, the maximum velocity (V_{max}) attainable with a given amount of enzyme is not affected, but the apparent K_m is increased.

NONCOMPETITIVE

In noncompetitive inhibition, the inhibitor usually is not structurally similar to the substrate and is assumed to bind to the enzyme at a point other than the substrate binding site. Sometimes, the enzyme remains active, but its activity is diminished. Increasing substrate concentration does not alter inhibition, since substrate and inhibitor bind at different sites. In noncompetitive inhibition, the maximum velocity (V_{max}) attainable with a given amount of enzyme is reduced, but no change is observed in apparent K_m.

The linerized form of the Michaelis–Menten equation (Lineweaver–Burk plot) is used to distinguish between competitive and noncompetitive inhibition.

OTHER INHIBITORS

The presence of metal ions may also act as inhibitors. In many enzymes, the presence of free cysteine sulfhydryl (—SH) groups is essential for activity. Alterations in these groups, such as oxidation with heavy metals, may cause conformational changes in the enzyme (denatures the enzyme), thus altering the enzyme activity. Heavy metals may also compete with an activator, resulting in loss of enzyme activity. To avoid metal contamination, glassware used in enzyme analysis should be thoroughly cleaned.

Denaturation of the enzyme may also occur with heat, acid, base, organic solvent, or high salt concentration. Denaturation is thought to result from rupture of weak ionic, salt, and nonpolar bonds, which are responsible for maintaining the enzyme in its tertiary configuration. Denaturation does not involve covalent bond cleavage.

ISOENZYMES

Isoenzymes may be defined as multiple molecular forms of the same enzyme. They represent different proteins (the amino acid content differs) present in the same individual with similar enzymatic activity (act on the same substrate to form the same products). It is believed that the catalytic or active center in isoenzymes is identical or similar. It is possible that all enzymes may occur in multiple isoenzyme forms.

Few enzymes in the body are tissue specific. Therefore, an elevated blood enzyme level is not particularly helpful in pinpointing the organ source or site of lesion. Since tissues vary in their isoenzyme content, isoenzyme patterns are used in clinical medicine to help identify the organ or tissue involved in a pathophysiologic process. For example, creatine kinase (CK or CPK) is present in high concentration in skeletal muscle, heart muscle, and brain; however, the distribution of CK isoenzymes present in each organ source is different. When total CK in the blood is elevated, the organ source is not specified; however, separation of the isoenzyme fractions may indicate the organ source responsible for the elevation. Since the heart CK isoenzyme is considered to be organ specific, the presence of this isoenzyme is helpful in the differential diagnosis of patients with chest pain.

Isoenzymes differ in their (1) rates of electrophoretic mobility; (2) stability to heat denaturation; (3) resistance to various chemical inhibiting agents; (4) affinity for substrates and coenzymes; and (5) immunogenic properties.

In the clinical laboratory, CK and LD isoenzymes are most often separated by electrophoresis. The isoenzymes of alkaline phosphatase do not separate well on electrophoresis, and heat or chemical separation in conjunction with electrophoresis is often employed. Heat denaturation has also been used to separate several of the isoenzymes of LD. The prostatic isoenzyme of acid phosphatase can be separated from red blood cell acid phosphatase using chemical inhibition (tartrate ion). However, immunochemical methods are now available that allow direct quantitation of this isoenzyme fraction.

ETIOLOGY OF ENZYMES IN THE BLOOD

Most enzymes serve no function in the plasma. The few functional plasma enzymes are synthesized in the liver and are then released into the plasma. Functional plasma enzymes include pseudocholinesterase, lipoprotein lipase, ceruloplasmin, and the blood-coagulation enzymes. These *plasma-specific* enzymes are present in higher concentration in plasma than in cells. The plasma-specific enzymes are clinically significant when the serum level is decreased below the reference range.

Nonfunctional plasma enzymes are derived from the cells of organs and tissues. These enzymes are present in high concentrations within cells, and the low levels of these enzymes in normal plasma are thought to be derived from those cells destroyed in normal tissue turnover. Nonfunctional or *non-plasma-specific* enzymes are clinically important when the serum level is increased above the reference range. Normally, non-plasma-specific enzymes are contained within cell membranes, the cell membrane being impermeable to enzymes as long as the cell is functioning normally. However, during periods of stress, such as periods of decreased oxygen or glucose supply, or in the presence of bacterial or viral infections, the cell membrane becomes permeable or the membrane ruptures, causing release of soluble enzymes into the extracellular fluid. These enzymes eventually reach the blood. Some of the enzymes found in cells are free in the cytoplasm, and some are found in cellular structures such as the mitochondria and lysosomes. When the enzymes associated with cellular structures are present in the plasma, the degree of cell damage is thought to be greater than when cytoplasmic enzymes alone are found.

When the cell membrane becomes permeable to enzymes, the degree of leakage is determined by the concentration gradient that exists between the intracellular and extracellular enzyme levels. Smaller enzymes also leak more readily than do enzymes with large molecular weights. The elevated enzyme levels associated with hepatitis and following myocardial infarction are thought to be derived from a change in cell permeability or from cell death causing release of enzymes into the extracellular fluid.

The exocrine enzymes are those that are secreted into ducts. The exocrine enzymes, amylase and lipase, are normally present in low and constant concentrations in the plasma. If the duct through which the enzyme passes becomes blocked, the plasma level of these enzymes increases rapidly. Ductal obstruction may result in an increase in intraductal pressure, which causes rupture of acinar cells. The released enzymes then enter extracellular fluid and diffuse into the plasma.

Excessive synthesis or induction of an enzyme by cells with overflow into the plasma is also thought to cause elevated enzyme levels. In obstructive liver disease, the increase in liver alkaline phosphatase is thought to be due to this process.

Increased synthesis can also occur when cells that normally produce an enzyme are increased in number, such as when tumor is present. Neoplastic processes may also elevate enzymes by causing structural changes that allow intracellular enzymes to leak into the circulation. Neoplasms may also induce normal cells to alter enzyme-related physiology, leading to increased enzyme production. In certain cases, tumor cells are thought to produce enzymes that are normally produced only in immature or fetal cells. It is believed that tumor cells revert to a more primitive cell form and produce enzyme patterns normally found only in fetal or immature cells. The Regan alkaline phosphatase isoenzyme is thought to be derived from this process.

Finally, vigorous exercise can elevate certain enzyme levels through the release of small quantities of muscle enzymes. If the route of enzyme clearance is via the kidney, renal disease, with decreased clearance, can also cause elevations of certain enzymes (i.e., amylase and CK).

ENZYME ASSAY METHODS

SPECIMEN REQUIREMENTS

Serum is the specimen of choice for enzyme assay unless otherwise specified. Anticoagulants inhibit the activity of many enzymes. EDTA chelates divalent metal ions and may therefore remove activators. Specimens should be free from hemolysis, because the ruptured cells will release cellular enzymes into the plasma. Hemoglobin, released from cells, may also inhibit enzyme activity. Serum will contain enzymes from lysed platelets, but with the exception of acid phosphatase, the majority of enzymes routinely determined are not present in platelets. Enzymes should preferably be assayed without storage. If this is not possible, the serum should be refrigerated or frozen for longer periods of storage.

There are several exceptions to this general rule. The isoenzymes of lactate dehydrogenase, LD-4 and LD-5, are inactivated at cold temperatures. Serum for LD can be stored at room temperature for two to three days with no loss of activity. Alkaline phosphatase activity appears to increase in specimens that are allowed to stand before analysis (either at room temperature or refrigerated temperatures). The increase in activity may be due to the increase in pH that occurs as serum is allowed to stand. Acid phosphatase is very unstable and will deteriorate even if refrigerated unless the pH of the serum is adjusted to 6.0 or below with citric acid.

ASSAY METHODOLOGY

Since the enzyme protein is present in a large matrix of other proteins, partial or total isolation is impractical. Under conditions where the enzyme concentration is the rate-limiting factor, (that is, where substrate concentration is high, product formation is low, pH remains favorable throughout the assay and the temperature is controlled), the activity rate of an enzyme may be used to determine enzyme concentration.

Fixed Time or End Point

Under ideal conditions, the rate of product formed should parallel the drop in concentration of the substrate. Therefore, in fixed-time or end-point methods, the enzyme activity rate is determined either (1) by measuring the increase in an end product concentration or (2) by measuring the drop in substrate concentration. Measurement of an end product concentration is preferred because, with analytical methods, it is easier to measure small changes from zero than to measure decreases from a large substrate concentration. After termination of the enzyme-catalyzed reaction, the product or residual substrate is color complexed with a reagent and the absorbance of the complex is then measured.

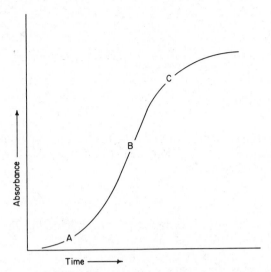

Figure 10.3. Enzyme reaction curve plotting absorbance against time. *A* represents the lag phase and is nonlinear; *B* represents zero-order kinetics and the linear portion of the curve; and *C* represents substrate depletion and reversion to nonlinearity.

Kinetic or Continuous Monitoring

In kinetic or continuous monitoring methods, the rate of change in the concentration of a substrate, product, or coenzyme is measured as a means of determining the rate of reaction. In kinetic methods the reaction is monitored as it proceeds. The serum sample is added to a cuvette, which contains prewarmed buffer substrate mixture. The reaction is then monitored in a spectrophotometer, which has a controlled-temperature cuvette compartment. The temperature in the cuvette compartment should not vary by more than $\pm0.1°C$. Absorbance readings are taken at timed intervals. When the change in absorbance per unit of time is the same on several measurements, the reaction is proceeding at zero-order kinetics and is in a linear portion of the curve (see Figure 10.3). The average of several consecutive readings can then be used to calculate enzyme activity. Continuous monitoring methods detect lag phases and substrate depletion because the reaction becomes nonlinear (change in absorbance per unit of time is not the same). The continuous monitoring or kinetic method is the preferred method of analysis.

The majority of the kinetic methods use the coenzyme NAD^+. The reduced form of this coenzyme (NADH) absorbs at a wavelength of 340 nm; the oxidized form (NAD^+) does not. When the reaction measures the formation of NADH from NAD^+ ($NAD^+ \rightarrow NADH$), the absorbance at 340 nm increases. When the reaction measures the formation of NAD^+ from NADH (NADH $\rightarrow NAD^+$), the absorbance at 340 nm decreases. The coenzyme system $NADP^+$/NADPH may also be used—the reduced form absorbs at 340 nm; the oxidized form does not. If the enzyme to be measured does not use NAD^+ or $NADP^+$, a product from the first reaction can usually be used as a substrate for a coupled reaction that uses NAD^+ or $NADP^+$. The coupled reaction is called the indicator reaction. By using coupled reactions, kinetic methods may be used to monitor enzymes that do not require coenzymes:

$$\text{substrate} \xrightarrow[\text{(enzyme under assay)}]{\text{enzyme 1}} \text{product 1} + \text{product 2}$$

$$\text{product 1 (substrate)} + \text{NADH} \xrightarrow[\text{(enzyme used as a reagent)}]{\text{enzyme 2}} \text{NAD}^+ + \text{product}$$

In the example given, the change in absorbance at 340 nm would decrease.

The reduced forms of these coenzymes (NADH and NADPH) also fluoresce when activated with ultraviolet light.

Immunoassay

Immunoassay methods were developed in the anticipation that these methods would offer greater sensitivity and specificity than conventional methods. Certain isoenzymes may be prepared in a purified form that will allow preparation of an antibody directed against specific antigenic determinants. However, because many isoenzymes are composed of combinations of polypeptide chains, the antibody produced may not be totally specific for only one isoenzyme fraction. For example, the isoenzymes of LD are tetramers composed of a combination of two polypeptide chains, H and M. An antibody could be prepared that would react with only H or only M antigenic determinants. However, three of the five isoenzymes of LD are composed of a combination of both polypeptide chains and would therefore cross-react with either antibody, reducing test specificity.

The prostatic acid phosphatase is antigenically unique from other acid phosphatases, therefore permitting determination of this isoenzyme by immunological techniques. A immunoassay method for CK-MB is also available that uses two monoclonal antibodies and employs a sandwich technique (see CK, isoenzymes).

The determination of other isoenzymes by immunoassay techniques have lacked specificity; therefore, they have not replaced conventional methods.

UNITS OF MEASURE

Enzymes, which are measured by reaction rate methods, are dependent on substrate used, pH, buffer, temperature, and concentration of coenzymes and activators, and these factors must be specifically stated in defining a unit of enzyme activity. Older procedures measured activity rates in units that were individually defined for a particular procedure. The Commission on Enzymes of the International Union of Biochemistry proposed a standardized unit known as the International Unit (U), which is defined as follows: One Unit (U) of any enzyme is that amount that will catalyze the transformation of 1 μmol of substrate per minute under standard conditions. The concentration is to be expressed in terms of units per milliliter (U/mL) of serum or milliunits per milliliter (mU/mL), which is the same as units per liter (U/L) whichever is the more convenient numerical value. The recommended temperature for analysis is 30°C:

$$U = \mu\text{mol/min/mL (or L) of serum}$$

By using a standardized definition, the amount of enzyme present can be compared among different procedures. However, comparison of enzyme activity rates with different procedures should be used with caution.

Enzymes measured by immunoassay procedures are quantitated and reported in the amount of enzyme protein found in the specimen or in weight/volume units (e.g., ng/mL).

OXIDOREDUCTASE ENZYMES

LACTATE DEHYDROGENASE (LD, LDH)
L-LACTATE: NAD⁺ OXIDOREDUCTASE; EC 1.1.1.27

Lactate dehydrogenase is an intracellular enzyme found in the cell cytoplasmic supernatant (cytosol) and functions in the glycolytic pathway.

Function

LD reversibly catalyzes the conversion of pyruvate to lactate with the oxidation of NADH to NAD⁺:

$$\text{pyruvate} + \text{NADH} + \text{H}^+ \underset{}{\overset{\text{LD}}{\rightleftharpoons}} \text{L-lactate} + \text{NAD}^+$$

In glucose metabolism, the LD-catalyzed reaction allows NADH to be converted to NAD⁺. NAD⁺ serves as the hydrogen acceptor when the glucose molecule is split into two three-carbon fragments (see Figure 5.9). If NAD⁺ could not be reoxidized after reduction, glycolysis would become stalled after all available NAD⁺ was reduced. The formation of lactate from pyruvate provides a mechanism for recylcling NAD⁺ and allows glycolysis to continue when oxygen supplies are low or absent. The lactic acid that accumulates in cells can be converted back to pyruvate when sufficient oxygen is again present. The equilibrium of the reaction (i.e., P → L or L → P) can be shifted by changing the pH.

Source

As a glycolytic enzyme, LD can be found in virtually all tissue. The liver and skeletal muscle have the greatest concentrations, followed by the heart. The isoenzyme pattern for various tissues is used to pinpoint organ source when total LD is elevated.

LD is a tetramer (composed of four polypeptide chains). Each tetramer is composed of a combination of two different subunits (polypeptide chains), the H (heart) type, which is abundant in heart, and the M (muscle) type, which is abundant in skeletal muscle. The five isoenzymes of LD are (1) LD-1 (H_4 or HHHH); (2) LD-2 (H_3M or HHHM); (3) LD-3 (H_2M_2 or HHMM); (4) LD-4 (H_1M_3 or HMMM); and (5) LD-5 (M_4 or MMMM). LD-1 and LD-2 are found in great abundance in heart, red blood cells, and kidney. LD-3 is found in many tissues (i.e., spleen, bowel, uterus, thyroid, bladder, and lung) and is considered nonspecific. However, LD-3 may be useful in the evaluation of those individuals with suspected pulmonary embolism. LD-4 and LD-5 are found in high concentration in liver and skeletal muscle. A sixth isoenzyme, LD-X, has been isolated. It occurs in human testis and sperm.

The activity of alpha-hydroxybutyrate dehydrogenase (HBDH), once considered a separate enzyme, is now recognized to represent that of isoenzymes of LD (largely LD-1) on another substrate (alpha-hydroxybutyrate). HBDH does not have a IUB number and is no longer routinely analyzed in most clinical laboratories.

Specimen Requirements

The specimen of choice is nonhemolyzed serum. Red blood cells contain LD concentrations 150 times that found in serum. The serum should, therefore, be separated from the clot as soon as possible to avoid any leakage of LD from cells. If storage is required, it is best to leave the sample at room temperature, since LD-4 and LD-5 are destroyed at refrigerated temperatures. If storage is required beyond two or three days, the sample should be kept at 4°C with NAD^+ or glutathione added. These additions decrease the rate of inactivation of LD-4 and LD-5.

Assay

Kinetic methods measure the rate of change in absorbance at 340 nm as NAD^+ is converted to NADH (L → P) or as NADH is converted to NAD^+ (P → L) as an indication of enzyme activity:

$$\text{L-lactate} + NAD^+ \overset{\text{LD}}{\rightleftharpoons} \text{pyruvate} + NADH + H^+$$

Two colorimetric methods are available, which use coupled reactions. Method A uses the product, pyruvate, which is reacted with 2,4-dinitrophenylhydrazine (2,4-DNPH) at an alkaline pH to form a golden brown phenylhydrazone. The intensity of the color is then measured by absorbance spectrophotometry:

$$\text{pyruvate (product)} + \text{2,4-DNPH} \overset{OH^-}{\rightarrow} \text{phenylhydrazone (colored)}$$

Method B uses the product NADH, an enzyme that acts as an intermediate electron carrier (diaphorase), and the tetrazolium dye INT, which is colorless in its oxidized form. The electron carrier transports its electron to INT, which forms a colored compound in its reduced form:

$$\text{NADH} + \text{INT} \xrightarrow{\text{diaphorase}} NAD^+ + \text{INTH}$$

Isoenzyme Separation

The isoenzymes of LD are most commonly separated by electrophoresis. Either cellulose acetate or agarose may be used as the electrophoretic support media. The isoenzymes are named according to their rate of migration; that is, the fastest migrating isoenzyme is designated LD-1, followed in order by the others, with LD-5 representing the slowest isoenzyme. After separation, the isoenzymes may be visualized by overlaying the electrophoresis medium with a substrate–buffer–indicator mixture. Each separated isoenzyme catalyzes a reaction with the substrate. A product of the reaction becomes visible through an indicator reaction (one of the two colorimetric reactions described above). The intensity of the color of each isoenzyme is proportional to its concentration. If needed, a densitometer can then be used to quantitate each isoenzyme. The fluorescence of the NADH produced in the basic LDH reaction can also be used to visualize and quantitate individual isoenzymes.

Clinical Significance

LD is primarily used to detect acute myocardial infarction (AMI). Following an AMI, LD begins to rise in approximately 12–24 hours, frequently reaching levels greater than two to three times the upper limits of normal. The peak activity usually occurs on the third day. LD returns to normal in 7–12 days if no complications occur. In the LD isoenzyme pattern of normal sera, LD-2 is greater than LD-1. A "flip" is said to occur when LD-1 exceeds LD-2.

To diagnose an AMI with a high degree of specificity (afforded by LD isoenzymes) and a high degree of sensitivity (afforded by CK isoenzymes), a combined isoenzyme pattern is used. The combined criteria required for the diagnosis of AMI are (1) a flipped LD pattern (i.e., LD-1 greater than LD-2) and (2) the presence of CK-2. The CK-2 must appear in a sample drawn in the first 48 hours following the onset of chest pain or other symptoms characteristic of AMI. In most individuals, the flipped-LD never precedes the appearance of CK-2 and occurs within 48 hours after an AMI.

More than one sample is drawn for isoenzyme studies. For greater specificity, it is recommended that serial sampling be obtained. The first sample should be obtained 6–8 hours after onset of pain, followed by collections at 12–16 hours, at 20–24 hours, and, optionally, at 36–48 hours. In practice, most hospitals collect three serial samples. One is obtained on admission, one collected 6–13 hours later, and the third obtained 24–36 hours after admission.

If CK-2 fails to appear within a 48-hour period following chest pain, AMI can be ruled out with 100% accuracy. A flipped-LD in the absence of CK-2 may indicate hemolysis or renal infarction. LD-1 will also increase in the megaloblastic anemias.

LD-5 will increase with hepatic damage. The presence of CK-2 and flipped-LD with an increase in LD-5 is consistent with the diagnosis of AMI complicated by congestive heart failure or shock. LD-5 will also increase in viral and drug-induced hepatitis, infectious mononucleosis, and neoplastic liver disease. Inflammatory myopathies will also produce elevations of LD-5.

Prominent LD-2 and LD-3 patterns are consistent with pulmonary embolism.

An increase in all LD isoenzymes indicates multisystem involvement such as in shock, anoxia, or trauma.

Carcinoma or sarcoma may produce isoenzyme patterns with bizarre multiple isoenzyme abnormalities.

ISOCITRATE DEHYDROGENASE (ICD)
THREO-D-ISOCITRATE : NADP⁺ OXIDOREDUCTASE
(DECARBOXYLATING); EC 1.1.1.42

Source

There are two forms of ICD: (1) a mitochondrial form that occurs only in the mitochondria, participates in hydrogen transfer in the citric acid cycle, and requires NAD^+ as the hydrogen acceptor and (2) the form found in serum that is present in the cytosol and mitochondria and requires $NADP^+$ as the hydrogen acceptor. The serum-type ICD is found in high concentration in liver, heart, skeletal muscle, kidney, adrenal tissue, platelets, and red blood cells. The ICD in serum is thought to originate primarily from the liver.

Clinical Significance

ICD is considered to be a very sensitive indicator of parenchymal liver disease. Elevated levels are found in toxic and viral hepatitis and in infectious mononucleosis. After a myocardial infarction, ICD is not elevated despite large concentrations in heart muscle unless the MI is accompanied by congestive heart failure resulting in hepatic ischemia. Rapid removal of the heart enzyme from the circulation has been postulated as an explanation for this apparent discrepancy.

GLUCOSE-6-PHOSPHATE DEHYDROGENASE (GPD)
D-GLUCOSE-6-PHOSPHATE : NADP⁺ OXIDOREDUCTASE; EC 1.1.1.49

Function

This enzyme catalyzes the transfer of hydrogen from glucose-6-phosphate to $NADP^+$, the initial reaction of the pentose phosphate or shunt pathway (see Figure 5.9). The purpose of the pentose phosphate shunt is to produce NADPH and pentose phosphate. NADPH is needed by red blood cells to maintain the integrity of the cell membrane structure. Defects in the cell membrane increase the susceptibility of cells to lysis. Pentose phosphate is needed as the starting material for nucleotide and nucleic acid synthesis. GPD is present in practically all cells. The enzyme present in red blood cells is the one of clinical interest.

Specimen Requirements

A whole blood sample collected in EDTA, heparin, or ACD (acid–citrate–dextrose) is required. A hemolysate is prepared from the sample.

Clinical Significance

A few individuals congenitally lack red cell GPD activity. Other individuals have variant forms of GPD. Only a fraction of individuals with the variant forms demonstrate depressed GPD activity. Some of these individuals are subject to drug-induced hemolytic disease. Primaquine (antimalarial), naphthalene, and fava beans precipitate severe hemolytic episodes. In other individuals, the precipitating factor may be some physiological crisis. After the acute attack is ended, the individual fully recovers and remains well until another attack is precipitated. During the hemolytic episodes, only the older cells hemolyze; the younger cells remain intact.

CERULOPLASMIN (FERROXIDASE I)
IRON (II):OXYGEN OXIDOREDUCTASE; EC 1.16.3.1

Function

Ceruloplasmin is a glycoprotein that binds 94–95% of copper in the plasma and acts as the copper-transport protein. Ceruloplasmin is also an oxidase enzyme that catalyzes the oxidation of ferrous iron (Fe^{2+}) to ferric iron (Fe^{3+}). Only the ferric form of iron can be incorporated into apotransferrin to form transferrin, the iron-transport protein. Transferrin can donate its iron directly to the RE cells of the bone marrow for hemoglobin synthesis. It is postulated that without ceruloplasmin, the synthesis of transferrin would be depressed, compromising iron delivery to sites of heme synthesis. Such compromised delivery would result in a deficiency of hemoglobin or other iron-containing proteins.

Clinical Significance

Ceruloplasmin is a member of the acute phase proteins and will increase in physiologically stressful conditions, such as in acute infections, after myocardial infarction or surgery, or in chronic conditions such as systemic lupus erythematosus or rheumatoid arthritis.

Decreased ceruloplasmin levels may occur in renal diseases associated with protein loss in the urine and in severe nutritional protein deficiencies. Ceruloplasmin is most frequently requested to evaluate Wilson's disease (see Chapter 8). In Wilson's disease, the serum copper and ceruloplasmin are both decreased.

TRANSFERASE ENZYMES

TRANSAMINASES

Function

Transaminase enzymes catalayze the transfer of an amino group ($—NH_2$) from an amino acid directly to another molecule without the liberation of free ammonia. The molecule receiving the amino group is a keto ($C{=}O$) acid with a carbon skeleton like an amino acid. The net result of transamination is that the molecule donating the amino group becomes a keto acid and the acceptor keto acid becomes a new amino acid. All of the 20 natural amino acids are capable of undergoing transamination, and each amino acid requires a different enzyme. All transamination reactions require vitamin B6 (pyridoxine) derivatives as coenzymes.

Transamination represent a major link between amino acid and carbohydrate metabolism. Amino acids, needed for protein synthesis, are obtained from dietary protein, tissue breakdown, and carbohydrates via the process of transamination. Since transamination reactions are reversible, keto acids, derived from trans-amination, may function in the synthesis of new carbohydrate (gluconeogenesis).

Two transaminase enzymes frequently measured in clinical laboratories are aspartate aminotransferase (AST), formerly called glutamate oxaloacetic trans-aminase (GOT), and alanine amino transferase (ALT), formerly called glutamate pyruvate transaminase (GPT).

Aspartate Aminotransferase (AST, GOT)
L-Aspartate : 2-Oxoglutarate Aminotransferase; EC 2.6.1.1

Source
AST is found in the cell cytosol and mitochondria, and normal serum contains only small amounts of this enzyme. When mild tissue damage occurs, the predom-inate form found in the serum is derived from the cytoplasm; when the damage is more severe, both forms will appear in the serum. AST is found in many tissues, with highest concentrations in heart, liver, skeletal muscle, and kidney.

Specimen Requirements
The specimen of choice is serum free from hemolysis and immediately separated from cells. Erythrocytes contain 10 times the AST content of serum; therefore, the presence of hemolysis will falsely elevate values. The enzyme is stable for a week at refrigerated temperatures. For long periods of storage, samples should be frozen.

Assay
In the presence of the coenzyme, pyridoxyl-5-phosphate (P-5-P), AST catalyzes the reaction seen in Figure 10.4. This coenzyme is not consumed during the course of the reaction but functions to accept the amino group. The amino group is then transferred to the keto acid; therefore, the amino form of the coenzyme does not appear as a reaction product.

In the laboratory, aspartic acid and α-ketoglutaric acid are used as substrates. The oxalacetic acid (product) is determined colorimetrically as a phenylhydrazone or diazonium salt.

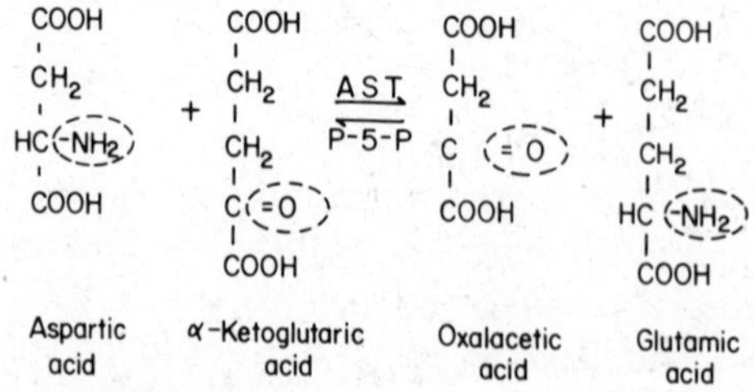

Figure 10.4. Reaction catalyzed by AST illustrating the transfer of an amino group from aspartic acid (amino acid donor) to alpha-ketoglutaric acid (acceptor keto acid with a carbon skeleton of an amino acid) resulting in the formation of a new amino acid (glutamic acid) and a new keto acid (oxalacetic acid).

A kinetic method has been devised using a coupled reaction. The enzyme, malate dehydrogenase (MDH), is used as a reagent:

$$\text{oxalacetic acid} + \text{NADH} + \text{H}^+ \overset{\text{MDH}}{\rightleftharpoons} \text{malate} + \text{NAD}^+$$

The reaction is monitored at 340 nm as NADH is oxidized to NAD^+; the decrease in absorbance per unit of time is directly proportional to AST activity.

Clinical Significance

AST was formerly one of a group of enzymes (AST, LD, CK) used to evaluate those patients with suspected myocardial infarctions. The LD and CK isoenzyme studies, which offer greater specificity and sensitivity, have now essentially replaced these total enzyme studies. Since AST is present in high concentration in heart muscle, AST may rise to very high levels following an AMI. The rise begins about 8–12 hours following the onset of pain and reaches levels two to three times the upper limit of normal. Peak activity usually occurs 24 hours after the symptoms begin. AST will return to normal in four to six days unless a new infarct occurs. The serum level is roughly proportional to the severity of the damage.

AST is now used primarily to evaluate hepatic disease. AST may elevate to very high levels in parenchymal liver disease, such as viral hepatitis, toxic hepatitis, and infectious mononucleosis. AST will also become elevated in cholestatic disease (both intrahepatic and extrahepatic cholestasis), but the extent of elevation is less than with those disorders involving necrosis (parenchymal disease). In cirrhosis, AST may be elevated or in the upper range of normal depending on the status of the cirrhotic process. Elevations of AST may also occur in either primary or metastatic carcinoma of the liver.

Skeletal muscle contains high levels of AST, and the enzyme is elevated in progressive muscular dystrophy and dermatomyositis, but generally remains normal in muscular disease of neurogenic origin. Muscle trauma may also elevate AST.

Alanine Aminotransferase (ALT, GPT)
L-Alanine : 2-Oxoglutarate Aminotransferase; EC 2.6.1.2

Source

The highest concentration of ALT occurs in the liver, followed by the kidney, heart, skeletal muscle, and pancreas. The concentration in the liver is over twice the concentration found in the kidney and six or more times that found in the heart, skeletal muscle, and pancreas.

Specimen Requirements

The specimen of choice is serum free from hemolysis and immediately separated from cells. Erythrocytes contain five to seven times the ALT content of serum; therefore, the presence of hemolysis will falsely elevate concentration values. The enzyme may be stored in the refrigerator but should be frozen for long periods of storage.

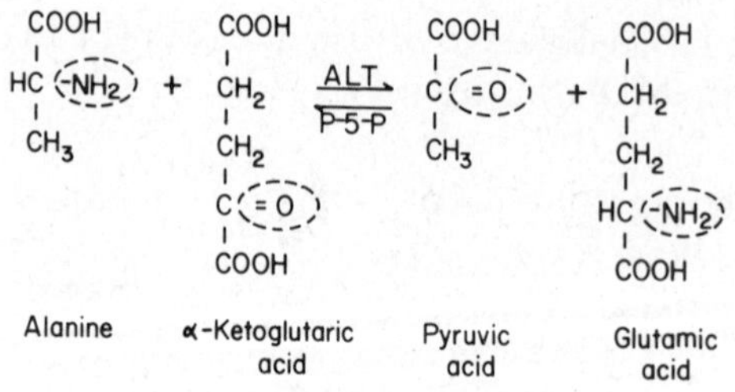

Figure 10.5. Reaction catalyzed by ALT.

Assay

In the presence of the coenzyme, pyridoxal-5-phosphate (P-5-P), ALT catalyzes the reaction seen in Figure 10.5. As in the AST reaction, the coenzyme is not consumed and is used over again.

For endpoint or colorimetric analysis, the product, pyruvic acid, can be coupled with dinitrophenylhydrazine to form a colored hydrazone.

A kinetic method has been devised using a coupled reaction. The enzyme lactate dehydrogenase (LD) is used as a reagent:

$$\text{pyruvate} + \text{NADH} + \text{H}^+ \overset{\text{LD}}{\rightleftharpoons} \text{lactate} + \text{NAD}^+$$

The reaction is monitored at 340 nm as NADH is oxidized to NAD^+; the decrease in absorbance per unit of time is directly proportional to ALT activity.

Clinical Significance

ALT is used primarily to evaluate hepatic disease. In parenchymal liver disease, ALT values frequently exceed AST values. ALT will be elevated in other forms of liver disease, such as obstructive liver disease and in cirrhosis; in these diseases, however, AST generally exceeds ALT levels. ALT is considered a more liver specific enzyme than is AST.

CREATINE KINASE (CK, CPK)
ADENOSINE TRIPHOSPHATE: CREATINE N-PHOSPHOTRANSFERASE; EC 2.7.3.2

Function

Creatine kinase or creatine phosphokinase catalyzes the reversible transfer of a phosphate group from ATP to creatine, forming creatine phosphate. Creatine phosphate serves to store energy in muscle. When energy is needed for muscle contraction, the phosphate from creatine phosphate can be transferred to ADP to form the high-energy compound ATP.

Source

Creatine kinase is a dimer (composed of two polypeptide chains). The subunits, designated M (muscle) and B (brain), combine to form three isoenzymes, which are (1) BB or CK-1; (2) MB or CK-2; and (3) MM or CK-3. CK-1 is found in high concentration in the brain, lung, bladder, and bowel. The CK-1 "brain form" is seldom found in the serum, because it cannot cross the blood brain barrier. CK-2, once considered specific for heart muscle, can also be found in varying small quantities in skeletal muscle. CK-3 is found in high concentration in skeletal muscle and heart.

Specimen Requirements

The specimen of choice is nonhemolyzed serum that has been separated from cells as soon as possible after collection. CK is heat labile, with CK-2 being the most fragile even at room temperature. If not analyzed immediately, the sample should be refrigerated; and if not analyzed within 24 hours, it should be frozen.

Assay

Creatine kinase catalyzes the following reaction; magnesium is required as an obligate activator:

$$\text{creatine phosphate} + \text{ADP} \underset{\text{Mg}^{2+}}{\overset{\text{CK}}{\rightleftharpoons}} \text{creatine} + \text{ATP}$$

Two additional coupled reactions are needed to measure the enzyme kinetically. A product of each reaction is used as a substrate for additional enzymes used as reagents:

$$\text{ATP} + \text{glucose} \overset{\text{hexokinase}}{\rightleftharpoons} \text{ADP} + \text{glucose-6-phosphate}$$

$$\text{glucose-6-phosphate} + \text{NADP}^+ \overset{\text{GPD}}{\rightleftharpoons} \text{6-phosphogluconate} + \text{NADPH} + \text{H}^+$$

The increase in absorbance is measured at 340 nm as NADP^+ is reduced to form NADPH. The absorbance increase is directly proportional to the activity of CK.

To determine the enzyme by end-point or colorimetric methods, an additional tetrazolium indicator reaction is needed. The tetrazolium salt is reduced to form a colored formazan dye:

$$\text{NADPH} + \text{tetrazolium (colorless)} \overset{\text{diaphorase}}{\rightleftharpoons} \text{NADP}^+ + \text{formazan (colored)}$$

Isoenzyme Separation

Electrophoresis is still considered the reference method for the separation of CK isoenzymes. Either cellulose acetate or agarose can be used as the electrophoretic support media. Following separation, the plate is overlaid with the reagents described above to form either a fluorescent product (NADPH) or a colored formazan. The fluorescent product can be visually inspected under a UV light source or scanned fluorometrically. The colored formazan can also be scanned without an ultraviolet scanner.

Other methods of separation include separation by ion-exchange chromatography, immunological inhibition, selective activation, and immunoassay methods.

The ion-exchange method requires stepwise elutions, is technique dependent, and cannot separate CK-2 and CK-1 well.

The immunological inhibition technique measures residual enzymatic activity of a sample after it has been incubated with anti-CK-3 antibody. The CK-3 antibody will inhibit any M subunit activity. This assay is more sensitive to CK-1 (BB) than to CK-2 (MB), and false positives are possible due to the presence of CK-1.

Selective activation relies on two sulfhydryl activators to differentiate CK isoenzymes. Total CK is activated with dithiothreitol (DTT); CK-3 is activated with reduced glutathione (GSH). The difference between the total CK minus CK-3 is equal to the CK-2 activity present, assuming that CK-1 is absent from the serum. A precise analytical system is needed to determine small differences between two large numbers.

A radioisotope method uses anti-BB antibody, which has specificity for BB (CK-1) and MB (CK-2) isoenzymes. It will not cross-react with MM. However, it cannot distinguish between MB and BB and would not be reliable when BB is present.

A immunoradiometric sandwich assay (IRMA) is also available that is specific for CK-MB. Two monoclonal antibodies are used, one specific for the B subunit and one for the M subunit. Anti-CK-B is attached to a solid phase (plastic bead). The sample is incubated with anti-CK-B. Both CK-MB and CK-BB will combine with the anti-CK-B and will be retained on the solid support. The liquid, containing mostly CK-MM, is removed. In the second step, ^{125}I-labeled antibody to CK-M reacts with the M subunit of CK-MB. CK-BB has no M subunit and will not bind labeled anti-CK-M. The unreacted liquid is then removed. The tubes are counted in a gamma counter. The amount of radioactivity is directly proportional to the concentration of CK-MB.

A immunoenzymetric sandwich procedure is also available. The test is similar to the immunoradiometric sandwich procedure; however, the antibody label is an enzyme rather than a radioisotope. In this procedure, both monoclonal antibodies are allowed to react with the B and M subunits. The unreacted material is then removed. Substrate is added and enzyme activity is measured. The enzyme activity is directly proportional to the concentration of CK-MB.

Clinical Significance

The presence of the isoenzyme CK-2 (MB) is most often used as an indicator of myocardial damage. In cases of acute myocardial infarction, CK-2 appears approximately 2–12 hours after the onset of chest pain and usually disappears after 30 hours but may persist for as long as 72 hours. Because of this transient rise, specimens for isoenzyme studies should be drawn within 24 hours after the onset of symptoms. The presence of CK-2 in the serum is highly specific for myocardium. Small quantities, however, may be found in skeletal muscle. When a patient presents with chest pain and CK-2 fails to appear within 48 hours, AMI can be ruled out with 100% accuracy. CK-3 is also found in high concentration in heart muscle (heart muscle contains 60% CK-3 and 40% CK-2) and will also become elevated following an AMI. Creatine kinase is rapidly cleared from the blood, and CK-3 levels return to normal in three to five days.

Table 10.2. Rise and Fall of the Heart Enzymes Following an Acute
Myocardial Infarction

Enzyme	Rise	Peak	Return to Normal
CK-2 (MB)	4–8 hours	18–24 hours	By 30 hours usually, but to 72 hours
CK-3 (MM)	4–8 hours		3–5 days
AST	8–12 hours	24 hours	4–6 days
LD	12–24 hours	72 hours (flip occurs 12–24 hours, usually)	7–12 days

Combined interpretation of CK and LD isoenzyme studies permit greater test
sensitivity and specificity. The diagnosis of an AMI can be made with 100%
confidence in those individuals who demonstrate CK-2 and a flipped-LD pattern in
samples drawn in the first 48 hours following an acute episode. The presence of
CK-2 with no flipped pattern may indicate an AMI, but coronary insufficiency and
myocardial ischemia also demonstrate this pattern. Table 10.2 summarizes the
rise and fall of enzymes following an acute myocardial infarction.

Increased CK-3 activity is present in a wide variety of neuromuscular diseases.
Several of these disorders (Duchenne muscular dystrophy, polymyositis) also
liberate CK-2 into the serum. CK-3 will also become elevated following vigorous
exercise or following intramuscular injections.

CK-1 (BB), if found in the blood, is believed to originate from the bowel wall,
lung, or tumor. CK-1 originating from the brain does not cross the blood brain
barrier. The appearance of CK-1 in the serum is not associated with any particular
disease state.

There are two other CK isoenzymes that have been reported, a mitochondrial
type, which migrates cathodically to the CK-3 band, called CK-MIT and a band
that appears between CK-3 and CK-2, called "macro" CK. The mitochondrial band
may occur when the extent of tissue damage is high. It has been found in cases of
malignant tumor or in association with cardiac abnormalities. Its appearance suggests a grave prognosis. The "macro" CK is CK-BB bound to IgG. It usually
occurs in elderly women and is not associated with any specific disease. Most
often, CK-MIT and "macro" CK are found when total CK is within normal limits
or is only slightly elevated.

Case History, Acute Myocardial Infarction

A 73 year-old white male was scheduled for an elective coronary artery bypass
graft surgery. Prior to his hospital admission date, he suffered acute chest pain
and was admitted to the hospital. Enzyme studies and isoenzyme patterns on
cellulose acetate are given in Figure 10.6.

Discussion
The presence of the MB band is determined from visual inspection of the electrophoresis strip, as well as by densitometry tracing. The admission CK isoenzyme
pattern was considered to be negative for the MB band; the small peak in the MB
area was considered to be an artifact. The LD-flip did occur with this patient (i.e.,

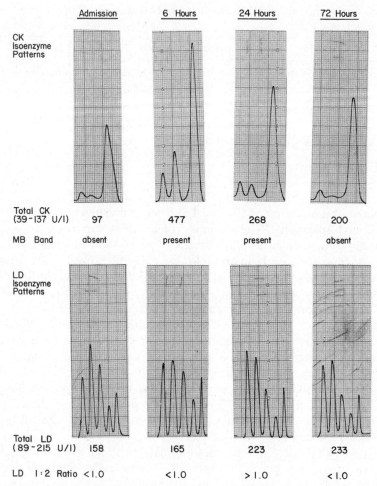

	Admission	6 Hours	24 Hours	72 Hours
CK Isoenzyme Patterns				
Total CK (39-137 U/l)	97	477	268	200
MB Band	absent	present	present	absent
LD Isoenzyme Patterns				
Total LD (89-215 U/l)	158	165	223	233
LD 1:2 Ratio	<1.0	<1.0	>1.0	<1.0

Figure 10.6. Myocardial infarction case-history CK and LD isoenzyme patterns demonstrating the presence of CK-2 (MB) band and flipped-LD (LD-1 greater than LD-2).

LD-1 greater than LD-2); however, an increase in LD-1 fraction is considered to be significant even if a flip does not occur. The ratio of LD-1/total LD has been proposed to replace the LD-1/LD-2 ratio in order to provide a better criterion for evaluation. CK is rapidly cleared from the blood. Note that by day three the total CK is returning to normal, and the MB band is considered absent by day three. The timing in the collection of samples is important in order to provide the physician with the maximum amount of information from these determinations.

In this patient, the elevated CK and the isoenzyme patterns (that is, the presence of MB and flipped-LD) suggest myocardial infarction. The LD patterns may also show an increase in LD-5, a finding associated with hepatic congestion, which may occur following compromised blood delivery to the liver. The electrocardiogram was abnormal with changes that correlated with severe ischemia. It was felt that the patient had experienced a definite infarct, and the elective surgery was postponed. A repeat cardiac catheterization was performed 11 days postadmission. The results showed little change from the previous catheterization performed during the initial pre-op work up. Triple saphenous vein coronary artery bypass graft surgery was performed. The patient tolerated the procedure well and was sent home.

GAMMA GLUTAMYLTRANSFERASE (GGT)
γ-GLUTAMYL-PEPTIDE: AMINO ACID γ-GLUTAMYLTRANSFERASE;
EC 2.3.2.2

Function

The primary physiological role of GGT appears to be the hydrolysis of glutathione (tripeptide) into the products L-glutamate and a dipeptide, cysteinyl glycine, which is further broken down into cysteine and glycine. These three amino acids are then available for resynthesis of glutathione. It is not known why the breakdown and resynthesis of glutathione is needed in the body. GGT is classified as a transferase enzyme, because it transfers the gamma-glutamyl group from peptides and compounds that contain it to some acceptor. When water is the acceptor, hydrolysis takes place. Other acceptors may include the substrate itself or some amino acid or peptide.

Source

GGT is found in many tissues. The highest levels of GGT are found in the brush borders of the proximal renal tubules, followed by the liver and then pancreas. GGT is found primarily on the outer surfaces of cell membranes. In the liver, it is strongly localized in the canilicular portions of the hepatocyte membrane. It is also present on the plasma membrane of epithelial cells that line the bile ducts, although to a lesser extent. The GGT in serum is thought to originate from the liver.

Specimen Requirements

Serum GGT is stable and can be stored in the refrigerator for several days without loss of activity. For long periods of storage, the samples should be frozen.

Assay

The reaction sequence of the reference method for GGT is given below. The gamma-glutamyl-p-nitroaniline is the gamma-glytamyl donor substrate; the glycylglycine is the acceptor substrate and also serves as the buffer:

$$\gamma\text{-glutamyl-}p\text{-nitroaniline} + \text{glycylglycine} \xrightarrow{\text{GGT}} \gamma\text{-glutamylglycylglycine} + p\text{-nitroaniline}$$

The end product, p-nitroaniline, is yellow in color. The rate of formation of the yellow end product is monitored at 405 nm, and the increase in absorbance is directly proportional to GGT activity.

Clinical Significance

GGT is elevated in all forms of liver disease, but is considered a cholestasis-indicating enzyme. The highest levels are associated with either intra- or extrahepatic cholestasis. It is considered more sensitive than either alkaline phosphatase or the transaminases in detecting cholestasis.

The measurement of GGT has also been proposed as a screening test to measure alcohol consumption. Chronic alcohol or drug consumption induces proliferation of the endoplasmic reticulum of liver cells. When this occurs, the microsomal enzymes that metabolize drugs or alcohol, as well as GGT, are induced. Some of the newly synthesized GGT is secreted into the circulation. It was found that GGT will increase in extended chronic alcohol consumption, but that normal social drinking is unlikely to produce abnormal GGT levels in healthy persons. When the alcohol consumption is reduced, GGT falls; therefore, GGT could also be used as a test to monitor treatment progress.

GGT is also considered to be a more reliable test to detect liver disease in children and adolescents than is alkaline phosphatase. In children up to six months of age, GGT levels are higher than adult levels, because immature livers have higher concentrations of GGT. However, after six months of age, the levels remain fairly constant. Alkaline phosphatase levels do not show stability in children and adolescents, because osteoblasts, which make alkaline phosphatase, are quite active during growth periods. As a result of this instability, alkaline phosphatase levels may vary and are higher throughout childhood than are the levels in adults. This variability may create difficulties in pediatric test interpretation.

GGT levels are sex dependent—males have higher levels than do females. Extreme obesity produces values above the reference range. Various medications (i.e., anticonvulsants, tricyclic antidepressants, tranquilizers, barbiturates) when taken chronically also cause elevations.

HYDROLASE ENZYMES

PHOSPHATASES

The phosphatases are a group of related enzymes that have low substrate specificity and are characterized by their ability to hydrolyze a large variety of organic phosphate esters, forming an alcohol and inorganic phosphorus (Figure 10.7). The phosphatases may also be called phosphomonoesterases. In the laboratory, two types are clinically significant: (1) alkaline phosphatase and (2) acid phosphatase.

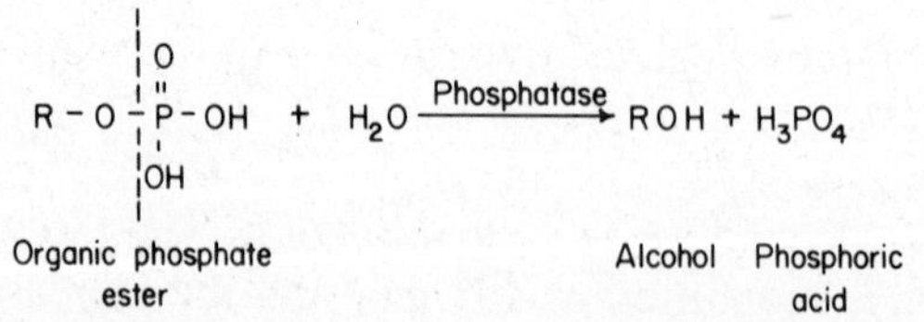

$$R-O-\overset{\overset{\displaystyle O}{\|}}{\underset{\underset{\displaystyle OH}{|}}{P}}-OH \ + \ H_2O \ \xrightarrow{\text{Phosphatase}} \ ROH + H_3PO_4$$

Organic phosphate   Alcohol   Phosphoric
ester   acid

Figure 10.7. Reaction catalyzed by phosphatase enzymes.

Alkaline Phosphatase (ALP)
Orthophosphoric Acid Monoester Phosphohydrolase; EC 3.1.3.1

Function

Except in bone, histochemical methods have demonstrated that alkaline phosphatase is located mainly at the surface or just below cell membranes. It is, therefore, believed that alkaline phosphatase may be involved with the transport of substances into or out of cells. In bone, alkaline phosphatase is believed to aid in the precipitation of calcium phosphate, since alkaline phosphatase activity is intensified during periods of active bone growth.

Source

Alkaline phosphatase is widespread in many tissues, but the highest concentrations occur in bone, liver, intestine, placenta, and kidney tubules. The alkaline phosphatase in normal serum is thought to be derived from the liver, bone, intestine, and, sometimes, placenta. Renal alkaline phosphate is found in urine but not in serum.

Specimen Requirements

Fasting serum, free from hemolysis, is the specimen of choice. Alkaline phosphatase activity appears to increase at both room temperature and under refrigeration, with increases of up to 30%. In some frozen serum, ALP activity will increase with time, but in others, there is a decrease in activity. If frozen serum is used, it is recommended that the sample be thawed and then stored in the refrigerator for 24 hours before assay. Alkaline phosphatase is unique, as it is the only enzyme that increases in activity upon standing. Persons of B or O blood group who are secretors may show increases in the intestinal ALP following a fatty meal. In these persons, this ALP may account for about 20% of total activity.

Assay

Since the natural substrate of the enzyme is not known, many substrates have been used in analytical methods. Alkaline phosphatase exhibits optimal activity at a pH of about 10.0. One colorimetric or end-point procedure uses thymolphthalein monophosphate as the substrate. After incubation, alkaline carbonate (color developer) is added, which stops the reaction by raising the pH and converts the liberated thymolphthalein into its blue form. The absorbance of the colored product is directly proportional to ALP activity; magnesium is needed as an activator:

$$\text{thymolphthalein monophosphate} \xrightarrow[\text{Mg}^{2+}]{\text{ALP}} \text{thymolphthalein}$$

$$+ \text{ inorganic phosphate}$$

$$\text{thymolphthalein (colorless)} \xrightarrow{\text{color developer}} \text{thymolphthalein (colored)}$$

A kinetic method uses *p*-nitrophenyl phosphate (PNPP) as a substrate. PNPP does not absorb at the wavelength chosen to read the test (405 nm). As ALP cleaves phosphate, free *p*-nitrophenol (PNP) is formed and converted to a yellow color at an alkaline pH. Since the optimum pH for ALP is about 10.0, the rate of formation of the yellow color can be used to measure enzyme activity:

$$PNPP + H_2O \xrightarrow[Mg^{2+}]{ALP} PNP + \text{inorganic phosphate}$$

Isoenzyme Separation

At least nine isoenzymes of alkaline phosphatase have been identified in human serum. These are fast liver (preliver), liver, bone, placental, Regan, Nagao, renal, intestinal, and PA. Methodologies for separating and identifying isoenzymes include electrophoresis, heat inactivation, use of chemical denaturation or inhibition, ion-exchange chromatography, and immunological techniques. The last two methods are not practical for routine laboratory use.

The properties of several of the ALP isoenzymes show little variation (especially liver and bone); therefore, separation is difficult, and more than one technique is frequently needed for discrimination.

Alkaline phosphatase enzymes do not separate well on electrophoresis. A schematic representation of separation of all nine isoenzymes on cellulose acetate is seen in Figure 10.8.

The fast liver is the most anodal fraction, followed by the liver isoenzyme. The bone isoenzyme is usually diffuse and may overlap the liver. Placental, Regan, Nagao, and renal isoenzymes also migrate in the bone area but do not exhibit the wide diffuse band exhibited by bone.

Serum from normal individuals contains small amounts of ALP derived from liver, bone, and intestine. Adults under the age of 50 usually have primarily liver ALP. Adults past the age of 50 will usually show increases in the bone isoenzyme due to the skeletal changes that occur with age. Individuals with B or O type blood will show an intestinal band following a fatty meal. Placental ALP may be seen in pregnant women as early as the latter part of the first trimester to as late as the termination of pregnancy. Following termination, ALP may remain elevated for up to one month longer. Children have a large bone band with little or no liver and intestinal forms.

Regan, Nagao, fast liver, and PA are all abnormal bands associated with neoplasms. None are specific tumor markers. These isoenzymes may be present when the total alkaline phosphatase is within the reference range. Fast liver has been suggested as a diagnostic tool that could be used to identify metastatic disease of the liver. Fast liver is also found in patients with viral hepatitis, alcoholic cirrhosis, and other liver diseases.

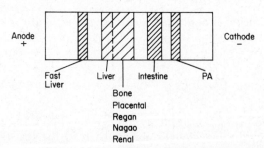

Figure 10.8. Schematic representation of the separation of the isoenzymes of alkaline phosphatase on cellulose acetate.

For valid interpretation of electrophoretic patterns, reference serum of known isoenzyme composition should be used in conjunction with the unknown specimens. The isoenzyme markers may then be used as an aid in band identification.

Starch, sugar, polyacrylamide gels, and cellulose acetate have all been used as electrophoretic support media.

Separation by heat is based on the variance in thermolabile properties of isoenzymes. Placenta, Nagao, and Regan isoenzymes are the most heat stable, followed in order by intestine and liver. Bone is the most heat labile (placenta, Nagao, Regan > intestine > liver > bone). Heat separation is most often performed to separate bone and liver isoenzymes. Separation by heat must be carefully controlled to achieve accurate results. The test tube size, sample size, water bath temperature, and timing are critical. Precisely at the end of heating, the sample should be immersed in ice. It is recommended that the test be performed in the following way to give reproducible results:

1. The serum sample is split into two aliquots.
2. A 0.5 mL aliquot is placed into a 12×75 mm test tube.
3. The test tube is immersed in a 56° water bath for 10 minutes. The temperature and timing should be carefully controlled.
4. After heating, the sample is immediately immersed in ice. The heated and unheated samples are then assayed for ALP activity or electrophoresis is performed on both samples (which are innoculated side by side on the same electrophoresis plate).

Bone alkaline phosphatase is destroyed at 56°C (90–100% inactivated), while liver is not affected as much (approximately 30% destroyed). If the alkaline phosphatase originated from the bone, the isoenzyme pattern of the heated sample should show a marked decrease or absence of the bone band as compared to the unheated sample. If electrophoresis is not used, the interpretation is based on the residual enzyme activity that remains after heating. If the residual activity is 20% or less of the total, the isoenzyme is probably of bone origin. If residual activity is 25–55%, the isoenzyme is probably of liver origin.

Heat separation may also be used to identify the presence of the Regan tumor marker. The Regan isoenzyme is extremely heat stable. When the sample is heated to 65°C for 30 minutes, intestine, liver, and bone isoenzymes will be destroyed, but the Regan isoenzyme remains stable. The placenta isoenzyme is also stable at this temperature. The Regan isoenzyme has been called the carcinoplacental isoenzyme because of its similarities to the placental isoenzyme. The Regan isoenzyme is produced by the tumor itself. This isoenzyme has been found in several types of malignancies, but the highest association is in ovarian and gynecologic cancers.

Urea denaturation has also been used in separation, since the isoenzymes of ALP are selectively denatured by urea. The bone isoenzyme is the most susceptible to denaturation; liver isoenzyme is of intermediate susceptibility; and intestine, Regan, and placenta isoenzymes are more resistant. No agreement has been reached as to the concentration of urea to use or to the duration of treatment. After treatment with urea, both samples (i.e., treated and untreated) are assayed for ALP activity or electrophoresis is performed (on both samples simultaneously) in order to determine residual remaining activity.

Other investigators have developed methods based on preferential inhibition of isoenzymes by chemicals. Inhibitors that have been used include L-phenylalanine, L-tryptophan, L-homoarginine, L-leucine, levamisole, and imidazole. Only the L-isomers of the amino acids are active, and the inhibition is noncompetitive. L-Phenylalanine inhibits intestinal, placental, Regan, and Nagao isoenzymes. To differentiate between Regan and Nagao, L-leucine is used. Nagao is sensitive to L-leucine, while Regan is not affected. The separation of ALP isoenzymes is summarized in Table 10.3.

Clinical Significance

Alkaline phosphatase is used primarily in the diagnosis of bone diseases associated with osteoblastic activity and in hepatic disease associated with biliary obstruction.

Paget's disease of the bone will produce the highest levels of serum alkaline phosphatase. Serum activity of this enzyme correlates well with hydroxyproline excretion in the urine, which is used as an indicator of bone turnover. Cancer of the bone of the blastic type is associated with high enzyme levels. Bony metastases of the blastic type are common with cancer of the prostate, breast, thyroid, and pancreas.

Individuals with primary hyperparathyroidism may or may not have increased serum alkaline phosphatase activity. The test is of little diagnostic value in this disease. Secondary hyperparathyroidism may produce elevated levels. The degree of elevation is dependent on the extent of skeletal involvement.

Moderate elevations of ALP are seen in osteomalacia and rickets, but bone changes may be present in the absence of serum enzyme changes. Thus, ALP is not recommended as a screening test for this disorder. In osteoporosis, ALP levels are normal.

ALP is a cholestasis-indicating enzyme. It is believed that obstruction of the biliary tree induces cells to increase the production of ALP, thus increasing the amount of enzyme that enters the blood. The increase is more marked in extrahe-

Table 10.3. The Separation of ALP Isoenzymes using Heat and Chemical Inhibition Techniques

Separation Techniques	Alkaline Phosphatase Isoenzymes				
	Placenta	Regan or Nagao	Intestine	Liver	Bone
56°C for 10 minutes	Stable	Stable	Stable	Stable	Destroyed
65°C for 30 minutes	Stable	Stable	Destroyed	Destroyed	Destroyed
Urea inhibition	Most resistant	Most resistant	Most resistant	Intermediate susceptibility	Most susceptible
L-Phenylalanine inhibition	Inhibits	Inhibits	Inhibits	Little effect	Little effect

patic obstruction (stones, stricture, tumor at head of pancreas) than in intrahepatic obstruction. Enzyme levels rise in proportion to the severity of the obstruction. Toxic agents whose principal target is the liver ductal system will also increase ALP. Space-occupying lesions of the liver are also associated with elevated levels.

ALP levels may be normal or only modestly elevated in liver diseases associated with parenchymal damage (i.e., hepatitis or toxic agents whose principal target is parenchymal cells). ALP may aid in differentiating hepatitis from inherited metabolic disease (i.e., Gilberts, Rotor's, Dubin–Johnson). The ALP activity will be normal in the metabolic diseases, unless hepatocellular or fibrosis is present, but may be increased in hepatitis.

Serum ALP levels are higher in children than in adults. Alkaline phosphatase levels also fluctuate in children in association with periods of rapid bone growth. There are two peak bone periods; the first occurring before the first postnatal year, and the second occurring during puberty. Peak serum activity during puberty correlates with sexual maturity rather than with chronological age.

Acid Phosphatase (ACP)
Orthophosphoric Monoester Phosphohydrolase; EC 3.1.3.2

Function
The acid phosphatases refer to a group of similar or related enzymes whose optimal activity occurs at a pH below 7.0. The acid phosphatases present in liver, serum, and prostate exhibit optimal activity at pH 4.9–5.0, while the optimal activity for the red blood cell isoenzyme occurs at pH 5.5–6.0.

The acid phosphatases are intracellular enzymes, most of which are located inside cell organelles, principally lysosomes, and are believed to function in digestive processes. ACP may function in the autodigestion of dying cells or it may aid in the breakdown of large molecules that enter the cell. Normally the enzymes in lysosomes are contained within the lysosomal membrane, and none escape into the blood. The phosphatase of prostatic origin (PAP) is found in the epithelial cells of the prostate gland and is secreted into prostatic secretions under the influence of male sex hormones. The prostatic isoenzyme is considered organ specific.

Source
Acid phosphatase is found in many cells with significant amounts present in prostate, liver, spleen, red blood cells, platelets, and bone marrow.

The acid phosphatase in normal serum is believed to be derived from platelets, osteoclasts, erythrocytes, and leukocytes. The prostatic acid phosphatase is normally confined to the prostate. When neoplasia is present, it is believed that the prostatic cells become leaky, permitting the enzyme to enter the blood. The acid phosphatase of diagnostic importance is the prostatic isoenzyme.

Specimen Requirements
The samples for acid phosphatase require special processing because ACP is very unstable. The serum should be separated from the cells as soon as possible and then refrigerated. If analysis is to be delayed, acidify the serum with disodium citrate or acetic acid to a pH below 6.5 and freeze. No hemolysis is permitted as red blood cells contain acid phosphatase.

Since RIA procedures measure only the immunological properties of the molecule, the problem with sample decay is obviated in these procedures. However, those procedures based on both immunological and enzymatic properties require the same attention to sample handling as described above.

Assay

Acid phosphatase levels are of diagnostic importance in the diagnosis of prostatic cancer and in monitoring the response of the patient to therapy. Therefore, it is important to differentiate between acid phosphatase of prostatic origin and that derived from other sources. A common differential procedure is to use chemical inhibition. Acid phosphatase of prostatic origin is inhibited by the L-tartrate ion (almost 99% of the PAP activity is inhibited by tartrate). Chemical inhibition is performed as follows:

1. Two tubes are prepared for each sample. One tube contains a substrate buffer mixture only, and the second tube contains the substrate buffer mixture plus the tartrate ion.
2. The sample is added to each tube, and the acid phosphatase activity is determined in each tube.
3. The tube without the tartrate ion will represent acid phosphatase activity from all forms of acid phosphatase, that is, platelets, red blood cells, and prostate. The tube with the tartrate ion will represent acid phosphatase activity from platelets and red blood cells. The difference between assay values is equal to the prostatic acid phosphatase:

$$\text{prostatic acid phosphatase} = \text{total acid phosphatase} - \text{tartrate assay}$$

An alternative method is to inhibit red blood cell acid phosphatase. Red blood cell acid phosphatase is inhibited by formaldehyde or cupric ions.

An end-point or colorimetric procedure using the substrate, *p*-nitrophenylphosphate (PNPP), is shown below. The substrate is the same as that used for the assay of alkaline phosphatase. The acid pH required for enzyme activity requires alkalinization after incubation for color development:

$$\text{PNPP} + H_2O \xrightarrow{\text{ACP}} \text{PNP} + \text{inorganic phosphate}$$

$$\text{PNP (colorless)} \xrightarrow{\text{OH}^-} \text{PNP (colored)}$$

Another end-point method employs the substrate, thymolphthalein monophosphate (TMP), which is considered to be a more specific substrate for the prostatic isoenzyme. At the wavelength chosen to read the test, TMP does not absorb, while the liberated product, thymolphthalein, does absorb. The reaction is terminated with NaOH, which increases the color intensity of the chromophore:

$$\text{TMP} \xrightarrow{\text{ACP}} \text{thymolphthalein} + \text{inorganic phosphate}$$

$$\text{thymolphthalein} \xrightarrow{\text{OH}^-} \text{thymolphthalein anion}$$

Measurement of Prostatic Acid Phosphatase

Three immunoassay tests have been developed to measure prostatic acid phosphatase: (1) radioimmunoassay (RIA); (2) counterimmunoelectrophoresis (CIEP); and (3) enzyme immunoassay (EIA).

Prostatic cancer is difficult to diagnose in its early and potentially curable stages. Immunochemical methodology was developed in the anticipation that methods which were more specific and sensitive than conventional colorimetric methods would be able to detect prostatic cancer in its curable stages. RIA methods measure only the immunoreactive portion of the PAP molecule and can directly quantitate the enzyme concentration. However, it has also been determined that the number of false positive and false negative test results obtained by RIA procedures render this test questionable in its ability to serve as a screening test for early detection of prostatic cancer.

Counterimmunoelectrophoretic methods measure both immunoreactivity and enzyme activity. It is believed that the false positive rate with CIEP methods are less than with RIA methods. Thus, CIEP is considered the preferred method by some chemists.

Enzyme immunoassay methods also measure both immunoreactivity and enzymatic activity.

Clinical Significance

In the United States, prostatic cancer is the second most common cancer in white males and is the most common cancer found in black males. The risk of developing prostatic cancer increases with age. Prostatic cancer is usually divided into four stages. In Stage A, the tumor is confined to the prostate and is not palpable; in Stage B, the tumor is localized in the prostate but can be detected by palpation; in Stage C, the tumor is invasive but confined to the pelvic area; in Stage D, the tumor has metastasized to bone and other distant organs. Staging the cancer is important because the proper treatment regimen is based on the extent of the disease. When the tumor is confined to the capsule (Stage A and B), the cancer may be surgically curable. Conventional colorimetric procedures lack sensitivity and can only detect elevations in acid phosphatase after the tumor is no longer confined to the capsule. Immunoassay methods, although more sensitive, are subject to false positives.

Presently, PAP immunoassay methods are used to confirm malignancy and to follow the response of the patient to therapy. Using PAP as a general screening test procedure is not recommended. The colorimetric or end-point methods are also used to follow the response of the patient to therapy.

Total acid phosphatase may also be increased in Paget's disease and other bone disorders, hepatobiliary disease, Gaucher's disease, and cancers of the breast and colon.

Since PAP occurs in high concentrations in prostatic secretions, it is frequently used in forensic medicine in the investigation of rape or similar offenses.

PSEUDOCHOLINESTERASE (CHS)
ACYLCHOLINE ACYL-HYDROLASE; EC 3.1.1.8

Function

The cholinesterases refer to a group of related enzymes that hydrolyze choline esters. Physiologically, the most important choline ester found in the body is acetylcholine, a neurotransmitter synthesized at nerve endings, which transmits nerve impulses from nerve to muscle fiber. Cholinesterase destroys acetylcholine after the impulse has been conducted, so that additional impulses may then be transmitted. If acetylcholine was not destroyed, the nerve would remain electrically charged and further conduction would not be possible.

Source

There are two types of cholinesterase enzymes found in the blood, each derived from different sources. The first type, acetylcholinesterase, specific cholinesterase, or true cholinesterase is found primarily in red blood cells, brain, lungs, spleen, and nerves, and only very small amounts occur in the serum. It is responsible for the degradation of neurotransmitted acetylcholine and reacts poorly with other choline esters.

The second type, butyrylcholinesterase, nonspecific cholinesterase, or pseudocholinesterase occurs in the serum as well as in brain, liver, pancreas, and heart. It has no known physiological role. Of the two enzymes, pseudocholinesterase is the enzyme of greater clinical interest and the one routinely measured in clinical laboratories.

Specimen Requirements

The required specimen for pseudocholinesterase is nonhemolyzed serum. For short periods of storage (up to one week), the serum may be stored under refrigeration. For long periods of storage, the sample should be frozen. Repeated freezing and thawing should be avoided.

Assay

Several methods are available for the measurement of cholinesterase activity. Most methods either estimate the amount of acid released by the hydrolysis of a choline ester, or measure the residual unhydrolyzed ester remaining following incubation with a cholinesterase source, or measure the release of choline or thiocholine following enzymatic hydrolysis of an amino alcohol ester (e.g., butyrylthiocholine). An example of the latter method follows.

Butyrylthiocholine is hydrolyzed by pseudocholinesterase releasing thiocholine. Thiocholine reduces a blue dye (2,6-dichlorophenolindophenol or DIP) to a colorless form. The decrease in absorbance at 600 nm is directly proportional to pseudocholinesterase activity:

$$\text{butyrylthiocholine} + H_2O \xrightarrow{\text{CHS}} \text{butyric acid} + \text{thiocholine}$$

$$2\ \text{thiocholine} + \text{DIP (colored)} \rightarrow \text{dithiocholine} + \text{DIP-}H_2 \text{ (colorless)}$$

Clinical Significance

Pseudocholinesterase activity is used as a presurgical screening test to detect those individuals who may be at risk following the use of muscle relaxants during surgical procedures. Susceptible individuals inherit genes that code for variant forms of pseudocholinesterase. These variant forms hydrolyze choline esters at a slower rate than does the cholinesterase found in normal persons. These individuals may exhibit unusual sensitivity to the drug succinylcholine, a muscle relaxant used in surgery that is similar to acetylcholine and is hydrolyzed by the cholinesterase enzymes. In homozygotes, the usual dose of succinylcholine may produce a paralytic apnea requiring artificial ventilation. Heterozygotes are not usually affected. Decreased pseudocholinesterase activity allows the anesthesiologist to identify those individuals at risk and to avoid the use of this muscle relaxant.

Inhibition studies may be used to confirm the presence of a variant form of pseudocholinesterase. Enzyme activity is determined in the presence of each of two inhibitors, dibucaine and fluoride ions. The degree of inhibition is expressed as a number. For example, the number 62 means that 62% of the enzyme activity is lost in the presence of the inhibitor. The response of the enzyme to the two inhibitors allows the enzyme phenotype to be determined.

The measurement of pseudocholinesterase may also be requested as a screening test to detect toxic exposure to organophosphate insecticides. Toxic effects with insecticides can occur by inhalation or by contact. After exposure to organophosphates, cholinesterase activity decreases and does not return to normal until new protein is synthesized by the liver.

Besides inhibition by organophosphates, blood cholinesterase enzymes are also inhibited by other substances. Cholinesterase inhibition by nonorganophosphorus compounds is generally reversible and disappears within a few hours following withdrawal of the inhibitor.

Pseudocholinesterase activity has also been used to measure hepatic parenchymal function; however, base values are needed for test interpretation (look for a decrease from base values). Other enzyme tests are available that measure parenchymal function as well or better.

Decreased pseudocholinesterase activity may also be found in pregnancy and a number of other unrelated conditions.

BIBLIOGRAPHY

Baillie, E. E. "CK isoenzymes: Part 1," *Laboratory Medicine* **10**(5), 267 (1979).

Baillie, E. E. "CK isoenzymes: Part 2," *Laboratory Medicine* **10**(6), 337 (1979).

Bates, H. M. "GGTP and Alcoholism: A sober look," *Laboratory Management,* 17–19 (Mar. 1981).

Bates, H. M. "Is your laboratory measuring prostatic acid phosphatase by immunoassay?" *Laboratory Management,* 11–16 (Feb. 1980).

Batsakis, J. G. "Serum alkaline phosphatase, refining an old test for the future," *Diagnostic Medicine* **5**(3), 25 (1982).

Biomedical Reference Laboratories. *Prostatic Acid Phosphatase.* Pamphlet, Burlington, N.C.

Choe, B. "New dimensions for prostatic acid phosphatase," *Diagnostic Medicine* **5**(3), 81 (1982).

Demers, L. M. "Prostatic acid phosphatase: Assay comparisons," *Laboratory Management,* 45–51 (Aug. 1981).

Galen, R. S. "Isoenzymes and myocardial infarction," *Diagnostic Medicine,* **1**(1), 40 (1978).

Galen, R. S. "Isoenzymes, which method to use," *Diagnostic Medicine* **1**(3), 42 (1978).

Galen, R. S., Gambino, S. R., et al. *Isoenzymes of CPK and LDH in Myocardial Infarction and Certain Other Diseases,* Cardiac Profile Booklet. Beaumont, TX: Helena Laboratories, 1975.

Hadden, D. M. and Prentiss, T. "Cardiac profiling by electrophoresis," *Laboratory Management,* 19–24 (May 1977).

Harper, H. A. *Review of Physiological Chemistry,* 15th ed. Los Altos, CA: Lange Medical Publications, 1975.

Lum, G. "Diagnostic value of CK isoenzymes," *Laboratory Management,* 14–16 (May 1977).

Pesce, M. A. "Interpreting CK isoenzyme patterns," *Laboratory Management,* 25–37 (Oct. 1982).

Shaw, L. M. "Keeping pace with a popular enzyme GGT," *Diagnostic Medicine* **5**(3), 59 (1982).

Shaw, L. M. "The GGT assay in chronic alcohol consumption," *Laboratory Management,* 56–63 (May 1982).

Sims, G. M. "Enzyme substrate optimization and the routine laboratory," *Lab World,* 11–14 (Jan. 1976).

Speicher, C. E. *Unexplained Elevation of Serum Alkaline Phosphatase.* Check Sample Program, CC-81-2, American Society of Clinical Pathologists, 1981.

Tietz, N. (ed). *Fundamentals of Clinical Chemistry,* 2nd ed. Saunders: Philadelphia, 1976.

Tomashefsky, P., Romas, N. A., Hsu, K. C. and Tannenbaum, M. "Prostatic acid phosphatase: Methods and their utility," *Laboratory Management,* 33–38 (May 1980).

Warren, B. M. *The Isoenzymes of Alkaline Phosphatase.* Beaumont, TX: Helena Laboratories, 1981.

Wills, J. H. "Blood cholinesterase: Assay methods and considerations," *Laboratory Management,* 53–64 (Apr. 1982).

POST-TEST

All Multiple choice questions can have one or more than one answer.

1. The following are statements regarding enzymes. Circle all true statements.
   a. Enzymes are proteins that catalyze reactions by supplying energy to activate a specific reaction.
   b. Enzymes are used up as the reaction proceeds.
   c. Most enzymes are reaction specific.
   d. Enzymes of clinical usefulness are organ specific.

2. Match the enzyme class with the reaction catalyzed.
   _____ 1. transferases
   _____ 2. hydrolases
   _____ 3. lyases

   a. nonhydrolytic enzymes that catalyze removal of specific groups from substrates

_____ 4. oxidoreductases
_____ 5. isomerases
_____ 6. ligases

b. enzymes that catalyze oxidore-
ductive reactions between two
substrates

c. enzymes that catalyze transfer
of a specific group other than
hydrogen between a pair of substrates

d. enzymes that catalyze the building of new compounds

e. enzymes that catalyze cleavage
of a variety of bonds by the addition of water

f. enzymes that catalyze the interconversion of isomers

3. When an enzyme is assayed under zero-order kinetics, this means that:
   a. the enzyme concentration exceeds substrate concentration
   b. the reaction rate is dependent on substrate concentration
   c. the reaction rate is dependent only on enzyme concentration
   d. the substrate concentration is in excess
   e. the reaction rate is proceeding at one-half maximal activity

4. K_m refers to:
   a. substrate concentration at which the reaction is proceeding at one-half
      maximal rate
   b. substrate concentration at which all enzyme is bound to substrate
   c. substrate concentration at which the amount of product formed is di-
      rectly proportional to enzyme concentration
   d. substrate concentration at which the reaction is proceeding at V_{max}

5. First-order kinetics means:
   a. the reaction rate is maximal
   b. the reaction rate is dependent on substrate concentration
   c. the reaction rate is proceeding at one-half maximal velocity
   d. the substrate concentration is not sufficient to bind all enzyme present

6. The Enzyme Commission recommends that enzymes be assayed at:
   a. 37°C
   b. 25°C
   c. 30°C
   d. 40°C

7. An enzyme procedure for LD follows: The LD substrate contains the following concentration of reagents:

Lithium lactate	125 mmol/L
NAD$^+$	4.2 mmol/L
Diaphorase	120 mmol/L
INT	0.6 mmol/L
tricine–carbonate buffer	50 mmol/L

The reactions are as follows:

$$\text{lactate} + \text{NAD}^+ \xrightarrow{\text{LD}} \text{pyruvate} + \text{NADH} + \text{H}^+$$

$$\text{NADH} + \text{INT} \xrightarrow{\text{diaphorase}} \text{INTH} + \text{NAD}^+$$

The product, NADH, produced in the first reaction, acts as a reducing agent. INT is reduced to form a colored end product. Diaphorase acts as an electron carrier agent. The procedure calls for a 10 minute incubation time at 37°C. Indicate how the following would affect test results. Use:

a. no change in activity
b. increased activity
c. decreased activity
d. change, but more information is required to determine the direction of change

Following each answer you choose, explain why the test results would be affected in the way indicated.

—————— 1. Water bath temperature at 38°C.
            Why? _____
—————— 2. Reaction stopped at 10½ minutes.
            Why? _____
—————— 3. NAD⁺ concentration 1.2 mmol/L.
            Why? _____
—————— 4. Lithium lactate prepared with a concentration of 130 mmol/L.
            Why? _____
—————— 5. Tricine–carbonate buffer concentration 25 mmol/L.
            Why? _____

8. Creatine kinase requires the presence of magnesium for activity. Magnesium is an example of:
   a. an activator
   b. a cofactor
   c. a coenzyme
   d. a cosubstrate

9. Lactate dehydrogenase requires the presence of NAD⁺ for activity. NAD⁺ is an example of:
   a. an activator
   b. a cofactor
   c. a coenzyme
   d. a cosubstrate

10. The following are statements regarding coenzymes. Circle all true statements.
    a. Coenzymes are water-soluble organic molecules.
    b. Vitamins may be components of coenzymes.

    c. ATP and pyridoxyl phosphate are examples of coenzymes that transfer groups other than hydrogen.

    d. NAD^+ and NADP are examples of coenzymes that transfer hydrogen.

    e. Coenzymes as cosubstrates are used up as the reaction proceeds.

    f. Enzymes that belong to reaction classes that transfer groups frequently require coenzymes.

11. Alkaline phosphatase enzymes catalyze the hydrolytic cleavage of a large variety of organic phosphate esters. This enzyme could therefore be said to exhibit:

    a. stero specificity

    b. absolute specificity

    c. group specificity

12. If an enzyme is able to catalyze a reaction with β-D-glucose but is not able to use α-D-glucose as a substrate, this enzyme could be said to exhibit:

    a. stero specificity

    b. absolute specificity

    c. group specificity

13. All isoenzymes of lactate dehydrogenase catalyze a single reaction, the reversible conversion of lactate to pyruvate. This enzyme could therefore be said to exhibit:

    a. stero specificity

    b. absolute specificity

    c. group specificity

14. Denaturation of an enzyme involves disruption of:

    a. primary structure

    b. secondary structure

    c. tertiary structure

    d. peptide bonds

15. An enzyme assay procedure was being evaluated. It was noted that the activity rate was inhibited by the presence of an end product of a coupled reaction. The inhibition was overcome by increasing the substrate concentration. From these data, you could conclude that the type of inhibition was:

    a. competitive inhibition

    b. noncompetitive inhibition

    c. inhibition by metal ions

16. When K_m is small, the affinity of the enzyme for substrate is _____ ; when K_m is large, the affinity of the enzyme for substrate is _____ .

    a. high

    b. low

17. Isoenzymes may be defined as:

    a. multiple molecular forms of the same enzyme

    b. enzymes that are found in both plasma and tissue

    c. enzymes that vary in amino acid content but catalyze the same reaction

    d. enzymes that migrate to the same position when electrophoresed at pH 8.6

18. Isoenzymes may be separated by:
    a. chemical inhibition
    b. stability to heat denaturation
    c. electrophoresis
    d. ion-exchange columns
    e. immunochemical methods

19. Match the enzyme with the cause for elevation. (Each may have more than one answer.)

    _____ 1. Alkaline phosphatase elevated in obstructive liver disease

    _____ 2. Creatine kinase increased after myocardial infarction

    _____ 3. Amylase elevated in persons with renal disease

    _____ 4. Alkaline phosphatase elevated in diseases associated with increased bone turnover

    _____ 5. Lactate dehydrogenase increased after myocardial infarction

    _____ 6. Acid phosphatase increased in prostatic carcinoma

    a. Failure to excrete an enzyme normally cleared by the kidney

    b. Blockage of a duct system through which an enzyme passes

    c. Release of enzymes from damaged cells

    d. Overproduction of an enzyme by a tumor

    e. Change in permeability allowing leakage of enzymes into circulation

    f. Increased synthesis of enzymes by normal cells with overflow into plasma

20. Plasma-specific enzymes are clinically significant when their blood level is _____ ; non-plasma-specific enzymes are clinically significant when their blood level is _____ .
    a. increased
    b. decreased

21. Indicate how the enzyme assay would be affected under the conditions described. Use ↑ (increased); ↓ (decreased); → (not affected).

    _____ 1. Sample for total acid phosphatase very slightly hemolyzed

    _____ 2. Alkaline phosphatase sample collected two days prior to assay and stored in the refrigerator

    _____ 3. Sample for total LD placed in the refrigerator overnight

    _____ 4. Sample for total LD collected in the evening, serum separated and left at room temperature for assay with the morning workload

    _____ 5. Sample for total acid phosphatase spun down, serum removed, pH adjusted to 6.0, and frozen

_____ 6. Sample for LD determination hemolyzed

_____ 7. Sample for AST hemolyzed

22. Three methods for assaying enzymes are listed below. Match the assay method with the procedure described.

a: continuous monitoring or kinetic

b: fixed time or end point

c: radioimmunoassay

_____ 1. In the determination of amylase, starch, as a substrate, and patient sample were incubated. The reaction was stopped and glucose, the end product, was then determined by a colorimetric procedure.

_____ 2. In the determination of lactate dehydrogenase, lactate and NAD^+ were incubated with patient sample. As HADH was formed, the change in absorbance per unit of time was monitored.

_____ 3. In the determination of prostatic acid phosphatase, patient sample, radioisotope-labeled prostatic acid phosphatase, and specific antibody were allowed to incubate. The radioactivity of the bound complex was then determined using a gamma counter.

23. Of the methods that measure activity rate, the assay method of choice is:
a. colorimetric or end point
b. kinetic or continuous monitoring

24. When reaction rates are monitored using coenzyme I, the form that absorbs at 340 nm is:
a. NAD^+
b. NADH
The form that fluoresces is:
a. NAD^+
b. NADH

25. An International Unit may be defined as the amount of product produced or substrate utilized at a given pH and temperature in terms of:
a. mmol/min/mL
b. mmol/mL/min
c. μmol/min/mL
d. mg/min/mL

26. Name the enzyme that catalyzes the reaction described.

_____ 1. This oxidoreductase enzyme, in the presence of NAD^+, reversibly catalyzes the conversion of lactate to pyruvate.

_____ 2. This hydrolase enzyme cleaves phosphate ester bonds at an acid pH.

      _____ 3. This transferase enzyme transfers an amino group from aspartic acid to a keto acid.

      _____ 4. This transferase enzyme transfers a phosphate group from ATP to creatine to form creatine phosphate and ADP.

      _____ 5. This hydrolase enzyme cleaves phosphate ester bonds at an alkaline pH.

      _____ 6. This transferase enzyme transfers an amino group from the amino acid alanine to a keto acid.

      _____ 7. This oxidoreductase enzyme transfers hydrogen from glucose-6-phosphate to $NADP^+$.

      _____ 8. This transferase enzyme transfers a gamma-glutamyl group to some acceptor.

27. A flipped-LD, that is LD-1 greater than LD-2, may be found:
    a. after a MI
    b. in hemolyzed samples
    c. in renal infarction
    d. in skeletal muscle disorders

28. The most abundant sources for LD-1 and LD-2 are:
    a. kidney
    b. RBC
    c. heart
    d. skeletal muscle
    e. liver

29. The most abundant sources for LD-4 and LD-5 are:
    a. kidney
    b. RBC
    c. heart
    d. skeletal muscle
    e. liver

30. The CK in normal persons is thought to be derived from:
    a. heart muscle
    b. brain
    c. liver
    d. skeletal muscle

31. A patient, with chest pain, entered the hospital. A cardiac profile was ordered. The total CK in the admission sample was normal; however, in subsequent samples, the total CK was elevated and the presence of the MB band was noted. The LD-flip did not occur in any of the samples. From these results, you conclude:
    a. these findings are consistent with an acute myocardial infarction
    b. these findings may be consistent with an acute myocardial infarction
    c. these findings may be consistent with cardiac ischemia
    d. these findings definitely rule out acute myocardial infarction

32. A patient entered the hospital complaining of symptoms of several days duration. A cariac profile was ordered. The total CK was elevated but no MB

band was present in any of the samples. A flipped-LD was noted on the admission sample. From these results you conclude:
a.   these findings are consistent with acute myocardial infarction
b.   these findings may be consistent with acute myocardial infarction
c.   these findings definitely rule out acute myocardial infarction
d.   these findings may be present in disorders not related to acute myocardial infarction

33.   ALT (GPT) is found in the highest concentration in the:
a.   heart
b.   muscle
c.   liver
d.   kidney

34.   A serum sample was assayed for alkaline phosphatase, which was found to be elevated. An aliquot of the sample was heated in a water bath at 56°C for 10 minutes and then assayed for alkaline phosphatase. The residual activity in the heated sample was found to be less than 20% of the original alkaline phosphate activity. From these results, you conclude that the origin of the alkaline phosphatase is probably from:
a.   bone
b.   liver
c.   intestine
d.   Regan isoenzyme

35.   An elevated serum alkaline phosphatase was found on a sample obtained from a 65-year-old male. An aliquot of this sample was heated at 65°C for 30 minutes. The alkaline phosphatase activity in the heated sample showed little change from the alkaline phosphatase activity in the unheated sample. From these results, you can conclude that the origin of the alkaline phosphatase is probably from:
a.   bone
b.   liver
c.   intestine
d.   Regan isoenzyme

36.   A mildly elevated alkaline phosphatase was found in a sample from a 14-year-old child. This finding:
a.   is normal for a child during periods of rapid bone growth
b.   indicates bone disease
c.   indicates liver disease.
d.   indicates a hemolytic disorder

37.   The most sensitive laboratory test presently available to aid in the diagnosis of chronic alcoholism is:
a.   AST
b.   GGT
c.   ALP
d.   ALT

38. Inherited variant forms of the enzymes, _____ and _____ , may be associated with clinical symptoms in the presence of precipitating factors.
    a. gamma-glutamyl transferase
    b. ceruloplasmin
    c. pseudocholinesterase
    d. glucose-6-phosphate dehydrogenase

39. Enzymes are clinically useful in assisting the physician in diagnosis. Match the enzyme with the disease state in which the enzyme has its greatest diagnostic usefulness.

    _____ 1. CK-2   a. parenchymal liver disease
    _____ 2. LD-1   b. cholestatic disease of the liver
    _____ 3. LD-5   c. acute myocardial infarction
    _____ 4. AST   d. skeletal muscle disease
    _____ 5. ALT
    _____ 6. ALP
    _____ 7. GGT
    _____ 8. CK-3

ANSWERS

1. c
2. (1) c; (2) e; (3) a; (4) b; (5) f; (6) d
3. c, d
4. a
5. b, d
6. c
7. (1) b; Enzyme activity increases with an increase in temperature until de-natured. The enzyme is not denatured at 38°C.
    (2) b; Under zero-order kinetics, increase in time will increase product formed.
    (3) d or c; The coenzyme has to be present in excess because it is con-sumed as the reaction proceeds. With depletion, reaction ceases. The NAD$^+$ concentration is less than procedural requirements.
    (4) a; Enzymes are assayed under zero-order kinetics, which means sub-strate concentration is in excess.
    (5) d; Buffer has to be present in sufficient concentration to buffer any pH changes that might occur with an increase in product or a decrease in substrate. With a pH shift, activity rate is altered.
8. a, b
9. b, c, d
10. all statements are true
11. c

12. a
13. b
14. c
15. a
16. a, b
17. a, c
18. a, b, c, d, e
19. (1) f; (2) c, e; (3) a; (4) f; (5) c, e; (6) d, e
20. b, a
21. (1) increased; (2) increased; (3) decreased; (4) not affected; (5) not affected; (6) increased; (7) increased
22. (1) b; (2) a; (3) c
23. b
24. b, b
25. c
26. (1) LD; (2) ACP; (3) AST; (4) CK; (5) ALP; (6) ALT; (7) GPD; (8) GGT
27. a, b, c
28. a, b, c
29. d, e
30. d
31. b, c
32. b, d
33. c
34. a
35. d
36. a
37. b
38. c, d
39. (1) c; (2) c; (3) a; (4) a; (5) a; (6) b; (7) b; (8) d

CHAPTER 11
LIVER, PANCREAS, INTESTINAL, AND GASTRIC

OBJECTIVES

On completion of this chapter, you will be able to:

1. Describe the anatomy of the liver.
2. List the major functions of the liver.
3. Trace the metabolism and excretion of bilirubin.
4. Define the following words as they apply to the liver:
   a. unconjugated bilirubin
   b. conjugated bilirubin
   c. jaundice
   d. kernicterus
   e. urobilinogen
   f. urobilin
   g. cholelithiasis
5. Classify the type of jaundice based on bilirubin test results, and list the disease states associated with each group.
6. Correlate laboratory findings in serum and urine in hemolytic disease, and in parenchymal and obstructive liver disease.
7. Discuss the Evelyn–Malloy and Jendrassik–Grof methodology for bilirubin.
8. Describe the reactions that occur in the brain during the process of ammonia detoxification.
9. Describe the clinical and laboratory findings associated with Reye's syndrome.
10. List the abnormal serum proteins and serum antibodies that may accompany liver disease.
11. Discuss the interpretation of the galactose tolerance test.
12. Distinguish porphyrias from porphyrinurias.
13. List the types of porphyrias with associated laboratory findings.
14. Discuss assay methods for porphyrins.
15. List the endocrine and exocrine functions of the pancreas.
16. Define the following words as they apply to the pancreas:
    a. acinar cell
    b. islets of Langerhans
    c. ampulla of Vater
    d. macroamylase
    e. secretin
    f. cholecystokinin–pancreozymin
17. Describe the pathophysiology of acute pancreatitis, chronic pancreatitis, carcinoma of the pancreas, and cystic fibrosis.
18. For amylase, discuss (a) enzyme sources; (b) test principle for amyloclastic, saccharogenic, turbidimetric, and dye-labeled substrate methods; and (c) clinical significance.

19. For lipase, discuss (a) enzyme sources, (b) test principle for original and modified Cherry–Crandall methods and turbidimetric method, and (c) clinical significance.

20. List the procedural steps for performing a sweat chloride test.

21. Discuss sweat chloride test interpretation.

22. Discuss test interpretation for the following: (a) D-xylose, (b) vitamin A, (c) fat absorption, and (d) secretin stimulation tests.

23. Describe the anatomy of the stomach.

24. List the functions of the stomach.

25. Discuss the function and regulation of the hormone gastrin.

26. Describe the mechanisms that initiate gastric secretion in the cephalic, gastric, and intestinal phases of gastric secretion.

27. List the physical characteristics of gastric residue, and describe the normal and abnormal aspects of each.

28. List the normal chemical constituents of gastric residue.

29. List abnormal chemical constituents of gastric residue, and explain the clinical significance associated with their presence.

30. Describe the conditions necessary for the collection of a suitable specimen for gastric analysis.

31. List substances used for gastric stimulation, and designate those which have the greatest stimulatory effect.

32. Describe the test principle, test procedure, and method of calculation for gastric analysis reported in meq/L.

32. Calculate gastric problems when given data.

33. Discuss test interpretation for the following tests: (a) gastrin, (b) intrinsic factor, (c) Hollander test, and (d) lactic acid test.

34. Describe laboratory findings associated with each of the following diseases of abnormal gastric function: (a) ulcers, (b) pernicious anemia, (c) Zollinger–Ellison syndrome, (d) carcinoma of the stomach, (e) pyloric obstruction, and (f) atrophy of the stomach.

GLOSSARY OF TERMS

Acinar cells.   The exocrine cells of the pancreas that secrete pancreatic fluid. In anatomy, acinus (from grape) is used to designate a small saclike structure.

Alpha-fetoprotein (AFP).   An alpha-1-globulin normally present in fetal and neonate blood. In adults, high levels of alpha-fetoprotein are associated with hepatocellular carcinoma. Alpha-fetoprotein may leak into amniotic fluid and maternal serum when the fetus is afflicted with anencephaly or open spina bifida. The measurement of AFP in amniotic fluid or maternal serum is of diagnostic value in these diagnoses.

Ampulla of Vater.   The opening through which the common bile duct and pancreatic duct enter the lumen of the duodenum.

Bilirubin. Breakdown product derived from the porphyrin rings of hemoglobin and other heme-containing proteins.

Cholecystokinin–pancreozymin. Small intestinal hormone which stimulates the pancreas to secrete a fluid high in enzymes, low in bicarbonate.

Cholelithiasis. The presence of stones in the gallbladder or bile duct.

Conjugated bilirubin. Bilirubin that has been conjugated in the liver (via enzymes) with glucuronate to form bilirubin diglucuronide. Conjugated bilirubin is water soluble.

Endocrine. A term used to denote that an organ or gland secretes substances directly into the blood without the intervention of ducts.

Exocrine. A term used to denote that an organ or gland secretes substances into ducts.

Free acid. Free acid (or acid in the dissociated form) is defined as material having a pH below 3.0. Töpfers reagent (dimethylaminoazobenzene in ethanol) is an indicator that changes color in the pH range of 2.8–3.5. If free acid is present, the indicator will be red, if free acid is absent, the indicator will be yellow.

Gastric residue. Contents of the stomach after a fast of approximately 12 hours.

Gastrin. Hormone of the stomach antrum and small intestine that stimulates the parietal cells to secrete HCl.

Heme. Heme is protoporphyrin in combination with ferrous iron. Heme is found in hemoglobin (contains four heme groups), myoglobin (contains one heme group), and some enzymes (catalase, peroxidase, and cytochromes).

Intrinsic factor. A glycoprotein secreted by parietal cells of the gastric mucosa that is necessary for the absorption of vitamin B12 from the gastrointestinal tract.

Islets of Langerhans. Clusters of endocrine cells of the pancreas that secrete insulin and glucagon and may also produce gastrin.

Jaundice. Clinical finding associated with deposition of bilirubin in skin and other tissue that produces a yellow discoloration of these tissues.

Kernicterus. Deposition of unconjugated bilirubin in brain tissue that is associated with widespread destructive changes in the brain.

Kupffer cells. The fixed macrophage cells of the liver that are found on endothelial surfaces of sinusoidal spaces in the liver. These cells are part of the reticuloendothelial system and perform phagocytic functions.

Macroamylase. Amylase that is complexed with immunoglobulins or other proteins resulting in a molecule that is too large for renal clearance.

Parietal cells. Cells of the gastric body that secrete HCl into the gastric lumen.

Pathognomonic. Term used to describe distinctive specificity of a test, symptom, or sign that is characteristic for a disease because the finding in diseased individuals does not overlap that observed in healthy individuals.

Porphyrin. A cyclic compound formed by four pyrrole rings linked by methylene bridges ($=$CH$-$).

Secretin. Small intestinal hormone that stimulates the pancreas to secrete a fluid high in bicarbonate, low in enzymes.

Unconjugated bilirubin.   The lipid-soluble form of bilirubin that circulates in the bloodstream bound to albumin. Binding to albumin increases water solubility, thus allowing transport in plasma.

Urobilin.   Oxidized urobilinogen that gives feces its characteristic red-brown color.

Urobilinogen.   A colorless reduction product of conjugated bilirubin that is formed in the intestinal tract by the action of intestinal microorganisms.

LIVER

The blood flow to the liver is substantial. The liver receives about 1500 mL of blood per minute, provided by a dual blood supply from the hepatic artery and the portal vein (GI tract to the liver). The liver also has an extensive lymphatic system.

The liver is composed of lobes made of parenchymal liver cells (hepatocytes) arranged in layers or sheets one cell thick (Figure 11.1). These layers are separated by a capillary system called the sinusoidal system. The sinusoidal spaces are lined with Kupffer cells (sinusoidal epithelial cells), a type of reticuloendothelial cell. The bile canaliculi have no separate structure of their own but are formed by the membranes of adjoining parenchymal cells. They carry bile to the ductules, which then combine to make up larger bile ducts. The interstitial fluid is formed in the space of Disse and is carried by the lymphatic system.

The liver consists of 60% parenchymal cells and 30% Kupffer cells. The remaining cells include vascular and supporting tissue and bile ducts.

FUNCTIONS OF THE LIVER

Excretory

Bilirubin, alkaline phosphatase, cholesterol, and catabolic end products of cholesterol (bile acids) are excreted by the liver via the bile. Substances withdrawn from the blood by hepatic activity, such as heavy metals and dyes, are excreted by the liver.

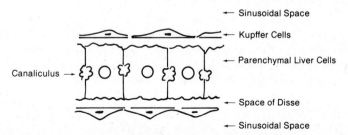

Figure 11.1.   Liver structure.

Hematological

Hematopoiesis occurs in the liver of the embryo; hematopoiesis may also occur in the adult liver in abnormal states. Aged red blood cells are destroyed in the liver by reticuloendothelial cells. Heparin, fibrinogen, prothrombin, and other coagulation factors are produced in the liver.

Metabolic

The liver is the site of synthesis and breakdown of glycogen, which helps to regulate blood glucose levels. It serves as the site for conversion of carbohydrates (galactose, fructose) to glucose.

Albumin and many alpha and beta globulins (other than the immune globulins) are synthesized in the liver. It is the site of amino acid metabolism.

Lipoproteins are synthesized in the liver. The liver is a site for cholesterol esterification and an excretory route for cholesterol via the bile.

Vitamin B12 is stored in the liver. The liver aids in promoting the absorption of fat-soluble vitamins (A, D, E, and K) through the emulsification properties of bile.

The liver is the first site of hydroxylation in the formation of activated vitamin D. Activated vitamin D is needed for calcium absorption and promotes increased phosphorus absorption.

Protective and Detoxification

Kupffer cells aid in the immune mechanism by removing foreign bodies from the blood via phagocytosis. The liver is the site of drug detoxification via conjugation, methylation, oxidation, and reduction reactions. Most steroid hormones are inactivated in the liver. They are conjugated into water-soluble products that are excreted in the urine. The liver is the site of removal of ammonia from the blood. Both the ammonia derived from the diet and carried to the liver via the portal vein and ammonia derived from protein catabolism are removed via the urea cycle.

Circulatory

The liver may act as a site for blood storage and helps to regulate blood volume. It may transfer blood from the portal to the systemic circulation.

BILIRUBIN METABOLIC PATHWAY

Red blood cells live approximately 120 days and are then destroyed. Bilirubin is a waste product derived from hemoglobin destruction (i.e., from prophyrin rings). The route of excretion is through the biliary tract. The excretory pathway is given in Figure 11.2. The key below is used to interpret each numbered step of Figure 11.2.

1. Bile pigments originate from RE cells (Kupffer cells in the liver or other RE cells where erythrocytes are destroyed). The globin is broken down to amino acids, which go to the amino acid pool; iron enters the body stores to be

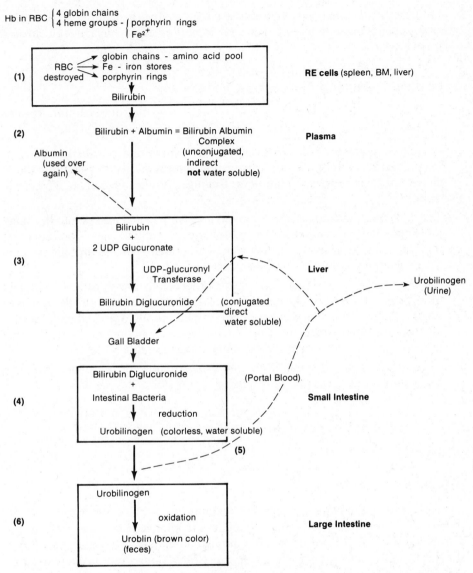

Figure 11.2. Bilirubin metabolic pathway. The numbers in parentheses refer to reactions described in the text. (Adapted with permission from Davidson, I. and Henry, J. (eds.). *Clinical Diagnosis by Laboratory Methods,* 15th ed. Philadelphia: Saunders, 1974.)

reused. The protoporphyrin ring is opened to form biliverdin. The biliverdin is reduced to form bilirubin within RE cells.

2. The bilirubin formed in RE cells (other than Kupffer cells) is transported to the liver sinusoidal spaces via a bilirubin–albumin complex. In the liver, the albumin is separated from bilirubin. Albumin may then be used again for further transport.

3. In liver parenchymal cells, bilirubin is conjugated with glucuronide through a series of enzymatic reactions. The most important enzyme is UDP-glucuronyl transferase:

bilirubin + 2 uridine diphosphate glucuronate

$$\xrightarrow{\text{UDP-glucuronyl transferase}} \text{bilirubin–diglucuronide}$$

Before conjugation, bilirubin is not water soluble and is called indirect or unconjugated bilirubin. After conjugation, bilirubin is water soluble and is called direct or conjugated bilirubin. Conjugated bilirubin can be excreted in urine.

4. The bilirubin–diglucuronide is excreted into the bile and passes into the small intestine where it is reduced by anaerobic bacterial enzymes. The reduction products consist of mesobilirubinogen, stercobilinogen, and urobilinogen (all are collectively known as urobilinogen).

5. A portion of the urobilinogen is reabsorbed from the intestinal tract and returned to the liver via the enterohepatic–portal system. Normally, the liver removes all but a small portion of the recycled urobilinogen and reexcretes it via the bile. Urobilinogen is water soluble. A small amount of the reabsorbed urobilinogen escapes hepatic removal and is excreted via the urine.

6. The urobilinogen remaining in the intestine is oxidized to urobilin, which is colored and gives feces its brown color.

JAUNDICE

Definition

When bilirubin in the blood is excessive, it deposits in tissues, which then become yellow. The condition is known as jaundice or icterus. Levels of total serum bilirubin above 2.5–3.0 mg/dL usually produce jaundice. In healthy individuals, all or most of the total bilirubin is unconjugated, resulting from the daily destruction of red blood cells.

Jaundice may be caused by (1) production of more bilirubin than the normal liver can excrete (this type of jaundice occurs in hemolytic disease); (2) failure of the damaged liver to excrete the bilirubin produced in normal amounts (this type of jaundice may occur in hepatitis or in liver disease associated with hepatoxins); and (3) obstruction of the excretory ducts of the liver. The obstruction may occur in the liver, and is then called intrahepatic cholestasis, or the obstruction may occur in the extrahepatic biliary tree and is called extrahepatic cholestasis or posthepatic obstruction. Intrahepatic obstruction occurs when there is defective transport of conjugated bilirubin into bile canaliculi. This can occur as a result of inflammation or swelling of liver cells, which then block excretory ducts in the liver. Extrahepatic obstruction may occur due to stones, tumor, or stricture causing obstruction of the biliary tree.

Kernicterus

Bilirubin is soluble in lipid solvents but almost insoluble in water. Since it is lipid soluble, it can cross cell membranes readily. The binding of bilirubin to albumin makes it more water soluble, so that it can be transported in the plasma for excretion in the bile. Bilirubin bound to albumin does not cross cell membranes readily. If the bilirubin is excessively high in the newborn, and there is insufficient albumin to bind all bilirubin, bilirubin may enter brain tissue producing "brain jaundice" or kernicterus.

CLASSIFICATION OF JAUNDICE

A laboratory classification of jaundice based on fractionated bilirubin is given below:

I. Unconjugated

In the unconjugated type of jaundice, at least 80% of the serum bilirubin is unconjugated or indirect.

A. Prehepatic

This jaundice is due to increased production of bilirubin secondary to increased destruction of red blood cells. Increased destruction of red blood cells may occur in (a) hemolytic anemia and (b) neonatal jaundice due to Rh or ABO incompatibilities.

B. Hepatic

This jaundice is due to defective removal of bilirubin from sinusoidal blood by the liver parenchymal cells or due to a conjugation defect in the liver.

1. Preconjugation transport failure

Preconjugation transport failure occurs in Gilbert's disease. Gilbert's disease is a familial type of nonhemolytic jaundice. It is relatively common and asymptomatic. The transport of bilirubin into liver cells from the sinusoidal space is impaired. The plasma bilirubin fluctuates and increases during illness.

2. Conjugation failure in the liver

Neonatal physiological jaundice is a common type of jaundice found in the newborn. This type of jaundice is due to an inefficient or immature fetal liver. The liver is not able to conjugate all of the bilirubin produced because of deficient enzymes. As the liver matures, enzyme function improves, and jaundice disappears in a few days. Neonatal physiological jaundice is treated with phototherapy. Crigler–Najjar disease is a second type of jaundice caused by conjugation failure. This type of jaundice is caused by a deficiency of the conjugating enzyme system in the liver. The infant is unable to conjugate bilirubin. Kernicterus follows with early death.

II. Conjugated

In the conjugated type of jaundice, the percentage of conjugated or direct bilirubin to total bilirubin is much higher than in the unconjugated type. The percentage of conjugated bilirubin may reach 40–50% or higher.

A. Hepatic
   1. Hepatocellular
      Diffuse hepatocellular damage or necrosis causes the conjugated
      bilirubin to regurgitate into sinusoidal blood. The damaged liver
      cells are also incapable of normal function and cannot efficiently
      remove the bilirubin normally produced; this derangement results in
      increased unconjugated bilirubin also. Hepatocellular damage may
      result from (a) viral hepatitis; (b) toxic hepatitis (drugs, toxins); and
      (c) cirrhosis.
   2. Hepatocanalicular
      Intrahepatic obstruction causes defective transport of the conju-
      gated bilirubin into the bile canaliculi. Intrahepatic obstruction may
      occur due to (a) viral hepatitis; (b) drugs; and (c) Dubin–Johnson
      disease. In Dubin–Johnson disease, a genetic defect results in post-
      conjugation transport failure. The conjugated bilirubin cannot be
      transported into bile canaliculi.
B. Posthepatic
   Posthepatic jaundice is caused by obstruction of the common bile duct.
   The obstruction may be caused by (a) stones; (b) stricture or spasm; and
   (c) neoplasms.

LABORATORY FINDINGS IN DISEASE STATES

Hemolytic Processes

The jaundice produced by hemolytic processes results from an increased production of bilirubin following accelerated destruction of red blood cells. The icteric
sera is derived from bilirubin that is in transit to the liver via the blood (unconjugated form).

The laboratory findings associated with hemolysis include (1) an increase in
total bilirubin, primarily of the unconjugated form and (2) increased fecal and
urine urobilinogen. A normal liver has large functional reserve and is able to
remove, conjugate, and excrete the increased load of bilirubin with no regurgitation into sinusoidal blood. Therefore, in the hemolytic state, conjugated bilirubin
is normal and urine bilirubin is negative. However, the amount of conjugated
bilirubin reaching the intestinal tract is increased; the amount of urobilinogen
formed from the conjugated bilirubin is increased, resulting in increased fecal
urobilinogen. Since a portion of the urobilinogen undergoes recycling, the amount
of urobilinogen entering the blood increases, resulting in increased urine urobilinogen.

Parenchymal Liver Cell Damage

The diseased liver is impaired in its ability to excrete the bilirubin produced from
normal destruction of aged red blood cells. It is deficient not only in its capacity to
remove bilirubin from the blood but also in its conjugation and excretory functions.

The laboratory findings associated with parenchymal cell damage include (1) increased total bilirubin composed of both the conjugated and unconjugated forms; (2) positive urine bilirubin and increased urine urobilinogen; and (3) slightly decreased fecal urobilinogen. The total bilirubin is increased because the damaged liver cells are ineffective in removing bilirubin from blood (increased unconjugated bilirubin) and because inflammation with associated swelling causes cell damage, necrosis, and blockage of bile canaliculi (intrahepatic obstruction) resulting in regurgitation of conjugated bilirubin into sinusoidal blood and general circulation. The urine bilirubin is positive, since conjugated bilirubin is water soluble and can be excreted in the urine. The amount of conjugated bilirubin reaching the intestinal tract is reduced, resulting in decreased fecal urobilinogen. The urine urobilinogen may be increased, because that portion of the urobilinogen reabsorbed for recycling is not efficiently removed by the diseased liver for reexcretion. Thus, the blood urobilinogen increases. This water-soluble urobilinogen is then excreted in the urine.

Early Posthepatic Obstruction

Any obstruction of the common bile duct leads to ineffective excretion of conjugated bilirubin with regurgitation into sinusoidal blood. The icteric serum is caused primarily by conjugated bilirubin that "leaks" from the liver.

The laboratory findings associated with posthepatic obstruction include (1) increased total bilirubin composed in large part of the conjugated form; (2) positive urine bilirubin; and (3) decreased to negative (depending on the extent of obstruction) fecal and urine urobilinogen. The urine bilirubin is positive, because conjugated bilirubin enters the blood stream via regurgitation from the liver. Due to the obstruction, less bilirubin enters the intestinal tract, resulting in decreased urobilinogen, hence, the decreased fecal and urine urobilinogen. If the obstruction is complete, the fecal and urine urobilinogen may be negative, and the feces will be clay-colored, since urobilin, formed from urobilinogen, gives feces the characteristic brown color. If the obstruction is not relieved, parenchymal cell damage may occur.

Table 11.1 summarizes the findings associated with hemolysis, parenchymal liver cell damage and posthepatic obstruction.

BILIRUBIN METHODOLOGY

Specimen Requirements

A fasting serum specimen is preferred for bilirubin assay. Hemolysis causes false low results because hemoglobin interferes with the diazo reaction. With newer methods (i.e., Jendrassik–Grof), slight hemolysis can be tolerated. If the serum is exposed to direct sunlight (UV light), unconjugated bilirubin is destroyed. The serum is stable in the refrigerator for several days if protected from light. The sample should be frozen for longer periods of storage.

Evelyn–Malloy Procedure

The Evelyn–Malloy procedure is an old procedure. However, the principle of the test is the same as in newer methodologies.

Table 11.1. Laboratory Findings in Jaundice

Jaundice Produced By	Disease	Serum Bilirubin			Urine		Feces	Miscellaneous Tests
		Unconj.[a]	Conj.	Total	Urobilinogen	Bilirubin	Urobilinogen	
Increased RBC destruction	Hemolysis	↑	normal to sl. ↑	↑	↑↑	negative	↑↑	↓ Haptoglobin ↑ LD-1 ↑ Reticulocyte count
Liver cell damage	Hepatitis (early)	↑↑	↑	↑	↑	positive	sl. ↓	↑ AST ↑ ALT ↑ LD-5
Obstruction	Extrahepatic biliary obstruction (early)	↑	↑↑	↑	↓ to negative	positive	↓ to negative	↑ ALP ↑ GGT ↑ 5' Nucleotidase ↑ Leucine aminopeptidase ↑ Cholesterol

[a] ↑ = increased; ↓ = decreased; sl. = slightly; unconj. = unconjugated; conj. = conjugated.

Principle

Bilirubin and the diazo reagent form an azobilirubin complex, which can be measured colorimetrically. The color of the azobilirubin varies with pH.

The Evelyn–Malloy test is based on the following principles: (1) conjugated (direct) bilirubin is water soluble and, therefore, will react with the diazo reagent in a water solution and (2) unconjugated bilirubin (indirect) is not water soluble, therefore, alcohol is necessary to put the unconjugated bilirubin in solution so that it can react in the diazo reaction.

Test

1. The total bilirubin is determined by:

   serum + alcohol + diazo reagent → "total bilirubin"

2. The direct (conjugated bilirubin) is determined by:

   serum + water + diazo reagent → "conjugated bilirubin"

3. The indirect (unconjugated) bilirubin is calculated:

   total bilirubin − conjugated bilirubin = "unconjugated bilirubin"

Sources of Error

Sometimes protein interferes, because alcohol can cause protein to precipitate. A protein-free filtrate cannot be used, because some of the bilirubin is coprecipitated with the protein. The alcohol concentration is adjusted to 50% to avoid precipitation of protein.

Jendrassik–Grof Procedure

Principle

Diazo is a prefix used in clinical chemistry to denote that a compound contains the nitrogen radical. Sulfanilic acid is diazotized with sodium nitrite and hydrochloric acid to form the diazo reagent. Bilirubin is coupled with diazotized sulfanilic acid to form a diazo product called azobilirubin, which behaves as a pH indicator. It appears blue at a strongly acidic or alkaline pH, and red near neutrality. Since pH affects the color produced, the pH must be carefully controlled.

Test

1. Total bilirubin (conjugated and unconjugated) forms a colored complex with diazo reagent in the presence of a catalyst:

   serum + sodium acetate + diazotized sulfanilic acid
   (bilirubin)   (buffer)   (diazo reagent)

   $$\xrightarrow[\text{(catalyst)}]{\text{caffeine and sodium benzoate}} \text{azobilirubin}$$

2. Ascorbic acid is added, which destroys the excess diazo reagent (terminates the reaction).

3. Alkaline tartrate is added, which converts the medium to an alkaline pH. The red-purple azobilirubin is converted to a blue azobilirubin.

4. Conjugated bilirubin (direct) is determined by the diazo reaction in the absence of the catalyst and at an acid pH. Under these conditions, only the conjugated bilirubin will react:

$$\underset{\text{(bilirubin)}}{\text{serum}} + \text{dilute HCl} + \text{diazo reagent} \rightarrow \underset{\text{(conjugated only)}}{\text{azobilirubin}}$$

The reaction is terminated by ascorbic acid. Alkaline tartrate is added to make the pH alkaline, as in the procedure for total bilirubin.

5. Both total bilirubin and conjugated bilirubin tests are read at a wavelength of 600 nm.

6. The unconjugated fraction is calculated:

$$\text{total bilirubin} - \text{conjugated bilirubin} = \text{unconjugated bilirubin}$$

The Jendrassik–Grof procedure is the method of choice for bilirubin assay.

Reference Range

Bilirubin (adults)	total 0.2–1.2 mg/dL
	conjugated 0.0–0.4 mg/dL
	unconjugated 0.0–0.7 mg/dL
Bilirubin (newborn)	total 1.0–12.0 mg/dL

BLOOD AMMONIA

The major source of blood ammonia is the gastrointestinal tract; a minor contributor is the kidney. In the GI tract, bacteria release ammonia from nitrogen-containing food. This ammonia, plus that ingested as ammonium salts, is absorbed by the enterohepatic portal vein and carried directly to the liver, where it enters the urea cycle (Krebs–Henseleit) to form urea. Normally, the level of circulating ammonia is very low. Elevation of blood ammonia can arise by two processes: (1) impaired liver parenchymal cell function and (2) shunting of the portal blood past the liver, which may occur in cirrhosis.

Ammonia is toxic to the brain; it may produce permanent brain damage. However, in the brain, ammonia reacts with glutamate (a detoxifying agent) to produce glutamine, a benign substance. In the brain, an increase in ammonia above the detoxification level causes cerebral dysfunction. When the blood ammonia level is increased, ammonia uptake by the brain increases. Thus, glutamine synthesis increases. It is postulated that when the synthesis of glutamate is accelerated, the concentration of alpha-ketoglutarate is reduced, resulting in impairment of the cerebral citric acid cycle, thus precipitating hepatic coma:

$$\text{alpha-ketoglutarate} + NH_4^+ + NADH \rightarrow \text{glutamate} + NAD^+ + H_2O$$

$$\text{glutamate} + NH_4^+ \rightarrow \text{glutamine}$$

Blood ammonia assay may be requested to establish a diagnosis of impending or existing hepatic coma. It may be useful in diagnosing Reye's syndrome.

Reye's Syndrome

Reye's syndrome is a form of encephalitis. It occurs most frequently in children (6–12 years), but can occur in individuals to late teens. The etiology is unknown but follows a viral infection. Reye's syndrome is associated with influenza B, and less commonly with influenza A and varicella infection. Reye's syndrome has also been associated with the ingestion of drugs (in particular aspirin), fungal toxins (aflatoxin), and exposure to pesticides. It is suggested that the injury in Reye's syndrome is toxic rather than infectious. The brain becomes edematous and coma follows.

The laboratory findings in Reye's syndrome may include (1) elevated blood ammonia; (2) elevated liver enzymes (AST, ALT); (3) hypoglycemia; (4) prolonged prothrombin time; (5) acid–base disturbances; (6) modest ketonemia; (7) lipid abnormalities (decreased total cholesterol and VLDL cholesterol and increased levels of free fatty acids); and (8) elevations in CK-MM and CK-MB (sometimes) but no CK-BB. The cerebrospinal fluid cell count, glucose, and protein are essentially normal; lactate, pyruvate, and glutamine are increased.

It is postulated that the major pathological disturbance in Reye's syndrome is a form of severe mitochondrial injury. In the liver, mitochondrial enzyme activity is reduced. The urea cycle cannot function properly, leading to an increase in blood ammonia. Abnormalities in gluconeogenesis produce hypoglycemia. Disturbances in the synthesis of clotting factors lead to abnormalities in the coagulation profile.

Mitochondrial abnormalities in the brain similar to those seen in the liver have been noted on cerebral biopsies. Mitochondrial abnormalities have also been found in skeletal muscle, myocardium, kidney, and pancreas.

Case History, Reye's Syndrome

A 4 year-old lethargic, semicomatose white female was admitted to the hospital. At the time of admission, she showed signs of seizure activity. Approximately one week prior to this admission, she had been treated for a febrile upper respiratory infection. Two days prior to this admission, she began vomiting "coffee-ground" material. Urine output was scanty. A lumbar puncture was performed and blood chemistries were ordered. The patient was taken to surgery for the insertion of a subdural intracranial pressure monitor. She was placed on medication to control seizure activity.

On days two and three, renal chemistry tests showed continued deterioration in renal function, attributed to acute tubular necrosis. Edema became marked. An EEG was performed; the results were markedly abnormal.

On day four, the urine output decreased, the patient lapsed into a deep coma and expired. The laboratory data are listed in Table 11.2. On admission, the blood pH was 7.16, and pCO_2 was 78 mm Hg. The cerebrospinal fluid cell count and protein were normal, and the glucose reflected the plasma glucose concentration. The CSF culture showed no growth. Cerebrospinal fluid glutamine was not ordered.

Table 11.2.   Reye's Syndrome Case-History Laboratory Data

	Ammonia (11–33)[a] μmol/L	AST (16–55) U/L	BUN (7–21) mg/dL	Creatinine (0.6–1.3) mg/dL	Na (135–145) mmol/L	K (3.5–5.0) mmol/L	Cl (98–106) mmol/L
Admission							
Day 1	840	475	49	1.3	125	6.3	91
Day 2	80		85	3.6	124	6.9	86
Day 4			104	4.6	136	5.4	94

[a] The reference ranges for components are indicated in parentheses.

Discussion

In a child, the history of a previous viral illness followed by lethargy, vomiting, and coma is typical of Reye's syndrome. The elevation in blood ammonia is thought to be due to deficient enzyme activity, possibly that of ornithine carbamyl transferase, in the liver. This mitochondrial enzyme functions in the urea cycle, where urea is formed from ammonia. If the level of ammonia in the brain rises above the detoxification level, brain damage may result. The prognosis appears to be grave in those individuals with extreme hyperammonemia (greater than 600 μmol/L). The glutamine level in the CSF rises. Ammonia stimulates the respiratory center in the brain, and the patient may hyperventilate with resultant respiratory alkalosis, upon which may be superimposed a metabolic acidosis. The severe acidosis in this patient appears to be a respiratory acidosis (unexplained). The degree of metabolic acidosis seems to correlate with the final outcome with survival being markedly decreased in those patients admitted with a blood pH less than 7.2.

Both liver transaminase enzymes, AST and ALT, increased early in the illness, reflecting hepatic involvement. The patient may develop hypoglycemia, reflecting decreased gluconeogenesis. The prothrombin time may be prolonged.

The etiology of the renal failure in this patient was unexplained. Serum and urine osmolalities were performed. The results indicated the inability of the kidneys to concentrate, a finding in acute tubular necrosis.

Blood Ammonia Methodology

The four methods available for blood ammonia assay include a Conway microdiffusion method, cation-exchange resin methods, ammonia electrode method, and enzymatic (kinetic) methods.

The microdiffusion procedure lacks precision and has been replaced by other methods. The cation-exchange resin methods are based on selective isolation of ammonia by ion-exchange resins, elution of the ammonia followed by colorimetric reaction with phenol–hypochlorite (Berthelot reaction). The ammonia electrode method measures a change in electrical potential as ammonia gas diffuses across a semipermeable membrane and comes into equilibrium with ammonium ions. The enzymatic methods for NH_3 are adaptable to automated instrumental systems and use either NADPH or NADH as coenzymes. NADPH is preferred as interferences from other NADH-consuming reactions are eliminated. This method follows.

Specimen
The preferred specimen for blood ammonia is a plasma (EDTA–Na or heparin–Na) sample, which must be submersed in ice immediately following collection. Arterial rather than venous collections are preferred by some laboratories; however, there is normally little difference in the ammonia level between venous and arterial blood. The test should be performed immediately because the ammonia level rises rapidly due to enzymatic deamination of labile amides. The collection tube should be filled to capacity and kept tightly stoppered until the analysis is performed.

Principle
As the enzyme, glutamate dehydrogenase (GLDH), catalyzes the condensation of ammonium and α-ketoglutarate, NADPH is oxidized to NADP$^+$. The decrease in absorbance at 340 nm (due to the disappearance of NADPH) is directly proportional to the ammonia concentration in the sample:

$$NH_4^+ + \alpha\text{-ketoglutarate} + NADPH \xrightarrow{\text{GLDH}} \text{L-glutamate} + NADP^+ + H_2O$$

Reference Range

Venous plasma   11–35 μmol/L

The reference range is influenced by the care taken in specimen collection and handling.

OTHER ASSOCIATED FINDINGS AND TESTS IN LIVER DISEASE

Abnormal Proteins in Liver Disease

Lipoprotein X
Lipoprotein X is an abnormal lipoprotein found in nearly all cases of cholestasis (both intra- and extrahepatic cholestasis). It is not present in healthy people and rarely present in patients with hepatocellular damage, unless there is also cholestasis.

Alpha-Fetoprotein (AFP)
Alpha-fetoprotein is synthesized by the fetus *in utero*. It reaches its peak at about 12 weeks gestation and drops to low levels by birth. Alpha-fetoprotein has been used as a tumor marker for hepatomas. When tumor is present, AFP may appear in the blood. It is postulated that liver cells revert to a more primitive form and produce AFP. AFP is not specific for hepatomas and is present in other malignant disorders (i.e., germinal cell testicular tumors, gastric cancer, cancer with liver metastases) and in other hepatic disorders not associated with malignancy.

Serological Abnormalities

Serological abnormalities may be present in liver disease. Antimitochondrial antibodies are those directed against mitochondrial membranes of cells from many tissues. These antibodies may be present in individuals with cirrhosis and active chronic hepatitis and occasionally in individuals with obstruction of the main bile duct. Anti-smooth muscle antibodies are found in about half of those individuals with active chronic hepatitis.

Galactose Tolerance Test

Principle
Galactose is a monosaccharide and no hydrolysis is required prior to absorption from the GI tract. Once absorbed, the galactose is transported to the liver by the enterohepatic portal system. If liver function is not impaired, galactose may be enzymatically converted to glucose and then utilized by the body. If galactose is not converted to glucose, it is excreted in the urine.

Test
An oral or intravenous dose of galactose is given. Blood and urine samples are collected, and galactose is measured.

Interpretation
The interpretation is based on the speed of removal of galactose by the liver. The galactose blood level remains high for a prolonged period of time in patients with liver disease as compared to those with normal liver function. This test is an insensitive liver function test.

Bile Acids

Bile acids are formed from cholesterol in the liver, and are transported through bile ducts to the gall bladder. During digestion, bile acids enter the intestinal tract and are reabsorbed into portal blood. These bile acids are then removed by hepatocyte extraction and reexcreted into the bile. Normally, only small amounts of bile acids are present in the blood. However, in liver disease, the hepatocyte is impaired in its ability to remove bile acids, and increased concentrations will be found in the blood and urine (in both parenchymal and cholestatic disease).

Since bile acids are synthesized from cholesterol, serum bile acid levels remain normal in hemolytic disease unless liver disease is also present. Serum bile acid levels are also normal in Gilbert's disease, Crigler–Najjar syndrome, and Dubin–Johnson syndrome. Bilirubin is elevated in all of these conditions.

Serum bile acids are assayed by RIA procedures. A fasting specimen is required to eliminate the slight rise in bile acids that may occur following a meal.

PORPHYRINS AND PORPHYRINURIAS

Porphyrins are complex ring compounds made of four pyrrole rings linked together (cyclic tetrapyrrole compounds) as seen in Figure 11.3.

Free porphyrins play no biological role in man. They are active in the form of

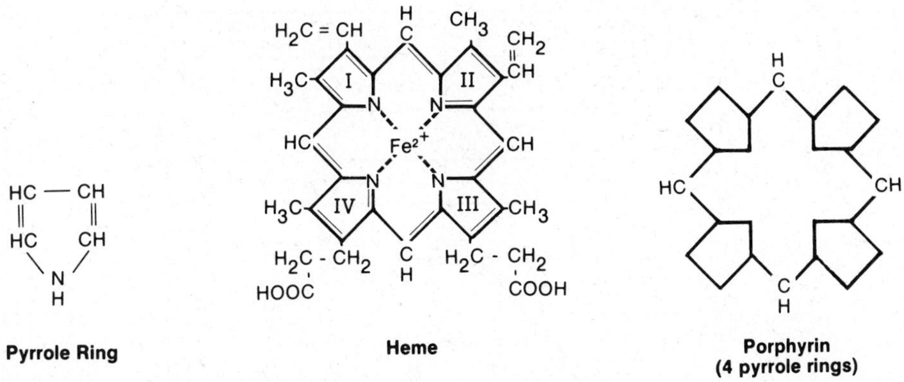

Figure 11.3. Structure of porphyrins and heme.

metal chelates; the iron chelate of porphyrin is called heme (Figure 11.3). The iron in heme is ferrous iron.

Hemoglobin Synthetic Pathway

In humans and other mammals, heme is found in hemoglobin, myoglobin, and the cytochrome enzymes (where heme proteins act as free-electron carriers in the final stage of glucose metabolism). All heme-containing compounds function in biological oxidation.

The porphyrin in heme is called protoporphyrin. Porphyrins are mainly synthesized in the liver and the marrow of the long bones by a series of enzymatic reactions as shown in Figure 11.4. The key below is used to interpret each numbered step of Figure 11.4.

1. Glycine from the amino acid pool and succinate from the Krebs cycle form delta-aminolevulinic acid (ALA) in the presence of the enzyme ALA synthetase.

2. Two ALA molecules combine in the presence of the enzyme, ALA dehydrase, to form porphobilinogen (PBG).

3. Under the influence of enzymes, four molecules of PBG combine to form uroporphyrinogen (a cyclic tetrapyrrole). There are four possible isomers of uroporphyrinogen, but only two have been identified in nature, isomer I series and isomer III series. Normally, the isomer III series is the predominant pathway. The III series leads to protoporphyrin IX synthesis, which is the immediate precursor to heme. The isomer I series has no known physiological function.

4. The uroporphyrinogens are then decarboxylated to form coproporphyrinogens of the same isomeric type. Uroporphyrinogen and coproporphyrinogen are not colored but may be oxidized (no enzyme required) upon standing to uroporphyrin and coproporphyrin, which are colored.

5. Coproporphyrinogen III is then oxidized to protoporphyrin IX, the porphyrin of heme.

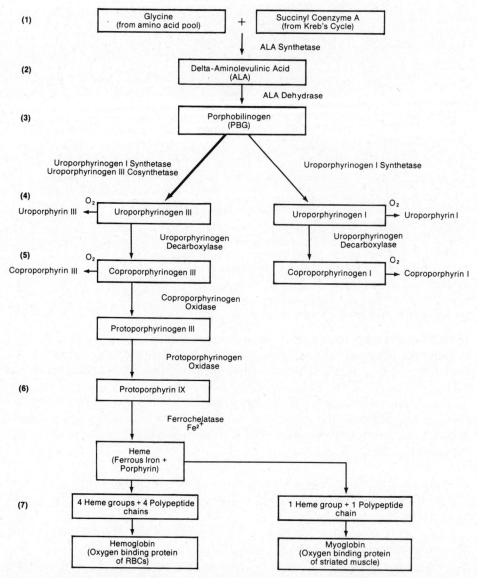

Figure 11.4. Hemoglobin synthetic pathway. The numbers in parentheses refer to reactions described in the text.

6. A molecule of ferrous iron is added to protoporphyrin IX to form heme.

7. Four heme groups and four polypeptide chains form hemoglobin. Myoglobin, a heme protein of striated muscle, contains one heme group and one polypeptide chain. The molecular weight of myoglobin is approximately one-fourth that of hemoglobin.

The metabolic pathway is efficient and, normally, very low levels of porphyrins and precursors are excreted via the urine and feces. The route of excretion depends on water solubility. ALA, PBG, and uroporphyrins are excreted primarily in the urine. Protoporphyrin is excreted primarily via the fecal route and coproporphyrin by either route.

Control of Porphyrin Synthesis

The control of porphyrin synthesis is believed to occur via a negative feedback system with heme and the enzyme ALA synthetase. When heme levels are decreased, ALA synthetase activity increases, which then restores adequate heme levels. When heme levels are adequate, ALA synthetase production is inhibited.

Other factors may affect ALA synthetase activity. Certain drugs, steroid hormones, and iron induce the enzyme. Glucose inhibits ALA synthetase activity.

Porphyrias

Porphyrias are inherited diseases associated with enzyme defects that involve the heme biosynthetic pathway. In these diseases, large amounts of porphyrins or precursors are excreted in the urine and feces. Porphyrias are primary abnormalities. When a primary abnormality occurs, the activity of other enzymes increase (usually ALA synthetase increases), and heme synthesis remains adequate.

Porphyrinurias

In porphyrinurias, the increase in porphyrin excretion is moderate. The increase is secondary to another disease process not related to inherited enzyme defects in porphyrin metabolism. Examples of acquired disorders of porphyrin metabolism occur in liver disease and in lead intoxication. Therefore, porphyrinurias are secondary abnormalities.

Porphyrias

Porphyrins are synthesized in many tissues of the body, but the porphyrin requirement of tissue varies. Erythropoietic tissue requires greater porphyrin synthesis than liver, and liver requires greater porphyrin synthesis than other tissues. Genetic enzyme defects will affect enzyme synthesis in all porphyrin-synthesizing tissue; however, the metabolic consequences are most often associated with the liver and erythropoietic system. Prophyrias are classified as either (1) erythropoietic in origin or (2) hepatic in origin depending on where excess porphyrin production occurs. For each porphyria, an enzyme defect is known or postulated. The diagnosis of the type of porphyria is made on the basis of clinical symptoms and type(s) of porphyrins excreted in the urine and/or feces.

Symptoms

The clinical symptoms of porphyrias are of three types: (1) sensitivity to sunlight (photosensitivity); (2) gastrointestinal complaints; and (3) neurological disturbances.

Porphyrins may accumulate in the skin. Upon exposure to UV light (i.e., sunlight), cutaneous eruptions occur. The degree of photosensitivity varies with the type of porphyria. The eruptions may be filled with fluid which fluoresces. They may ulcerate and heal slowly, thus producing scarring and deformity. In some lesions, no ulceration or scarring occurs.

Gastrointestinal disturbances usually occur intermittently with symptoms of intense abdominal pain, constipation, and vomiting. Neurological changes usually occur in conjunction with gastrointestinal disturbances.

Erythropoietic Porphyrias

Congenital Erythropoietic Porphyria (CEP) (Gunther's Disease)
Congenital erythropoietic porphyria becomes evident at birth or during infancy. Porphyrins are deposited in skin and bones.

Clinical Symptoms. These individuals exhibit extreme photosensitivity. Upon exposure to sunlight, bullous skin eruptions occur. The lesions ulcerate and heal slowly, leading to deformity. Teeth are often discolored and will fluoresce when exposed to UV light. These individuals may experience hemolytic episodes. No gastrointestinal or neurological disturbances are associated with CEP.

Laboratory Findings. The urine appears red and turns dark brown on standing. It contains large quantities of uroporphyrin and small amounts of coproporphyrin, predominately of the isomer type I. The feces contain elevated coproporphyrin, predominately type I. Erythrocytes have high levels of uroporphyrin I.

Enzyme Defect. It is postulated that the enzyme defect is decreased uroporphyrinogen III cosynthetase or an increase in uroporphyrinogen I synthetase or both.

Erythropoietic Protoporphyria (EPP)
Erythropoietic protoporphyria begins in early childhood or adolescence.

Clinical Symptoms. Erythropoietic protoporphyria is characterized by solar eczema. Protoporphyrins are deposited in the liver, sometimes leading to chronic hepatic disease. EPP may eventually be classified under the hepatic types, because hepatic consequences are more severe.

Laboratory Findings. There is no abnormal urinary excretion of porphyrins or precursors. Protoporphyrin IX is increased in the plasma, red blood cells, and stool.

Enzyme Defect. The enzyme defect associated with EPP is a ferrochelatase deficiency in red blood cells.

Hepatic Porphyrias

Acute Intermittent Porphyria (AIP)
Acute intermittent porphyria rarely occurs before adolescence and usually becomes evident in the third decade of life. The attacks are intermittent and may be induced by drugs or estrogens.

Clinical Symptoms. Acute intermittent porphyria is characterized by both gastrointestinal and neurological symptoms, which may occur only during acute attacks. These individuals are not photosensitive. AIP is more prevalent in women than in men.

Laboratory Findings. The urine is colorless but turns dark brown on standing. During an acute attack, the urine contains large amounts of PBG and lesser amounts of ALA (prophyrin precursors). Urinary excretion of uroporphyrin and coproporphyrin is also elevated.

Enzyme Defect.   The defect appears to be a decrease in uroporphyrinogen I synthetase, leading to decreased heme production and a compensatory increase in ALA synthetase.

Hereditary Coproporphyria (HCP)

In hereditary coproporphyria, the attacks are intermittent and may be drug induced.

Clinical Symptoms.   HCP may be associated with both gastrointestinal and neurological symptoms and is rarely associated with photosensitivity.

Laboratory Findings.   The urinary excretion of ALA, PBG, coproporphyrin, and uroporphyrin is increased. In HCP, fecal coproporphyrin excretion is greatly increased. The measurement of fecal coproporphyrin is a test that is used to differentiate HCP from AIP.

Enzyme Defect.   The enzyme defect is postulated to be a deficiency of coproporphyrinogen oxidase with a compensatory increase in ALA synthetase.

Variegate Porphyria (VP)

The onset of variegate porphyria is usually in the fourth or fifth decade of life. VP may be drug induced (easily induced by barbiturates and sulfonamides).

Clinical Symptoms.   VP is associated with gastrointestinal and neurological symptoms, photosensitivity, and cutaneous lesions.

Laboratory Findings.   During acute attacks, the urine contains large amounts of ALA, PBG, uroporphyrin, and coproporphyrin. The fecal excretion of coproporphyrin and protoporphyrin are increased in both the acute and latent stages.

Enzyme Defect.   The enzyme deficiency is postulated to be a deficiency in the protoporphyrinogen oxidase or ferrochelatase enzyme.

Porphyria Cutanea Tarda (PCT)

Porphyria cutanea tarda usually occurs between the ages of 40 to 60. PCT is the most common form of porphyria.

Clinical Symptoms.   Only photo cutaneous lesions are present in PCT. Liver disease is usually present as a consequence of chronic alcoholism.

Laboratory Findings.   Urine uroporphyrin is increased, the type I isomer predominating.

Enzyme Defect.   The enzyme defect is a deficiency in uroporphyrinogen decarboxylase in both the red blood cells and liver.

Porphyrinurias

Porphyrinurias are acquired or are secondary to a disorder that leads to elevated prophyrin excretion. They may be acquired due to (1) liver disease (hepatitis or cirrhosis) or (2) heavy metal or organic solvent toxicity (lead, gold, mercury, arsenic, benzene, carbon tetrachloride). The secondary prophyrinurias are much more common than primary prophyrias. Porphyrinuria due to lead poisoning is the most common type of secondary porphyrinuria. It may occur in factory workers who handle lead or in children who chew on paint that contains lead. In suspected lead poisoning, urine is analyzed for ALA. Coproporphyrin III is also increased. Lead blocks the enzyme ALA dehydrase.

Laboratory Evaluation

Specimen

To quantitate urinary porphyrins, a twenty four hour urine specimen is collected. During collection, the specimen should be protected from light and stored in a cool place. The porphyrin precursors, porphobilinogen (PBG) and delta-aminolevulinic acid (ALA), are more stable in an acid pH; porphyrins are more stable in alkaline urine. For porphyrin screening procedures, a random urine specimen may be used. The screening test should be performed immediately; however if this is not possible, the pH should be adjusted to 6–7 and the specimen may then be stored in the refrigerator.

Other suitable specimens include heparinized plasma or washed red blood cells (for erythrocyte porphyrins) or fecal samples. Fecal specimens should also be protected from light and stored in a cool place.

Assay

Porphyrins are pigments or precursors to pigments and will color the urine pink to deep wine to almost black. Porphyrinogens are colorless but are rapidly oxidized to porphyrins. Some individuals will excrete urine that will be normal in color but will turn dark after exposure to light.

Porphyrins fluoresce when exposed to long UV light (397–408 nm), which gives an intense orange–red color. Porphyrinogens do not fluoresce. Screening tests for porphyrins are based on this naturally fluorescent property.

ALA and PBG will react with Ehrlich's reagent (p-dimethylaminobenzaldehyde in hydrochloric acid) to form a magenta-colored complex. Urobilinogen will also react with Ehrlich's reagent producing the same color. The Watson–Schwartz test can be used to distinguish between urobilinogen and porphobilinogen when Ehrlich's reagent is used.

Column chromatography with fluorometric detection may be used to quantitate individual porphyrins.

PANCREAS

The pancreas has large functional reserve. It is estimated that 50% or more of the cells are destroyed before pancreatic insufficiency can be clearly demonstrated.

PANCREATIC FUNCTIONS

Pancreatic functions may be divided into two groups: (1) endocrine (ductless) functions and (2) exocrine (through ducts) functions.

Endocrine

Insulin is secreted by beta cells of the islets of Langerhans. Insulin is needed by adipose and muscle cells for cellular uptake of glucose.

Glucagon is secreted by alpha cells of the islets of Langerhans and helps to regulate blood glucose between meals through glycogenolysis.

Exocrine

Exocrine pancreatic functions include the production and secretion of pancreatic fluid. Pancreatic fluid is produced in acinar cells then transported through ducts into the main pancreatic duct that empties into the intestine via the ampulla of Vater. Frequently, the bile duct joins the pancreatic duct just proximal to or at the ampulla of Vater. An obstruction to the outflow of pancreatic fluid (by obstruction or edema or spasm), decreases the flow of pancreatic fluid to the intestine and causes an increase of enzymes in the serum. Clinically, the two most important enzymes in the serum regarding pancreatic function are amylase and lipase.

DISEASES OF THE PANCREAS

Acute Pancreatitis (Acute Hemorrhagic Pancreatitis)

Pathophysiology
Acute pancreatitis is a serious disease with a high mortality rate. Causative factors include (1) gall stones, (2) alcoholism, (3) trauma (including abdominal surgery); and (4) infection (e.g., mumps virus). It is believed that the obstruction (due to stones, edema, or spasm) causes an increase in intraductal pressure leading to rupture of acinar cells or ductals. The leaking enzymes somehow become activated and initiate cell necrosis. Hemorrhage is usually present, probably due to necrosis of blood vessels by enzymes. Shock may ensue. The diagnostic problem is to distinguish acute pancreatitis from an acute surgical abdomen that would require surgical intervention (e.g., acute appendicitis, bowel obstruction, perforated ulcer).

Elevations in serum and/or urine amylase and serum lipase are used to diagnose acute pancreatitis. After an attack, increased amylase values may be demonstrated as early as 2–12 hours after the onset of symptoms. These values may return to normal in three to four days. The maximum peak is usually reached on the first day of illness. If little functioning pancreatic tissue is left, values may be normal or borderline. In acute pancreatitis, lipase activity essentially parallels that of amylase. Elevated amylase levels may be present in other nonpancreatic disorders. When the measurement of both amylase and lipase are used together, the ability of these test procedures to aid in the diagnosis of acute pancreatitis is increased.

Case History, Acute Pancreatitis
A 21-year-old white female was seen in the emergency room. She stated that she had been awakened from sleep with sharp epigastric pain that radiated to the lower back. Approximately two weeks previous to this visit, she had experienced

a similar (but less severe) type of pain, but did not seek medical advice. This pain dissipated after three days.

She had been vomiting but denied vomiting blood. Bowel sounds were good, and she denied tenderness elsewhere in the abdomen except over the epigastrium. X-ray studies revealed gall stones. Laboratory data are given in Table 11.3.

Discussion.   Pancreatitis is often associated with gallbladder disease. The etiology in these situations is unclear. It has been postulated that the passage of a stone causes obstruction of the pancreatic duct that drains the pancreatic fluid into the intestine; however, other factors are probably involved. Untreated gallbladder or biliary tract disease often leads to relapsing acute pancreatitis. The pancreatitis in this patient was thought to be secondary to cholecystitis and cholelithiasis.

Note the acute onset and the very high serum amylase level, which fell rapidly. The timing in the collection of the specimen following the onset of pain may be a critical factor; if the time interval is too long, the elevated amylase may be missed. However, in this patient, the amylase level fell, but was still abnormal on her fourth day of illness. The liver enzymes and bilirubin fell abruptly after the initial rise. The rise in total bilirubin was due primarily to an increase in the conjugated fraction, a finding in obstructive liver disease. The increased alkaline phosphatase enzyme is also an indicator of cholestatic disease.

The patient's serum calcium is on the lower limits of the reference range. Calcium may be abnormally low following an acute pancreatic attack. It is believed that calcium forms insoluble soaps with fatty acids released from tissue by enzymatic breakdown.

Other intra-abdominal diseases such as gastric or duodenal ulcers and intestinal obstruction may also cause amylase values to elevate, but these possibilities were ruled out by the physician. The amylase value in these diseases usually does not reach the very high levels seen in acute pancreatitis.

Laboratory Diagnosis of Acute Pancreatitis

Amylase Methodology.   Two types of amylase are found in the serum. The P-type isoamylase is derived only from the pancreas. The S-type isoamylase is derived from multiple organs including salivary glands, ovaries, uterus, lactating

Table 11.3.   Acute Pancreatitis Case History Laboratory Data

	Amylase (5–75)[a] U/L	AST (16–55) U/L	ALT (3–35) U/L	ALP (17–60) U/L	Bilirubin (Total: 0.1–1.0) mg/dL	Ca (8.5–10.2) mg/dL
Admission Day 1	5680	675	506		Total—2.2 Conj.—1.6 Unconj.—0.6	
Day 2	1527	216		232	Total—0.9	8.6
Day 3	217	68		184	Total—0.6	8.9

[a] The reference ranges for components are listed in parentheses.

breast, lung, and fallopian tubes. In normal serum, the ratio of S-type isoamylase to P-type isoamylase varies from 1 : 1 to 3 : 1. Amylase has a molecular weight of 45,000 g/mole and is cleared by the kidneys; therefore, amylase is normally present in both urine and serum. Amylase is usually absent in the serum of the newborn. It appears at approximately two months of age and reaches adult levels at the end of the first year.

Alpha amylase catalyzes the hydrolysis of starches, glycogen, oligosaccharides, and polysaccharides. Amylase belongs to the hydrolase class of enzymes. Substrates commonly used for analysis include potato, corn, and Lintner's soluble starch; other substrates include pure amylose, amylopectin, and glycogen.

1. *Specimen Requirements.* A serum sample is used for amylase assay. With the exception of heparin, anticoagulants inhibit amylase activity. Some physicians prefer urine as the specimen of choice. It has been claimed that urine amylase increases before serum amylase rises, and urine amylase remains elevated longer than serum amylase. Amylase is activated by calcium (metalloenzyme) and chloride. The enzyme is stable for several days at room temperature and longer if refrigerated or frozen.

2. *Amyloclastic (Iodometric) Method.* Starch will bind iodine to form a deep blue iodine complex. The hydrolysis of starch is paralleled by a gradual loss in its ability to bind iodine. In amyloclastic methods, a decrease in substrate is measured, based on the decrease in the blue starch iodine color.

3. *Saccharogenic (Reductometric) Method.* As starch is hydrolyzed, glucose and dextrins are released. In saccharogenic methods, the carbohydrate end products are measured. The test is performed in the following way:

1. The specimen is incubated with a starch solution.
2. The reaction is stopped.
3. Glucose is then measured in two samples, one before incubation and one following incubation.
4. The difference in glucose values is an indication of amylase activity.

The end products of starch hydrolysis consist largely of maltose and some glucose. Some assay methods employ an additional enzyme (maltase) as a reagent to change all end products to glucose:

$$starch \xrightarrow{amylase} mostly\ maltose + glucose$$
$$maltose \xrightarrow{maltase} glucose$$

Glucose is then determined by any glucose method (see Chapter 5).

4. *Turbidimetric method.* Starch is cloudy. As it is hydrolyzed, the cloudiness decreases. In turbidimetric methods, the decrease in turbidity of a starch suspension is measured.

5. *Dye-Labeled Amylase Substrate Method.* When dye is bound to starch, it is insoluble. As amylase attacks the starch, small, water-soluble dye-containing

fragments are produced. The fragments are measured colorimetrically after being separated by centrifugation or filtration from the insoluble unreacted substrate.

6. *Clinical Significance*. Amylase may be increased in (1) acute pancreatitis; (2) intra-abdominal disease (e.g., perforated peptic ulcers, gastric or duodenal ulcers, intestinal obstruction, strangulation of the intestine, and after abdominal surgery); (3) renal disease (due to decreased clearance of amylase by the kidney); (4) salivary gland disorders; and (5) macroamylasemia. The amylase levels in intra-abdominal disorders are generally not as high as in acute pancreatitis.

Macroamylasemia is characterized by the presence of amylase complexed with IgA and IgG globulins. Other binding molecules have also been reported. The large size of the complexes reduces the renal clearance of amylase from the serum. Thus, the serum amylase appears to be elevated while the urine amylase is normal.

The amylase clearance/creatinine clearance ratio (C_{am}/C_{cr}) is used as a test to screen for macroamylasemia. The ratio is determined by measuring the concentrations of amylase and creatinine in simultaneously collected samples of serum and urine. The urine specimen can be a random specimen. There is no need to collect a 24-hour specimen or other timed collection. The same urine specimen must be used for both the amylase and creatinine determination:

$$\frac{C_{am}}{C_{cr}} (\%) = \frac{\text{urine amylase}}{\text{serum amylase}} \times \frac{\text{serum creatinine}}{\text{urine creatinine}} \times 100$$

The clearance ratio reference range varies depending on the methodology used in amylase assay.

In macroamylasemia, the ratio percentage will be low due to the inability of the kidney to clear the large amylase-carrier molecule. In pancreatitis, the ratio percentage will be elevated, but other conditions may also exibit elevated ratios, thus limiting the ability of this test to diagnose pancreatitis.

Lipase Methodology. Lipase activity is found in pancreas, leukocytes, adipose tissue, and milk. The lipase activity in normal serum may not entirely originate from the pancreas. There are four known lipases present in the serum. The lipase of clinical significance is "pancreatic lipase."

The lipase enzyme catalyzes the hydrolysis of esters of long-chain fatty acids. Substrates used for analysis include corn oil, olive oil, or triolein.

1. *Specimen Requirements*. Unhemolyzed serum is the specimen of choice. Hemoglobin inhibits pancreatic lipase activity. Lipase is stable at room temperature for several days and longer if refrigerated or frozen. Lipase activity may increase with bacterial contamination due to microbial lipases.

There is no lipase activity in urine. Either the enzyme is not cleared by glomerular filtration, or it may be bound and inactivated, or its activity is inhibited by substances present in the urine.

2. *Cherry–Crandall Method*. As triglycerides are hydrolyzed by action of lipase, fatty acids are liberated. The quantity of fatty acids liberated is determined by titrating with 0.05 *N* NaOH. The following conditions were used in the original Cherry–Crandall method:

1. The substrate used was olive oil buffered with phosphate.
2. The incubation time was 24 hours at 37°C.
3. The pH indicator used was phenophthalein.
4. The titrating agent was 0.05 N NaOH.

The liberated fatty acids were titrated to a red color. The color change, using phenophthalein as an indicator, occurred at approximately pH of 8.8. In using phenophthalein as an indicator, only 70% of the fatty acids were actually titrated.

Modifications of the Cherry–Crandall method are still used. These modifications include:

1. The incubation time is shorter with times of 6, 4, 3, and 1 hour. It is now known that over 50% of the hydrolytic activity occurs in the first few hours.
2. Different buffers are used.
3. Different indicators for end-point detection are used. Thymophthalein is recommended because the end point (pH 10.5) is considered more accurate. Also, the end point may be measured potentiometrically with a pH meter.

3. *Turbidimetric Method.* Emulsions of fats in water are milky. As lipase hydrolyzes fats, the turbidity decreases. The change in turbidity is measured photometrically as an indication of lipase activity.

4. *Clinical Significance.* In general, lipase values parallel amylase values, except in mumps where lipase is normal. Lipase is less affected by intra-abdominal disease than is amylase.

Chronic Pancreatitis (Pancreatic Cirrhosis)

In chronic pancreatitis, recurrent bouts of inflammation of the pancreas occur and pancreatic tissue is gradually destroyed. Normal pancreatic tissue is replaced by fibrous (scar) tissue. The onset is insidious. Causative factors include the presence of gall stones and chronic alcohol consumption. When the damage is sufficiently extensive, lack of pancreatic enzymes causes malabsorption problems.

Amylase values are of little use in diagnosing chronic pancreatitis. Serum amylase activity may be slightly increased but is often normal. The most common finding in chronic pancreatitis is malabsorption associated with decreased secretion of digestive enzymes.

Carcinoma of the Pancreas

Tumors of the pancreatic gland can arise from the head, body, or tail of the gland. The tumor site determines associated symptoms. If the head of the gland is involved, obstruction of the bile duct may cause jaundice, and obstruction of the pancreatic duct may cause malabsorption problems. Glucose tolerance tests may be abnormal. Serum amylase may be elevated but is of little value for diagnostic purposes.

Cystic Fibrosis (Fibrocystic or Mucoviscidosis)

Cystic fibrosis (CF) is a familial disease characterized as a disorder of exocrine glands. The exocrine glands secrete a viscous material which obstructs exocrine ducts. The exocrine glands most prominently affected include pancreas, lung and sweat glands.

In the United States, approximately 1 in 1600 white infants will be afflicted with this disease. Approximately 1 in every 20 Caucasians is a carrier. The major clinical problems of CF are chronic lung disease and malnutrition secondary to malabsorption. In the pancreas, the ducts become plugged, and pancreatic tissue is replaced with fibrotic tissue. It has been found that the sodium and chloride content of sweat in individuals with cystic fibrosis is significantly elevated even before clinical symptoms develop. These sweat electrolytes remain elevated throughout life.

Sweat Test

The sweat test is the most valuable single test in the diagnosis of cystic fibrosis. In afflicted infants, the test becomes positive between 3 and 5 weeks of age. The test is equally diagnostic in adults. In 1959, Gibson and Cooke developed a simplified method to measure the electrolyte content of sweat.

Test

1. Pilocarpine is applied to the skin and a weak current is used to induce sweating.
2. The sweat is collected on a preweighted gauze sponge. After collection, the sponge is reweighed to determine the amount of sweat collected.
3. The sweat is analyzed for sodium and chloride.

The measurement of sodium and chloride is equally diagnostic for cystic fibrosis. Analysis of both sodium and chloride is recommended in order to evaluate the test better. If there is a wide difference between the sodium and chloride concentration, the test is probably not reliable and should be repeated.

An iontophoresis instrument has also been used to measure chloride on skin surfaces. After sweat is induced, the chloride-selective electrode is applied to the skin, and the chloride is measured directly.

Clinical Significance. A positive test is generally regarded as a sodium or chloride value above 60 mmol/L. Values between 50 and 60 mmol/L are borderline and should be repeated. A positive test is pathognomonic for cystic fibrosis.

GASTRIC–PANCREATIC–INTESTINAL DIGESTIVE PATHWAY

The gastric–pancreatic–intestinal digestive pathway is shown in Figure 11.5. The key below is used to interpret each numbered step of Figure 11.5.

1. Carbohydrate digestion begins in the mouth and is initiated by salivary amylase (ptyalin).

2. The vagus nerve stimulates the stomach to increase the secretion of gastric

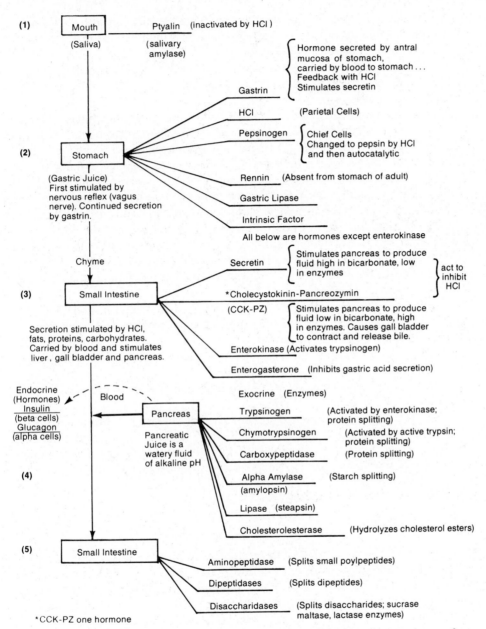

Figure 11.5. Gastric–pancreatic–intestinal digestive pathway. The numbers in parentheses refer to reactions described in the text.

juice. HCl is secreted from parietal cells and gastrin is secreted from the antral portion of the stomach. Gastrin, a hormone, acts on the stomach to allow continued secretion of gastric juice.

Gastrin secretion is also stimulated by food intake (principally protein, and calcium salts) and distension of the stomach. Gastrin also stimulates the small intestine to begin the secretion of secretin, a hormone that acts on the pancreas.

Gastrin secretion is regulated by a negative feedback mechanism with HCl. As the pH of stomach contents decreases, gastrin secretion is inhibited. This safety mechanism prevents oversecretion of acid.

The acid pH inhibits the enzymatic activity of salivary amylase. Pepsinogen, secreted by the stomach, is changed to pepsin by HCl, and protein digestion is initiated.

3. As the stomach contents enter the small intestine, small-intestinal hormones are secreted. These enter the portal blood (and general circulation) and act on the pancreas, liver, and gallbladder. The small-intestinal hormones initiate the following responses: (a) Secretin stimulates the pancreas to produce a fluid high in bicarbonate and low in enzymes; (b) Cholecystokinin–pancreozymin (CCK-PZ) (one hormone) stimulates the pancreas to produce a fluid low in bicarbonate and high in enzymes. It causes the gallbladder to contract and to release bile. Secretin and CCK-PZ also function in a control mechanism which inhibits the secretion of HCl. (c) Fat in the duodenum stimulates the secretion of enterogasterone, a hormone that inhibits the activity of gastric muscle and gastric acid secretion.

An enzyme, enterokinase, has no digestive function but activates trypsinogen to form trypsin.

4. Pancreatic fluid is secreted in response to gastrin, secretin, and CCK-PZ. Pancreatic fluid contains large amounts of bicarbonate and the following protein-splitting enzymes: trypsinogen, chymotrypsinogen, and carboxypeptidase. The protein-splitting enzymes are all secreted in an inactive form and are activated in the intestinal tract.

Pancreatic amylase hydrolyzes carbohydrates, and fats are hydrolyzed by pancreatic lipase.

5. Any small polypeptides and dipeptides that remain are broken down to amino acids by aminopeptidase and dipeptidase enzymes of the small intestine. The disaccharidases of the small intestine hydrolyze the disaccharides to monosaccharides. The amino acids and monosaccharides are then absorbed.

TESTS TO EVALUATE MALABSORPTION PROBLEMS

Absorption Tests

Intestinal absorption tests are based on the fact that hydrolysis of fats, proteins and carbohydrates must occur before absorption is possible. The pancreas is the main source of enzymes for fat, protein, and carbohydrate hydrolysis. Malabsorption may occur due to (1) deficiency of pancreatic enzymes; (2) deficiency of disaccharidases of the bowel (this type would affect carbohydrate absorption); (3) lack of bile salts caused by obstruction of the bile duct (this type would affect fat absorption); and (4) impaired transport across the intestinal wall. Therefore, abnormal intestinal absorption may occur in diseases that are not associated with pancreatic dysfunction. The pancreas has large functional reserve and intestinal absorption tests may not be abnormal until late in the disease. Careful interpretation of test results is necessary.

D-Xylose

Principle. D-Xylose is a pentose. No enzymes are needed for absorption. Given orally, it is passively absorbed in the small intestine, passes unchanged through the liver, and is excreted in the urine.

Test

1. An oral dose of D-xylose is given to the patient.
2. Urine and serum samples are collected at timed intervals.
3. D-Xylose is measured in both serum and urine.

Interpretation. The test is used to determine the site of malabsorption, that is, pancreatic or intestinal in origin. The D-xylose test is normal in pancreatic insufficiency. In intestinal malabsorption, D-xylose absorption is impaired and blood D-xylose levels are low. Both urine and serum values are needed for proper test interpretation.

Vitamin A
Principle. Vitamin A is a fat-soluble vitamin. The enzyme lipase is needed for fat hydrolysis, which will then allow vitamin A to be absorbed. Serum vitamin A levels in children with cystic fibrosis of the pancreas and in patients with pancreatic insufficiency often show decreased vitamin A levels.

Test

1. Vitamin A is determined in the serum for baseline comparison.
2. An oral dose of vitamin A is given.
3. The vitamin A in the serum is again measured.

Interpretation In patients with pancreatic insufficiency (lacking lipase), there is only a slight increase in vitamin A despite the oral dose of vitamin A. Also if there is insufficient bile, vitamin A will not be absorbed. Hence any test that depends on fat hydrolysis has to be interpreted in light of the availability of both enzyme and bile.

Fat Absorption
Principle. Absorption of fat from the intestinal tract requires adequate amounts of lipase and bile. Radiolabeled fatty acids are given as a high fat meal. Serum fats are determined after the meal (4 and 6 hour specimens).

Test

1. Two different radiolabeled fats are given orally, ^{131}I-labeled triolein and ^{125}I-labeled oleic acid. (Since two different isotopes are used, the measurement of both fats are possible in the same sample.)
2. Blood samples are obtained 4 and 6 hours after ingestion of the labeled fats.
3. The isotopes are measured in the samples.

Interpretation. If lipase is present, ^{131}I triolein may be hydrolyzed, and absorption can occur thus ^{131}I will appear in the blood. If only the ^{125}I is present in the blood, the ^{131}I triolein could not be hydrolyzed due presumably to lack of pancreatic enzymes. The ^{125}I oleic acid absorption is independent of lipase activity. If the malabsorption is intestinal in origin, the absorption of both isotopes is impaired and blood levels of either isotope will be low.

Secretin Stimulation Test

Secretin is a hormone secreted by cells in the upper small intestine. The major stimulus for secretion is HCl. Hence as the stomach contents enter the small intestine, HCl stimulates the secretion of secretin. Secretin then acts on the pancreas, causing secretion of a fluid high in bicarbonate content. As the pancreatic juice enters the intestinal tract, the gastric acid is neutralized, removing the stimulus for secretin production.

Test
1. The patient should be fasting. A double radio-opaque gastroduodenal tube is passed and positioned so that duodenal contents may be aspirated through one opening and gastric contents through the other.
2. Secretin is given. The injection may be intramuscular, but the intravenous route is preferred. Timed duodenal secretions are collected.
3. The volume is measured and the bicarbonate content is determined.

Interpretation
In general, decreased bicarbonate with relatively normal fluid volume characterizes the diffuse inflammation of chronic pancreatitis. Localized duct obstruction, characteristic of tumor, is indicated by decreased fluid volume and normal bicarbonate concentration. If steatorrhea is present, the test may be helpful in identifying the site of abnormality. In steatorrhea of pancreatic origin, the bicarbonate concentration is low; in steatorrhea of intestinal origin, the bicarbonate is normal.

GASTRIC

The stomach can be divided into four zones (Figure 11.6). The *cardia zone* is located at the esophageal end of the stomach, and surrounds the cardia sphincter. The *fundus zone* is located in the upper portion of the stomach close to the cardia zone. The cells in the fundus secrete mucous and intrinsic factor. The *body zone* comprises the main portion of the stomach. Four types of cells are found here: (1) surface epithelial cells that secrete mucous, (2) parietal cells that secrete HCl, (3) chief or peptic cells that secrete pepsinogen; and (4) neck chief cells that secrete mucous and pepsinogen. The *pyloric zone* is the narrow portion of the stomach that adjoins the duodenum. The pyloric zone is divided into three subdivisions: antrum, pyloric canal, and sphincter. Pyloric cells secrete mucous, some pepsinogen and gastrin.

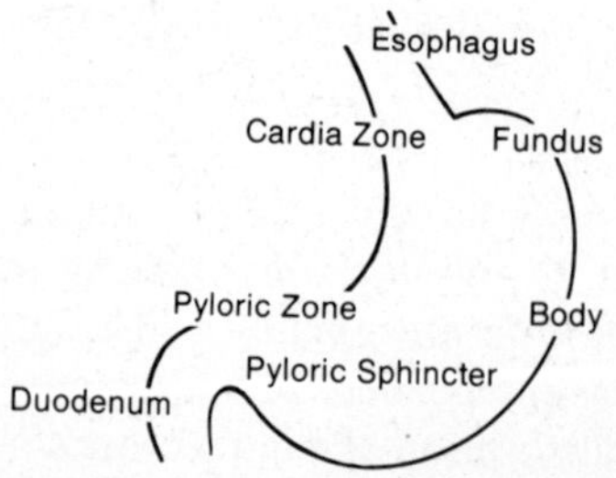

Figure 11.6. Stomach anatomy, anterior view.

The functions of the stomach include the following: (1) acts as a container for the acceptance of food; (2) mixes the food with gastric juice; (3) temporarily stores food and regulates food entry into the small intestine; (4) secretes HCl and intrinsic factor; and (5) provides some digestive enzymes (pepsinogen for protein digestion and gastric lipase for lipids).

PHASES OF GASTRIC ACID SECRETION

Gastric secretion is initiated by vagal stimulation and by the hormone gastrin. Gastrin is secreted by specialized cells called G-cells located in the antrum of the stomach and duodenum. Gastrin initiates the following responses: (1) stimulates the secretion of HCl from parietal cells; (2) stimulates the secretion of pancreatic enzymes from pancreatic cells; (3) stimulates release of secretin by duodenal mucosa; (4) weakly stimulates pepsinogen secretion; (5) stimulates antral motility; and (6) promotes gastric and mucosal growth. A negative feedback mechanism exists between gastrin and HCl. At pH 2.5, there is an 80% reduction in gastrin release; maximal suppression occurs at pH 1.0.

Gastric secretion may be divided into three phases. The *cephalic* or neurogenic phase is initiated by the sight or smell of food, or by chewing and swallowing. The vagus nerve stimulates the secretion of HCl from parietal cells and the secretion of gastrin from the stomach antrum. The *gastric* or chemical phase is initiated by gastrin. In addition to vagal stimulation, gastrin is secreted in response to alcohol, by the presence of calcium salts, by distention of the stomach, and by proteins and breakdown products of proteins. The last phase, *intestinal* phase, is the least important of the three phases. In response to the entry of food into the duodenum, humoral substances are secreted into the blood. Intestinal gastrin is the principal hormone initiating gastric secretion. Some intestinal hormones inhibit gastric secretion. Enterogastrone inhibits gastric secretion and motility.

PHYSICAL CHARACTERISTICS OF GASTRIC RESIDUE

Color

Gastric fluid is usually colorless. If bile is present, the color may be slightly yellow or green. A red or brown color indicates the presence of blood.

Volume

The fasting residual total volume is between 20 and 100 mL, but is usually below 50 mL. A volume above 100 mL is abnormal. An increase in volume may be due to (1) delay in emptying (pyloric obstruction); (2) increase in gastric secretion (duodenal ulcer or Zollinger–Ellison syndrome); and (3) regurgitation of duodenal material (bile will be present).

Odor

The odor is sour normally. A foul odor indicates putrification or fermentation of gastric contents.

Consistency

The consistency of the gastric fluid is non-viscous in nature, unless there is increased amounts of mucous present.

CHEMICAL CONTENT OF GASTRIC RESIDUE

Normal Contents

HCl (Secreted by Parietal Cells)
Free HCl is present in most normal individuals without stimulation. About 4% of healthy young individuals have no free HCl. By age 60, 25% of the population has no free HCl. The words, anacidity and achlorhydria, have no universally accepted definitions. Anacidity has been defined as failure of the gastric fluid pH to fall below 6.0 or 7.0 after maximal stimulation. Achlorhydria has been defined as gastric secretion above pH 3.5 with failure of the pH to fall more than one unit with maximal stimulation. Anacidity and achlorhydria are considered synonymous by some investigators and indicates absence of free HCl.

Enzymes
Pepsinogen, a precursor to pepsin, is secreted by the stomach. Pepsin is an enzyme that digests protein. A small amount of pepsinogen enters the blood directly from peptic cells and is excreted in the urine as uropepsinogen.

Rennin is found in the stomach of infants, but is absent from the stomach of adults. Rennin acts as a casein-coagulating or curd-forming enzyme. Casein is a milk protein.

Gastric lipase contributes only slightly to the digestion of fats. Some nondigestive enzymes, such as LD and alkaline phosphatase, may also be found in the stomach.

Mucous
Mucous is normally present in only small amounts. Mucous may be increased in gastritis, gastric carcinoma, or as a result of mechanical irritation during passage of a gastric tube.

Intrinsic Factor
Intrinsic factor is a protein binder specific for vitamin B12 and is secreted by mucosal cells located in the fundus of the stomach. Intrinsic factor is absolutely necessary for vitamin B12 absorption.

Other
Small amounts of albumin, globulins, and electrolytes are also present.

Abnormal Contents

Blood
Blood is not normally present in gastric fluid. If fresh blood is observed, it may be due to trauma (as occurs in tube passage), carcinoma, or ulcer disease. When

blood remains in the stomach over a period of time, the acid pH converts it to acid hematin, which resembles coffee grounds. Coffee-ground blood may be found in patients with cancer, peptic ulcers, and gastritis.

Food and Organic Acids

Organic acids (lactate, butyrate) in gastric fluid are formed by bacterial action on food retained in the stomach for long periods of time at neutral or alkaline pH. The retained food neutralizes the acid, resulting in the alkaline pH. If free acid is present, organic acids will not be formed. Carcinoma of the stomach or phyloric stenosis may cause food to be retained.

GASTRIC ANALYSIS

Diagnostic Indications for Gastric Analysis

Quantitative tests of gastric acid secretion are now rarely used. Reasons for performing a gastric analysis include (1) to aid in the evaluation of a patient with recurrent ulcer disease; (2) to aid in the diagnosis of Zollinger–Ellison (ZE) syndrome by demonstrating the hypersecretory state of ZE; (3) to determine the completeness of vagotomy after gastric surgery; and (4) to determine if the individual can secrete HCl. If HCl can be secreted, pernicious anemia (PA) may be ruled out. However, in rare cases of junvenile PA, acid may be present.

Assay Method

Collection

The patient should have no food or drink for 12 hours prior to the collection of the gastric specimens. The patient is not permitted to smoke the morning of the test or during the test and should avoid exercise. A gastric tube is inserted and gastric contents are aspirated. After the fasting gastric residue is withdrawn, a gastric stimulus is given. Several stimuli have been used:

1. *Histamine.* Histamine is a powerful stimulus for gastric secretion. Reactions to histamine require administration of an antihistamine prior to injection of histamine. The antihistamine does not significantly interfere with the stimulatory effect on gastric secretion.
2. *Histalog.* Histalog, an analogue of histamine, stimulates the secretion of HCl from parietal cells. Injection of antihistamine prior to histalog injection is recommended by some, although others feel it is not necessary. Until recently, histalog was the most common type of stimulus used.
3. *Gastrin.* Gastrin is a hormone of the stomach antrum and duodenum. It is the most potent stimulator of HCl secretion. On a molar basis, it is 500 times more potent than histamine.
4. *Pentagastrin.* Pentagastrin is a synthetic gastrin but not quite as potent as gastrin. Pentagastrin is now the recommended stimulus.

After stimulation, samples are collected at timed intervals.

Gastric Analysis Reported in meq/L (Titratable Acidity)

The gastric acid is determined by titration of an aliquot of gastric contents with 0.1 N NaOH to pH 7.0–7.4. This pH endpoint may be determined by a pH meter (preferred) or colorimetrically using phenol red as an indicator (endpoint pH; 6.8–8.4).

Specimen. The patient should fast for 12 hours with no liquids and no exposure to olfactory or visual stimuli. A tube is inserted and residual gastric residue is removed and discarded. Samples are then collected at timed intervals. This test measures the *rate* of gastric secretion.

Basal Acid Output (BAO). Basal acid output is reported in terms of output of acid/hour. Specimens are collected at 15 minute intervals. If samples are collected for only 30 minutes, the answer is corrected to reflect output of acid/hour.

Maximal Acid Output (MAO). A stimulus is given to the patient. Samples are collected at 15 minute intervals. The number collected depends on the type of stimulant used. MAO is reported as maximal acid output/hour. When more than four 15-minute samples are collected, the four highest *consecutive* values are used to calculate the output of acid/hour.

Calculations. The normality of each gastric collection is determined. The definition of normality is the number of equivalents per liter. The equivalents per liter are then changed to milliequivalents per liter. The formula, volume 1 × concentration 1 = volume 2 × concentration 2 ($V_1 \times C_1 = V_2 \times C_2$), may be used. An example is shown below.

EXAMPLE 11.1   Twenty milliliters of gastric contents required 3.2 mL of 0.1 N NaOH. Find the acidity in meq/L.

Answer. $V_1 = 3.2$ mL; $C_1 = 0.1$ N NaOH; $V_2 = 20$ mL; $C_2 =$ unknown

$$(3.2)(0.1) = (20)(C_2)$$

$$C_2 = 0.016 \ N$$

The gastric material is 0.016 N or contains 0.016 equivalents per liter. Now, change eq/L to meq/L:

$$\frac{0.016 \ \cancel{eq}}{L} \times \frac{1000 \ meq}{1 \ \cancel{eq}} = 16 \ meq/L$$

Once the meq/L value is known, the answers may be converted to meq/specimen based on the volume collected, then converted to meq/hr. Use the four highest consecutive specimens to calculate meq/hr. The four highest MAO consecutive values were (15 minute intervals):

meq/L Acid in Specimen	Volume of 15-minute Specimen in Liters	meq per Specimen
20	25 mL = 0.025 L	0.500
35	60 mL = 0.060 L	2.100
50	55 mL = 0.055 L	2.750
45	45 mL = 0.045 L	2.025
		Total 7.375 = 7 meq/hr

OTHER TESTS OF GASTRIC FUNCTION

Gastrin

Radiommunoassay methods are available for the measurement of serum gastrin. Plasma samples are unsuitable for analysis. Proteolytic enzymes of the serum will destroy gastrin, therefore, the specimen has to be frozen if not analyzed immediately. The sample should be obtained after a 10 hour fast because protein meals stimulate gastrin secretion.

Gastrin is used in the diagnosis of Zollinger–Ellison syndrome. It may also be used to evaluate individuals with unusual, severe, or recurrent peptic ulcers and those with severe duodenal ulcer disease. Individuals with pernicious anemia have elevated gastrin levels because they are unable to secrete HCl (no feedback mechanism). Gastrin is excreted by the kidney thus, gastrin may be elevated in individuals with renal disease.

Intrinsic Factor (Schilling Test)

The absorption of vitamin B12 requires the presence of intrinsic factor, which is secreted by mucosal cells of the fundus of the stomach. The Schilling test measures the absorption of radiolabeled vitamin B12 to determine if intrinsic factor is present.

Test
1. An intramuscular injection of unlabeled vitamin B12 is given prior to the test to saturate binding sites for B12 in the liver and plasma.
2. An oral dose of radiolabeled (^{57}Co or ^{58}Co) B12 is given.
3. Urine samples are collected and the radioactivity is determined.

Interpretation
If intrinsic factor is present, the radiolabeled vitamin B12 is absorbed, enters the blood, and is then excreted in the urine, since all binding sites for B12 were previously saturated with the unlabeled B12 that was given prior to the test.

Hollander Test (Insulin Test)

The Hollander test is used to determine the completeness of vagotomy. Hypoglycemia stimulates the vagus nerve. The vagal stimulation then increases gastric acid secretion.

Test
1. A basal gastric specimen is collected and the acid content is determined. A basal blood sample is collected and blood glucose determined.
2. Insulin is given intravenously.
3. Gastric samples and blood samples are collected at timed intervals. HCl is determined in gastric samples and glucose is determined in blood samples.

Interpretation
If vagotomy is complete, there should be no increase in HCl after insulin injection because the vagus is unable to stimulate secretion in hypoglycemia. The test is valid only if the glucose value falls below 45 mg/dL at some point in the test.

Detection of Lactic Acid

Normally, lactic acid is not present in gastric contents. When food is retained in the stomach or when free HCl is absent, lactic acid may be produced by fermentation of gastric contents by bacteria. Carcinoma of the stomach is associated with gastric retention and hypochlorhydria; this test is used to evaluate this disease.

LABORATORY FINDINGS IN DISEASES ASSOCIATED WITH ABNORMAL GASTRIC FUNCTION

Ulcers

The measurement of gastric acid in the evaluation of ulcer disease is of little diagnostic value since there is much variation in gastric secretion, possibly due to the etiology of the ulcer.

Peptic implies that the ulcer is produced by gastric juice high in pepsin content and HCl concentration. Peptic ulcers are associated with an increased rate of gastric acid secretion. Peptic ulcers, if found in the stomach, are most often in the antral region. Most peptic ulcers occur in the duodenum. The etiology of many ulcers is not known. Approximately 70% of individuals with duodenal ulcer disease exibit increased gastric volume with hyperchlorhydria.

Pernicious Anemia

Pernicious anemia (PA) results from lack of vitamin B12. In the GI tract, vitamin B12 cannot be absorbed if intrinsic factor is not available. In PA, the laboratory findings consist of decreased gastric volume with no HCl or pepsinogen content. Serum gastrin is elevated. Rare cases of juvenile PA may have normal gastric acid secretion.

Zollinger–Ellison Syndrome

Zollinger–Ellison (ZE) syndrome is caused by excess gastrin production. The gastrin may be produced by neoplasia or hyperplasia of gastric gastrin-producing cells or by tumors of the pancreas involving nonbeta islet cells. The tumors are called gastrinomas. Gastrin secretion is continuous and prolonged with no feedback regulation between gastrin and HCl (i.e., secretion is autonomous). The volume of gastric residue is high. The overnight fluid volume may be greater than 1 L. The acid secretion is high. Fasting gastrin levels are high.

Patients with ZE show symptoms of diarrhea and poor fat absorption, probably secondary to the low pH in the intestinal lumen; they will have recurrent ulcers. Death may occur secondary to gastric complications. The treatment may include gastrectomy to remove the target organ. Drugs are also available that block histamine receptors on parietal cells (H_2 antagonists) and reduce acid secretion from all forms of stimuli, (i.e., food, gastrin, or histamine). Patients with ZE require continual treatment.

Carcinoma of the Stomach

Achlorhydria occurs in about 50% of patients with carcinoma of the stomach. The remainder show hypochlorhydria. A small percentage may show hyperacidity.

Blood may be present in gastric contents and the lactic acid test may be positive. Gastric cytology is the most useful test in the evaluation of carcinoma of the stomach.

Pyloric Obstruction

The volume of gastric residue is large since the exit into the duodenum is obstructed. Food will be present in the gastric residue, which neutralizes the acid. Lactic and butyric acid and yeast cells may be present if the pH is high enough.

Atrophy of the Stomach (Atrophic Gastritis)

Atrophy refers to progressive degeneration and loss of function. In atrophic gastritis, chronic inflammation of the stomach causes degeneration of the mucous membrane. The laboratory findings consist of achlorhydria and low fluid volume. Serum gastrin is elevated due to lack of HCl feedback.

BIBLIOGRAPHY

Bates, H. M. "Macroamylasemia," *Laboratory Management,* 33–37 (June 1981).

Bates, H. M. "Why laboratorians must educate physicians on the use of serum bile acid measurements," *Laboratory Management,* 20–21 (Feb. 1979).

Breuer, R. I. "Stalking the elusive pancreas: A clinician's viewpoint," *Laboratory Medicine* **8** (1), 7 (1977).

Corless, J. K. and Middleton, H. M. "Normal liver function," *Archives of Internal Medicine* **143,** 2291 (1983).

Couch, W. D. *Macroamylasemia and Pancreatitis.* Check Sample Program CC-144, ASCP, 1983.

Demers, L. M. "Diagnostic use of serum bile acid measurements," *Laboratory Management,* 43–50 (May 1979).

De Vivo, D. C. and Nicholson, J. F. "Pathologic and clinical findings in Reye syndrome," *Laboratory Management,* 41–45 (Mar. 1981).

Glasgow, A. M. "Clinical application of blood ammonia determinations," *Laboratory Medicine* **12** (3), 151 (1981).

Henry, J. B. (ed.). *Clinical Diagnosis and Management by Laboratory Methods,* 16th ed. Philadelphia: Saunders, 1979.

Noel, S. and Byers, J. M.: *Laboratory Evaluation of Reye's Syndrome.* Check Sample Program CC-127, ASCP, 1981.

Ratliff, R. C. and Hall, F. F. "Blood ammonia returns to the laboratory," *Laboratory Management,* 16–23 (Aug. 1979).

Shwachman, H. "Overnight sweat test for cystic fibrosis diagnosis," *Lab World,* 41–44 (Apr. 1981).

Tietz, N. (ed.). *Fundamentals of Clinical Chemistry,* 2nd ed. Philadelphia: Saunders, 1976.

Van Lente, F. "Diagnosing acute pancreatitis the enzymatic way," *Diagnostic Medicine* **5** (5), 50 (1982).

Yancy, E. and Schreiner, R. L. "Evaluation and Management of the Newborn with Jaundice," *Journal of the Indiana State Medical Society,* 353–359 (May 1979).

Zinterhofer, L. *Amylase and Lipase.* Check Sample Program CC-118, ASCP, 1979.

POST-TEST

1. The phagocytic cells in the liver are called the _____ and are part of the reticuloendothelial system.

2. Draw a diagram explaining bilirubin metabolism. Explain or show:
    a. What happens to the RBC in the RE system.
    b. How bilirubin is transported in the blood (bound to what?).
    c. The process of conjugation in the liver and name the most important enzyme involved in the conjugating process.
    d. What happens to the bilirubin after it enters the intestinal tract.

3. Match the following:
    1. _____ indirect bilirubin   a. conjugated bilirubin
    2. _____ direct bilirubin   b. unconjugated bilirubin
    3. _____ water soluble
    4. _____ water insoluble
    5. _____ type which may be found in the urine
    6. _____ forms urobilinogen in the intestinal tract

4. Reye's syndrome is typically associated with all of the following *except:*
    a. increased blood ammonia
    b. increased AST, ALT
    c. encephalitis
    d. acid–base disturbances
    e. increased CSF protein

5. Kernicterus is a term used to describe:
    a. deposition of bilirubin in the skin to produce jaundice
    b. deposition of bilirubin in the sclera to produce "yellow eyes"
    c. the icteric color of serum when bilirubin is elevated
    d. deposition of bilirubin in brain tissue of newborns

6. Differentiate between intrahepatic cholestasis and extrahepatic cholestasis.

7. All of the following laboratory findings could be associated with hemolytic anemia *except:*
    a. elevated total bilirubin
    b. decreased haptoglobin
    c. serum unconjugated bilirubin, high
    d. positive urine bilirubin
    e. increased fecal urobilinogen

8. All of the following findings could be associated with viral hepatitis *except:*
    a. elevated total bilirubin
    b. decreased haptoglobin
    c. positive urine bilirubin
    d. elevated urine urobilinogen
    e. icteric skin

9. All of the following findings could be associated with posthepatic obstruction *except:*
    a. elevated total bilirubin
    b. positive urine bilirubin

    c.   icteric skin
    d.   increased fecal urobilinogen
    e.   serum conjugated bilirubin, high

10.   The pigment that gives the orange brown color to feces is:
    a.   urobilin
    b.   urobilinogen
    c.   both
    d.   neither

11.   All or most of the bilirubin in normal individuals is:
    a.   conjugated
    b.   unconjugated
    c.   haptoglobin bound
    d.   hemopexin bound

12.   A 12-year-old boy was noted to be jaundiced but had no other symptoms. Upon physical examination, no abnormalities were found other than the jaundice. Laboratory tests were performed and results were:
Initial exam:

Bilirubin	total 6.1 mg/dL
	direct 0.4 mg/dL
AST	normal

Two-weeks later:

Bilirubin	total 5.2 mg/dL
	direct 0.3 mg/dL
AST	normal
Alkaline phosphatase	normal for age
Reticulocyte count	0.9%
Haptoglobin	normal
Hemoglobin	14.0 g/dL

What is your diagnosis based on the laboratory findings?
    a.   hemolytic anemia
    b.   Dubin–Johnson disease
    c.   Gilbert's disease
    d.   viral hepatitis

13.   A stat bilirubin determination was performed using a standard of 6.2 mg/dL, a control serum, and the patient's serum. The following results were obtained:

Bilirubin control assay value	6.9 mg/dL
Acceptable limits	(± 2 SD units) 1 SD = 0.2 mg/dL
Standard absorbance	0.420
Control absorbance	0.483
Patient's absorbance	0.130

The data may best be interpreted as follows:
a. The control value is within acceptable limits, the patient value is normal, and the patient value can be reported to the physician.
b. The control value is within acceptable limits, the patient value is abnormal, and the patient value can be reported to the physician.
c. The control value is outside acceptable limits, the patient value is normal, and the patient value cannot be reported to the physician.
d. The control value is outside acceptable limits, the patient value is abnormal, and the patient value cannot be reported to the physician.

14. Ammonia is normally removed from the blood by:
a. glutamine synthesis in the brain
b. formation of urea
c. excreted by glomerular filtration in the urine
d. excreted via the respiratory system

15. Match the following:
1. _____ Dubin–Johnson
2. _____ Crigler–Najjar
3. _____ Gilbert's
4. _____ neonatal physiological jaundice

a. conjugation failure due to deficiency of liver enzymes (familial type)
b. failure of transport system that transports bilirubin into liver cells
c. failure of transport system that transports bilirubin into caniliculi
d. type of jaundice associated with immature liver

16. Match the following:
1. _____ chronic active hepatitis
2. _____ Gilbert's syndrome
3. _____ hepatoma
4. _____ primary biliary obstruction

a. elevated gamma globulin, antimitochondrial and smooth muscle antibodies
b. elevated cholesterol and alkaline phosphatase and conjugated bilirubin
c. elevated alpha-fetoprotein
d. elevated unconjugated bilirubin

17. Match the following:
1. _____ hemolytic jaundice
2. _____ intrahepatic jaundice
3. _____ extrahepatic obstructive jaundice

a. elevated serum bilirubin, mostly indirect; urine urobilinogen increased; urine bilirubin negative; fecal urobilinogen increased
b. elevated serum bilirubin, direct, higher; urine urobilinogen negative;

urine bilirubin positive; fecal urobilinogen nega- tive

c. elevated serum biliru- bin, direct may be higher than indirect; urine urobilinogen in- creased; urine bilirubin positive; fecal urobilino- gen decreased

18. What is lipoprotein X?

19. Conjugated bilirubin will react with Ehrlich's diazo reagent (diazotized sul- fanilic acid):
    a. in water
    b. in alcohol
    c. only after proteins are precipitated
    d. a and b

20. Sunlight (UV light) destroys:
    a. conjugated bilirubin
    b. unconjugated bilirubin

21. The final color of the azobilirubin in both the Evelyn–Malloy and Jendras- sik–Grof procedures is determined by the _____.

22. When total bilirubin is 6.1 mg/dL and the conjugated bilirubin is 4.2 mg/dL, the unconjugated bilirubin should be reported as _____. (Give units.)

23. How would the following affect the bilirubin determination? (Use ↑ for increased; ↓ for decreased; NE for no effect.)
    1. _____ exposure to light, especially sunlight
    2. _____ slight hemolysis, Jendrassik procedure
    3. _____ gross hemolysis
    4. _____ specimen left on counter top all day before analysis
    5. _____ specimen stored in refrigerator overnight, wrapped

24. Match the following:
    1. _____ severe hemolytic disorder
    2. _____ hepatocellular disease
    3. _____ early, complete biliary obstruction
    4. _____ normal

Serum Bilirubin		Fecal Urobilinogen	Urine	
Unconjugated	Conjugated		Urobilinogen	Bilirubin
a.	0.9	5.8	not detected	not detected	positive
b.	2.9	5.8	decreased	increased	positive
c.	0.7	0.1	normal	normal	negative
d.	6.3	0.8	greatly increased	increased	negative

25. All of the following are heme-containing compounds *except:*
    a. myoglobin
    b. chlorophyll
    c. hemoglobin
    d. cytochrome enzymes

26. The type of prophyrin in hemoglobin and myoglobin is called:
    a. coproporphyrin
    b. porphobilinogen
    c. urophorphyrin
    d. protoporphyrin

27. Match the following, each has three answers, one from each group:
    1. _____ _____ _____ porphyrias   a. primary abnormality
    2. _____ _____ _____ porphyrinurias   b. secondary abnormality

                                              c. acquired
                                              d. inherited

                                              e. most common
                                              f. least common

28. In suspected lead poisoning, urine is analyzed for:
    a. coproporphyrin I
    b. delta-amino levulinic acid
    c. porphobilinogen
    d. uroporphyrin III

29. A property of porphyrins used to identify and quantitate porphyrins is:
    a. insolubility in chloroform
    b. natural pigment properties
    c. ability to form a color complex with Ehrlich's reagent
    d. ability to fluoresce when exposed to long UV light

30. Match the disease with associated clinical and laboratory findings:
    1. _____ porphyria cutanea tardia
    2. _____ acute intermittent porphyria
    3. _____ congenital erythropoietic porphyria

    a. evident at birth or infancy, extreme photosensitivity with bullous skin
       eruptions, may have hemolytic episodes; urine positive for uroporphyrin
       and coproporphyrin (predominately type I)
    b. attacks may be drug or estrogen induced, gastrointestinal and neurologi-
       cal symptoms, more prevalent in third decade women; urine contains
       porphyrin precursors; defect in uroporphyrinogen I synthetase
    c. Most common type porphyria, onset usually middle years, photocuta-
       neous lesions only; urine positive for uroporphyrin, type I predomi-
       nating

31. Exocrine functions of the pancreas include production of:
    a. insulin and glucagon
    b. bicarbonate

    c. amylase and lipase
    d. b and c
    e. a and b

32. Amylase is cleared:
    a. in the bile
    b. by the liver
    c. by the kidney
    d. by RE cells

33. Amylase and lipase determinations are used to evaluate:
    a. pancreatic disease
    b. liver disease
    c. renal disease
    d. endocrine disturbances

34. In chronic pancreatitis, the most common finding is:
    a. elevated lipase
    b. elevated amylase
    c. malabsorption
    d. elevated urine amylase

35. Amylase may be elevated in all of the following *except:*
    a. mumps
    b. pneumonia
    c. acute pancreatitis
    d. renal disease
    e. intra-abdominal disease

36. Match the following:
    1. _____ amylase   a. hormone secreted by small intes-
    2. _____ lipase   tine, which stimulates the pan-
    3. _____ secretin   creas
    4. _____ trypsin   b. enzyme that catalyzes protein
    5. _____ mucoviscidosis   breakdown
    6. _____ insulin and   c. enzyme that catalyzes carbohy-
                     glucagon   drate breakdown
    7. _____ acinar cells   d. enzyme that catalyzes fat break-
                                        down
                               e. synonym for cystic fibrosis
                               f. endocrine pancreatic function
                               g. cells that produce pancreatic hor-
                                  mones
                               h. no match

37. Match the following for amylase methodology:
    1. _____ saccharogenic methods   a. measures decrease in ab-
    2. _____ dye-labeled substrate   sorbance
    3. _____ turbidimetric methods   b. measures increase in end
    4. _____ amyloclastic methods   product

        c. colorimetric procedure, measures soluble dye fragments

        d. measures drop in substrate concentration

38. The original procedure used to determine lipase based on titration of liberated fatty acids with NaOH is called _____ .

39. Cystic fibrosis is associated with all of the following *except:*
    a. elevated Na and Cl in sweat
    b. malabsorption
    c. increased excretion of Na and Cl in the urine
    d. chronic lung disease

40. The test that is considered pathognomonic for cystic fibrosis is:
    a. vitamin A absorption test
    b. sweat chloride test
    c. stool trypsin
    d. D-xylose absorption test

41. Malabsorption due to lack of pancreatic enzymes could be associated with all of the following *except:*
    a. steatorrhea
    b. increase in amylase activity in blood
    c. deficiency of fat-soluble vitamins
    d. malnutrition
    e. deficiency of vitamin-K-dependent clotting factors

42. How would the following affect the determination (use ↑ increased; ↓ decreased, NE no effect):
    Hemolysis:
    1. _____ amylase
    2. _____ lipase
    Specimen stored in the refrigerator for three days:
    3. _____ amylase
    4. _____ lipase

43. D-Xylose is a test used to measure:
    a. carbohydrate tolerance
    b. fat absorption
    c. pancreatic insufficiency
    d. the functional status of the small intestine

44. The cells in the stomach that secrete HCl are called:
    a. acinar cells
    b. islet cells
    c. parietal cells
    d. peptic cells

45. The most powerful stimulus for gastric secretion is:
    a. histamine
    b. histalog

   c. pentagastrin
   d. gastrin
   e. alcohol

46. The Zollinger–Ellison syndrome is characterized by all of the following *except:*
    a. large volumes of gastric juice
    b. low to negative serum gastrin level
    c. high acid secretion in gastric juice
    d. clinical symptoms of malabsorption and diarrhea
    e. recurrent ulcers

47. Titrable acidity is reported in:
    a. meq/L
    b. meq/100 mL

48. Match the following:

1.	_____	pernicious anemia	a. no acid or pepsinogen, decreased gastric volume, increased gastrin
2.	_____	cancer of the stomach	b. very high gastrin levels, large increase in gastric volume, greatly increased acid
3.	_____	Zollinger–Ellison syndrome	c. decreased acid, lactic acid positive, blood in gastric contents

49. Five milliliters of gastric juice required 2.5 mL of 0.1 N NaOH to reach neutrality. The acidity in meq/L is _____ .

50. The enzyme in the stomach that digests protein is called:
    a. gastrin
    b. rennin
    c. trypsin
    d. pepsin

51. The nerve that is stimulated by the sight or smell of food and by chewing and swallowing and causes HCl and gastrin secretion is called the _____ nerve.

52. Gastrin has a negative feedback system with HCl. This means that in a normal person, as gastric HCl increases, the blood gastrin level will:
    a. increase
    b. decrease

ANSWERS

1. Kupffer cells
2. See Figure 11.2
3. (1) b; (2) a; (3) a; (4) b; (5) a; (6) a
4. e

5. d

6. Intrahepatic—obstruction of the bile ducts that occurs within the liver that results in ineffective transport of bilirubin into caniliculi; extrahepatic—obstruction of the bile duct that occurs in the biliary tract distal to the liver. Posthepatic obstruction may occur due to stones, stricture, or tumors.

7. d

8. b

9. d

10. a

11. b

12. c

13. b

14. b

15. (1) c; (2) a; (3) b; (4) d

16. (1) a; (2) d; (3) c; (4) b

17. (1) a; (2) c; (3) b

18. Abnormal lipoprotein found in persons with cholestasis.

19. d

20. b

21. pH

22. 1.9 mg/dL

23. (1) decreased; (2) no effect; (3) decreased; (4) decreased; (5) no effect

24. (1) d; (2) b; (3) a; (4) c

25. b

26. d

27. (1) a, d, f; (2) b, c, e

28. b

29. d

30. (1) c; (2) b; (3) a

31. d

32. c

33. a

34. c

35. b

36. (1) c; (2) d; (3) a; (4) b; (5) e; (6) f; (7) h

37. (1) b; (2) c; (3) a; (4) d

38. Cherry–Crandall

39. c

40. b

41. b

42. (1) no effect; (2) decreased; (3) no effect; (4) no effect

43. d
44. c
45. d
46. b
47. a
48. (1) a; (2) c; (3) b
49. $(5)(C_1) = (2.5)(0.1)$: $C_1 = 0.05$ eq/L = 50 meq/L
50. d
51. vagus
52. b

CHAPTER 12
ENDOCRINOLOGY

I. HORMONES
   A. Hormone Functions

II. HORMONES OF THE HYPOTHALAMUS AND PITUITARY
   A. Hypothalamic "Hormones"
   B. Pituitary Hormones
      1. Anterior Lobe
      2. Posterior Lobe
   C. Assay Methods for Hormones of the Hypothalamus and Pituitary
      1. Radioimmunoassay
      2. Bioassay

III. CONTROL OF HORMONE PRODUCTION AND RELEASE
   A. Hypothalamus and Pituitary Control
   B. Negative Feedback
      1. Long and Short Loops
   C. Positive Feedback
   D. Diseases Caused by the Absence of Feedback

IV. MECHANISMS OF HORMONE ACTION
   A. Secondary Messenger
   B. Binding of Receptor Proteins
   C. Thyroid Hormone Action
   D. Characteristics of Target Cell Disorders
      1. Hormone Spillover

V. CHEMICAL STRUCTURE OF HORMONES

VI. HORMONES WITH THE STEROID STRUCTURE
   A. Chemical Structure
   B. Types of Steroidal Hormones
   C. Steroid Classes
   D. Nomenclature of Steroids
      1. Sterioisomers
   E. Assay Methods for Steroids
      1. Cortisol
         a. Urine Measurements for Cortisol

OBJECTIVES

On the completion of this chapter you will be able to:

1. Define, illustrate, or explain the following terms:
   a. negative feedback
   b. positive feedback
   c. androgen
   d. estrogen
   e. mineralocorticoid
   f. glucocorticoid
   g. steroid hormones
   h. catecholamines
   i. Cushing's disease
   j. thyroxine

  k. transcortin
  l. TBG
  m. T_3 resin uptake
2. Discuss the major hormones of the hypothalamus and the pituitary.
3. Discuss the major aspects of hormone stimulation and regulation with emphasis on negative feedback mechanisms.
4. Discuss the endocrinology of the adrenal cortex including glucocorticoids, mineralocorticoids, androgens, and estrogen with emphasis given to adrenal hypo- and hyperfunction.
5. List and describe the various assay methods for steroid hormones in the urine and blood.
6. List the various laboratory methods used for the assessment of thyroid function and give the principle behind each method.
7. Describe the renin–angiotensin–aldosterone system and its relationship to electrolyte balance and blood pressure regulation.
8. Describe the endocrinology of the menstrual cycle along with major hormones associated with the maintenance of pregnancy.

GLOSSARY OF TERMS

Adrenogenital syndromes.   Inherited enzymatic blocks in the formation of adrenal corticosteroids that feedback to the pituitary and hypothalamus. Thus, the adrenal gland is overstimulated with the increased production of certain steroids and the absence of others.

Androgens.   Steroid hormones produced in the testes and, to a lesser extent, in the adrenal cortex and ovary, responsible primarily for secondary male sex characteristics.

Catecholamines.   Hormones having the structure of hydroxylated benzene derivatives synthesized primarily in the adrenal medulla and the postganglionic sympathetic nerves.

Estrogens.   Steroid hormones produced primarily in the ovaries (nonpregnant females) and are responsible for secondary female sex characteristics.

Hormone.   Chemical products of organs or tissues of the endocrine system that are transported in the blood and elicit a regulatory effect on other organs and tissues (target cells).

Mineralocorticoids.   Steroid hormones primarily involved in the regulation of electrolyte levels.

Pituitary gonadotrophins.   Pituitary hormones that are primarily involved in the regulation of ovarian and testicular function.

Secondary messenger.   The release of cyclic-AMP due to binding of a hormone to a specific cell receptor on the surface of the target cell.

HORMONES

Hormones are specific chemical products of organs or tissues of the endocrine system that are transported by the blood or other body fluids, and elicit a specific regulatory effect on certain other target tissue or organs. Hormone regulation tends to maintain the chemical integrity of the cell environment.

HORMONE FUNCTIONS

Specific actions of hormones may include (1) a whole body response to hormone stimulation (e.g., when epinephrine is released due to stress); (2) a regulatory action where, for example, salt and water balance are controlled by aldosterone; (3) a morphogenic action where hormones control the rate and type of body growth as in sex hormones and their influence on secondary sex characteristics; and (4) a permissive or complementary action of hormones where certain cells may be only partially or totally unable to respond to a hormone unless the second hormone is present.

The major glands of the endocrine system include the hypothalamus and pituitary, located at the base of the brain; the thyroid and parathyroid glands located in the neck; and the adrenal glands (outer cortex and central medulla areas) capping the upper pole of each kidney. Also included in the endocrine system are the islets of Langerhans of the pancreas, the ovaries (follicles and corpora lutea), and testicles. The pituitary and hypothalamus are connected by the hypophyseal stalk allowing the delivery of hypothalamic neurohumors or peptides to the pituitary. The pituitary gland, often called the "master gland," consists of anterior and posterior lobes, each with separate structures and functions.

The four parathyroid glands are located in the posterior portion of the two thyroid lobes, two parathyroid glands per lobe.

Other glands and tissue have endocrine function, such as the production of digestive hormones by the mucous membranes of the gastrointestinal tract. The placenta also functions as a member of the endocrine system when present.

HORMONES OF THE HYPOTHALAMUS AND PITUITARY

HYPOTHALAMIC "HORMONES"

The hypothalamus, located at the base of the brain, releases hormones (or neuro-humors) that both stimulate and inhibit the release of hormones from the pituitary. These neurohumors all have a polypeptide structure and act on the anterior lobe of the pituitary.

Hypothalamic hormones include thyrotropin-releasing hormones (TRH); growth-hormone-releasing factor (GRF) and -inhibiting factor (GIF); prolactin-inhibiting factor (PIF) and -releasing factor (PRF); gonadotropin-releasing factor (GRF); and melanocyte-releasing factor (MRF) and -inhibiting factor (MIF). Growth-hormone-inhibiting factor (GIF) is also called somatostatin.

PITUITARY HORMONES

Anterior Lobe

The anterior pituitary has seven specific hormones, that stimulate (or inhibit) the various other "target" glands or tissue associated with the endocrine system, including: (1) adrenocorticotropin (ACTH)—adrenal cortex; (2) thyrotropin (thyroid-stimulating hormone, TSH)—thyroid; (3) follicle stimulating hormone (FSH)—ovary and testis; (4) lutenizing hormone (LH)—ovary and testis; (5) somatotropin (growth hormone, GH)—entire body; (6) prolactin—mammary glands; and (7) melanocyte-stimulatory hormone (MSH)—skin. FSH and LH are also called pituitary gonadotrophins. Later in this chapter, we will discuss how these target glands or tissue of the endocrine system, in response to stimulation by the pituitary, release hormones themselves that are directed toward other organs and tissue systems.

Posterior Lobe

The posterior pituitary has two major hormones with the following functions: (1) vasopressin (antidiuretic hormone, ADH)—elevates blood pressure; and (2) oxy-tocin—stimulates uterine contraction in parturition.

Antidiuretic hormone controls blood pressure by (1) regulating renal tubular reabsorption of water and (2) acting as a vasoconstrictor.

ASSAY METHODS FOR HORMONES OF THE HYPOTHALAMUS AND PITUITARY

Radioimmunoassay

In general, most protein and peptide hormones are quantitated by radioimmunoas-say (RIA) where the specimen containing the "unknown" hormone level is mixed with radioactively labeled hormone and antibody. In RIA, the higher the hormone level in the specimen, the less the radioactive hormone is bound to the antibody. Hence, using a standard curve, the unknown hormone level can be determined by counting the radioactivity of the antibody-bound fraction.

Bioassay

A more laborious assay method for protein hormones is called "bioassay" where the specimen containing the "unknown" level of hormone is injected into an animal with some known physiological response being recorded (e.g., uterine weight). This response is then compared to responses in control animals. This method is now obsolete in clinical laboratories.

CONTROL OF HORMONE PRODUCTION AND RELEASE

Overall control of hormone release by the glands of the endocrine system is maintained by an intricate combination of hormone stimulation with modulation via "feedback" mechanisms. Hormone release from an endocrine gland can be stimulated by either a change in a level of circulating substance (e.g., Ca^{2+}, glucose) or the increase in blood concentration of a trophic (or pituitary) hormone (e.g., ACTH, TSH, FSH, etc.).

HYPOTHALAMUS AND PITUITARY CONTROL

As previously described, the hypothalamus can both stimulate and inhibit the release of pituitary hormones through the action of hypothalamic "releasing factors" or "inhibiting factors." These factors, called neurohumors, are released into the pituitary portal system from hypothalamic tuberoinfundibular neurons. The pituitary then stimulates other endocrine glands and tissue by release of its hormones.

NEGATIVE FEEDBACK

The level of pituitary hormone production and secretion is, in general, regulated by negative feedback mechanisms, as illustrated in Figure 12.1. As shown, hormone production and release by target endocrine glands such as the thyroid and adrenal cortex are stimulated directly by the pituitary (and indirectly by the hypothalamus) by release of TSH and ACTH, respectively. When the resultant hormones produced by these glands (i.e., thyroxine and cortisol) reach a high blood concentration, stimulation of the target endocrine glands by both the hypothalamus and pituitary is decreased.

Long and Short Loops

The "long-loop" mechanism is where the blood concentration of hormones feeds back to the hypothalamus or pituitary (as described above); the "short loop" is

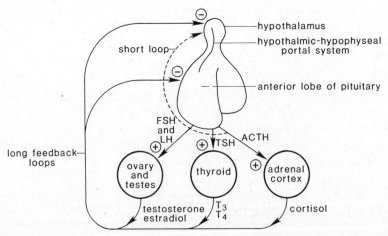

Figure 12.1. Negative feedback mechanisms. (Adapted with permission from Catt, K. J. *Lancet* **I**, 827, 1970.)

where trophic hormones of the pituitary act on the hypothalamus in a feedback mechanism. The T_4 feedback acts primarily on the pituitary, inhibiting the response of the pituitary to TRH stimulation, thus decreasing the amount of TSH produced and released.

Cortisol, on the other hand, has a feedback mechanism that acts primarily on the hypothalamus rather than the pituitary.

POSITIVE FEEDBACK

A positive hormone control system can also influence hormone release and can be present as part of a hormone control system. For example, in the female menstrual cycle, the increased estrogen concentration from the ovaries stimulates LH release.

Normally, estrogen production exhibits negative feedback to the hypothalamus and pituitary for FSH and LH release; however, at midcycle in the female, increased estrogen levels trigger a surge in LH, an example of *positive feedback*.

DISEASES CAUSED BY THE ABSENCE OF FEEDBACK

Many endocrine-related diseases are caused by the absence of normal feedback. Examples are as follows: (1) goiter formation—TSH overstimulation of the thyroid due to decreases in circulatory T_4 and T_3; (2) hypertension and virilization—ACTH overstimulation due to diminished blood levels of cortisol; and (3) hot flashes—pituitary or hypothalamic factors causing menopause symptoms because of decreased levels of estrogen.

MECHANISMS OF HORMONE ACTION

It is generally believed that hormone acts on target tissue by two major mechanisms.

SECONDARY MESSENGER

Some hormones, with the chemical structure of peptides (or proteins) and low-molecular-weight amines, act on target tissue through a secondary messenger, cyclic-AMP. The stimulating hormone attaches to receptor sites on the target cell membrane. This attachment stimulates the cellular production of the enzyme adenylate cyclase, which in turn stimulates the cellular conversion of adenosine triphosphate (ATP) to cyclic-AMP. Cyclic-AMP then produces the intended physiological response through activation of a kinase enzyme system. Cyclic-AMP activates a kinase enzyme that converts certain proteins by reaction with ATP to protein phosphate derivatives. These phosphorylated proteins may lose their ability to repress specific genes. Hence, by this mechanism, we could see increases in

DNA and RNA synthesis, increased cell growth and division, influence of carbohydrate metabolism, and steroidogenesis.

In general, hormones that act by the secondary messenger mechanism include
protein and polypeptide hormones, catecholamines, prostaglandins, and various
releasing factors.

BINDING OF RECEPTOR PROTEINS

A second major mechanism of hormone action involves the combination of the
stimulating hormone with specific receptor proteins inside the cell. The hormone
must move through the cell membrane into the cytoplasm and be lipid soluble.
Steroid hormones (to be discussed below) follow this mechanism.

Once the hormone–receptor complex is formed, it migrates to the nucleus of
the cell where the complex associates with nuclear chromatin to facilitate the
transcription of genes and the formation of messenger-RNA (mRNA). The mRNA
is coded for the production of proteins (enzymes, proteins or protein precursors,
etc.), which are responsible for the target cell response to stimulation. Steroid
hormones follow this mechanism.

THYROID HORMONE ACTION

Thyroid hormones penetrate the cell and bind directly to receptors located in
nuclear chromatin. These receptors are always present regardless of the presence
of thyroid hormone, a feature contrasting the proposed mechanism of steroid
hormones. However, thyroid hormones stimulate protein formation through
mRNA in a manner similar to steroid hormones.

CHARACTERISTICS OF TARGET CELL DISORDERS

Endocrine diseases are most often associated with hypofunction or hyperfunction
of an endocrine gland or tissue with the production of either too much or too little
of a particular hormone. However, the "target cell" of the hormone may have
defective hormone receptors, blunting the response of the target cell to hormone
stimulation. For example, in diabetes type II (or non-insulin-dependent diabetes),
the insulin blood level is often elevated in hyperglycemia due to ineffective cell
receptor binding to insulin. Nephrogenic diabetes insipidus is due to kidney cell
resistance to vasopressin.

Hormone Spillover

Some endocrine diseases are related owing to *hormone spillover* of one hormone
reacting with cell receptors normally "reserved" for other hormones. For example, in female patients with pituitary tumors, HGH elevations with acromegaly is
often seen along with glactorrhea. The HGH apparently reacts with cell receptors

in the breast for prolactin (an example of hormone "spillover"). Also, elevated ACTH in Addison's disease reacts with cell receptors on melanocytes to cause skin darkening.

CHEMICAL STRUCTURE OF HORMONES

The chemical structure of hormones plays a significant role in (1) the mechanism of hormone action in the cell and (2) the method of laboratory analysis, which is based on unique chemical groupings in the chemical structure of different hormones. Hence, it is useful to classify hormones based on their chemical structure, especially in view of the fact that hormones produced in the same glands often have closely related structures. For example, hormones of the anterior pituitary are proteins, glycoproteins, or peptides. Hormones with a steroid chemical structure are produced in the adrenal cortex, while the adrenal medulla produces hormones with the chemical structure of aromatic amines. Also, the neurohumors of the hypothalamus are low-molecular-weight polypeptides. Hormones of the thyroid are iodine derivatives of aromatic amino acids. The estrogens produced by the ovaries are uniquely structured steroids in that one of the rings of the steroid has the chemical structure of phenol.

HORMONES WITH THE STEROID STRUCTURE

CHEMICAL STRUCTURE

Steroid hormones have the basic steroid nucleus consisting of the saturated phenanthrene cyclic rings with an additional five-membered ring attached. The steroid structure along with carbon atom numbering are shown in Figure 12.2.

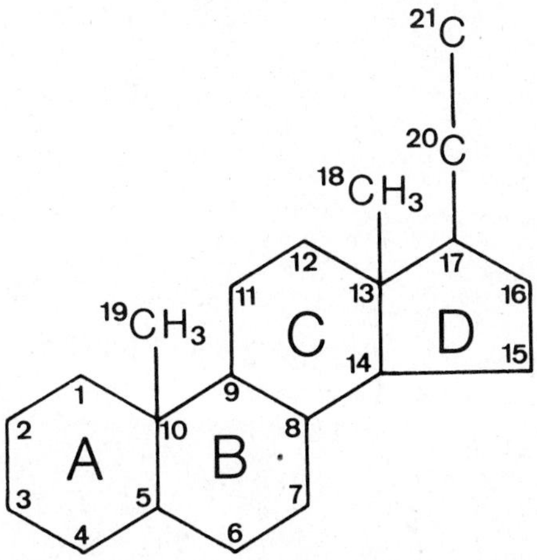

Figure 12.2. Carbon ring systems and atom numbering for steroid hormone structures.

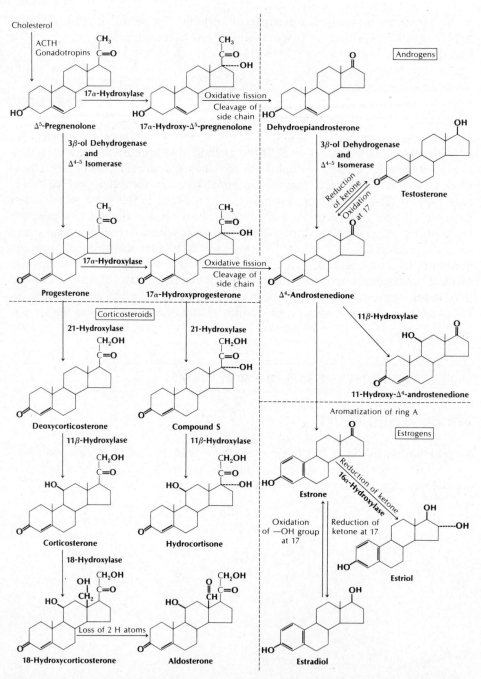

Figure 12.3. The synthesis of steroid hormones. (Reproduced with permission from Freeman, M. and Freeman, S. *Introduction to Steroid Biochemistry and Its Clinical Applications.* New York: Harper and Row, 1970.)

TYPES OF STEROIDAL HORMONES

Steroid hormones are produced and released primarily by the ovary, testis, placenta, and adrenal cortex. In the adrenal cortex, steroid hormones are derived from cholesterol through a series of enzyme reactions.

Although the biosynthesis of the steroid hormones shown in Figure 12.3 is complicated, there are only three main types of steroids of the adrenal cortex. These types include (1) steroids with 18 carbon atoms (C-18, the estrogen group), which stimulate development of secondary female sex characteristics; (2) steroids with 19 carbons (C-19, the androgen group), which primarily stimulate development of secondary male sex characteristics; and (3) steroids with 21 carbons (C-21) consisting of glucocorticoids, mineralocorticoids, and progesterone, which are mainly involved in carbohydrate metabolism, electrolyte balance, and reproductive physiology, respectively. The basic structural differences of each type of steroid is shown in Figure 12.4.

The ovaries and placenta release C-18 steroids (estrogen) and a C-21 steroid, progesterone; the testes release a C-19 steroid, testosterone, a very potent androgen.

STEROID CLASSES

As seen in Figure 12.4, the C-18 steroids (called estrane) has an 18-carbon steroid nucleus with a methyl group (—CH$_3$) bonded at carbon number 13. The C-19 steroids (called androstane) has a total of 19 carbons with two methyl groupings bonded at carbons 10 and 13. The C-21 steroids (called pregnane) have methyl groups at 10 and 13 and, in addition, C-21 steroids have a two-carbon side chain bonded at carbon 17.

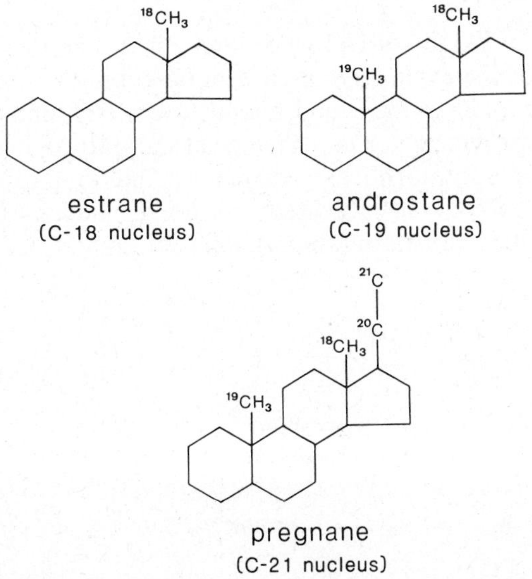

estrane
(C-18 nucleus)

androstane
(C-19 nucleus)

pregnane
(C-21 nucleus)

Figure 12.4.   The basic steroid nucleus associated with estrane, androstane, and pregnane.

NOMENCLATURE OF STEROIDS

Systematic nomenclature of steroids uses the base name for the three classes of compounds described above. The suffix on the base name varies to describe the compound where (1) the "ane" suffix means a fully saturated steroid ring (i.e., no double bonds, as in pregn*ane*); (2) the "ene" suffix denotes one carbon–carbon double bond in the steroid ring, as in pregn*ene*; and (3) an "ol" and "one" suffixes indicate the presence of a hydroxyl group or ketone group in the steroid structure.

After the appropriate base name is discerned, locations of double bond(s) and functional group(s) are indicated with carbon numbers of the steroid nucleus (see Figure 12.2).

Stereoisomers

Also, β is used to denote whether the substituent on the ring is on the same planar side of the ring as the methyl groups at carbons 10 and 13, and α is used to indicate that the substituent is on the opposite side of the ring as the methyl groups at carbons 10 and 13. Substituent bonds opposite to the carbon-10 and carbon-13 methyl groups are shown with dashes (see Example 12.1).

EXAMPLE 12.1.   Name the steroid compounds shown in Figure 12.5.

Answer:
(A)   3 β-hydroxyandrost-5-en-17-one (dehydroepiandrosterone, DHEA)
(B)   11 β,17,21-trihydroxypregn-4-ene-3,20-dione (cortisol)
Note: Trival names for the steroids are listed in parentheses.

ASSAY METHODS FOR STEROIDS

Cortisol

The measurement of blood cortisol levels and cortisol and its metabolites in the urine is essential for the evaluation of adrenal function. Cortisol, a major adrenal hormone, is transported in the blood bound to cortisol-binding globulin (called CBG or transcortin) where it has a biological half-life of 90 minutes. Free (or unbound) cortisol is the metabolically active form and is rapidly catabolized by the liver; CBG-bound cortisol is "protected" from liver metabolism. Major metabolites of cortisol include (1) tetrahydrocortisol, (2) tetrahydrocortisone, (3) cortol,

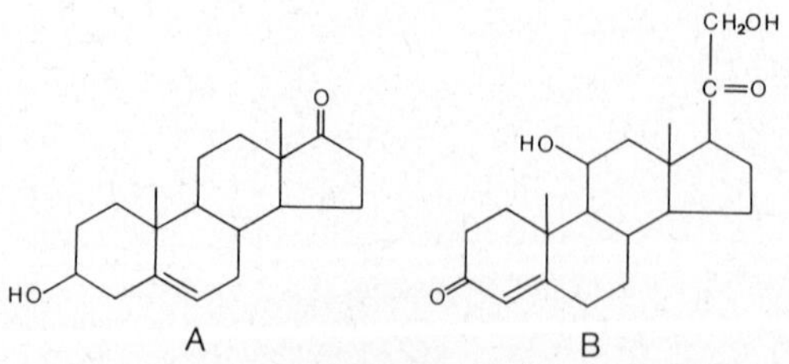

Figure 12.5

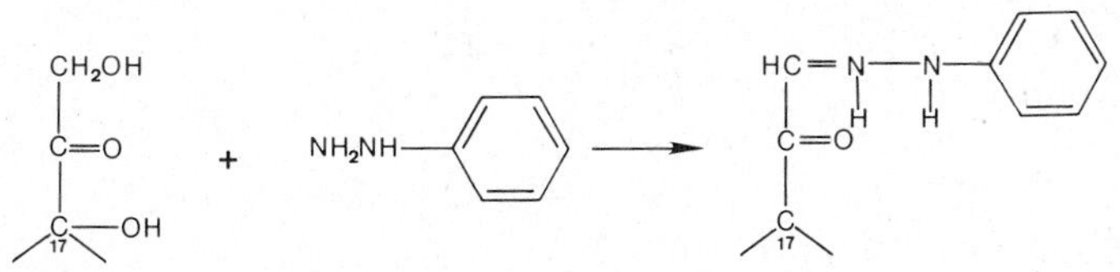

Figure 12.6. Formation of a hydrazone in the Porter–Silber reaction.

and (4) cortolone; these metabolites are conjugated as glucuronides or sulfates and are excreted in the urine. Free cortisol in the urine, an excellent indicator of adrenal hyperfunction, normally comprises only 1% of the total 17-OH steroids excreted.

Urine Measurements for Cortisol

A 24-hour urine collection is required due to the diurnal variation of cortisol. The amount of cortisol released into the blood peaks at about 8:00 a.m. and reaches its lowest blood level at 4:00 p.m. The urine should be stabilized by acidification with boric acid during the collection. A urine creatinine assay should be performed on the specimen to verify the completeness of collection.

Hydrolysis and Extraction of Urinary Corticosteroids

In general, the steps for analysis of steroids in urine include (1) hydrolysis—sulfates and glucuronic steroid conjugates are hydrolyzed by either treatment with acid or by enzymatic hydrolysis; (2) extraction—the steroid analyte is extracted in an organic solvent; (3) purification—where coextracted impurities are removed using additional extractions at different pHs; and (4) quantitation—where the extracted and purified steroid analyte is measured using colorimetric or fluorimetric methods. Other quantitative methods may employ high-pressure liquid chromatography (HPLC) or gas chromatography (GC) after hydrolysis and organic solvent extraction steps.

Colorimetric Reactions for Hydrolized and Extracted Corticosteroids

17-Hydroxycorticosteroids

Approximately 50% of the urinary metabolites of cortisol (having hydroxyl groups at C-21 and C-17 and a keto group at C-20) undergo a condensation reaction with phenylhydrazine to form a yellow hydrazone. This reaction is often called the Porter–Silber reaction for steroids (see Figure 12.6). The 17-hydroxycorticosteroids measured by this method have a normal range of 3–10 mg/24 hr for males and 2 to 8 mg/24 hr for females.

17-Ketogenic Steroids

Approximately 80% of the urinary metabolites of cortisol can be measured as 17-ketogenic steroids. The method involves treating the 17-hydroxy metabolites of cortisol with sodium borohydride ($NaBH_4$, see Figure 12.7). This reaction breaks the C-17–C-20 steroid bond and generates a keto group at carbon number 17. This

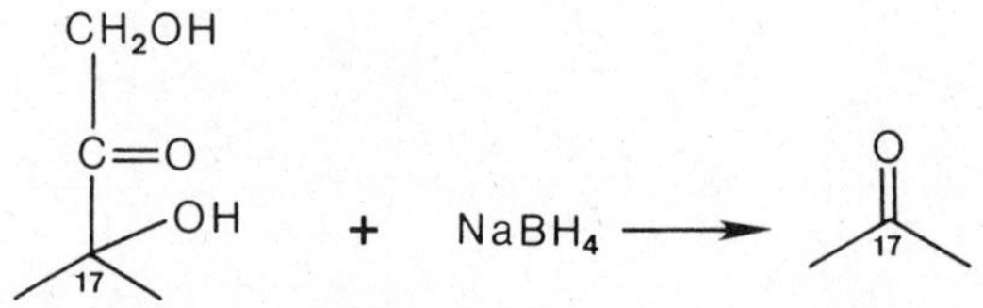

Figure 12.7. Conversion of 17-hydroxysteroids to 17-ketosteroids (used for the ketogenic steroid assay).

"ketogenic" steroid is then reacted with meta-dinitrobenzene to form a purple compound (see Figure 12.8). This reaction is called the "Zimmerman reaction" (see Figure 12.8). The normal range for ketogenic steroids is 5–23 mg/24 hr for males and 3–15 mg/24 hr for females.

17-Ketosteroids

The measurement of 17-ketosteroids in the urine is generally regarded as a "screening test" for androgens and not a measurement of cortisol levels since only 10% of the metabolites of cortisol are 17-keto steroids. Also, the value of ketosteroids as an indication of androgen production is questioned since testosterone, the most potent of the androgen family, is not included in the 17-ketosteroid measurement. The method is based on the treatment of the hydrolyzed and extracted steroid metabolites with meta-dinitrobenzene in alcoholic KOH to form a purple derivative (see Figure 12.6).

The normal range for 17-ketosteroids is 8–20 mg/24 hr for men and 6–15 mg/24 hr for females.

Radioimmunoassay for Steroids

Using radioimmunoassay (RIA) for the measurement of steroid hormones directly in the serum is now much preferred because of the increased sensitivity and specificity of RIA over the older and more time consuming urinary testing procedures. Also, the accurate collection and preservation of a 24-hour urine specimen is often difficult to manage due to involvement of the patient and nursing staff.

Cortisol

Cortisol is commonly measured by RIA using a solid-phase or "coated-tube" radioimmunoassay procedure. The antibody to cortisol is coated on a polypropylene tube. Patient's specimen containing cortisol along with radiolabeled cortisol (^{125}I) is added to the antibody tube, incubated, then decanted. The total radioactivity bound to the tube is then counted in a gamma counter. By comparing radioactivity in patient's unknowns with that of a standard curve, the concentration of cortisol in the patient's specimen can be determined.

The normal range for serum cortisol for an 8 a.m. specimen is 7–25 μ/dL. For specimens collected at 4:00 p.m., the normal range is 4–18 μg/dL. Urine-free cortisol can also be measured in urine by RIA after extraction of the cortisol in an organic solvent.

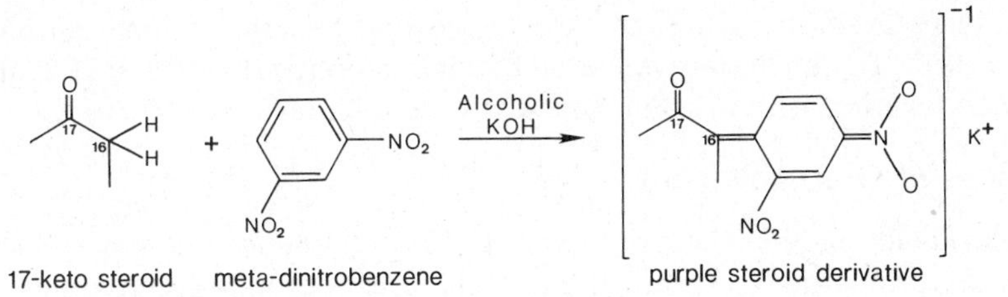

Figure 12.8. Zimmerman reaction for 17-keto steroids.

Other Steroids
Serum levels of most steroid hormones can be readily obtained using radioimmunoassay. Commercial "kit" RIA methods are available for the important androgens, estrogens, mineralocorticoids, and glucocorticoids.

Peptide Hormones
Radioimmunoassays are now available in "kit" form for pituitary peptides that stimulate steroid hormone production; these include assays for ACTH, LH, and FSH. A nonpituitary peptide hormone, human chorionic gonadotrophin (HCG), also stimulates steroid production and can be measured by RIA.

ENDOCRINOLOGY OF THE ADRENAL CORTEX

The adrenal cortex secretes a wide variety of steroid hormones with more than 40 being identified. The adrenal cortex is composed of three distinct layers: outer layer—glomerulosa; middle layer—fasiculata; and innermost layer—reticularis. As mentioned earlier, all hormones of the adrenal cortex are synthesized from the cholesterol as shown in Figure 12.3. To be discussed later relative to Figure 12.3 are major enzyme blocks that cause adrenogenital syndromes. As mentioned above, adrenal cortex steroids of major interest include glucocorticoids, mineralocorticoids, androgens, estrogens, and progesterone. Also, the adrenal cortex is stimulated by pituitary ACTH with the major feedback inhibitory effect coming from cortisol, a glucocorticoid. ACTH has little controlling effect on aldosterone, a potent mineralocorticoid. This hormone is regulated by the renin–angiotensin system to be discussed later.

GLUCOCORTICOIDS

The two major glucocorticoids are hydrocortisone (cortisol) and corticosterone (see Figure 12.3).

Diurnal Variation

There is significant diurnal variation in the secretion of cortisol. For example, after midnight, the blood level rises to reach a maximum at early morning; this level gradually declines to lowest levels, usually in the early evening. In certain diseases like Cushing's syndrome (adrenal cortical hypersecretion), diurnal variation is absent. For an initial evaluation of adrenocortical function, an 8 a.m. blood specimen together with a 4 p.m. specimen are both tested for cortisol.

Functions of Glucocorticoids

Glucocorticoid functions include enhancement of fat lipolysis and protein catabolism while promoting the production of glucose from noncarbohydrate sources (gluconeogenesis). Accordingly, abnormally high levels of these steroids cause "muscle wasting" together with the redistribution of fat. Elevated glucocorticoids also (1) decrease peripheral utilization of glucose, which increases levels of blood glucose and can cause an abnormal glucose tolerance test; (2) inhibit the immune response by decreasing circulating lymphocytes; (3) elevate neutrophiles but depress eosinophils and basophils; (4) increase erythropoesis; (5) induce osteoporosis; (6) increase gastric acid secretion; and (7) are associated with psychotic behavior.

ADRENAL HYPOFUNCTION

Addison's Disease

Low levels of cortisol (the major glucocorticoid) are associated with either primary adrenal hypofunction (Addison's disease) or anterior pituitary hypofunction. Addison's disease is caused by the progressive destruction of the adrenal cortex, usually by infectious agents, especially fungal. Also, amyloidosis of the adrenals or idiopathic adrenal atrophy can produce Addison's disease. The latter has been associated with autoimmune diseases.

Clinical Findings
Clinically, the patient with Addison's disease will have fatigue, weakness, diarrhea, anorexia, nausea, vomiting, and dizziness. The skin often shows increased pigmentation due to hypersecretion of ACTH, which has some melanin-like activity.

Laboratory Findings
Laboratory findings may include: (1) hypoglycemia; (2) hyperkalemia and hyponatremia with hypotension due to decreased aldosterone (a potent mineralocorticoid); (3) eosinophilia; (4) hypercalcemia; and (5) low plasma cortisol levels and associated low urinary 17-hydroxysteroids and ketogenic steroids.

Also, in Addison's disease, there is no stimulation of the adrenals by intramuscular injection of ACTH (or synthetic polypeptide similar to ACTH). Treatment of patients with Addison's disease involves replacement therapy with both glucocorticoids and mineralocorticoids.

ADRENAL HYPERFUNCTION

Cushing's Disease and Syndrome

Adrenal hyperfunction results in increased blood levels of cortisol and is described as Cushing's disease (where bilateral adrenal hyperplasia is present due to excessive pituitary stimulation with ACTH) or Cushing's syndrome. Cushing's syndrome includes (1) adrenal stimulation by ectopic production of ACTH by tumor(s) outside the pituitary (most commonly tumors of the lung, thymus, and pancreas); (2) benign adrenal adenomas where small tumors synthesize cortisol under essentially no ACTH control (many adrenal adenomas do respond to ACTH stimulation); or (3) adrenocortical carcinomas with virulent tumors, which are often metastatic at the time of detection or surgical removal.

Clinical Findings
Patients with adrenal hyperfunction often exhibit (1) central redistribution of fat with a "buffalo hump" torso and a rounded face described as a "moon face;" (2) muscle wasting in limbs; (3) loss of elastic fibers in the skin; (4) bone loss due to osteoporosis; (5) masculinization of females (hirsute appearance often with amenorrhea); (6) hypertension due to increased levels of aldosterone; and (7) hyperpigmentation where increases in ACTH are involved.

Laboratory Findings
Laboratory findings in adrenal hyperfunction include (1) hypokalemia and hypernatremia; (2) alkalosis; (3) hyperglycemia; (4) elevated red blood cell count; (5) neutrophilia; and (6) elevated cortisol level with associated urinary increases in 17-hydroxycorticosteroids and 17-ketogenic steroids. The treatment for patients with adrenal hyperfunction may include the following: (1) irradiation of the pituitary; (2) bilateral adrenalectomy; (3) surgery to remove ACTH-secreting tumors; and (4) chemotherapy.

ADRENOGENITAL SYNDROMES (AGS)

21-Hydroxylase Deficiency

Adrenogenital syndromes (AGS) are associated with enzyme deficiencies along the adrenal cortical pathways of steroid production (see Figure 12.3). The most common block is *21-hydroxylase* deficiency along the cortisol pathway (or "simple virilizing" AGS) where the lack of cortisol feedback to the pituitary causes overstimulation of the gland with resultant overproduction of androgenic hormones. Marked elevations of serum 17-hydroxyprogesterone (a cortisol precursor) and its urinary metabolite, pregnanetriol, are consistent with this form of AGS. A 21-hydroxylase block in both the cortisol and aldosterone pathways is incompatible with life unless treated with appropriate steroids and salt replacement.

11-Hydroxylase Deficiency

Figure 12.9 summarizes the interrelationship between the pituitary and adrenal glands and various endocrine disorders. A "hypertensive" AGS form is seen in the *11-beta-hydroxylase* block where increases in mineralcorticoids, 11-deoxycortisol, and deoxycorticosterone are observed. Treatment with cortisol supplements is generally recommended in AGS.

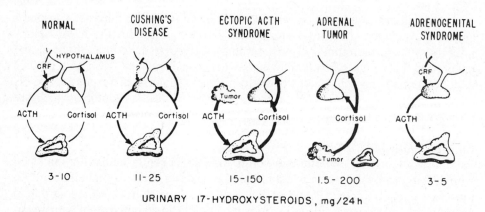

Figure 12.9. The adrenal–pituitary axis in disorders of the adrenal cortex. [Reproduced with permission from Lipsett, M. B., et al. "Humoral syndromes associated with nonendocrine tumors,"*Annals of Internal Medicine*, **61,** 733 (1964).]

FUNCTIONAL STUDIES FOR ADRENAL FUNCTION

ACTH Stimulation

As mentioned earlier, functional tests can be used to diagnose disease relative to adrenal hypo- or hyperfunction. The ACTH stimulation test is useful in distinguishing between hypofunction due to adrenals (Addison's disease) versus hypopituitarism.

To evaluate adrenal hypofunction plasma cortisols and 17-hydroxycorticosteroids are measured before and after ACTH administration. If no rise in these steroids is recorded, adrenal hypofunction is indicated. If cortisol and urinary metabolites rise over baseline levels, hypopituitarism may be indicated. In hypersecretory states, ACTH stimulation in adrenal hyperplasia will cause increases in cortisol and metabolites of three to five times control levels, while in adrenal carcinoma, little or no change over baseline levels is observed. Note that in carcinoma, baseline cortisol levels are often elevated prior to ACTH stimulation.

Dexamethasone Suppression

Another function study for adrenal endocrinology is the dexamethasone suppression study. Dexamethasone is a fluorinated steroid that mimics the feedback of cortisol. When injected, it will lower the level of ACTH stimulation and, in turn, should lower the level of cortisol (and urinary metabolites thereof) if the adrenal glands are under the control of the pituitary. Patients with adrenal hyperplasia will show a decrease in cortisol upon dexamethasone administration, while patients with adrenal carcinoma will maintain an elevated cortisol level throughout the suppression study indicating no pituitary control. The dexamethasone test has also been used to evaluate patients with depression and may help discriminate those patients needing antidepressant drug therapy.

Metyrapone Stimulation

Metyrapone is a drug that blocks the conversion of a precursor steroid to cortisol, allowing little or no cortisol feedback to the pituitary. In this event, increased levels of ACTH stimulate the production of 17-hydroxy and 17-ketogenic steroids. Hence, this function test is used to measure pituitary ACTH reserve.

ACTH Levels

The availability of specific radioimmunoassays for ACTH has enabled the evaluation of the pituitary–adrenal axis without function tests in some patients. For example, an increased serum ACTH together with a low plasma cortisol is consistent with adrenal hypofunction. If both ACTH and cortisol are decreased, then pituitary hypofunction is indicated. Also specific radioimmunoassay tests are available for the various androgens, estrogens, and mineralocorticosteroids, making diagnosis of adrenal dysfunction less complicated. However, it should be noted that androgens and estrogens are, in most circumstances, produced in other endocrine glands to be discussed later.

MINERALOCORTICOSTEROIDS

The major mineralocorticosteroids are, in order of potency, aldosterone, deoxycorticosteroid (DOC), and 11-deoxycorticosteroid.

Aldosterone

Function

Aldosterone is formed from progesterone in the zona glomerulosa layer of the adrenal cortex and has the physiological effect of regulating the transport of ions across cellular membranes, particularly those of the renal tubules. Increases in aldosterone cause the kidney (1) to retain sodium in the plasma (by increasing the reabsorption of Na^+ from the tubular filtrate) and (2) excrete K^+ and H^+ in the urine. The resultant increase in sodium in the plasma leads to an expansion of the extracellular fluid volume which elevates blood pressure. The combined effects of increased blood pressure and decreased levels of plasma K^+ cause a decrease in the production of renin (see below). Hence, patients with increased aldosterone levels often have elevated blood pressure along with increased blood sodium, decreased blood potassium, and alkalosis. The alkalosis is due to the hypokalemia because of (1) the unavailability of K^+ for K^+-Na^+ exchange in renal tubules with enhanced reliance on the H^+-Na^+ exchange mechanism and (2) the shift of H^+ into cells to replace lost potassium ions.

Renin–Angiotensin–Aldosterone System

The renin–angiotensin–aldosterone system provides the major physiological control of extracellular fluid. It is generally believed that a decrease in blood pressure activates "stretch receptor" in the juxtaglomerular cells (JG cells) located in the wall of the afferent arteriole of the kidney. This event stimulates the release of a proteolytic enzyme called *renin* from the JG cells. Renin act on angiotensinogen (which is produced in the liver and released in the plasma) to form angiotensin I, an inactive intermediate. In the lung, angiotensin I is acted on by a converting enzyme (ACE) to form angiotensin II, which, in addition to being a vasoconstrictive agent, stimulates the adrenal cortex to produce and release aldosterone. Hence, decreased blood flow to the kidney as in renovascular stenosis will produce hypertension due to activation of the renin–aldosterone system. In general, ACTH plays a minor role in the stimulation of aldosterone release and maintenance of physiological blood levels. However, pharmacological doses of ACTH can stimulate aldosterone production and release. Also, plasma loading with potassium will stimulate aldosterone as will sodium depletion.

Primary Aldosteronism (Conn's Disease)

Primary aldosteronism (or Conn's disease) is most often due to a single benign adrenal adenoma. Primary aldosteronism, more common in 30–50 year-old women, can also be caused by hyperplasia of the zona glomerulosa. Patients often have increased sodium with increased plasma volume, which results in hypertension. Moreover, serum potassium is reduced, while the bicarbonate is increased. The resulting alkalosis often causes decreases in ionized calcium (due to increased Ca–protein binding) with the development of tetany. Laboratory results should reveal an increased serum aldosterone with a decreased (or normal) plasma renin activity (PRA). PRA is determined by RIA where the amount of angiotensin I produced in a given time is measured using angiotensinogen as substrate.

Secondary Aldosteronism

Secondary aldosteronism is associated with stimulation of aldosterone secretion through the renin–angiotensin–aldosterone system described above. Secondary aldosteronism can occur in (1) changes in effective plasma volume (as seen in hemorrhage or salt depletion); (2) potassium loading; or (3) Bartter's syndrome (or renal juxtaglomerular hyperplasia). The plasma renin activity will be elevated along with the aldosterone level in secondary aldosteronism.

Normal Range
The normal range for plasma aldosterone is 0.3–2.5 ng/dL for the recumbent patient and 4–19.6 ng/dL for the upright patient.

ADRENAL ANDROGENS

In the male, testosterone, the most potent androgen, is secreted by the testes under control of LH (or interstitial-cell-stimulating hormone, ICSH) with only small amounts produced in the adrenals. The major androgens produced by the adrenals from pregnenolone (see Figure 12.3) are dehydroepiandrosterone (DHEA) and androstenedione; these hormones have one-twentieth to one-fortieth the masculinizing effects of testosterone, but rather exhibit an anabolic effect.

In females, only the weak androgens (DHEA and androstendione) are normally produced by the adrenals and the ovaries. Testosterone is produced in the female by liver metabolism of DHEA (20%) and androstendione (65%) with only 20% due to ovarian secretion.

ADRENAL ESTROGENS

Estrogens are female sex hormones and are produced in only minute quantities in the adrenals and testes.

ANDROGENS

The C-19 steroids have significant influence on sexual and nonsexual tissue, including seminal vesicles, prostate, penis, scrotum, vas deferens, bones, muscle, kidney, larynx, and hair growth. In the female, excess androgen production in the adrenals or ovaries can cause hirsutism (increase in body hair) and other forms of virilization. Major androgens in order of potency include testosterone, androstendione, and dehydroepiandrosterone (DHEA).

TESTOSTERONE

As mentioned above, in the male, the synthesis and secretion of testosterone by the interstitial Leydig cells in testes are under the influence of ISCH. Increased testosterone levels "feedback" and suppress ISCH levels. Spermatogenesis in the seminiferous tubules of the testes is regulated by follicle-stimulating hormone (FSH). Both ISCH and FSH are termed pituitary gonadotrophins; their release from the anterior pituitary is regulated by releasing factors of the hypothalamus.

To measure serum levels of androgens, RIA methods are available and, in most cases, preferred over the urinary 17-ketosteroid assay, which does not measure testosterone. Testosterone levels are highest around 8:00 a.m. and lowest between midnight and 4:00 a.m. Normal levels of testosterone in males range from 500 to 860 ng/dL.

Testosterone Levels and Disease

In males, low plasma testosterone levels are valuable for the diagnosis of hypogonadism. In females, elevated plasma testosterone levels are seen in ovarian tumors, adrenogenital syndrome, virilizing tumors, polycystic ovary syndrome, and hirsutism. Also, amniotic fluid testosterone levels may be of value in determining the sex of the fetus.

ENDOCRINOLOGY OF THE FEMALE MENSTRUAL CYCLE

GONADOTROPINS

Pituitary gonadotrophins, FSH and LH, in the female play a major role in the menstrual cycle, which can be broken down into (1) the proliferative follicular phase, when the growth of the ovarian follicle occurs; (2) ovulation roughly at day 14; and (3) the secretory luteal phase, where growth of the corpus luteum occurs. Implantation of the fertilized egg (or ovum) occurs approximately nine days after fertilization in pregnancy and thickening of the uterine lining continues. Otherwise, menstruation occurs and the entire cycle repeats.

FOLLICULAR PHASE

FSH stimulates ovarian follicle development during the follicular phase (see Figure 12.10). LH must be present in order for FSH to function, an example of "permissive hormone action." Both pituitary gonadotrophins stimulate C-18 estrogen (mostly estradiol) production by the ovary, whereupon, at midcycle, an abrupt surge of LH induces ovulation (see Figure 12.10). The increased level of estradiol stimulates the LH surge, an example of "positive feedback."

LUTEAL PHASE

After ovulation, the follicle becomes lutenized (or becomes a secretory body) to form the corpus luteum, which produces progesterone and estrogen. When the egg (or ovum) reaches the uterus, if not fertilized, it degenerates and is sluffed along with the uterine lining (or endometrium) during menstruation. The menstrual cycle repeats every 28 days; hence, as soon as the luteal phase ends, the follicular phase begins anew.

FERTILIZATION OF THE EGG

Trophoblast

If fertilization occurs in the oviduct, the fertilized egg, called a trophoblast, develops invasive capacities that allow it to implant into the uterine wall, a process requiring six to nine days. The trophoblast secretes chorionic gonadotrophin (HCG), which maintains the life of the corpus luteum (six to eight weeks). The corpus luteum releases estrogen and progesterone. Progesterone blood levels are later maintained by progesterone released by the placenta. The elevated levels of

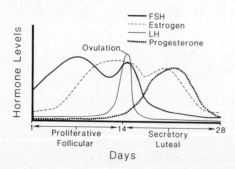

Figure 12.10.   Various hormone levels in the menstrual cycle.

progesterone during pregnancy inhibit ovulation. Pregnancy ends after approximately 270 days and, with parturition, progesterone levels drop as relaxin is released from the placenta. Relaxin acts on the hip joints to increase motility.

PREGNANCY TESTS

Human Chorionic Gonadotrophin (HCG)

HCG is a polypeptide hormone, molecular weight of 38,000 g/mole which consists of an "alpha" and "beta" chain. The alpha chain is quite similar to the alpha chain of the pituitary gonadotrophins, FSG and LH. The beta chain or subunit is unique for HCG.

Serum Pregnancy Tests

Most pregnancy tests are based on the measurement of HCG in either the urine or serum. RIA procedures for serum levels of HCG measure the "beta-chain subunit" of HCG. Beta HCG tests are more reliable than urine pregnancy screens and provide for early detection of pregnancy even in tubal pregnancies and threatened abortions, where HCG levels are somewhat lower. High levels of beta-HCG are also useful for the determination of multiple pregnancies. Beta-HCG also serves as a "tumor-marker" for gestational trophoblastic disease (i.e., choriocarcinoma, molar pregnancy, and others).

HCG AS A TUMOR MARKER

Choriocarcinoma

In choriocarcinoma, the beta-HCG is quite elevated (often 10,000 mIU/mL or higher); however, upon effective removal or chemotherapy, the beta-HCG falls below 2.5 mIU/mL. Thus, beta-HCG serves as an effective "tumor marker" for trophoblastic malignancies.

HCG and Testicular Cancer

HCG is elevated in testicular tumors, however, it returns to low levels when the cancer is removed. Hence, HCG serves as a valuable "tumor marker" in testicular cancer.

ESTROGEN

Estrogen Production

Estrogen levels in pregnancy are elevated due to placental production of estriol while, in nonpregnant females, the ovaries produce estrogen (mostly estradiol). Oral contraceptives contain estrogen and progesterone and suppress ovulation by reducing pituitary gonadotropin levels by the feedback mechanism.

Actions of Estrogens

Actions of estrogens include (1) ovary—increases the sensitivity of granulosa cells to FSH; (2) uterus—increases tissue protein synthesis of the myometrium (muscle wall of uterus); (3) increases motility of fallopian tubes; and (4) produces secondary female sex characteristics (breast and reproductive system development, fat distribution, etc.). Major estrogens are (1) estradiol—major ovarian estrogen and most potent; (2) estrone—ovarian estrogen; and (3) estriol—placental estrogen, which generally reflects the integrity of the fetal–placental unit.

Measurements

Estrogens are measured in the serum by RIA methods. Urinary measurements of estrogens are based on the Kober reaction where the phenolic ring of C-18 steroids together with oxygen groups at C-3 and C-17 cause the formation of a pink color when the estrogen steroids react with sulfuric acid.

Hypo- and Hyperestrogenism

Hypoestrogenism is characterized by (1) sterility related to the absence of ovulation and corpus luteum function; (2) inadequate sexual maturation; and (3) amenorrhea. Increased estrogen levels in males can result from tumors of the adrenal cortex causing gyenecomastia, testicular atrophy, impotency, and azospermia. In young girls, increased estrogen levels may cause female hair patterns to develop early.

Estriol and Fetal Distress

Increased estriol levels during pregnancy indicate good placental function and fetal "well being." In pathological pregnancies such as those complicated by toxemia, hypertension, and diabetes mellitus, the estriol often drops precipitiously prior to fetal death *in utero*. Combined measurements of serum and urinary estriols, human placental lactogen (a placental hormone), and amniotic fluid surfactant (which relates to fetal lung maturity) helps clinicians decide when to perform a caesarean delivery in cases of fetal distress.

OTHER PITUITARY PEPTIDE HORMONES

HUMAN GROWTH HORMONE

With the possibilities of producing significant amounts of protein hormones by genetic engineering, where, through gene-splicing techniques, bacteria can be "programmed" to produce specific proteins, interest in laboratory testing for pituitary hormones has increased. Growth hormone (GH) is a pituitary hormone that increases the growth in bony, connective, and epithelial tissue and also raises blood glucose levels (as an antagonist to insulin) while stimulating the mobilization of fats. Increases in GH after puberty, possibly from a hypothalamic or pituitary adenoma, can cause acromegaly, a disease characterized by abnormal skeletal growth of the bones and fat. Giantism is observed if the increase in GH occurs *before* puberty (or before epiphyseal plates are closed). In instances where pituitary gonadotropins are low or absent, epiphyseal plates do not close and growth may continue for 30 years or longer. Dwarfism is a disease caused by the lack of GH. Dwarfs have large heads and are of short stature but have normal intelligence. In pituitary hypofunction with decreased GH and TSH, a mentally retarded dwarf is often seen.

MELANOCYTE-STIMULATING HORMONE

Melanocyte-stimulating hormone (MSH) acts on melanocytes to stimulate production of melanin pigment of the skin. MSH and ACTH have similar peptide structures and, accordingly, ACTH, when elevated, may have melanin skin action. ACTH is also used in adrenocortical "function tests" (discussed earlier in this chapter).

PROLACTIN

Pituitary prolactin (otherwise called lactogenic hormone, mammotropin or leutotropin) release is controlled by inhibitory and releasing hypothalamic neurohumors. The prolactin inhibitor (PIH) is suppressed in the mother after delivery and, as a result, prolactin stimulates the secretion of milk in mammary glands. Another hormone of the posterior pituitary, oxytocin, is necessary for the release of milk from mammary glands. Prolactin is measured by radioimmunoassay.

Dopamine acts to decrease prolactin production. Prolactin release is primarily inhibited rather than stimulated. Elevated prolactin levels in infertile women with menstrual cycle disturbances (due to FSH elevations) are suggestive of space-occupying lesions in the pituitary–hypothalamic region. Prolatinomas or pituitary tumors secreting prolactin (sometimes called "chromophobe tumors" because resistance to certain tissue-staining procedures) have been shown to shrink in response to bromocryptine. Galactorrhea disappears in many female patients treated with bromocryptine.

In patients with hyperprolactinemia, a TRH stimulation study helps differentiate patients with pituitary adenomas versus other causes of prolactinemia. Patients with pituitary adenomas show no change in prolactin levels on stimulation with TRH; patients with prolactinemia from other causes show an increase in prolactin levels on TRH administration.

THYROID ENDOCRINOLOGY

PITUITARY THYROID STIMULATING HORMONE (TSH)

Pituitary thyrotropin or thyroid stimulating hormone (TSH) stimulates the growth and function of the follicular cells of the thyroid gland, which causes an increase in production of thyroglobulin and thyroid hormones (T_3 and T_4).

Measurements

TSH is analyzed by radioimmunoassay and is an excellent test for the diagnosis of hypothyroidism. For example, in primary hypothyroidism, TSH is elevated. In secondary or tertiary hypothyroidism (hypofunction of the pituitary or hypothalamus, respectively), TSH will be low. With the administration of thyroid-releasing hormone (TRH) to a hypothyroid patient and subsequent monitoring of TSH, lesions of the pituitary can be distinguished from hypothalamic lesions.

THYROID HORMONES

Functions

The thyroid gland produces and releases iodoamino acid hormones, which influence the rate of metabolism of body cells including growth, cerebral activity, alertness, etc. If the thyroid gland is hyperactive, increased levels of thyroid hormones are produced and released causing an increased rate of body metabolism (basal metabolic rate, BMR). A hypofunctioning thyroid results in low levels of thyroid hormones and low metabolic rate. The thyroid is controlled directly by pituitary TSH and indirectly by hypothalamic TRH.

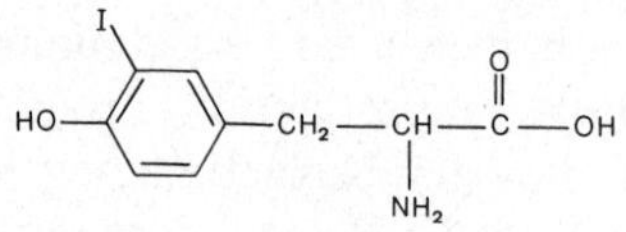

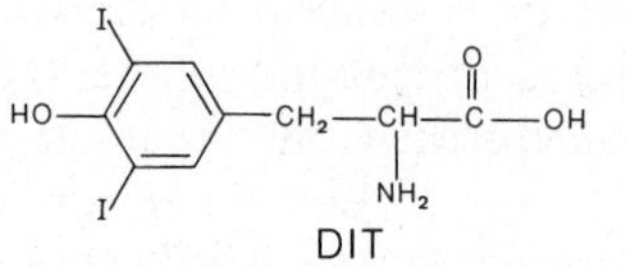

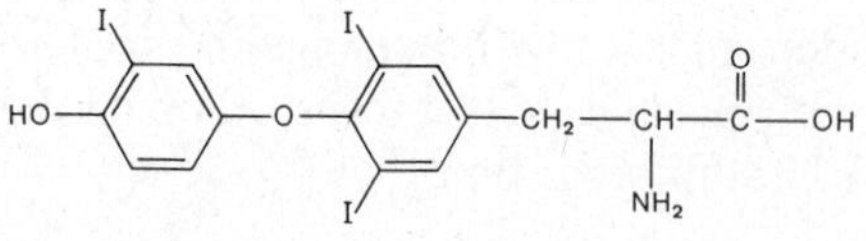

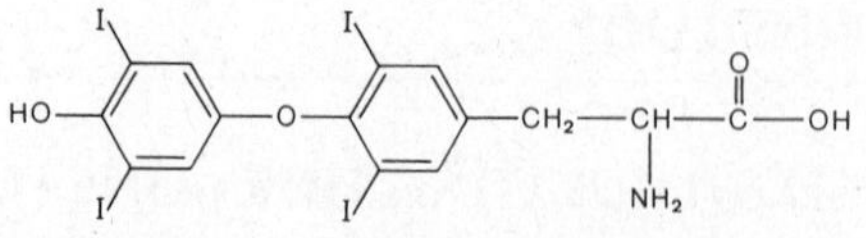

Figure 12.11. Structure of thyroxine (T_4) and triiodothyronine (T_3).

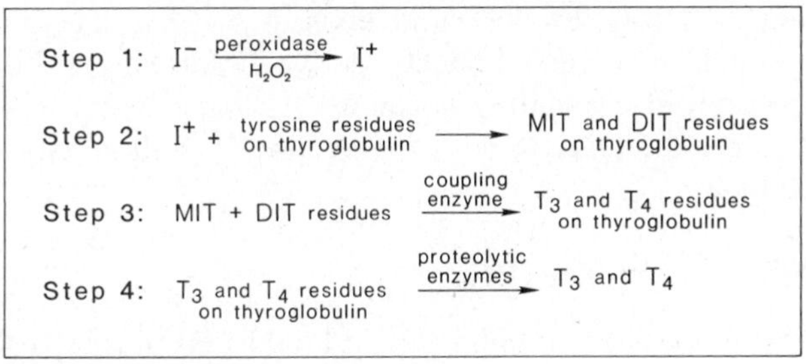

$$\text{Step 1:} \quad I^- \xrightarrow[H_2O_2]{\text{peroxidase}} I^+$$

$$\text{Step 2:} \quad I^+ + \underset{\text{on thyroglobulin}}{\text{tyrosine residues}} \longrightarrow \underset{\text{on thyroglobulin}}{\text{MIT and DIT residues}}$$

$$\text{Step 3:} \quad \text{MIT + DIT residues} \xrightarrow[\text{enzyme}]{\text{coupling}} \underset{\text{on thyroglobulin}}{T_3 \text{ and } T_4 \text{ residues}}$$

$$\text{Step 4:} \quad \underset{\text{on thyroglobulin}}{T_3 \text{ and } T_4 \text{ residues}} \xrightarrow[\text{enzymes}]{\text{proteolytic}} T_3 \text{ and } T_4$$

Figure 12.12.   The synthesis of T_3 and T_4.

Thyroxine (T_4) and Triiodothyronine (T_3)

The thyroid produces two major hormones, T_3 and T_4, which "feedback" and reduce levels of TRH and TSH. The chemical structures of T_3 and T_4 and precursors monoiodotyrosine and diiodotyrosine are shown in Figure 12.11.

Formation of T_3 and T_4

Iodine for the formation of T_3 and T_4 is reduced in the intestinal tract to iodide and absorbed (primarily in the small intestine). Circulating iodide is "trapped" by the thyroid to be used later for hormone synthesis. Iodide trapping by the thyroid is used for radioactive iodine uptake studies (RAI, ^{131}I) using X-ray scans. At higher dosages, ^{131}I can be used to therapeutically destroy thyroid tissue.

Once iodide is trapped, enzyme systems in the thyroid (usually under control of TSH) facilitate the synthesis of T_3 and T_4. Initially, peroxidase converts iodide (I^-) to reactive electrophilic iodinium (I^+), which, in the presence of iodinase enzyme, bonds to the aromatic ring of tyrosine residues, forming monoiodotyrosine (MIT) and diiodotyrosine (DIT) residues. These reacting tyrosine residues are located on thyroglobulin, a follicular thyroid protein.

Coupling enzymes then combine the iodinated tyrosine residues to form 3,5,3'-triiodothyronyl and 3,5,3',5'-tetraiodothyronyl residues, which, when proteolytically cleaved from thyroglobulin, yield triiodothyronine (T_3) and tetraiodothyronine (T_4), respectively. These reaction steps are summarized in Figure 12.12. Thyroglobulin does not normally leave the thyroid and, if present in the blood or lymph, may stimulate autoantibody formation. Also, the presence of thyroglobulin in the blood may indicate thyroid carcinoma.

"Free" and Protein Bound T_3 and T_4

There is 70 times more plasma T_4 circulating than T_3, with about 50% of the latter derived from peripheral cellular deiodinization of T_4. T_3 is more metabolically active than T_4 and is loosely bound to a plasma-carrier protein, thyroxine binding globulin (TBG). Approximately 99.7% of T_3 is bound to TBG with the remaining 0.3% active (or "free"), being unbound. It is the "free" (or unbound) hormone that binds to cell surface receptors on target tissue and provides "feedback" to the hypothalamus and pituitary. T_4 is strongly bound to TBG (60–75%); thyrobinding prealbumin (TBPA), 15–30%; and albumin (5–10%). The remaining 0.04% is "free" or unbound T_4, the metabolically active T_4 fraction. T_4 comprises about 95% of the total thyroid circulating hormone. The approximate biological half-lives of T_4 and T_3 are 6.5 and 1.5 days, respectively, with metabolic breakdown occurring in the liver.

Thyroxine-Binding Globulin (TBG)

Levels of thyroxine binding globulin in the blood play a vital role in the total amount of T_4 measured by radioimmunoassay. For example, if the level of TBG is increased in a patient (as seen in the pregnancy "estrogen effect" or women taking birth control pills containing estrogen), the level of "free" hormone feeding back to the hypothalamus and pituitary is decreased. Hence, the reaction below is shifted to the right:

$$T_4 + TBG \rightleftarrows TBG \cdot T_4$$
$$\text{"free } T_4\text{"} \qquad \text{"bound } T_4\text{"}$$

This event results in increased TRH and TSH, which stimulate the thyroid to increase the total thyroid level to a point where the "free" hormone is again within normal limits (or levels prior to estrogen in this case) as shown in Figure 12.13.

Other hormones such as anabolic steroids lower the TBG levels and decrease the total amount of circulating T_4 by shifting the above reaction to the left. The resulting increased levels of free T_4 lowers TRH and TSH, and thyroid output of T_4 is reduced until the amount of free hormone is again physiologically normal. Therefore, a measurement of binding sites on plasma TBG is required to accurately interpret total T_4 levels (see "T_3 resin uptake" test later in this chapter).

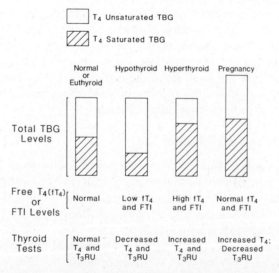

Figure 12.13. The relationship of TBG levels and levels of free T_4.

Thyroxine-Binding Globulin Levels
The level of thyroxine-binding globulin can be accurately and directly measured by RIA. The test can be particularly valuable in evaluating T_4 levels especially in neonates. The normal range for TBG in adults is 1.5–3.4 μg/dL. In pregnancy, the expected range is 5.6–10.2 μg/dL.

THYROID LABORATORY TESTS

Thyroxine (T_4)

General Procedure
Radioimmunoassay for T_4 involves the reaction of a three-component system: (1) the patient's serum containing T_4 treated with a chemical compound to displace T_4 from TBG; (2) radiolabeled ^{125}I thyroid hormone ($*T_4$); and (3) an animal antibody (Ab, usually rabbit) with IgG immunoreactivity for T_4. These three components undergo a chemical reaction as described below:

$$T_4 + {}^*T_4 + Ab \rightleftarrows Ab \cdot T_4 + Ab \cdot {}^*T_4$$

The amount of added antibody is insufficient to bind all of the T_4 in the test tube (patients' and added radiolabeled T_4); hence, the patient's hormone and radiolabeled hormone compete for the limited binding sites on the antibody. Accordingly, the higher concentration of T_4 in the patient's serum, the lower the amount of radiolabeled T_4 bound by the antibody and vice versa. Therefore, the measured total radioactivity of the bound radiolabeled T_4 is inversely related to concentration. Hence, after the competition for the binding sites reaches equilibrium, the antibody bound T_4 (Ab $\cdot$ T_4 and Ab $\cdot$ $*T_4$) can be separated by centrifugation and decanted from the unbound T_4 by addition of polyethylene glycol, ammonium sulfate, or goat antirabbit antisera:

$$Ab \text{ (rabbit)} \cdot {}^*T_4 + Ab \text{ (goat)} \rightleftarrows Ab \text{ (goat)} \cdot Ab \text{ (rabbit)} \cdot {}^*T_4$$
insoluble complex

The test tube must be centrifuged at high speed (1500g for 15 minutes) to precipitate the bound fraction. Alternatively, dextran-coated charcoal can be added to bind the free (or unbound) T_4. Centrifugation and decantation of the test tube results in isolation of the unbound T_4 hormone on the charcoal.

Standard Curve
After the addition of the antirabbit antisera followed by centrifugation and decanting, the radioactivity of the patient's test tube is measured in a gamma counter. The concentration of the T_4 in the patient's specimen is determined by a standard curve where radioactivity is plotted versus known concentrations throughout the range (see Figure 12.14). The normal range for serum T_4 is 4.5–11 μg/dL.

T_3 Resin Uptake

As previously discussed, the metabolically active T_4 that feeds back to the pituitary and hypothalamus is the "free" or unbound T_4. The amount of thyrobinding

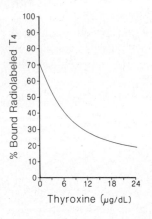

Figure 12.14. Sample standard curve for thyroxine (T_4) by radioimmunoassay.

globulin in the patient's blood then greatly affects both the total amount of T_4 and the free T_4 present.

$$T_4 + TBG \rightleftharpoons TBG \cdot T_4$$
$$\text{"free"} \qquad \text{"bound"}$$

For example, any increase in TBG will shift the above reaction to the right, lowering the free T_4 and stimulating TSH. Additional T_4 will be produced until the free T_4 is again within normal limits. The equilibrium expression for the above reaction reveals

$$[T_4] = \frac{K_{eq}[TBG \cdot T_4]}{[TBG \text{ binding sites}]}$$

where K_{eq} = equilibrium constant; $[T_4]$ = concentration of free T_4, which is inversely related to TBG concentration (or binding sites). The T_3 resin uptake (T_3RU) test is an indirect measurement of binding sites on TBG.

Principle of T_3RU

Excess radioactive $^*T_3$ (^{125}I) is added to a patient's specimen. The $^*T_3$ rapidly occupies the unsaturated binding sites on the TBG. T_3 has less binding affinity for TBG than T_4, hence no T_4 is displaced:

$$TBG \cdot T_4 + {}^*T_3 \rightleftharpoons TBG \cdot T_4 \cdot {}^*T_3$$

Excess $^*T_3$ (or the $^*T_3$ that did not bind to TBG) is removed by the addition of a solid resin:

$$^*T_3 + \text{resin} \rightleftharpoons \text{resin} \cdot {}^*T_3$$

The test tube is centrifuged and decanted and counted in a gamma counter. The counts of the patient's specimen are converted to percentage uptake based on counts obtained on a standard:

$$\text{T}_3\text{RU}(\%) \text{ of unknown} = \frac{\text{counts patient}}{\text{counts standard}} \times (\text{T}_3\text{RU of standard})$$

Interpretation of T₃RU

The higher the $\text{T}_3\text{RU}(\%)$, the lower the available binding sites on TBG; hence, T_3RU is inversely related to the available binding sites on TBG:

$$\text{T}_3\text{RU} \propto \frac{1}{\text{TBG binding sites}}$$

A hyperthyroid patient has more T_4 occupying binding sites on TBG, hence more of the $^*\text{T}_3$ binds to the resin and the T_3RU is increased. A hypothyroid patient has less T_4 occupying binding sites on TBG, hence more of the radioactive $^*\text{T}_3$ would attach to the TBG and less to the resin, causing a low T_3RU (see Figure 12.13).

Normal Ranges

The euthyroid range for T_3RU is 25–35%. In hyperthyroid patients with normal levels of TBG, T_3RU is usually greater than 35%. In hypothyroid patients with normal levels of TBG, the T_3RU is less than 25%.

Free Thyroid Index

Since the free T_4 is directly related to the total amount of TBG-bound T_4 and inversely related to TBG binding sites, the product of the total T_4 and the T_3RU gives an approximation of the level of "free" T_4 called the free thyroid index (FTI):

$$\text{FTI} = \frac{\text{T}_4 \times \text{T}_3\text{RU}}{100}$$

Since the FTI takes into account the variation of TBG binding sites, the FTI is a better indicator of thyroid disease than either T_4 or T_3RU used separately. The normal range for FTI is 1.3 to 3.8.

Direct Free T₄ Measurements

The free T_4 in a patient's specimen can be measured by (1) a laborious equilibrium dialysis procedure (where the free T_4 is allowed to pass through a semipermeable membrane prior to measurement) or (2) using a direct immunoextraction technique for extracting the free T_4 prior to analysis.

EXAMPLE 12.2.   A pregnant patient has hypothyroid symptoms but her T_4 of 4.8 μg/dL was in the normal range. Her T_3RU was 21%.

Answer: First calculate the FTI:

$$\text{FTI} = \frac{\text{T}_4 \times \text{T}_3\text{RU}}{100}$$

$$\text{FTI} = \frac{4.8 \times 21}{100} = 1.0$$

The FTI is low, so the patient probably is hypothyroid. Pregnancy increases the production of TBG, hence her total T_4 was in the normal range. However, her free T_4 was low. An elevated TSH in this patient would confirm primary hypothyroidism.

Reverse T_3

Reverse T_3 (rT_3) is a biologically inactive metabolite of T_4 (3,3',5'-triiodothyronine) formed by monodeiodination of T_4. Reverse T_3 may play a role in the assessment of thyroid function in the chronically ill patient and may be helpful in the diagnosis of "borderline" hyper- and hypothyroidism.

Reverse T_3 is measured by RIA with an expected normal range of 7.2–26.0 ng/dL.

Thyroglobulin

Thyroglobulin, a 19S glycoprotein, is a normal secretory product of the thyroid gland and, under the stimulation of TSH, is produced and transported to thyroid cellular lysosomes for digestion and subsequent release of T_4 and T_3.

Patients with differentiated thyroid tumors often show elevated thyroglobulin in the serum as do patients with Graves' disease. Thyroglobulins are assayed by RIA with normal levels being less than 38 ng/mL.

THYROID DISEASES

Thyroid diseases are associated with (1) thyroid hypofunction or hypothyroidism and (2) thyroid hyperfunction or hyperthyroidism. Myxedema is an edematous condition associated with hypothyroidism, while thyrotoxicosis is a condition associated with hyperthyroidism. A patient with no thyroid disease is said to be euthyroid.

Hypothyroidism

Disease Types
While patients' physical signs and symptoms are helpful in the diagnosis of hypothyroidism, measurements of thyroid hormones and TSH are required to confirm the diagnosis and locate the lesion as to (1) primary—lesion in the thyroid; (2) secondary—lesion in the pituitary; or (3) tertiary—lesion in the hypothalamus. Hypothyroidism is usually primary and rarely due to secondary and tertiary gland failure.

Symptoms
Some major symptoms of hypothyroidism include (1) mental dullness; (2) possible anemia with iron, B12, or folate deficiencies; (3) edema of face, eyelids, and extremities; (4) dry skin; and (5) dry, coarse hair.

Specific Diseases or Conditions
A severe thyroid necrosis is seen in Hashimoto's thyroiditis (or chronic lymphocytic thyroiditis). This disease is a possible inherited autoimmune disorder where

patients fail to suppress a clone of thyroid-directed T and B lymphocytes. Peak incidence of Hashimoto's thyroiditis is in adolescence, and it occurs more frequently in females and blacks. Hypothyroidism may also result from treatment of hyperthyroidism by surgery or radioactive iodine or hypothyroidism may be idiopathic.

Neonatal Hypothyroidism (Cretinism)
Neonatal hypothyroidism can result in irreversible brain damage (called cretinism) if not detected and treated shortly after birth with replacement thyroid hormone. Thyroid hormones are required for brain growth and development, which is essentially complete at 6 months. The incidence of neonatal hypothyroidism is 1 in 7000 live births. Most medical centers now screen all neonates for T_4 levels. If the T_4 is low, neonatal hypothyroidism is confirmed by an elevated serum TSH.

Laboratory Findings
Hypothyroid patients usually have decreased serum T_4 levels of 4.5 μg/dL or lower (normal T_4 ranges from 4.5 to 11.0 μg/dL). The T_3 resin uptake test is usually also decreased in hypothyroidism with levels being less than 25% (normal ranges from 25 to 35%). The FTI is also decreased. In primary hypothyroidism, the TSH is elevated, usually above 10 mIU/mL. In secondary and tertiary hypothyroidism, the TSH falls within normal limits.

Hyperthyroidism

Specific Diseases
Hyperthyroidism is more prevalent than hypothyroidism and is seen in (1) Graves' disease, where an autoimmune reaction results in the production of an IgG immunoglobulin with "TSH-like" properties [called long acting thyroid stimulator (LATS) or thyroid stimulating immunoglobulins (TSI)]; (2) toxic multinodular goiter; (3) toxic thyroid nodule (toxic adenoma); (4) T_3 thyrotoxicosis; or (5) exogenous thyroxine ingestion.

Symptoms
Exophthalmos is often associated with Graves' disease with other symptoms of hyperthyroidism being (1) diarrhea with possible anemia either from malabsorption of B12 or iron due to hyperperistalsis; (2) amenorrhea or irregular menstrual cycle or impotency in males; (3) CNS disturbances; (4) heat intolerance; (5) increased cardiac output; (6) increased appetite; and (7) thinning of hair and skin.

Laboratory Findings
Hyperthyroid patients usually have elevated serum T_4 levels of 11 μg/dL or higher (normal thyroid ranges from 4.5 to 11.0 μg/dL). The T_3 resin uptake test is also usually elevated in hyperthyroidism being greater than 35% (normals range from 25 to 35%). The T_7, or FTI should be greater than 3.9 (normals for T_7 range from 1.3 to 3.8). The TSH should be normal and toward the low end of the normal range. The normal range for TSH is less than 10 mIU/mL, with 7–10 mIU/mL being a "gray zone."

In thyrotoxicosis, a patient may be clinically hyperthyroid yet have a normal T_4

level. The serum total T_3 by RIA is usually elevated in thyrotoxic patients; hence, a serum T_3 is the recommended test in this disease.

Goiter

Thyroid enlargement or goiter is caused by overstimulation of the gland with TSH due to low circulating levels of free T_4 and T_3. Before and during the 1920s, symmetric diffuse goiter was caused by a dietary deficiency of iodine. The enlarged gland was more efficient in trapping iodine, hence the goiter was sometimes called "euthyroid" goiter. Euthyroid suggests that the patient's blood thyroid hormones were at normal levels. An asymmetric goiter may be due to single or multiple nodules in the thyroid, which may either be malignant (uncommon) or benign.

ENDOCRINOLOGY OF THE ADRENAL MEDULLA

CATECHOLAMINES

The major catecholamines are epinephrine (adrenaline), norepinephine (noradrenaline), and dopamine.

Synthesis

The chromaffin cells of the adrenal medulla produce epinephrine and, to a lesser extent, norepinephrine. Norepinephrine is also produced in the postganglionic sympathetic nerves and brain. Catecholamines are also synthesized by the chromaffin cells of other organs including the liver, heart, prostate, and lungs. Catecholamines are aromatic amines synthesized from tyrosine through a series of enzymatic reactions:

$$\text{tyrosine} \xrightarrow{\text{tyrosine hydroxylase}} \underset{\text{(Dopa)}}{\text{dihydroxyphenylalanine}} \xrightarrow{\text{dopa decaboxylase}}$$

$$\text{dopamine} \xrightarrow{\text{dopamine beta oxidase}} \text{norepinephrine} \xrightarrow[\text{N-methyl transferase}]{\text{phenylethanolamine}} \text{epinephrine}$$

The structures of epinephrine and norepinephrine are shown in Figure 12.15.

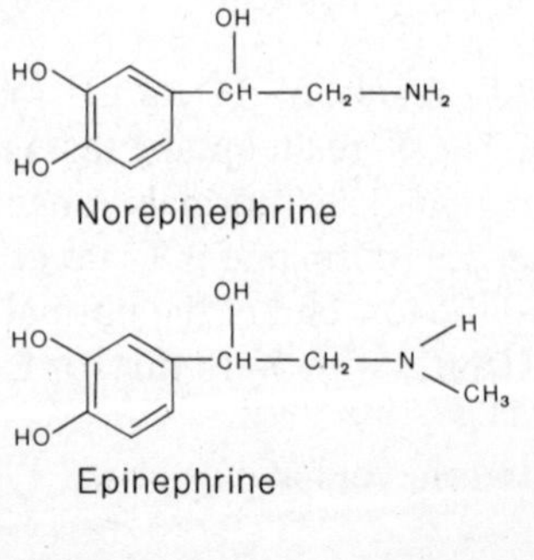

Norepinephrine

Epinephrine

Figure 12.15. Catecholamine structures.

Function

Once synthesized, catecholamines are stored in vesicles. The catecholamines are then released into circulation in response to stress (i.e., fear, exercise, hypoglycemia, hypotension). Release of catecholamines is stimulated by histamine, insulin, or angiotensin.

Epinephrine, when released, stimulates glycogen breakdown in the liver and hence raises the blood glucose level. Epinephrine also stimulates lipolysis in adipose tissue. Norepinephrine, on the other hand, is a more potent agent than epinephrine in raising blood pressure. Both catecholamines dilate the coronary arteries, while epinephrine dilates arteries supplying blood to the skeletal muscles.

Metabolism of Catecholamines

Two enzyme systems are responsible for the inactivation of catecholomines.

Metanephrines
Catechol-O-methyl transferases, an extra neural enzyme, converts catecholamines via O-methylation to metanephrines (metanephrine and normetanephrine). The measurement of urine levels of methanephrines is useful in the evaluation of hyperfunction of the adrenal medulla (see below).

VMA
Monoamine oxidase converts the metanephrines to vanilmandelic acid (VMA). Assays for VMA in the urine are also useful in the evaluation of adrenal medulla hyperfunction.

Catecholamine Measurement
Catecholamines can be quantitated on 24-hour urine specimens by fluorometric methods. This method is laborious and involves hydrolysis, selective adsorption, and elution on a column prior to generation of the fluorophore.

VMA is oxidized when treated with $NaIO_4$ to vanillin, which can be measured spectroscopically at 360 nm.

Plasma Catecholamines
Recently, plasma levels of catecholamines have been measured by radioenzymatic techniques; this assay may replace assays for urinary catecholamine and catecholamine metabolites. High-pressure liquid chromatography and gas chromatography–mass spectrometry have also been used to measure catecholamines in blood and urine.

Pheochromocytoma
Tumors of the chromaffin cells of the adrenal medulla, called pheochromocytoma, cause hyperfunction of the gland with markedly elevated levels of catecholamines being produced and released. Although cases of pheochromocytoma are rare, the hypertension produced is potentially lethal if left untreated. Urinary catecholamines, metanephrines, and VMA are elevated in pheochromocytomas. An acid-stabilized 24-hour urine collection is required for catecholamines.

Neuroblastoma and Dopamine

Tumors of neural origin, neuroblastomas and ganglineuromas, are usually associated with an increase in dopamine and its metabolite, homovanillic acid (HVA). Measurements of HVA in the urine are used to support the diagnosis of neuroblastoma.

PROSTAGLANDINS

First discovered in human semen and animal seminal tissue, prostaglandins have been found in a wide variety of body fluids where they exhibit a wide diversity of biological effects.

CHEMICAL STRUCTURE

Prostaglandins are synthesized primarily in cell membranes (in response to nerve stimulation) and have a fatty acid structure (see Figure 12.16). Changes in this basic structure (by forming double bonds on oxygenation) yield different prostaglandins including PGE_1, PGE_2, PGE_3, PGF_1-alpha, PGA_1, PGA_2, PGB_1, 19 hydroxy-PBG_1, etc.

FUNCTIONS OF PROSTAGLANDINS

Prostaglandins can help regulate the activity of smooth muscles including (1) stimulating the contraction of the uterus and (2) opening airways to the lung. Some prostaglandins raise blood pressure, while others lower it. Prostaglandins also may have potential for regulating secretions of certain endocrine glands.

SEROTONIN

Serotonin is not classified as a hormone, however it acts as a powerful muscle stimulant and vasoconstrictor. A derivative of the amino acid tryptophan, serotonin is formed primarily in the "argentaffine" tissue.

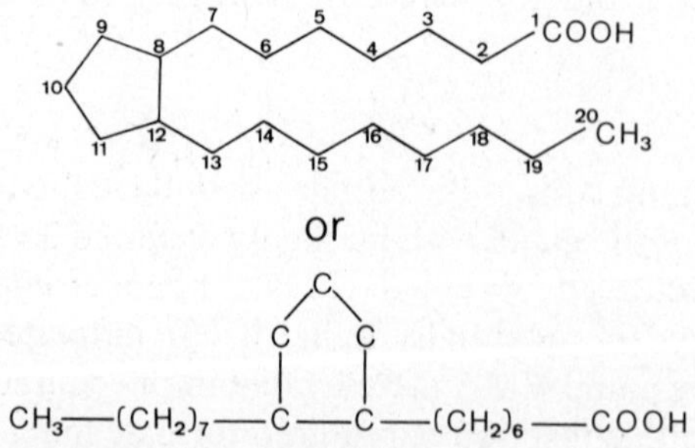

Figure 12.16. Structure of prostaglandins.

CLINICAL SIGNIFICANCE

In abdominal carcinoid tumors occurring in the wall of the small intestine or appendix, serotonin levels are elevated and are therefore responsible for many of the patient's symptoms. Symptoms include tachycardia, asthma, diarrhea, hyperperistalsis, and cardiac changes.

5-HYDROXYINDOLACETIC ACID (5-HIAA)

Serotonin is metabolized to 5-HIAA. This metabolite is measured in the urine as a test for metastatic carcinoid tumors (argentaffinoma).

HISTAMINE

Histamine is the potent vasoactive substance released from mast cells when antibody–antigen reactions take place on the cell surface. Increased histamine levels cause (1) vasodilation (can cause severe anaphylactic shock); (2) contraction of bronchi; and (3) stimulation of mucus glands.

KININS

BRADYKININ

This "kinin hormone" has a peptide structure and functions as a vasodilator. In addition, bradykinin plays a role in increasing vascular permeability and acts as a smooth muscle stimulant. Kinins are produced when kallikrein enzymes act on certain alpha globulins. Stimulation of kinin production is probably due to antibody–antigen reactions.

BIBLIOGRAPHY

Bates, H. M. "Ectopic pregnancy: RIA for the beta-HCG subunit," *Laboratory Management,* 27 (Feb. 1981).

Burke, M. D. and Rock, R. C. *Detection of Pheochromocytoma: Plasma Catecholamines Versus Urinary Catecholamines and Their Metabolites.* Check Sample Program ACC 81-1, ASCP, 1981.

Burke, M. D. "Hypertension: Test strategies for diagnosis and management," *Diagnostic Medicine,* 73 (May 1979).

Engel, R. H. "Diagnosing and monitoring virilizing congenital hyperplasia," *Laboratory Management,* 38 (Oct. 1979).

Fisher, D. A. "Screening for congenital hypothyroidism," *Hospital Practice,* 73 (Dec. 1977).

Gold, E. M. "The Cushing syndromes: Changing views of diagnosis and treatment," *Annals of Internal Medicine* **90,** 829 (1979).

Lange, P. H. "Testicular tumor markers," *Laboratory Management,* 27 (Oct. 1980).

Mitchell, M. L., et al. "Screening for congenital hypothyroidism," *Journal of the American Medical Association,* **239**(22), 2348 (1978).

Raphael, S. S. *Lynch's Medical Laboratory Technology.* Philadelphia: Saunders, 1983.

Rasor, J. L. and Braunstein, G. D. "A rapid modification of the beta-HCG radioimmunoassay," *Obstetrics and Gynecology* **50**(5), 553 (1977).

Slag, M. F. "Free thyroxine levels in critically ill patients," *Journal of the American Medical Association* **246**(23), 2702 (1981).

Teitz, N. (ed.). *Fundamentals of Clinical Chemistry,* 2nd ed. Philadelphia: Saunders, 1976.

Travis, J. C. (ed.). *Estriol and Human Choriomammotropin Tests of Fetoplacental Integrity.* Anaheim, CA: Scientific Newsletters, Inc., 1977.

Travis, J. C. (ed.). *Tumor Markers II.* Anaheim, CA: Scientific Newsletters, Inc., 1977.

Watts, N. B. *Thyrotopin-Releasing Hormone.* Check Sample Program ACC 81-3, ASCP, 1981.

Williams, R. H. (ed.). *Textbook of Endocrinology,* 6th ed. Philadelphia: Saunders, 1981.

Wyk, J. J. and Underwood, L. E. "Growth hormone, somatomedins and growth failure," *Hospital Practice* **13**(8), 57 (1978).

POST-TEST

1. Thyroid hormones are secreted in the following way. Thyroid-releasing hormone from the hypothalamus stimulates thyroid-stimulating hormone release from the anterior pituitary. TSH stimulates the thyroid gland to produce T_3 and T_4. As the blood level of thyroid hormones increases, TRH and TSH production decreases. This is an example of:
   a. positive feedback mechanism
   b. negative feedback mechanism

2. In the female menstrual cycle, FSH and LH stimulate estrogen secretion from the ovarian follicle. The rise of estradiol causes an abrupt surge of LH inducing ovulation. This is an example of:
   a. positive feedback mechanism
   b. negative feedback mechanism

3. There are three general mechanisms of hormonal action. Match the hormone with its mechanism of action.
   a. Hormone binds to cell surface receptors with the release of adenylate cyclase and the second messenger system.
   b. Hormone moves through cell membrane and binds to intracellular receptors; a conformational change occurs that allows the complex to bind to the nuclear chromatin; DNA is transcribed to mRNA with translation to specific proteins.
   c. Hormone penetrates cell and binds directly to receptors located in the chromatin; DNA is transcribed to mRNA with translation to specific proteins.
   _____ 1. estrogens
   _____ 2. androgens

_____ 3. TSH
_____ 4. ACTH
_____ 5. epinephrine
_____ 6. progesterone
_____ 7. T_3 and T_4
_____ 8. FSH

4. In the hormonal mechanism of action associated with the second messenger concept, the second messenger is frequently:
   a. adenylate cyclase
   b. the hormone itself
   c. cyclic-AMP
   d. a kinase enzyme

5. The immediate precursor to steroid hormone synthesis is:
   a. cholesterol
   b. ascorbic acid
   c. tyrosine
   d. cortisol

6. Draw the basic nucleus for steroid compounds, number the carbon atoms, and give the name of the compound.

7. Match the steroid compound with its basic nucleus.
   _____ 1. progesterone   a. C-18 compounds
   _____ 2. estrogens   b. C-19 compounds
   _____ 3. androgens   c. C-21 compounds
   _____ 4. mineralocorticoids
   _____ 5. glucocorticoids

8. Match the compounds on the left with the correct classification.
   _____ 1. estriol   a. mineralocorticoid
   _____ 2. testosterone   b. glucocorticoid
   _____ 3. cortisol   c. androgen
   _____ 4. estradiol   d. estrogen
   _____ 5. dehydroepiandosterone
   _____ 6. corticosterone
   _____ 7. deoxycorticosterone

9. The major regulatory effect of ACTH is in association with:
   a. aldosterone
   b. cortisol
   c. testosterone
   d. estradiol

10. Samples for cortisol determinations were obtained from a patient at 8 a.m. and 4 p.m. The technologist noted that the cortisol concentration in both samples showed little variation. From this information, you can conclude that (may be one or more answers):

a. The patient is probably normal since cortisol levels do not show daily fluctuation except in disease states.
b. The patient may have Cushing's disease.
c. The patient may have an adrenal tumor.
d. The patient has lost the diurnal rhythm of secretion.

11. A patient with *Cushing's disease* (pituitary in origin) may show all of the following *except:*
    a. hypercalcemia
    b. decreased number of circulating lymphs
    c. increased plasma cortisol
    d. loss of diurnal variation in cortisol secretion
    e. hypokalemia
    f. normal suppression with dexamethasone
    g. increased ACTH

12. A patient with Addison's disease may show all of the following *except:*
    a. increased plasma cortisol
    b. increased ACTH
    c. hyponatremia
    d. hyperkalemia
    e. hypoglycemia
    f. eosinophilia

13. The transport protein for cortisol is:
    a. albumin
    b. ceruloplasmin
    c. transcortin
    d. transferrin

14. Steroid hormones are produced in all of the following sites *except:*
    a. adrenal cortex
    b. adrenal medulla
    c. ovary
    d. testes
    e. placenta

15. In a person with normal renal function, the measurement of cortisol in the urine measures:
    a. the biologically active free cortisol
    b. free and protein bound cortisol
    Justify your answer.

16. Tests used to measure adrenal cortical function include all *except:*
    a. measurement of plasma cortisol
    b. measurement of 17-hydroxycorticosteroids
    c. measurement of 17-ketogenic steroids
    d. measurement of 17-keto steroids

17. Match the colorimetric reaction with the group measured.
    _____ 1. Zimmerman reaction   a.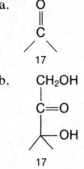
    _____ 2. Porter–Silber reaction

18. Two samples, collected at 8 a.m. and 4 p.m., were received in the laboratory for cortisol determination. The assay value for the sample collected at 4 p.m. was significantly higher than the one obtained at 8 a.m. You could interpret these results in the following way (choose the *best* answer):
    a. The patient results show the normal diurnal pattern of cortisol secretion.
    b. The results should be rechecked since the diurnal pattern of secretion is not the usual pattern seen.
    c. The samples were probably mislabeled when collected since the pattern of secretion is reversed from that normally seen.
    d. The patient results are probably normal since the sleep/awake pattern influences the diurnal pattern of secretion.

19. A patient, diagnosed as having Addison's disease, was admitted to the hospital. ACTH was given and cortisol measurements performed. The cortisol blood level increased after the ACTH was given. From this information, you can conclude that the site of hypofunction was:
    a. the adrenal gland
    b. the pituitary

20. A patient, with Cushing's syndrome, entered the hospital for diagnostic studies. Baseline cortisol measurements were performed and dexamethasone was given. A small dose of dexamethasone did not suppress the blood cortisol levels. A large dose of dexamethasone was given but there was little change in the blood cortisol values from baseline values. From this information, you can conclude:
    a. the patient is normal
    b. the patient has bilateral adrenal hyperplasia
    c. the patient probably has an adrenal tumor

21. A young man, 24 years old, has complaints of gradually increasing weakness, weight loss, and loss of appetite. On physical examination, he was observed to have bronzed skin; however, he reported no exposure to the sun. He was hypotensive and showed evidence of muscle wasting. Laboratory results included the following results:

    Serum sodium   125 mmol/L
    Serum potassium   6.2 mmol/L

Plasma cortisol (8:00 a.m.)   4 mg/dL (decreased)
17-OH-CS   below normal
Plasma ACTH   increased above normal

An ACTH stimulation test failed to stimulate an increase in plasma cortisol. The most likely diagnosis is:
a. Conn's disease
b. Cushing's syndrome
c. Addison's disease
d. Adrenal cortex hypofunction secondary to pituitary failure

22. A 40 year-old woman complains of amenorrhea and emotional disturbance perhaps partially due to her increasing obesity, which concentrated around the chest and abdomen. Her X-ray studies shows evidence of mineral bone loss (osteoporosis). Laboratory results:

Serum potassium   3.2 mmol/L
Fasting plasma glucose   140 mg/dL
Plasma cortisol (8:00 a.m.) 40 μg/dL (elevated)

(Note: A large dose of dexamethazone did not suppress the elevated cortisol.)
The most likely diagnosis is:
a. Cushing's (due to an adrenal adenoma)
b. Cushing's (due to bilateral adrenal hyperplasia)
c. pituitary failure
d. Addison's disease

23. *Secondary* aldosteronism may be caused by all of the following *except:*
a. potassium loading
b. salt depletion
c. hemorrhage
d. renal juxtaglomerular hyperplasia
e. adrenal adenoma

24. Excessive aldosterone may cause all of the following *except:*
a. hypernatremia
b. hyperkalemia
c. alkalosis
d. hypertension

25. The test used to measure androgen function is:
a. 17-ketogenic measurement
b. 17-hydroxycorticosteroid measurement
c. 17-ketosteroid measurement
d. free cortisol measurement

26. Fill in the blanks with the proper word from the list that follows.

```
        ┌─────────────────┐
        │ Low Salt Diet   │
        │ Upright Position│
        └─────────────────┘
                 ↓
         Decreased Plasma
              Volume
                 ↓
```

_____ 1. Name of enzyme released from the kidney
                 ↓
_____ 2. Name of protein released from the liver
                 ↓
_____ 3. Protein (released from the liver cleaved by enzyme) to:
                 ↓
_____ 4. Lung enzyme (ACE) cleaves molecule to:
                 ↓
_____ 5. Molecule released from adrenal cortex

a.   angiotensinogen
b.   angiotensin II
c.   renin
d.   angiotensin I
e.   aldosterone

27. Reproductive physiology (female) menstrual cycle—match the following:
_____ 1. Peak level of luteinizing hormone (LH)
_____ 2. Highest level follicle-stimulating hormone (FSH)
_____ 3. Highest level of urinary pregnanediol

a.   follicular
b.   midcycle
c.   luteal phase

28. The major estrogen secreted by the ovaries is:
a.   estriol
b.   estradiol
c.   estrone

29. The estrogen associated with pregnancy is:
a.   estriol
b.   estradiol
c.   estrone

30. The hormone that is used to determine pregnancy is:
a.   HCG
b.   LH
c.   FSH
d.   ICSH

31. A couple with infertility problems were to be evaluated. The physician requested pregnanediol determinations on samples obtained in the follicular and luteal stages of the woman's menstrual cycle. The results showed no

change in the pregnanediol concentration in these samples. From these results you can conclude:
a. ovulation is not occurring
b. ovulation is occurring
c. the estrogen concentration is low
d. the estrogen concentration is high
e. the LH concentration is high

32. The adrenogenital syndrome is usually associated with (may be one or more answers):
a. excess cortisol
b. increased ACTH
c. 21-hydroxylase enzyme deficiency
d. increased androgen production
e. 17-alpha hydroxylase enzyme deficiency

33. Growth hormone exhibits a feedback mechanism with:
a. insulin
b. blood sugar
c. fatty acids
d. amino acids

34. Catecholamines enable the body to react to stress. They prepare the body for
_____ , _____ , and _____ .

35. The major catecholamine secreted by the adrenal medulla is:
a. epinephrine
b. norepinephrine
c. dopamine
d. metanephrine

36. In the diagnosis of pheochromocytoma, the *screening* test of choice is:
a. urinary metanephrines
b. urinary VMA
c. plasma metanephrines
d. plasma VMA

37. The precursor for prostaglandin synthesis is:
a. ascorbic acid
b. tyrosine
c. arachidonic acid
d. cholesterol
e. tryptophane

38. 5-Hydroxyindolacetic acid is a metabolite of:
a. prostaglandins
b. serotonin
c. epinephrine
d. norepinephrine

39. Match the following:
_____ 1. pregnanetriol
_____ 2. 17,21-dihydroxy-20-ketosteroids
_____ 3. plasma cortisol

_____ 4. urinary 17-ketosteroids
_____ 5. vanilmandelic acid
_____ 6. 5-hydroxyindolacetic acid
_____ 7. phenolic A ring
_____ 8. progesterone

a. pheochromocytoma
b. androgens
c. maintains pregnancy
d. diurnal variation
e. Porter–Silber reaction
f. adrenogenital syndrome
g. serotonin
h. estrogens

40. The starting material for T_3 and T_4 synthesis is:
   a. thyroxine
   b. reverse T_3
   c. tyrosine
   d. iodine

41. While stored in the thyroid follicle, T_3 and T_4 are bound to a large protein called _____ .

42. Match the following:
   _____ 1. MIT   a. tetraiodothyronine or T_4
   _____ 2. DIT   b. triiodothyronine
   _____ 3. T_3 c. monoiodotyrosine
   _____ 4. thyroxine   d. diiodotyrosine

43. Match the following:
   _____ 1. most active thyroid hormone   a. T_3
   _____ 2. thyroid hormone present in highest concentra-   b. T_4
                 tion in circulating blood
   _____ 3. may be derived from tissue conversion as well
                 as from the thyroid gland
   _____ 4. exists free and bound to TBG, albumin and
                 TBA in the blood
   _____ 5. exists free and bound only to TBG

44. Match the following:
   _____ 1. primary hypothyroidism   a. defect in hypothala-
                                                  mus
   _____ 2. secondary hypothyroidism   b. defect in pituitary
   _____ 3. tertiary hypothyroidism   c. defect in thyroid
                                                  gland

45. Match the following:
   _____ 1. cretinism   a. hyperthyroid condition
   _____ 2. myxedema   associated with LATS
   _____ 3. Hashimoto's thyroiditis b. normal thyroid condition
   _____ 4. Graves' disease   c. hypothyroid condition
   _____ 5. goiter   believed autoimmune in
   _____ 6. euthyroid   origin

    d. clinical sign of hypothy-
        roidism
    e. enlarged thyroid gland
        associated with iodine
        deficiency
    f. hypothyroidism at birth

46. The T_3 uptake test measures:
    a. T_3
    b. T_4
    c. free TBG binding sites
    d. TSH

47. Complete the following table using arrows as follows, ($\uparrow$) increased; ($\downarrow$) decreased; ($\rightarrow$) normal.

Clinical Problem	T_4	RT_3U	FTI
Hyperthyroidism			
Hypothyroidism			
Primary TBG increase			
Primary TBG decrease			

48. Complete the following table using arrows as follows, ($\uparrow$) increased; ($\downarrow$) decreased.

Clinical Condition	TSH
Primary hypothyroidism	
Secondary hypothyroidism	

Tertiary hypothyroidism
  TRH given:
  The TSH before injection is _____.
  The TSH after injection is _____.	

49. A 42-year-old woman complained of "being cold all the time." She also noted that her skin and hair were dry and she "was tired all the time." She also noted a gain of six pounds without increasing her food intake. Thyroid function tests were ordered. The results of these tests were as follows: T_4—decreased; TSH—decreased; RT_3U—decreased.

    TSH was administered and T_4 was measured again. The T_4 increased over previous levels. Based on these results, one could rule out the _____ as the etiology of the hypothyroid condition. TRH was then given and the TSH measured. TSH did not increase after TRH administration. The etiology of the hypothyroid condition is _____.
    a. thyroid gland
    b. pituitary gland
    c. hypothalamus

50. A T_4 test on a 23-year-old pregnant female was elevated. The doctor ordered thyroid function tests. The result of these tests were as follows: T_4—increased; RT_3U—decreased. From these results, the patient is probably:
   a. hyperthyroid
   b. hypothyroid
   c. euthyroid

51. Laboratory results were as follows: T_4—decreased; RT_3U—decreased; TSH—increased. These results may be interpreted as:
   a. primary hypothyroidism
   b. secondary hypothyroidism
   c. tertiary hypothyroidism
   d. fails to diagnose any of the above

52. T_4 was measured by RIA. After separation, the radioactivity bound is determined to be greater in sample A than in sample B. These results may be interpreted as (two answers):
   a. the concentration of T_4 in sample A is higher than sample B
   b. the concentration of T_4 in sample A is less than sample B
   c. there is a direct relationship between radioactivity and sample concentration
   d. there is an inverse relationship between radioactivity and sample concentration

ANSWERS

1. b
2. a
3. (1) b; (2) b; (3) a; (4) a; (5) a; (6) b; (7) c; (8) a
4. c
5. a
6. cyclopentanoperhydrophenanthrene nucleus
7. (1) c; (2) a; (3) b; (4) c; (5) c
8. (1) d; (2) c; (3) b; (4) d; (5) c; (6) b; (7) a
9. b
10. b, c, d
11. f
12. a
13. c
14. b
15. a: Protein bound cortisol cannot be filtered by the glomerulus—only the free or biologically active forms will be found in the urine. Cortisol, once filtered, is passively reabsorbed, hence concentration independent. Urinary free cortisol is directly related to plasma free cortisol.
16. d
17. (1) a; (2) b

18. d
19. b
20. c
21. c
22. a
23. e
24. b
25. c
26. (1) c; (2) a; (3) d; (4) b; (5) e
27. (1) b; (2) a; (3) c
28. b
29. a
30. a
31. a
32. b, c, d
33. b
34. fight, flight, fright
35. a
36. a
37. c
38. b
39. (1) f; (2) e; (3) d; (4) b; (5) a; (6) g; (7) h; (8) c
40. c
41. thyroglobulin
42. (1) c; (2) d; (3) b; (4) a
43. (1) a; (2) b; (3) a; (4) b; (5) a
44. (1) c; (2) b; (3) a
45. (1) f; (2) d; (3) c; (4) a; (5) e; (6) b
46. c
47. Hyperthyroidism—increased, increased, increased; Hypothyroidism—decreased, decreased, decreased; Primary TBG increase—increased, decreased, normal; Primary TBG decrease—decreased, increased, normal
48. Primary hypothyroidism—increased; Secondary hypothyroidism—decreased; Tertiary hypothyroidism—before injection, decreased; after injection, increased
49. a, b
50. c
51. a
52. b, d

CHAPTER 13
NPN SUBSTANCES, RENAL FUNCTION, AND BODY FLUIDS

      C. Serous Fluids
          1. Pleural Fluid
          2. Pericardial Fluid
          3. Peritoneal Fluid
      D. Amniotic Fluid
          1. Isoimmunization
          2. Fetal Maturity
              a. Lecithin/Sphingomyelin (L/S) Ratio
              b. Bubble Stability or Shake Test
              c. Creatinine

OBJECTIVES

On completion of this chapter, you will be able to:

1. List the functions of the kidney.
2. Draw and label a nephron and discuss the purpose of each nephron segment.
3. Discuss the hormonal role of the kidney in the control of red blood cell volume and in the control of aldosterone production.
4. List the NPN substances.
5. Define the words (1) azotemia, (2) prerenal azotemia, (3) renal azotemia, (4) postrenal azotemia, and (5) uremia.
6. For each NPN substance listed below, discuss (1) source; (2) sample requirements; (3) assay procedures; (4) clinical significance; and (5) reference range.
   a. BUN
   b. Creatinine
   c. Uric acid
7. Convert BUN to urea concentration when given data.
8. Define the words (1) total renal blood flow; (2) effective renal plasma flow; and (3) glomerular filtration rate. List the normal volume range for 24 hour urine collections in adults.
9. Calculate clearance problems when given data.
10. Discuss the procedure for the performance of clearance tests that measure glomerular filtration rate.
11. Discuss the procedure for the performance of clearance tests that measure the secretory ability of the tubules.
12. Discuss the procedure for the performance of tests that measure the concentrating ability of the tubules.
13. In regard to the laboratory examination of cerebrospinal fluid, discuss (1) color and appearance; (2) cell count and differential; (3) total protein; (4) glucose; (5) glutamine; (6) lactic acid; and (7) lactate dehydrogenase in normal and abnormal specimens.

14. Discuss multiple sclerosis and describe the associated CSF laboratory findings.

15. In regard to the laboratory examination of synovial fluid, discuss (1) color and appearance; (2) cell count; (3) total protein; (4) glucose; and (5) microscopic findings in normal and abnormal specimens.

16. Describe the appearance of monosodium urate crystals and calcium pyrophosphate dihydrate crystals in synovial fluid using compensated polarized light microscopy.

17. Define transudate and exudate, and discuss the test procedures utilized to differentiate between the two types of effusions.

18. In regard to the laboratory examination of pleural fluid, pericardial fluid, and peritoneal fluid, discuss (1) appearance; (2) cell count; and (3) glucose in normal and abnormal specimens. For peritoneal fluid, discuss why amylase, hematocrit, alkaline phosphatase, BUN, and creatinine might be included among the tests performed.

19. Discuss the significance of the presence of bilirubin in amniotic fluid and the laboratory test procedure that is used to measure the degree of isoimmunization.

20. In regard to amniotic fluid, discuss (1) the etiology of respiratory distress syndrome; (2) the laboratory test procedures that measure fetal maturity; and (3) the interpretation of L/S ratios.

GLOSSARY OF TERMS

Amniotic fluid. The fluid that surrounds the fetus in the amniotic sac. It serves to protect the fetus and helps to maintain a constant body temperature. Fetal urine contributes to the amniotic fluid.

Azotemia. Azotemia is a term used to denote increases in serum nonprotein nitrogenous compounds, primarily urea, but may include other NPN substances as well.

Blood urea nitrogen (BUN). BUN actually measures the nitrogen content of urea. Urea (MW of 60) contains two nitrogen atoms per molecule, which is 46.7% of urea mass. Urea nitrogen is most often measured in serum or plasma rather than whole blood, but the term blood urea nitrogen is still used.

Cerebrospinal fluid. An ultrafiltrate of plasma that surrounds and fills the spaces in the central nervous system.

Clearance tests. Clearance tests measure the ability of the kidney to remove or clear a substance from the blood. Test results are reported in milliliters of blood cleared per minute or mL/min.

Effective renal plasma flow (ERPF). Volume of plasma that perfuses functional renal tissue, expressed in milliliters per minute. The average ERPF is about 660 mL/min.

Erythropoietin. A hormone synthesized in the kidney and released in response to anemia or hypoxia. Erythropoietin induces committed erythroid stem cells, primarily in the bone marrow, to differentiate into developing erythrocytes.

Exudate.   An effusion of fluid into a serous cavity as a result of damage to the mesothelial lining of a cavity. The damage may result from infection, inflammation, trauma, or neoplasm. The protein content is high as compared to a transudate.

Glomerular filtration rate (GFR).   The volume of filtrate that passes into Bowman's capsule, expressed in milliliters per minute. The GFR is about 120 mL/min in a man with two normal kidneys.

Mesothelial cells.   Cells derived from the epithelial lining of the serous membranes (peritoneum, pericardium, and pleura).

Nephron.   The functional unit of the kidney. Each kidney is composed of approximately 10^6 nephrons.

Nonprotein nitrogen substances (NPN).   The NPN substances consist of all nitrogen-containing substances of the blood with the exception of protein. NPN substances include urea, creatinine, creatine, uric acid, ammonia, and amino acids (and peptides).

Peritoneal (ascitic) fluid.   The serous fluid that lubricates the abdominal organs.

Pleural (thoracic) fluid.   The serous fluid that lubricates the lungs and thoracic wall during ventilation.

Postrenal azotemia.   Azotemia resulting from obstruction to the urinary flow.

Prerenal azotemia.   Azotemia produced by inadequate blood delivery to the kidneys because of blood loss, shock, dehydration, or congestive heart failure resulting in decreased glomerular filtration rate.

Renin.   A proteolytic enzyme (hydrolase class) of the kidney that catalyzes the conversion of angiotensinogen to angiotensin I in the plasma. Angiotensin I is then converted to angiotensin II by a converting enzyme of the lungs. Angiotensin II (a hypertensive agent) stimulates aldosterone production from the adrenal cortex.

Renal azotemia.   Azotemia produced by renal disease as a result of decreased clearance of NPN substances.

Total renal blood flow (TRBF).   Volume of blood that passes into renal arteries expressed in milliliters per minute. The average TRBF is about 1200 mL/min.

Transudate.   An effusion of fluid into a serous cavity that is not associated with inflammation but is produced by mechanical factors such as changes in osmotic or hydrostatic pressure which influences the formation or resorption of fluid. The protein content is lower when compared to an exudate.

Xanthochromic.   Xanthos means yellow; therefore, xanthochromic means yellow color. Xanthochromic cerebrospinal fluid most often indicates a previous intracranial hemorrhage.

The kidney is a complex organ that performs many functions. These functions can be divided into three general groups: (1) those that control the concentration of most of the constituents of the body; (2) those that aid in the control of blood pH; and (3) those that function to excrete end products of metabolism. The kidney's role in constituent concentration control (Chapters 5, 8, and 9) and acid–base balance (Chapter 9) have already been discussed. Its excretory functions include its ability to excrete water-soluble end products of metabolism. For example, urea, derived from protein catabolism, uric acid, derived from purine catabolism, and creatinine, derived from creatine phosphate (an energy storage source in muscle), are all excreted in the urine. The laboratory test procedures that are used to evaluate renal function measure the blood concentration of end products of metabolism normally excreted in the urine. When the kidney is diseased, the ability to clear or remove metabolic end products from the blood is deficient and the blood concentration of these components increases.

NEPHRON ANATOMY

The kidney has large functional reserve or in essence there is more tissue present than is needed to carry out the function of the organ. It is estimated that at least one-half of the kidney parenchyma (functional tissue of the kidney as opposed to support and connective tissue) is destroyed by the time renal function tests are abnormal. When man is at rest, survival is possible with as little as one-tenth functional renal tissue. Man can continue an active life with about 25% functional tissue.

The nephron is the functioning unit of the kidney. Each kidney contains approximately 10^6 nephrons, each capable of forming urine by itself. There are two types of nephrons: (1) those that lie principally in the cortical area (seven-eighths are cortical nephrons) and (2) those that originate near the border between the cortex and medulla which have long loops of Henle that extend deep into the medulla and are called juxtamedullary nephrons (one-eighth are juxtamedullary nephrons).

The blood supply to the kidney is abundant; about 25% of the blood volume perfuses renal arteries each time the heart beats. The renal arteries divide and eventually form the afferent arterioles that supply blood to individual nephrons. Each afferent arteriole divides to form two capillary systems: (1) the glomerular capillary system that converges to form the efferent arteriole which then forms the (2) peritubular capillary system that lies in close proximity to the nephron tubule. A schematic representation of a nephron is illustrated in Figure 13.1.

As blood flows through the glomerulus, water and selected substances enter Bowman's capsule. This glomerular filtrate then passes through the nephron tubule, where reabsorption and secretory processes alter the composition of the filtrate eventually forming urine. Reabsorption refers to the passage of substances from the tubular lumen (glomerular filtrate) to the peritubular capillary system. Secretion refers to the passage of substances from the peritubular capillaries to the tubular lumen (into the glomerular filtrate).

GLOMERULUS

The glomerulus is a network of approximately 30–40 capillaries which acts as a blood filter and allows water and small (low molecular weight) substances to pass freely into Bowman's capsule. If the molecular weight is large, the substances are excluded from the glomerular filtrate and remain in the blood.

The anatomy of the glomerular capillaries allows them to act as graded molecular sieves. The glomerular wall is composed of three layers: (1) an endothelial

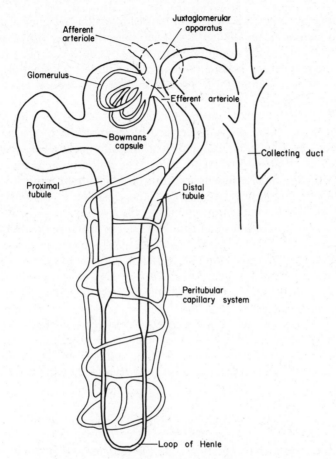

Figure 13.1. Schematic diagram of a kidney nephron.

layer (layer next to the blood); (2) basement membrane or middle layer; and (3) an epithelial layer, which consists of epithelial cells with foot processes called podocytes. The endothelial or innermost layer contains many openings or fenestrations, each about one-hundredth the size of a red blood cell. The basement membrane layer is composed of a complex of glycoproteins, some of which are negatively charged. The foot processes of the podocytes of the epithelial layer are embedded in the basement membrane and are covered by a thin layer of mucoprotein, which also bears a strong negative charge. Adjacent foot processes are also covered at their bases by a slit membrane that acts as a second selective barrier to particles that have penetrated the basement membrane. Solutes with low molecular weights such as electrolytes, sugars, amino acids, urea, polypeptides, and some low molecular weight proteins pass freely from the blood through the glomerular walls into Bowman's space. Both the size of the openings and the negative charge of the capillary walls retard the passage of intermediate or larger size proteins that are negatively charged at pH 7.4 (normal body pH), since negative charges repel one another. This allows for conservation of proteins in the plasma. Blood cells are blocked because the fenestrations or openings are normally too small to allow passage. All molecules with molecular weights less than 5000 g/mole easily penetrate the glomerular membranes and enter Bowman's capsule; substances with molecular weights up to 68,000 g/mole can penetrate to some extent, but the majority of plasma proteins are excluded. This action allows for a filtrate that is almost identical in composition to plasma except it is cell free and nearly protein free.

Each time the heart beats, approximately one-fourth of the blood goes to the kidneys. The hydrostatic pressure created by contraction of the heart and the resistance to blood flow created by afferent and efferent arterioles are the major forces causing a filtration process to occur. The dilitation and restriction of the arterioles are under humoral and neural control. This filtration force is opposed by (1) the collodial osmotic pressure (COP) of the plasma proteins that cannot pass through the walls of the capillaries and (2) by the hydrostatic pressure of the fluid in Bowman's space, which is small compared to the pressure created by the heart beat. The driving force creating filtration is the difference between the hydrostatic pressure created by the heart beat (about 45 mm Hg) and the hydrostatic pressure of the fluid in Bowman's space (about 10 mm Hg) plus the collodial osmotic pressure of the plasma proteins (about 20 mm Hg). At the beginning of the afferent arteriole, the filtration pressure is equal to $45 - (20 + 10) = 15$ mm Hg. As the blood in the glomerulus nears the efferent arteriole, the concentration of the plasma proteins increase (due to loss of water into Bowman's space) and the collodial osmotic pressure increases. At this point, filtration ceases since the hydrostatic pressure of the heart beat is exactly balanced by the two forces opposing filtration.

TUBULE

The tubule is a long coiled tube that converts the glomerular filtrate to urine. In one day, the kidney filters approximately 40 gallons of fluid but less than one-half gallon of fluid will finally be excreted. The volume of the glomerular filtrate is reduced to about one-hundredth of its original volume during its passage through the tubule. The tubule is divided into four parts: (1) Bowman's capsule or space; (2) proximal tubule; (3) loop of Henle; and (4) distal tubule. Each section of the tubule has specific functions.

Bowman's Capsule or Space

This is the expanded portion of the renal tubule, which envelopes the glomerulus. It serves as the receptacle for the glomerular filtrate.

Proximal Tubule

The majority of the reabsorptive and secretory mechanisms occur in the proximal tubule.

In this section of the nephron, approximately 60–70% of the water and salt (NaCl) is reabsorbed and at least 80% of the potassium is reabsorbed. Glucose, proteins, amino acids, and vitamins are completely reabsorbed in the proximal tubule so that normally very little of these solutes remain in the fluid entering the loop of Henle. Calcium, phosphate, urea, bicarbonate, and uric acid are also reabsorbed in the proximal tubule.

A variety of organic acids and bases, including many substances foreign to the body, are secreted into the tubular lumen. Hydrogen ion secretion occurs until the urinary buffers are titrated to their limiting pH 4.5. When this urinary pH is reached, hydrogen ion secretion ceases. The secreted hydrogen ions are buffered by filtered bicarbonate and phosphate, and by nonfiltered ammonia. Ammonia is formed in the renal tubular cells and secreted into the tubular lumen. The proximal tubule delivers about 30–40% of the glomerular filtrate to the loop of Henle.

Loop of Henle

These thin segments of the tubule called the loop of Henle neither reabsorb nor secrete large quantities of fluids and solutes. The fluid that enters Henle's loop is isoosmotic with plasma. A countercurrent exchange mechanism occurs in the loop of Henle that influences the osmotic concentration of the urine which flows on to the distal tubule. The medullary interstitial fluid surrounding the descending loop of Henle has a high osmotic concentration. Water moves out of the filtrate so that the tubular fluid becomes more concentrated. It is thought that the descending loop may be impermeable to sodium and chloride. In the ascending loop, sodium and chloride are actively reabsorbed. The ascending portion of the loop of Henle apparently has low permeability to water. The end result of these mechanisms is the production of the fluid that is hypoosmotic when compared to the cortical interstitial fluid. About 15–20% of the glomerular filtrate fluid is reabsorbed in the loop of Henle and 20–30% of the filtrate passes on to the distal tubule.

Distal Tubule

In the distal tubule, approximately 10% of the glomerular filtrate fluid is reabsorbed. Under the influence of antidiuretic hormone (ADH), free water (water without solutes) is reabsorbed. ADH increases pore size. Both sodium and potassium are reabsorbed, and potassium is also secreted under the influence of aldosterone. Hydrogen ions may also be secreted in this section of the nephron as long as the urinary pH is above 4.5.

COLLECTING DUCTS

ADH also influences water reabsorption in the collecting ducts. The final concentration of the urine occurs here. In the absence of ADH, the collecting ducts become impermeable to water and the volume of urine increases. Sodium is also reabsorbed in the collecting ducts, and this reabsorption is enhanced by aldosterone.

HORMONAL CONTROL EXERTED BY THE KIDNEY

ERYTHROPOIETIN

The kidney is the site of production of the hormone erythropoietin, which stimulates the bone marrow to produce red blood cells. The hormonal negative feedback mechanism is as follows. Decreased arterial oxygen saturation stimulates the kidney to produce erythropoietin. Erythropoietin stimulates the bone marrow to produce red blood cells (and hemoglobin), which facilitates the arterial oxygen saturation. The increase in arterial oxygen saturation removes the initiating stimulus, and production of erythropoietin is thus decreased.

A hemoglobin concentration of 7–8 g/dL can be maintained without erythropoietin, therefore, it is postulated that erythropoietin may also be formed at a site other than the kidney or that another mechanism may be involved in red blood cell production.

RENIN–ALDOSTERONE

Renin is an enzyme produced in specialized granular cells (also called juxtaglomerular cells) located in the wall of the afferent arteriole. These granular cells lie in close proximity to the macula densa cells found in the walls of the distal tubule. This close proximity occurs because the distal tubule of the nephron loops and lies close to the glomerulus of its own nephron (Figure 13.1). This region is known as the juxtaglomerular apparatus. The stimuli for renin release may be (1) from within the arteriole, that is, when the volume of blood perfusing the kidney is low; (2) from within the distal tubule due to a decrease in urinary sodium caused by a decrease in serum sodium; or (3) through sympathetic activity operating on beta receptors in the juxtaglomerular apparatus.

Renin, through the renin–angiotensin system, regulates aldosterone secretion from the adrenal gland. Aldosterone then acts on the distal tubule and collecting duct of the nephrons to increase reabsorption of sodium (and water) and to increase potassium excretion by increasing potassium secretion into the tubular filtrate. See Chapter 12 for the feedback mechanism.

NONPROTEIN NITROGEN SUBSTANCES

Nonprotein nitrogen (NPN) substances may be defined as all nitrogen-containing constituents of the blood with the exception of protein. The NPN constituents consist of (1) urea (45–55%); (2) amino acids (20%); (3) uric acid (20%); and (4) other substances (5%) including creatinine, creatine, and ammonia. The NPN constituents are end products of nitrogen metabolism and may be excreted in the urine.

Azotemia refers to an increase in the plasma concentration of NPN constituents, principally urea and creatinine.

Prerenal azotemia implies that the increase in NPN constituents is due to pathophysiologic processes that are not directly related to the kidney. Rather, prerenal azotemia results from inadequate delivery of blood to the kidney. When blood does not reach the kidney, the NPN constituents increase in the plasma as a result of depressed glomerular removal. Dehydration, shock, diminished blood volume, or congestive heart failure may all cause prerenal azotemia.

Renal azotemia implies that the increase in NPN constituents is caused by lesions in the kidney. The pathogenesis of renal azotemia primarily involves diminished glomerular filtration secondary to chronic or acute renal disease.

Postrenal azotemia implies that the increase in NPN constituents in the blood is secondary to pathophysiologic processes that also affect the glomerular filtration rate but are not directly related to lesions in the kidney. Urinary tract obstruction, which may be due to stones, tumor, or prostate enlargement, may increase the hydrostatic pressure in the kidney which reduces the filtration pressure in the glomeruli resulting in diminished removal of constituents from the blood.

Uremia is a clinical syndrome occurring with prolonged severe azotemia that includes metabolic acidosis, nausea, vomiting, water and electrolyte imbalances, neurological changes, and coma.

The measurement of individual NPN substances is performed routinely in the laboratory. The principal clinical application of each constituent is as follows. The measurement of the serum level of urea (BUN) and creatinine are used to evaluate renal function. Serum uric acid is used primarily to evaluate gout, and serum ammonia is used as an indicator of Reye's syndrome. Urine creatine excretion was once used to evaluate skeletal muscle disease but has been replaced by enzyme tests (LD-5, CK-3).

UREA

Source

Urea is the principal end product of protein and amino acid catabolism. This small molecule is water soluble and is readily excreted in the urine. Urea constitutes approximately one-half of total urinary solutes (about 25 g/day).

As amino acids (or protein) are deaminated in the liver, ammonia is liberated. Even small amounts of ammonia in the blood will adversely affect acid–base balance and cell function. Under normal conditions, the ammonia does not enter the blood but is converted into urea by a series of reactions that require specific enzymes found only in the liver. This formation of urea occurs via a complex cyclic process referred to as the Krebs–Henseleit or urea cycle. Ammonia, derived from the diet, is also prevented from reaching the circulation because it is carried directly to the liver via the enterohepatic portal system where it is then removed via the urea cycle.

Urea is freely filtered by the glomerulus, but a portion (40–80%) is passively reabsorbed with water and returned to the blood. The amount of reabsorption is dependent on the state of hydration. When the rate of urine flow is high, less urea is reabsorbed; when the rate of urine flow is low, more is reabsorbed.

Sample Requirements

Serum, plasma, urine, amniotic fluid, or other body fluids are all acceptable samples for the analysis of urea. Any of the common anticoagulants may be used for blood specimens. Blood samples collected in tubes which contain sodium fluoride cannot be used in procedures that utilize the enzyme urease since sodium fluoride inhibits this enzyme. Whole blood may be used, but deproteinization is necessary. Urea is stable but may be destroyed by bacteria, therefore refrigeration of samples is necessary if the assay is to be delayed for long periods. If a urine sample is to be assayed, refrigeration is especially important to inhibit bacterial destruction; freezing is required for long periods of storage.

Assay Procedures

Most urea assay methods do not measure urea directly but measure the nitrogen content of urea (hence the name BUN or blood urea nitrogen). The enzyme urease is used in hydrolysis. The reaction is as follows:

$$H_2N-\underset{\text{urea}}{\overset{\overset{\textstyle O}{\|}}{C}}-NH_2 + 2H_2O + H^+ \xrightarrow{\text{urease}} (NH_4)CO_3 + H^+ \rightarrow 2NH_4^+ + HCO_3^-$$

Upon alkalinization, ammonia is formed.

$$NH_4^+ \xrightarrow{OH^-} NH_3 + H_2O$$

The ammonia is then used in assay procedures.

Nesseler's Reaction

The ammonia formed in the urease reaction is reacted with Nesseler's reagent to form a colored end product that can be measured spectrophotometrically. Nesseler's reagent is an iodide salt of mercury and potassium ($HgI_2 \cdot 2KI$).

Berthelot Reaction

The ammonia formed in the urease reaction reacts with phenol and sodium hypochlorite to form a blue indophenol that can be measured spectrophotometrically. Sodium nitroprusside serves as the catalyst.

Kinetic Method

This method uses two enzymes as reagents. In addition to urease, glutamate dehydrogenase (GLDH) is needed. Ammonia and alpha-ketoglutarate serve as substrates. A coenzyme, NADH (reduced nicotinamide adenine dinucleotide), is oxidized to NAD^+ as alpha-ketoglutarate is reacted with ammonia:

$$\text{urea} + H_2O \xrightarrow{\text{urease}} 2NH_3 + CO_2$$

$$NH_3 + \text{alpha-ketoglutarate} + NADH \xrightarrow{\text{GLDH}} \text{L-glutamate} + NAD^+$$

The rate of change of the coenzyme is directly proportional to the urea concentration.

Electrode Methodology

As urease hydrolyzes urea, the conductivity in the reaction vessel increases due to the formation of NH_4^+ and HCO_3^-. A conductivity electrode and electronics are used to measure this change in conductivity, which is scaled to provide a number corresponding to the BUN concentration.

Diacetyl Monoxime

This method determines urea directly and does not use the enzyme urease. In an acid solution, the reagent diacetyl monoxime is hydrolyzed to diacetyl. The diacetyl reacts with urea to form a colored end product that can be measured spectrophotometrically. The liberated hydroxylamine, an interfering agent that accompanies the formation of diacetyl, is eliminated with an oxidizing agent such as ferric ammonium sulfate:

$$\text{diacetyl monoxime} + H_2O \xrightarrow{H^+} \text{diacetyl} + \text{hydroxylamine}$$

$$\text{diacetyl} + \text{urea} \xrightarrow{H^+} \text{yellow chromogen}$$

Conversion of BUN to Urea

In the United States, urea is reported as blood urea nitrogen. If the concentration of urea is desired, the urea nitrogen may be converted to an equivalent amount of urea by a conversion factor. This factor, 2.14, is derived by dividing the molecular weight of urea (60) by the molecular weight of the nitrogen content of urea (28), since there are two nitrogen atoms per molecule of urea (see molecular formula for urea under urease assay procedures). This conversion factor means that there is 2.14 parts of urea for every 1 part of urea nitrogen. Hence, to convert urea nitrogen to urea, multiply urea nitrogen by 2.14:

$$\text{BUN (mg/dL)} \times 2.14 = \text{urea (mg/dL)}$$

Clinical Significance

Increased

Since urea is an end product of metabolism normally excreted via the urine, any condition causing a decrease in the rate of clearance of urea from the blood or any condition that causes an excessive production of urea will elevate BUN .

The measurement of BUN is the most frequently requested test in the evaluation of renal function. Diseases of the kidney cause elevation of blood urea nitrogen because of decreased clearance. Failure to deliver blood to the kidney, as in cardiac decompensation or heart failure, also leads to decreased clearance and elevated blood levels, but the elevation is not related to renal disease. Dehydration may also cause elevated BUN levels due to the concentration of blood components which occurs with dehydration, and due to the decrease in renal blood flow which allows more filtered urea to be reabsorbed from the glomerular filtrate.

Excessive protein destruction within the body (such as occurs in burns, fever, or starvation) may elevate BUN as a result of excessive production of urea. In gastrointestinal bleeding, the ammonia produced as a catabolic product of blood enters the enterohepatic circulation. Then in the liver, the ammonia forms urea which may cause elevated BUN levels.

Decreased

Low BUN levels may occur in the presence of liver disease due to the inability of the liver to synthesize urea via the urea cycle. High fluid intake or excessive administration of intravenous fluids will result in decreased BUN levels due to an increase in renal blood flow, which allows less filtered urea to be reabsorbed from the glomerular filtrate.

In normal persons, the BUN may vary due to dietary factors. Normal persons on high protein diets will have higher values than those on vegetarian diets.

Reference Range

Serum or plasma BUN   7–21 mg/dL.

CREATININE

Source

Creatine is formed in the liver and pancreas, enters the blood stream, and is distributed to many cells, especially muscle cells. Within cells and in the presence of the enzyme creatine kinase, creatine is phosphorylated to form creatine phosphate. Creatine phosphate is then available to muscles as an immediate energy source (see Chapter 10, CK enzyme). The body content of creatine is proportional to muscle mass.

Creatinine (an anhydride of creatine) is a waste product of creatine formed during normal muscle metabolism. Creatinine is formed at a constant rate in an individual; the amount formed is based on muscle mass. Creatinine is water soluble and is readily excreted by the kidneys. In contrast to urea, once creatinine enters the glomerular filtrate, it is not reabsorbed to any significant extent. Creatinine levels are not appreciably affected by normal diets. Since the amount of creatinine excreted per day is based on muscle mass, daily urine excretion remains constant in the same individual in the absence of renal disease. The measurement of urine creatinine has been used to determine the completeness of a 24-hour urine collection. If the 24-hour urine collection is complete, the urine creatinine should fall within the normal range of 1–2 g/24 hr, provided the individual is free of renal disease. This test, as a predictor of the completeness of collection, is best utilized when multiple 24-hour collections are conducted on the same individual. In the same individual in the absence of renal disease, the amount of creatinine excreted per day should remain constant. For example, in high-risk pregnancy, multiple urine samples are collected for the determination of estriol. A fall in estriol indicates fetal distress necessitating termination of the pregnancy. Since the estriol assay requires 24-hour urines, an error in collection could provide the physician with false information resulting in the termination of pregnancy at an earlier time than would be desirable. By measuring the creatinine excretion in each serial collection before estriol assay, the laboratory may be able to detect an error in sample collection.

When kidney function is impaired, blood levels of creatinine increase. Creatinine may be secreted into the glomerular filtrate; the amount secreted is dependent on creatinine blood level. That is, as levels increase, more is secreted. At normal plasma levels, the amount secreted is considered negligible. Because creatinine is not appreciably affected by diet, degree of hydration or protein metabolism, the serum creatinine test is considered to be more specific but less sensitive than the BUN test for evaluation of renal function.

Sample Requirements

Any biological fluid such as serum, plasma, urine, or amniotic fluid may be used for creatinine assay. Whole blood contains many noncreatinine chromogens, so it is not considered an ideal specimen. Specimens may be preserved in the refrigerator for several days; if longer storage is needed, the samples should be frozen. Creatinine is formed from creatine at either an acid or alkaline pH, therefore specimens should be maintained at pH 7.0 during storage to minimize this interconversion.

Assay Procedures

Jaffe Reaction

The most common methodology used to determine creatinine is based on the Jaffe reaction. In the Jaffe reaction, creatinine is reacted with alkaline picrate solution (NaOH and picric acid) to form an alkaline creatinine picrate complex which is red in color. Under the usual assay conditions, the picric acid is present in excess and end point methods are not affected by minor variations in picric acid but are sensitive to variations in the concentration of hydroxide.

The reaction is sensitive to many interferences. The presence of noncreatinine chromogens or "Jaffe-positive" substances produces falsely elevated creatinine values. These Jaffe-positive substances include acetoacetic acid, pyruvic acid, and hydantoin. Some methods require deproteinization of the sample prior to the Jaffe reaction because protein is reactive with alkaline picrate and will form colored complexes.

A modified Jaffe procedure, which does not require a protein-free filtrate, measures the absorbance of the reaction mixture at two different pH levels (Jaffe, dual pH). The absorbance of the reaction mixture is recorded; acetic acid is then added to the reaction vessel. The acidification eliminates the color produced by creatinine while the color resulting from interfering substances such as protein is not eliminated. The absorbance, due to creatinine, is then calculated by subtracting the absorbance of the acidified tube from the initial absorbance.

Lloyd's reagent, an aluminum silicate, is used to isolate creatinine from the noncreatinine chromogens. In an acid medium, the creatinine is adsorbed by Lloyd's reagent, then desorbed in an alkaline solution. The creatinine is then determined by the Jaffe reaction. This method has been used as the reference method for creatinine. Cation exchange resins may also be used to separate creatinine from interfering agents. The eluted creatinine fraction may then be determined directly by ultraviolet spectrophotometry.

A kinetic or continuous monitoring method using the Jaffe reaction has been developed to be used in automated analyzers. Since noncreatinine chromogens produce color at slower or faster rates than does creatinine, the spectrophotometer readings are taken at a selected time interval to avoid the interferences of noncreatinine chromogens.

Enzyme Procedure

The Kodak Ektachem system uses the enzyme creatinine iminohydrolase, which catalyzes the hydrolysis of creatinine to form N-methylhydantoin and ammonia. The ammonia reacts with bromphenol blue to form a blue dye that is measured by reflectance photometry. A blank is utilized to eliminate any color due to endogenous ammonia in the patient sample. The reagents in the blank are the same as in the reaction slide except the creatinine iminohydrolase enzyme is absent.

Clinical Significance

The determination of the serum level of creatinine is used as a test to evaluate renal function. Creatinine, once filtered, is not reabsorbed but may be secreted by kidney tubules when blood levels are high. Blood levels, therefore, do not increase until renal function is substantially impaired. A creatinine value above 2–5 mg/dL is interpreted to mean moderate to severe kidney damage. Since creatinine is so readily excreted, blood levels of creatinine rise more slowly than do BUN levels in renal disease. Since BUN levels are affected by diet and state of hydration while creatinine values are independent of these factors, the creatinine test is considered to be more useful in following the progress of renal disease than is the BUN test.

For clinical interpretation, a ratio of BUN/creatinine may be calculated. In normal persons, the ratio is between 10:1 and 20:1. In prerenal azotemia, the ratio will rise because the decrease in blood delivery to the kidneys with decreased urine flow augments tubular reabsorption of urea, thereby increasing blood BUN values, while the creatinine value is less affected, since it is not reabsorbed once filtered. Also, in individuals with decreased muscle mass, the creatinine production is subnormal, and the BUN/creatinine ratio may also be high. Absorption of blood from the gastrointestinal tract following an episode of bleeding may elevate the ratio, because ammonia produced as a catabolic product of blood is converted to urea in the liver. Fever or burns will also increase the ratio as a result of increased protein catabolism.

A decreased ratio may be seen in states of overhydration, decreased protein intake, or severe liver disease.

In postrenal azotemia, both creatinine and BUN will be similarly affected because the mechanical obstruction suppresses the glomerular filtration rate.

Reference Range

Serum or plasma creatinine (nonspecific method)   0.6–1.3 mg/dL
Urine creatinine   1.0–2.0 g/24 hr

The determination of creatinine in the urine is performed to evaluate the completeness of a 24-hour collection or in conjunction with the creatinine clearance test to be discussed. Urine contains only small amounts of noncreatinine chromogens.

URIC ACID

Source

Uric acid (UA) is the end product of purine metabolism in man, anthropoid apes, and the Dalmation dog, since these species lack the enzyme uricase. In other mammals, uric acid is metabollized to allantoin, a more soluble end product:

$$\text{purines} \rightarrow \text{hypoxanthine} \rightarrow \text{xanthine} \xrightarrow{\text{xanthine oxidase}} \text{uric acid} \xrightarrow{\text{uricase}} \text{allantoin}$$

Most uric acid is formed in the liver, which contains high levels of xanthine oxidase. With the exception of the intestinal mucosa, most other tissues contain only traces of xanthine oxidase.

In man, uric acid is derived from the diet (exogenous nucleoproteins), from the breakdown of body nucleoproteins (endogenous nucleoproteins), and from direct transformation of purine nucleotides. Purine-rich foods include liver, kidney, thymus, herring, yeast, sardines, and leguminous vegetables.

Uric acid is freely filtered by the glomerulus but greater than 90% is subsequently reabsorbed in the proximal tubules. Tubular secretion of uric acid is then followed by a second tubular reabsorption. Uric acid is also presumed to be excreted through biliary, pancreatic, and gastrointestinal secretions with subsequent degradation by intestinal microorganisms.

Like man, birds and reptiles excrete uric acid in urine.

Sample Requirements

Plasma, serum, or urine may be used for uric acid assay. Any of the common anticoagulants may be used with the exception of potassium oxalate. Potassium forms insoluble potassium phosphotungstate resulting in turbidity in assay methods using phosphotungstic acid. Uric acid is stable for several days at room temperature but is susceptible to destruction by bacteria. Thymol may be used as a preservative to retard bacterial degradation. For long periods of storage, the sample should be frozen.

Assay Procedures

Phosphotungstic Acid

The phosphotungstic acid method is based on the reducing property of uric acid. Phosphotungstic acid is reduced to form a colored end product, tungsten blue:

$$\underset{\text{(reducing agent)}}{\text{uric acid}} + \text{alkaline phosphotungstate} \rightarrow \underset{\text{(tungsten blue)}}{\text{phosphotungstite complex}}$$

$$+ \underset{\text{(oxidized UA)}}{\text{allantoin}} + CO_2$$

The original procedure used a protein-free filtrate. Newer methods are modified to provide direct determination of UA without removal of protein. In the Caraway method, nonspecific reducing substances (ascorbic acid, glucose, tyrosine, and cysteine) are inhibited by the addition of sodium carbonate.

Uricase

In the uricase method, uric acid is converted to allantoin in the presence of the uricase enzyme. At the test wavelength (293 nm), uric acid absorbs but allantoin does not. The decrease in absorbance is proportional to the uric acid initially present:

$$\text{uric acid (absorbs)} \xrightarrow{\text{uricase}} \text{allantoin (does not absorb)} + CO_2$$

Since an enzyme is used as a reagent, this method is highly specific for uric acid.

Clinical Significance

Uric acid is most frequently assayed to evaluate gout. Gout is a disease of unknown etiology with a much higher incidence in males than females (95% of patients are males). In gout, urates tend to deposit in cartilage (ears and joints), epiphyseal bone, other articular structures, and kidneys, but rarely elsewhere. The crystalline urates (tophi) produce local necrosis with inflammation and destruction of cartilage and bone. In the kidney, the urates are deposited in the tubules, medulla, and interstitial tissue, resulting in inflammation and destruction of kidney tissue.

The first attack of gout usually occurs suddenly, lasts 3–14 days, and then disappears completely. The affected area is often the great toe or ankle, instep, or knee. Subsequent attacks follow with increasing frequency. Years later, after the first attack (5–14 years), chronic gouty arthritis develops. The hands and feet may become misshapen by tophaceous deposits. In some individuals, the sequence of events never progresses beyond the acute attack stage. The chief features of gout include arthritis, hyperuricemia, tophi, and late renal and cardiovascular complications.

Susceptibility to gout is believed to be transmitted by a single autosomal dominate gene with low penetrance. Persons who inherit the gene are considered to be carriers of latent gouty hyperuricemia. The hyperuricemia develops after puberty in males and after menopause in females. Three mechanisms have been proposed to explain the etiology of hyperuricemia: (1) diminished destruction of urates; (2) deficient excretion by the kidney; and (3) increased production of purines. The inherited form is called primary gout.

Secondary gout is not inherited, but the hyperuricemia is due to exogenous causes. Any disease with high tissue turnover, such as lymphatic leukemia or neoplasms of soft tissue, or after use of chemotherapeutic agents, or excessive exposure to radioactive radiation, may elevate uric acid levels. Certain drugs such as salicylates interfere with the renal handling of uric acid. Low-dose salicylates inhibit uric acid excretion (i.e., inhibit tubular secretion), but in high doses increase uric acid excretion (i.e., inhibit tubular reabsorption).

Uric acid will also be elevated in renal disease as a result of the inability of the kidney to excrete uric acid, but the uric acid assay is not used as a test to evaluate renal function.

There is a physiological rise in uric acid at the end of pregnancy, during and after labor, when the enlarged uterus is returning to normal size.

Decreased levels of uric acid are found in association with either congenital or acquired renal tubular reabsorption defects (i.e., Fanconi syndrome, Wilson's disease, toxic tubular damage).

Reference Range

Serum or plasma uric acid, male   3.7–7.9 mg/dL
                         female   2.7–6.6 mg/dL

Age, sex, race, social, and geographical factors affect reference ranges.

CLEARANCE AND CONCENTRATION TESTS

The tests that are used to evaluate renal function (BUN and creatinine) measure the blood level of an end product of metabolism normally excreted in the urine. When renal function is impaired, these end products are not effectively excreted and the blood concentration increases. Because the kidney has large functional reserve, it is estimated that 50% or more of the kidney parenchyma has been destroyed before BUN and creatinine become elevated in the blood. These tests, therefore, lack sensitivity.

The aim of clearance and concentration tests is to detect renal impairment as early as possible. These types of renal function tests show greater sensitivity to renal impairment than do the standard tests that measure blood urea nitrogen and serum creatinine.

FACTORS THAT ALTER TEST RESULTS

All clearance and concentration tests will be affected by prerenal, renal, and postrenal causes.

Prerenal

In prerenal disease, the decrease in kidney function occurs as a result of impaired delivery of blood to the kidney. Decreased blood delivery to the kidney occurs when the plasma volume is decreased (i.e., dehydration) or when the blood flow to the kidney is decreased (i.e., excessive blood loss, shock, cardiac failure).

Renal

Any disease affecting the kidney, such as glomerular disease, tubular disease, or disease involving the vascular system of the kidney altering blood flow, will influence renal function tests.

Postrenal

In postrenal disease, the decrease in renal function occurs as a result of decreased filtration pressure in the glomeruli caused by obstruction of the urine flow in the kidney. The obstruction may be due to renal stones, tumors, or prostate enlargement.

BLOOD FLOW IN RENAL TISSUE AND FORMATION OF URINE

With each heart beat, the renal tissue receives 20–25% of the cardiac output. The volume of blood that passes into renal arteries per minute is defined as total renal blood flow (TRBF) and is approximately 1200 mL/min.

The nonfunctional renal tissue (structural and supporting tissue) receives about 10% of the TRBF; functional renal tissue (glomeruli and peritubular capillary system) receives about 90% of the TRBF. The volume of plasma (note plasma, not blood) that perfuses functional renal tissue per minute is called the effective renal plasma flow (ERPF) and is approximately 660 mL/min. The ERPF may be measured by the paraaminohippurate clearance test when there is no impairment of renal function.

About 20% of the ERPF is extracted by the glomerulus. The volume of filtrate that passes into Bowman's capsule per minute is called the glomerular filtration rate (GFR) and is approximately 120 mL/min in a man with two normal kidneys. The GFR may be measured by the inulin clearance or creatinine clearance test.

Finally, the glomerular filtrate is altered as it progresses through the nephrons to form urine, which passes into the bladder at an average rate of 1 mL/min, or when awake, at 2 mL/min. The urine formation can range from as little as 0.3 mL/min to 15 mL/min depending on the state of hydration. In an adult, a normal 24-hour urine volume is 600–2000 mL.

The formation of urine is summarized in Figure 13.2.

RENAL FUNCTION TESTS

Kidney function tests may be divided into three major groups: (1) those that measure glomerular filtration rate (i.e., inulin and creatinine clearance); (2) those that measure secretory or proximal tubular function [i.e., paraaminohippurate (PAH) and phenolsulfonphthalein (PSP) clearance]; and (3) those that measure distal tubular function (i.e., dilution and concentration tests).

Mathematical Formula for Clearance Tests

The clearance tests all measure the ability of the kidney to remove or clear a substance from the plasma and are expressed as the number of milliliters of blood cleared of a substance per unit of time, usually expressed in mL/min. The formula for clearance tests is

$$\text{Clearance (mL/min)} = \frac{U}{P} \times V$$

where U = concentration of the substance in the urine
      P = concentration of the substance in the plasma
      V = volume of urine expressed in mL/min

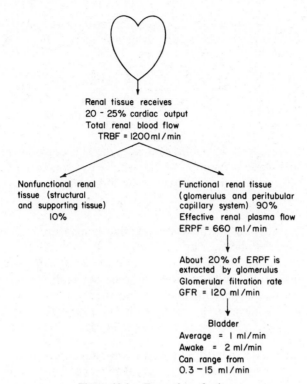

Renal tissue receives
20 - 25% cardiac output
Total renal blood flow
TRBF = 1200 ml / min

Nonfunctional renal
tissue (structural
and supporting tissue)
10%

Functional renal tissue
(glomerulus and peritubular
capillary system) 90%
Effective renal plasma flow
ERPF = 660 ml / min

About 20% of ERPF is
extracted by glomerulus
Glomerular filtration rate
GFR = 120 ml /min

Bladder
Average = 1 ml/min
Awake = 2 ml/min
Can range from
0.3 - 15 ml /min

Figure 13.2. Formation of urine.

The concentration of U and P may be expressed in any unit (mg/dL, g/dL) but both have to be expressed in the same units so that the units will cancel in the formula.

The clearance of a substance is proportional to renal parenchymal mass. The kidney mass is roughly proportional to the body surface of the individual; therefore, clearance tests are corrected for differences in individual size. The clearance formula with correction for body surface then becomes:

$$\text{Clearance (mL/min/m}^2) = \frac{U}{P} \times V \times \frac{1.73 \text{ m}^2}{\text{body surface (m}^2)}$$

The generally accepted average body surface in square meters is 1.73. The body surface of the patient expressed in square meters is usually determined by nomograms based on the patient's height and weight. Two examples of clearance problems are given below:

1. Calculate the creatinine clearance using the data: urine creatinine = 90 mg/dL; serum creatinine = 3.0 mg/dL; urine volume = 1000 mL/24 hr; patient height = 5.0 ft; patient weight = 90 pounds:

$$\text{Cl}_{cr} = \frac{90}{3.0} \times \frac{1000}{1440} \times \frac{1.73}{1.32} = 27 \text{ mL/min}$$

2. Calculate the creatinine clearance using the data: urine creatinine = 120 mg/dL; serum creatinine = 1.2 mg/dL; urine volume = 300 mL/4 hr.

$$Cl_{cr} = \frac{120}{1.2} \times \frac{300}{240} = 125 \text{ mL/min}$$

Tests That Measure Glomerular Filtration Rate

In tests that measure glomerular filtration rate, the substance that is measured is eliminated by the kidney predominately or wholly by glomerular filtration. In all clearance tests, the patient must be well hydrated before the test is initiated.

Exogenous

In exogenous clearance tests, the substance that is measured is not a substance normally present in the body. The substance to be cleared has to be given by intravenous injection. Inulin, a polysaccharide (MW of 5,100 g/mole), is freely filtered by the glomerulus and is neither secreted nor reabsorbed by the tubules. The inulin clearance test is considered to be the most accurate test available (reference method) to measure glomerular filtration rate.

The inulin clearance test is performed as follows:

1. A blood sample is taken before the test to be used as a control. The bladder is emptied before the test, and the urine saved to be used as a control.
2. A priming dose of inulin is injected intravenously and then inulin is infused intravenously throughout the test to maintain a constant blood level.
3. Three accurately timed urine samples are collected and blood samples are collected at the beginning and end of the urine collections.
4. Inulin is measured in blood and urine samples.
5. The mean value of the two blood samples is used in the clearance formula. The clearance for each urine sample is calculated. The average of the three values is used as an indication of the inulin clearance.

Since a physician has to be present throughout the inulin clearance test and the test is uncomfortable for the patient, endogenous clearance tests are the type routinely used in clinical settings.

Endogenous

In endogenous clearance tests, the substance that is measured is normally present in the blood and is not given by injection. Substances measured in endogenous clearance tests include urea and creatinine.

The urea clearance test was the first type of endogenous clearance test introduced but has been replaced by the creatinine clearance test. Urea clearance tests were difficult to interpret because the blood level of urea varies with dietary intake of protein and because urea, once filtered by the glomerulus, is passively reabsorbed. The amount reabsorbed is dependent on the rate of urine flow. The slower the rate of urine flow, the greater is the reabsorption, thus the clearance, in the same individual, appears lower when the rate of urine flow is low than when the rate of urine flow is maximum.

Creatinine clearance tests are more accurate than urea clearance tests because once creatinine is filtered by the glomerulus, it is not reabsorbed by the tubules and blood values are relatively stable from day-to-day in the same individual because the amount of creatinine formed is dependent on muscle mass.

The creatinine clearance test is performed as follows:

1. A precisely timed urine collection is obtained. The urine is usually collected over a 24-hour period but a 4-, 6-, 12-hour or any precisely timed collection can be used.
2. A blood sample is collected at some time during the urine collection.
3. Creatinine is measured in the blood and urine sample.
4. The clearance formula, with correction for body surface, is used to calculate the creatinine clearance.

Sources of Error
The greatest source of error in the creatinine clearance test is associated with faulty collection of the urine specimen. Incorrect timing of the collection or failure to completely empty the bladder at the beginning and end of collection results in an incorrect urine volume. Long periods of collection minimize errors due to failure to completely empty the bladder; therefore, 24-hour collections are preferred to shorter timed urine collections.

Reference Range

Inulin clearance, average, males	125 mL/min
average, females	115 mL/min
Creatinine clearance, males	85–125 mL/min
females	75–115 mL/min

When the creatinine clearance remains consistently below 10 mL/min, most patients die within a year. In individuals with creatinine clearances below 30 mL/min, electrolyte disturbances frequently develop. Individuals with creatinine clearance values above 30 mL/min seldom have electrolyte disturbances that can be attributed to glomerular filtration impairment. Most patients can remain active and work if the creatinine clearance remains above 30 mL/min.

Tests That Measure Proximal Tubular Function

Tests that measure proximal tubular function measure the secretory ability of the tubules. For these tests, substances are injected into the blood that are cleared either exclusively or predominately by the tubules.

Paraaminohippurate (PAH) Test
PAH, a substance foreign to the body, is removed from the plasma (up to 90%) in a single passage through the kidney. The PAH test is performed in much the same way as an inulin clearance test. A priming dose is given and continuous intravenous drip is needed to maintain a continuous blood level. Blood and urine samples are collected and clearance values are calculated. The PAH test is a research-type test, not a routine clinical one.

Since the ability to clear PAH from the blood is dependent on renal blood flow, PAH may be used to measure the effective renal plasma flow if there is no impairment of tubular function.

Phenolsulfonphthalein (PSP) Test

Phenolsulfonphthalein, a pH indicator, is predominately removed by tubular excretion. A small portion is removed by the liver and by glomerular filtration. The test is traditionally a 2-hour test; however, some laboratories collect only the first 15-minute specimen because this provides the closest distinction between normal and abnormal renal function.

The PSP test is performed as follows:

1. A standard dose of PSP dye is injected intravenously.
2. Urine samples are collected at 15, 30, 60, and 120 minutes after dye injection.
3. Each urine sample is measured, alkalinized, and diluted to 1 L. Upon alkalinization, the colorless PSP is converted to a colored form. The absorbance of the colored dye is measured spectrophotometrically. The percentage of PSP excretion is determined from a calibration curve.

Normally, at least 25% of the dye is excreted in the first 15 minutes and 60–85% is excreted by the end of 2 hours.

Reference Range

Paraaminohippurate clearance	600–750 mL/min
Phenosulfonphthalein clearance, first 15-minute specimen	25–50%
total excretion, end of 120 minutes	60–85%

Tests That Measure Distal Tubular Function

In diseases affecting the renal tubules, one of the first functions to be lost is the ability to concentrate the urine. When healing occurs, the ability to concentrate is the last function to return to normal.

The degree of urine concentration can be measured by specific gravity, by the refractive index principle, or by measurement of osmolality.

In the past, urine concentration tests were performed in the following way. Fluid intake was restricted, followed by serial collections of urine. The degree of concentration of each urine collection was determined either by measurement of the specific gravity or refractive index. If the ability of the kidney to concentrate urine was intact, the specific gravity or refractive index of the urine would increase as the kidneys attempted to conserve water under conditions of fluid restriction. Reference ranges were established for interpretation purposes. In the Fishberg test, the fluid intake was restricted beginning with the evening meal. No fluids were allowed during the night. Three urine samples were collected in the morning at 6, 8, and 10. In the Mosenthal test, the patient was allowed a controlled fluid intake. Then urine samples were collected over a 24-hour period.

The ability of the kidneys to concentrate urine is now measured by the simultaneous determination of urine and serum osmolality. The ratio of the urine osmolality to serum osmolality expresses the actual degree to which the kidneys have concentrated the glomerular filtrate. The normal ratio is between 1 and 3. The ratio is more meaningful when fluids are restricted whereupon the normal ratio is 3 or above. (See Chapter 9 where osmolality and test interpretation are discussed in detail.)

BODY FLUIDS

CEREBROSPINAL FLUID

The cerebrospinal fluid (CSF) is the fluid that fills the nontissue spaces of the brain and spinal column. It functions to maintain the constancy of the intracranial pressure, to protect the brain and nerve tissue, to maintain a stable chemical environment for these tissues, and to remove waste products of cerebral metabolism. CSF is formed by the ultrafiltration of plasma by a specialized structure of the brain called the choroid plexus of the cerebral ventricles (ventricles are the hollowed out cavities of the brain). The molecular sieve action of the choroid plexus decreases the protein content of CSF as compared to plasma.

Laboratory Examination

Color and Appearance
Normal CSF is clear and colorless with a viscosity similar to water. Xanthochromia (meaning yellow color) refers to yellow or pale pink to orange CSF that is associated with subarachnoid hemorrhage. Pale pink color indicates that free hemoglobin is present associated with recent brain hemorrhage while a yellow color indicates that the hemorrhage is not as recent, since the blood has undergone metabolic changes resulting in the formation of bilirubin. High serum levels of bilirubin with leakage into the CSF or CSF protein levels in excess of 150 mg/dL may also produce xanthochromic CSF. The presence of methemoglobin, an oxidized form of hemoglobin, may produce a brown-colored CSF.

A traumatic tap will also allow intact red blood cells to enter the CSF. In CSF collection, three tubes are generally used: the first for chemical and serological studies, the second for microbiological study, and the third for cell counts. If the number of cells in the third tube drops significantly from those seen in the first tube, the cells were probably derived from a traumatic tap.

Hazy or cloudy spinal fluid may be caused by the presence of bacteria, fungi, or ameba; leukocytes (greater than 200 cells/μL); erythrocytes (greater than 400 cells/μL), or aspiration of epidural fat during lumbar puncture.

Clotting is abnormal and is associated with an increase in CSF fibrinogen. A traumatic tap or markedly elevated CSF protein may result in clot formation.

Cells
Cell counts should be performed within 1 hour to avoid the error introduced by cell lysis. In normal adult CSF, the generally accepted reference interval is 0–5

mononuclear WBC/μL; in neonates, 0–15 mononuclear WBC/μL. Most of the cells are lymphocytes (more than two-thirds); the remaining cells are monocytes. The presence of neutrophils in CSF indicates acute inflammation and is generally associated with pyogenic organisms. In cases of untreated bacterial meningitis, neutrophils will constitute more than 90% of CSF cells. An increased number of neutrophils may be found in early viral meningitis (first 1–2 days), but the neutrophils rarely persist.

An increased number of mononuclear cells is associated with viral meningitis, tuberculous meningitis, fungal meningitis, syphilitic meningoencephalitis, and parastic disease as well as other noninfectious entities.

Protein

CSF, an ultrafiltrate of plasma, contains protein components present in serum but in lesser amounts, with predominance of low molecular weight proteins. The reference range for CSF total protein is 15–45 mg/dL. The electrophoretic pattern of concentrated CSF shows a prominent prealbumin band and proportionately more albumin and less gamma globulin than the serum pattern. Increases in CSF total protein are due to (1) changes in the permeability of the blood–CSF barrier, (2) local synthesis of central nervous system (CNS) immunoglobulins, (3) obstruction to the circulation of CSF, (4) decreased removal of protein molecules at the arachnoid villi, or (5) tissue degeneration. The blood–CSF barrier represents the interface between the choroid plexus epithelium and the endothelium of all capillaries in contact with the CSF. This is different from the blood-brain barrier, which is composed of the capillary endothelium in contact with astrocyte foot processes. Total protein may be increased in all types of infectious diseases that cause injury to the cerebral tissue, in intracranial hemorrhage, and in the presence of brain tumors.

In multiple sclerosis, a demyelinating disorder, the total protein may be normal, but the gamma globulin is increased due to synthesis of IgG within the CNS. Multiple sclerosis (MS) is a disease of unknown etiology and is the most common chronic neurologic disease affecting young adults. It most frequently begins between the ages of 15 and 45 with ambiguous clinical symptoms. The lesions may be located anywhere in the white matter of the CNS. The disease is progressive with periods of remission and may be immune in origin, but the antigenic stimulus is not known. MS may represent an autoimmune disorder with antibodies directed against myelin components or may be an immune response to a persistent CNS viral infection.

The CSF laboratory findings in MS are as follows. The appearance of the CSF, cell count, total protein, and glucose are all usually normal, but the total protein and cell count may be mildly elevated. The CSF IgG is increased in most cases due to IgG synthesis within the CNS. In order to discriminate between local (CNS) IgG production and IgG elevation secondary to changes in the blood–CSF barrier, the ratio of CSF IgG/CSF albumin is determined. The albumin serves as a reference protein since albumin is synthesized only in the liver. Therefore, the albumin in CSF is derived only from blood. When the origin of the IgG is due to changes in permeability of the blood–CSF barrier, both IgG and albumin will increase simultaneously and proportionately. Thus, the ratio should not be affected. If the increase in IgG originated in the CNS, the ratio will be elevated, since there should be no corresponding increase in albumin. For proper interpretation, the ratio should be performed on both the serum (serum IgG/serum albumin) and CSF (CSF IgG/CSF albumin) and the ratios compared to assure no changes have occurred in the serum that might affect the CSF protein constituents.

The IgG–albumin index is currently a popular method that is used for comparison. The formula for the index is as follows:

$$\text{IgG index} = \frac{\text{IgG (CSF)} \times \text{albumin (serum)}}{\text{IgG (serum)} \times \text{albumin (CSF)}}$$

The IgG index reference range should be established for each laboratory, but generally ranges between 0.25 and 0.85. The ratio will be elevated when IgG production occurs within the CNS.

The electrophoretic pattern of a concentrate of the CSF reveals oligoclonal bands (i.e., two or more discrete bands) in the gamma area in at least 90% of MS patients. These bands usually appear within the first year of the disease, are constant, and are not usually affected by exacerbations or remissions or therapeutic measures. The bands are considered to be products of multiple small clones of immunocytes in the CNS. High-resolution electrophoresis using agarose gel as the electrophoretic support media is required to demonstrate the oligoclonal bands. Isoelectric focusing, a technique whereby proteins are separated according to their isoelectric points in a pH gradient, may also be used to demonstrate the oligoclonal bands.

When electrophoresis is performed, both patient serum and CSF are plated side by side on the support media. A comparison of the serum pattern to the CSF pattern is needed to confirm that the oligoclonal bands originate from the CNS and are absent from the serum pattern. Positive specimens will demonstrate from 2 to 10 discrete bands in the gamma area. The preferred stain is 0.2% solution of Coumassie brilliant blue in 50% methanol. The patterns should be read visually, because densitometer tracings may not reveal the oligoclonal pattern. Serum and CSF samples for electrophoresis should be centrifuged, separated, and stored at 4°C, if the analysis is to be performed within two days. For longer periods of storage, the samples should be stored at −20°C. Repeated freezing and thawing should be avoided since the resolution of bands is affected.

Oligoclonal banding is not specific for MS and may be found in neurosyphilis, progressive multifocal encephalopathy, and some cases of bacterial or aseptic meningitis. Oligoclonal banding may occur transiently in Guillain–Barré syndrome and in acute herpes simplex encephalitis. Dense oligoclonal banding is present in subacute sclerosing panencephalitis (SSPE) and in progressive rubella panencephalitis; however, in these disorders, these bands should also appear in the serum.

Myelin, a major constituent of brain white matter, is composed of lipids (70%) and protein (30%). One of the protein constituents, myelin basic protein (MBP), appears to be specific for myelin and is released as myelin is destroyed. Myelin basic protein is found only in the active stage of demyelinating disorders and is not present in remission; therefore, it cannot be used as a diagnostic test. Myelin basic protein can be determined by RIA and is used to follow the activity of MS where its presence indicates active demyelination; it is also present in a variety of non-MS clinical entities.

Glucose

Glucose enters the spinal fluid both by passive diffusion and by active transport. Normally, CSF glucose is about 60–70% of blood levels. With high blood levels (800 mg/dL or above), the CSF glucose may be only about 30–40% of blood levels because the active transport system becomes saturated. For proper interpretation, a blood glucose should be drawn when the CSF is obtained. In a fasting patient with normal blood glucose, a CSF glucose value below 40 mg/dL is considered to represent a decreased CSF glucose.

CSF glucose may be decreased in bacterial, tuberculous, or fungal meningitis and in neoplasms involving the meninges. The disease is attributed to utilization of glucose by the CNS, leukocytes, and microorganisms, or impaired transport of glucose from plasma to the CNS.

Glutamine

The ammonia present in the CSF is approximately one-half of arterial blood levels. Since ammonia is toxic to brain tissue, the brain detoxifies ammonia by forming glutamine. When blood ammonia levels (hence CSF ammonia levels) are elevated, the glutamine concentration in the CSF increases. Elevated CSF glutamine levels are associated with hepatic encephalopathy. CSF glutamine levels are often used to evaluate Reye's syndrome where injury to hepatocytes affects the ability of the liver cells to synthesize urea from ammonia (urea cycle). Blood ammonia levels increase leading to elevated CSF glutamine levels (see Chapter 11, ammonia).

Lactic Acid

Lactic acid levels in CSF do not reflect blood levels. The lactic acid in CSF is probably derived from within the CNS through anaerobic metabolism. Inadequate blood or oxygen delivery to brain tissue may elevate CSF lactic acid levels. Increased lactic acid levels are also associated with decreased CSF glucose and are thought to reflect increased glucose utilization. The measurement of CSF lactic acid has been proposed as a test for meningitis. Lactic acid levels will increase in bacterial and tubercular meningitis, but should be within the reference range with most cases of viral meningitis.

Lactate Dehydrogenase

The lactate dehydrogenase activity in the CSF normally represents about 5–10% of plasma activity, since the majority of the enzyme protein is excluded by the normal barrier mechanisms. Lactate dehydrogenase in CSF is thought to be derived from: (1) diffusion across the blood–CSF barrier; (2) from the CNS tissue by diffusion across the brain–CSF barrier; and (3) from cellular elements in CSF, which may include bacteria, leukocytes, and tumor cells. The measurement of CSF lactate dehydrogenase isoenzymes has been proposed as a test to distinguish between bacterial and viral meningitis. The LD derived from brain tissue is predominately LD-1 and LD-2; the LD derived from lymphocytes is predominately

Table 13.1. Bacterial Meningitis Case-History CSF Laboratory Data

Appearance	Glucose (approximately two-thirds blood level)[a]	Protein (15–45) mg/dL	Cell Count (0–5 WBC/μL)	Gram Stain
Grossly purulent	None detected	1720	WBC—19,162 93% neutrophils 7% lymphocytes	Gram-positive diplococcus

[a] Reference ranges are given in parentheses.

LD-2 and LD-3; and that derived from leukocytes is predominately LD-4 and LD-5. In bacterial meningitis, the LD isoenzyme pattern reflects that derived from leukocytes with LD-4 and LD-5 present; in viral meningitis, the LD isoenzyme pattern reflects both a CNS and lymphocytic pattern with LD-1, LD-2, and LD-3 present. In either type of meningitis, high levels of isoenzymes derived from brain tissue (LD-1 and LD-2) indicate a grave prognosis.

Case History, Bacterial Meningitis

A 51-year-old black male was brought to the emergency room. Approximately one year ago, a partial mastoidectomy had been performed on the right ear because of chronic mastoiditis. Two days before this admission, the right ear began to drain clear fluid. Suddenly, he was unable to stand or speak and at the time of admission, the patient was responsive only to painful stimuli. The patient's neck was stiff. The blood BUN, glucose, and electrolytes were all normal. The CBC revealed a 20,300 white count with 70% polys and 13% band forms. A lumbar puncture was performed. The CSF laboratory data are given in Table 13.1.

Discussion

The CSF findings revealed an elevated white count composed almost totally of neutrophils, a decreased glucose, and an elevated protein; all are characteristic findings in bacterial meningitis. The CSF cultures were positive for *Streptococcus pneumoniae*. Blood cultures were also positive for the same organism.

The patient responded to high-dose penicillin. At the time of discharge, there was still tenderness over the right mastoid thought to be due to recurrent mastoiditis. The patient was referred to an otolaryngologist as a candidate for a complete mastoidectomy to prevent recurrent meningitis.

SYNOVIAL FLUID

Synovial fluid (SF) is an ultrafiltrate of plasma with an added hyaluronate–protein complex, which is manufactured by specialized cells lining the synovial membrane. Hyaluronate is a polysaccharide composed of repeating units of glucuronic acid and glucosamine. SF lubricates and nourishes articular cartilage. SF examination may be requested to evaluate joint infections, to determine the presence of gout or pseudogout, or to evaluate other forms of arthritis. Removal of SF from a joint is called arthrocentesis.

Laboratory Examination

Color and Appearance

Normal SF is clear and pale yellow in color, with a high viscosity. The degree of viscosity is related primarily to the concentration and polymerization of the SF hyaluronate. In inflammatory conditions, the viscosity decreases due to a decrease in the production of hyaluronate. The viscosity may be estimated by noting the "stringing" qualities of the aspirated fluid. This test is performed by placing a drop of SF on the thumb, touching a finger to the drop and noting the length of the string that forms as the fingers are separated. The length of the string should be longer than 4 cm in normal SF.

The mucin clot test measures the degree of polymerization of the SF hyaluronate. Normal SF will form a firm clot, surrounded by a clear solution when in contact with weak acetic acid. In inflammatory conditions, the clot formation is poor.

Cloudy SF may indicate an increase in leukocytes, which may occur in both septic and nonseptic inflammation.

Cells

Normal SF contains not more than 200 WBC/μL, primarily composed of lymphs and monocytes. The percentage of neutrophils should not exceed 25%. Very high leukocyte counts (100,000/μL or more) are suggestive of septic infection. In these conditions, the percentage of neutrophils is usually greater than 75%. Occasionally, active gout or rheumatoid arthritis will produce very high white counts in SF.

A search for LE cells and RA cells should also be conducted. RA cells or ragocytes are neutrophils with cytoplasmic granules. The granules are composed of immunoglobulins (IgG, IgM, rheumatoid factor) and complement. RA cells are found in about 95% of patients with rheumatoid arthritis but are also found in other conditions such as gout and septic arthritis.

If erythrocytes are present and are not derived from a traumatic tap, the cells should be counted and reported. Hemorrhagic effusions may be associated with joint trauma, fracture, tumor, hemophilia, pigmented villonodular synovitis, or septic arthritis.

Protein

The average total protein in normal SF is about 2 g/dL, but can range from 1 to 3 g/dL. As an ultrafiltrate of plasma, the types of proteins present reflect the plasma proteins, with the lower molecular weight proteins predominating. Inflammation increases permeability of the plasma membrane allowing larger molecular weight proteins to enter, increasing the total protein content. Fibrinogen has a large molecular weight and is normally excluded from the ultrafiltrate. If fibrinogen enters the joint space, the SF may clot on standing. Local synthesis of gamma globulins may also occur within the synovial tissue, elevating the total protein value. Normally, total protein is not included as a test in the routine examination of SF.

Glucose

For proper interpretation of the SF glucose, the sample should be obtained in a fasting state (6–12 hours) to allow glucose to equilibrate between plasma and SF. The glucose in SF is normally identical to or slightly less than the plasma glucose. In a nonfasting individual, a SF glucose level under 40 mg/dL is suggestive of decreased SF glucose.

Crystal Identification

Crystal identification in SF offers specificity for a single type of arthritis. However, two types of arthritis may coexist simultaneously in the same joint, decreasing its specificity.

To identify crystals, the SF should be examined by compensated polarized light microscopy. Since the crystals of interest are doubly refractive or birefringent, they will appear light against a dark background. The monosodium urate crystals (MSU) of gout appear as strongly birefringent needle-shaped crystals, which may vary from 1 to 20 μm in length. During acute attacks of gouty arthritis, most crystals are found within the cytoplasm of leukocytes or macrophages and may distend the shape of the cell. Between attacks, the majority of crystals are extracellular. With the use of a first-order red compensator, the crystals will have characteristic colors depending on their plane of orientation. They will be colored either blue or yellow depending on the alignment of the compensator.

Calcium pyrophosphate dihydrate crystals (CPPD) of pseudogout are weakly birefringent crystals and may appear as rhomboids, rods, rectangles, or needle-shaped crystals varying from 1 to 20 μm in length and up to 4 μm in width. When using a first-order red compensator, the colors seen with CPPD will be opposite to those of MSU.

Crystalline corticosteroids, which may appear following intraarticular injection, may polarize as long needles or rhomboids resembling MSU or CPPD. Cholesterol crystals may also appear in SF and are strongly birefringent, typically appearing as flat plates with one corner notched. However, in chronic effusions, cholesterol crystals may assume a form that resembles MSU or CPPD. SF may also contain birefringent debris, but in contrast to crystalline material, the debris has jagged edges and irregular shapes.

SEROUS FLUIDS

Serous (meaning derived from serum) fluids are formed by the ultrafiltration of plasma; no secretory processes are involved in their formation. Serious fluids are normally found in small amounts in the pleural (lung or chest), peritoneal (abdominal), and pericardial (heart) cavities, where they function to lubricate moving parts. These serous cavities are formed by a membrane; one surface lines the body wall, and the second surface lies in contact with the viscera. The area between the surfaces of the membrane contains the serous fluid. The cells that compose the membrane lining are called mesothelial cells. Removal of fluids from a cavity is called paracentesis. For example, abdominal paracentesis means removal of fluid from the abdominal cavity, thoracentesis means removal of fluid from the pleural cavity, and so on.

The formation and removal of these fluids by the body is a continuous process. The fluids are formed primarily through the balance of (1) hydrostatic pressure and (2) colloidal osmotic pressure. The hydrostatic pressure drives fluid into the tissue spaces and serous cavities, while the colloidal osmotic pressure pulls the fluid into the vascular space. The serous fluids are normally removed by the lymphatic system and returned to the blood. A disturbance in any of the normal mechanisms of fluid formation or removal results in effusions. For example, in liver disease or renal disease, the serum albumin may decrease, reducing the serum colloidal osmotic pressure. Fluids then tend to leave the vascular system producing tissue edema and effusions. In congestive heart failure, the hydro-static pressure increases, because blood tends to accumulate in organs. The asso-ciated poor circulation results in effusions. Decreased removal of serous fluids may occur when the lymphatic drainage is obstructed by a tumor, resulting in effusions.

Serous fluids are usually classified as either transudates or exudates. A transu-date is defined as a protein-poor fluid that is noninflammatory in origin. Transuda-tes, formed by diseases outside the body cavity, are frequently accompanied by generalized edema. An example of a transudate is a fluid that accumulates due to congestive heart failure or due to decreased serum albumin levels. Exudates are protein-rich fluids, considered to be inflammatory in origin. Exudates are formed by diseases that directly involve the organ or surfaces of the body cavity. Exam-ples of exudates include fluids that accumulate as a result of malignancies, infarc-tion, inflammation, and infection. In these processes, the accumulation of fluid occurs as a result of damage to the mesothelial linings.

Three tests may be used to provide a distinction between transudates and exudates. Exudates contain more protein than transudates. Pleural fluid, with a total protein greater than 3.0 g/dL is considered to represent an exudate. It has been suggested that a ratio of pleural fluid total protein/serum total protein pro-vides an improved method for distinction; a ratio value greater than 0.5 indicates an exudate. In peritoneal fluid, a total protein greater than 2.0–2.5 g/dL is consid-ered to represent an exudate. In pericardial fluid, the value needed to separate a transudate from an exudate has not been well established.

Exudates may also contain increased LD levels, which are thought to be de-rived from neutrophils. Pleural fluid LD greater than 60% of serum upper limit of normal or the ratio of pleural fluid LD/serum LD greater than 0.6 suggests a high probability that the fluid is an exudate.

Measurement of specific gravity provides the poorest method of distinction. Exudates are considered to have a specific gravity greater than 1.015.

Laboratory examination of serous fluids may include appearance, specific grav-ity, total protein, red blood cell and white blood cell count, differential and cytol-ogy examination. Additional testing may include hematocrit, glucose, amylase, LD, pH, BUN, creatinine, and microbiological cultures.

Pleural Fluid

Normal pleural fluid is clear and pale yellow in color. Cloudy or turbid fluid is usually due to increased numbers of leukocytes. Milky fluids are associated with lymphatic obstruction with leakage of thoracic duct contents into the serous fluid and the presence of chylomicrons (chylous fluid). If the cloudiness is due to chylomicrons, the cloudiness should disappear when mixed with ether (1 volume pleural fluid to 9 volumes ether). Milky fluid may also be found in effusions that contain lipids that are derived from cellular destruction (pseudochylous fluid). Pseudochylous fluids may be associated with any chronic effusion of long duration. Chylomicrons will be scanty or absent from pseudochylous fluids. Hemorrhagic fluids, if not due to a traumatic tap, are often caused by intrapleural malignancy but may also occur in association with many other disease entities.

About 50% of the effusions associated with pneumonia will have leukocyte counts over 10,000/μL with a predominance of neutrophils. In tubercular effusions, the leukocyte count generally exceeds 1000/μL, with over 50% lymphs and less than 1% mesothelial cells. However, these findings are not specific for tuberculosis and may have other etiologies. Mesothelial cells are normally sluffed into the serous fluid and generally make up at least 5% of the cellular elements. In tuberculosis, a fibrous exudate coats the cavity lining, possibly preventing the cells from entering the fluid or destroying underlying mesothelial cells.

The glucose in normal pleural fluid is approximately equal to that in whole blood. After meals, a 2–4-hour delay occurs before pleural fluid and blood glucose equilibrate. The serous fluid glucose is considered decreased if the difference between serous fluid and plasma glucose exceeds 40 mg/dL. If plasma glucose is not available for comparison, a serous fluid glucose under 60 mg/dL is considered decreased. Low glucose levels are associated with inflammatory diseases.

The measurement of pleural fluid pH may be helpful to determine if esophageal rupture has occurred. A pleural fluid pH under 6.0 is suggestive of rupture. Elevated amylase may also characterize a ruptured esophagus, the amylase being derived from the salivary gland. Pleural fluid amylase will also be elevated in effusions associated with pancreatitis.

Pericardial Fluid

Normal pericardial fluid is clear and pale yellow in color. Cloudiness is associated with an increase in neutrophils or the presence of chylous or pseudochylous effusions.

An increased white cell count (over 1000/μL) with a predominance of neutrophils is associated with bacterial pericarditis. The microscopic findings in tuberculous pericarditis are similar to those found in tuberculous pleural effusions.

The pericardial glucose parallels the serum whole blood glucose level. Decreased pericardial glucose occurs in bacterial pericarditis, rheumatoid disease, and malignancy.

Peritoneal Fluid

Peritoneal fluid may also be called abdominal or ascitic fluid. Normal peritoneal fluid is clear and pale yellow in color. Cloudy fluid is associated with bacterial infection, ruptured bowel, strangulated or infarcted intestine, peritonitis associated with appendicitis, or pancreatitis. Milky fluid, which is rare, may be due to chylous or nonchylous effusions. Green fluid contains bile, which may result from a perforated gall bladder or perforated duodenal ulcer or in association with acute pancreatitis. Red fluid indicates the presence of blood. Following abdominal trauma, whole blood may enter the peritoneal cavity. If the peritoneal fluid hematocrit is similar to the peripheral blood hematocrit, intraperitoneal injury is assumed to exist requiring exploratory laparotomy.

For undiluted, sterile peritoneal fluid, a leukocyte count exceeding $300/\mu L$ is considered abnormal. The differential is considered abnormal if the number of neutrophils exceeds 25%. The differential count in tuberculous peritonitis contains numerous mesothelial cells, which contrasts with tuberculous pleural fluid effusions.

Peritoneal fluid amylase activity may be elevated above normal blood levels in acute pancreatitis, in pancreatic pseudocyst, and following pancreatic trauma. Alkaline phosphatase activity may be elevated in individuals with perforation or strangulation of the small intestine. Following abdominal trauma, the measurement of peritoneal fluid creatinine and urea may be helpful to determine if bladder rupture has occurred.

AMNIOTIC FLUID

In utero, the fetus is immersed in a fluid-filled sac, which forms a protective cushion. The amniotic fluid that fills the sac is derived from both the amniotic membrane and secretions from fetal cells (gastrointestinal, respiratory, umbilical cord, and, in late pregnancy, fetal urine). There is also a fetomaternal exchange of water between amniotic fluid and maternal fluids.

The removal of amniotic fluid is called amniocentesis. Amniocentesis is usually performed to determine isoimmunization, to determine fetal maturity in high-risk pregnancies, or for genetic screening.

Isoimmunization

Isoimmunization is the result of blood group incompatibilities between mother and fetus. The IgG antibodies produced by the mother can cross the placenta causing destruction of the infant's red blood cells. The incompatibility may result from Rh or other blood group antigens.

The presence of hemolysis is detected by spectrophotometric scan of amniotic fluid, noting the presence of bilirubin at 450 nm. In normal fetuses, the bilirubin peaks at 23–25 weeks gestation and should disappear by the end of the 36th week. Deviation from this normal pattern gives evidence that hemolysis may be occurring. For the evaluation of hemolysis, amniotic fluid is usually collected after 28 weeks gestation.

Amniocentesis is indicated if maternal antibody titers show a pattern of increase. To determine maternal antibody titers, serial serum samples are collected from the mother. Prior to 26 weeks gestation, samples are obtained once per month, then every two weeks thereafter.

To scan amniotic fluid, the fluid is centrifuged, then filtered and scanned against a water blank with a double-beam narrow-bandpass recording spectrophotometer at wavelengths between 350 and 650 nm. These results, when plotted on one-cycle semilog paper, wavelength on the x-axis (linear scale) absorbance on the y-axis (logarithmic scale) show an inverse relationship between absorbance and wave-length. A normal curve shows a gradually decreasing absorbance as the wave-length increases, yielding essentially a straight line between 365 and 550 nm. An abnormal curve shows an absorption peak at 450 nm due to the presence of bilirubin. The magnitude of the peak at 450 nm is determined by measuring the height of the peak from the baseline of a straight line drawn between 365 and 550 nm. The straight line connecting the two points, 365 and 550 nm, is considered to represent the background absorbance of normal amniotic fluid. See Figure 13.3 for a schematic representation of a scan. Sometimes an additional peak occurs at 410 nm, called the Sorbet band, which is thought to be due to the presence of unconju-gated tetrapyrrole pigments. In practice, three absorbance points are plotted on the semilog paper, the absorbance at 365 nm, 450 nm, and 550 nm.

Amniotic fluid specimens for spectrophotometric scanning must be protected from light as the bilirubin pigment is destroyed by light exposure. Contamination of the fluid with blood, either fetal or maternal, may falsely elevate values. Cord blood bilirubin is 10–100 times greater than amniotic fluid bilirubin. The presence of oxyhemoglobin may change the baseline of the bilirubin scanning curve. The presence of meconium may also introduce error owing to changes in the baseline slope. Meconium causes a steep slope.

Fetal Maturity

Lecithin/Sphingomyelin (L/S) Ratio
In the United States, infant respiratory distress syndrome (IRDS) affects approxi-mately 40,000 infants per year resulting in 12,000 deaths.

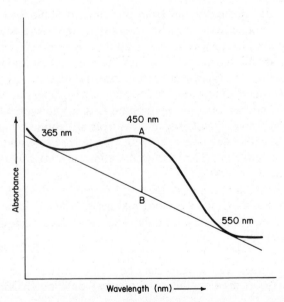

Figure 13.3. Amniotic fluid absorbance scan plotted on semilog paper (wavelength, abscissa, linear scale; absorbance, ordinate, log scale) to determine the degree of isoimmunization. Absorbance at 450 nm (A) minus expected absorbance at 450 nm (B) is equal to the absorbance contributed by bilirubin.

Infant respiratory distress syndrome is caused by a deficiency of pulmonary surfactant at birth. Pulmonary surfactant is a phospholipid protein complex that coats the alveoli preventing collapse of the air spaces on expiration. When surfactant is lacking, the infant cannot exchange oxygen and carbon dioxide. Surfactant is synthesized by type II pneumocytes that line the alveoli.

The major surface-active constituent of surfactant is dipalmitoyl phosphatidyl choline, which is called lecithin; phosphatidyl glycerol (PG) is the second most abundant surface-active component. Both are phospholipids and are secreted into amniotic fluid by the fetal lung. The determination of lecithin and PG levels in amniotic fluid is used as an aid in determining fetal lung maturity. Another phospholipid, sphingomyelin, also present in amniotic fluid, is used as an internal control (corrects for differences in the volume of amniotic fluid), since it remains at an essentially constant level throughout embryonic development.

In normal development up to the 26th week of gestation, the sphingomyelin concentration is greater than the concentration of lecithin, resulting in an L/S ratio of less than 1.0. After the 26th week, the lecithin concentration slowly increases until the 35th week after which there is a surge in lecithin resulting in a L/S ratio of 2.0 or more.

PG normally appears around 36–37 weeks of gestation and increases as the pregnancy progresses. It is believed that the appearance of PG signals the first biochemical stage of lung maturation and that it acts to stabilize lecithin.

The L/S ratio is determined by thin-layer chromatography (TLC). The amniotic fluid should be free from blood and meconium, since both contribute their own phospholipids. If the sample cannot be analyzed immediately, it should be centrifuged and the supernatant separated and stored at $-20°C$. The phosphodiesterase activity of the amniotic fluid destroys the lecithin if the sample is allowed to stand at room temperature.

The phospholipids are extracted from the amniotic fluid with a chloroform–methanol mixture. The organic phase is evaporated, and the residue is redissolved in chloroform. Some procedures include a cold acetone precipitation step, which separates saturated and unsaturated lecithin. Acetone precipitation has been suggested to increase test specificity. If acetone precipitation is performed, the chloroform residue is cooled and acetone is added dropwise. The sample is centrifuged and the acetone is poured off. The precipitate is then redissolved in chloroform.

Aliquots of the sample, standards, and controls are applied to a TLC plate and allowed to migrate with a developing solvent. The spots are visualized by charring or by use of lipid stains. The TLC patterns are interpretated by visual inspection and by densitometric scanning.

The presence of PG and a L/S ratio of 2 : 1 or greater using acetone precipitation are considered to indicated fetal lung maturity. When acetone precipitation is not used, it has been suggested that the L/S ratio cutoff should be raised to prevent false positives.

In certain pregnancies involving fetal stress (diabetes, preeclampsia, premature rupture of membranes), the L/S ratio may be falsely raised. Even in the presence of a L/S ratio of 2.0 or greater, these infants may develop IRDS.

Bubble Stability or Shake Test

Bubble stability in amniotic fluid is primarily due to the concentration of dipalmitoyl lecithin. The shake test is performed by shaking amniotic fluid with 95%

ethanol for 15 seconds and then observing the stability of the bubble pattern at the air–liquid interface after the sample has remained undisturbed for 15 minutes. The ethanol inhibits bubble formation by other constituents of the amniotic fluid permitting only the active surfactant phospholipids to influence the bubble pattern. A positive test is interpreted as a continuum of bubbles around the circumference of the liquid surface at the end of 15 minutes.

This test is difficult to interpret, but may be used as a screening procedure to determine fetal maturity.

Creatinine

In the latter part of gestation, fetal urine contributes to the amniotic fluid. The amniotic fluid creatinine level is related to fetal muscle mass, to fetal renal function, and to the maternal serum creatinine level. Normally, the amniotic fluid creatinine increases as the fetus matures but does not indicate fetal lung maturity or surfactant production. An amniotic creatinine value greater than 2.0 mg/dL is predictive of fetal maturity only if the maternal creatinine does not exceed 0.9 mg/dL.

This test should be used only as an assessment of fetal size and gestational age.

BIBLIOGRAPHY

Epstein, D. J. *CSF Protein Analysis in Multiple Sclerosis.* Check Sample Program CC 82-1, ASCP, 1982.

Garg, A. K. and Nanji, A. A. "The anion gap and other critical calculations," *Diagnostic Medicine* **5** (2), 32 (1982).

Gerson, B. *Laboratory Tests as Aids in Establishing the Diagnosis of Multiple Sclerosis: Oligoclonal Bands and Myelin Basic Protein.* Check Sample Program ACC 82-3, ASCP, 1983.

Glasser, L. "Body fluid evaluation: Serous fluids," *Diagnostic Medicine* **3** (5), 79 (1980).

Glasser, L. "Cells in cerebrospinal fluid," *Diagnostic Medicine* **4** (2), 33 (1981).

Glasser, L. "Reading the signs in synovia," *Diagnostic Medicine* **3** (6), 35 (1980).

Glasser, L. "Tapping the wealth of information in CSF," *Diagnostic Medicine* **4** (1), 23 (1981).

Glasser, L. and Finley, P. R. "Amniotic fluid and the quality of life," *Diagnostic Medicine* **4** (6), 31 (1981).

Henry, J. B. (ed.). *Clinical Diagnosis and Management by Laboratory Methods,* 16th ed. Philadelphia: Saunders, 1979.

Johnson, K. P. and Hosein, Z. "The laboratory identification of multiple sclerosis," *Laboratory Management,* 36–40 (May 1981).

Killingsworth, L. M. *High Resolution Protein Electrophoresis Booklet.* Beaumont, TX: Helena Laboratories, 1983.

Killingsworth, L. M., Cooney, S., Tyllia, M., and Killingsworth, C. "Protein analysis: Deciphering cerebrospinal fluid patterns," *Diagnostic Medicine* **3** (2), 23 (1980).

Rock, R. C. *Determination of Creatinine in Serum.* Check Sample Program CC 81-4, ASCP, 1982.

Smith, E. B. and Libb, S. "Recommendations for a fetal lung maturity profile," *Laboratory Management,* 27–30 (Sept. 1979).

Statland, B. E. and Freer, D. E. "Second-guessing Mother Nature: assessing fetal health and maturity." *Diagnostic Medicine* **2** (4), 72 (1979).

Sun, T. and Tria, L. *Amniotic Fluid Analysis.* Check Sample Program ST 83-3, ASCP, 1983.

Sun, T. *Laboratory Diagnosis of Multiple Sclerosis.* Check Sample Program ST 80-2, ASCP, 1980.

Tietz, N. (ed.). *Fundamentals of Clinical Chemistry*, 2nd ed. Philadelphia: Saunders, 1976.

Torday, J. S. "A new assay for fetal lung maturity," *Laboratory Management*, 45–51 (Oct. 1980).

Valitin, H. *Renal Function: Mechanisms Preserving Fluid and Solute Balance in Health.* Boston: Little Brown and Co., 1973.

Velander, R. K. and Pickering, N. A. "Fetal maturity—Clinical biochemical evaluation," *Laboratory Medicine* **12** (10), 604 (1981).

Warren, B. M. *L/S Ratio Booklet.* Beaumont, TX: Helena Laboratories, 1980.

Wenk, R. E. *Amniotic Fluid Phospholipids.* Check Sample Program ACC 80-4, ASCP, 1980.

POST-TEST

1. Draw a nephron, label its parts, and describe its function.

2. When substances are reabsorbed, the substances pass from the _____ ; when substances are secreted, the substances pass from the _____ .
   a. tubular lumen to peritubular plasma
   b. peritubular plasma to tubular lumen

3. The *major* force(s) that move(s) the filtrate across the glomerular capillary wall is(are): (may be more than one answer)
   a. colloidal osmotic pressure exerted by plasma proteins
   b. hydrostatic pressure created by the heart beat
   c. hydrostatic pressure of fluid in Bowman's space
   d. difference in lumen size of afferent and efferent arteriole

4. In a person with severe anemia, one would expect the erythropoietin level to be:
   a. increased
   b. decreased
   c. normal

5. The enzyme produced in the juxtaglomerular apparatus of the kidney is:
   a. ADH
   b. renin
   c. erythropoietin
   d. aldosterone

6. When renin is increased above normal values, one would expect the serum sodium to be _____ ; the blood volume to be _____ ; the serum potassium to be _____ .
   a. increased
   b. decreased
   c. normal

7. The NPN substances include all of the following *except:*
   a. urea
   b. creatinine
   c. ammonia
   d. amino acids
   e. albumin

8. Match the condition with its etiology.
    _____ 1. acute glomerulonephritis   a. prerenal azotemia
    _____ 2. enlarged prostate   b. renal azotemia
    _____ 3. nephrosis   c. postrenal azotemia
    _____ 4. acute blood loss
    _____ 5. dehydration
    _____ 6. obstructive renal stone
    _____ 7. congestive heart failure
    _____ 8. shock

9. Ammonia is removed from the body primarily by:
   a. glutamine synthesis in the brain
   b. urea synthesis in the liver
   c. excretion as free ammonia in the urine
   d. oxidation to nitrogen and water

10. All of the following will affect serum BUN levels *except:*
    a. dietary intake of protein
    b. state of hydration
    c. state of liver function
    d. state of kidney function
    e. muscle mass

11. In the same individual, when the rate of urine flow is 3.0 mL/min, you would expect the urinary BUN to be _____, as compared to the urinary BUN when the rate of urine flow is 0.5 mL/min.
    a. higher
    b. lower
    c. unaffected

12. In the methodology for blood urea nitrogen, all methods listed below utilize the enzyme urease as a reagent *except*:
    a. conductivity electrode method
    b. Berthelot method
    c. diacetyl monoxime method
    d. kinetic method using GLDH and NADH

13. A BUN was reported as 50 mg/dL. The concentration of urea in this same individual would be _____ .

14. Serum creatinine values are affected by all of the following *except:*
    a. muscle mass
    b. state of kidney function
    c. normal diet
    d. degree of tubule secretion

15. To check for the completeness of a 24-hour urine collection, the 24-hour urine _____ is measured.
    a. creatine
    b. uric acid
    c. urea
    d. creatinine

16. In the Jaffe reaction, noncreatinine chromogens may include all *except:*
    a. acetone
    b. acetoacetic acid
    c. pyruvic acid
    d. bilirubin

17. Of the two tests most frequently used for kidney function, the test that offers greatest sensitivity is _____; the test that offers greatest specificity is

    _____.
    a. serum BUN
    b. serum creatinine
    c. urine BUN
    d. urine creatinine

18. In gout, the crystalline material in trophi will be composed of:
    a. calcium oxalate
    b. calcium pyrophosphate
    c. xanthine
    d. monosodium urate

19. When the uricase enzyme is present, the end product of purine metabolism will be _____. When the uricase enzyme is absent, as in man, the end product of prine metabolism will be _____.
    a. uric acid
    b. allantoin
    c. xanthine
    d. hypoxanathine

20. Match the following. Each has two answers, choose one answer from each group.

    _____ 1. TRBF   a. volume of blood that passes into renal ar-
    _____ 2. ERPF   teries expressed in mL/min
    _____ 3. GFR   b. volume of plasma that perfuses functional
                              renal tissue expressed in mL/min
                           c. Volume of glomerular filtrate that passes
                              into Bowman's capsule expressed in mL/
                              min

                           d. 1200 mL/min
                           e. 120 mL/min
                           f. 660 mL/min

21. Given: Serum creatinine = 1.5 mg/dL; urine creatinine = 100 mg/dL; 24-hour urine volume = 1500 mL; patient's body surface = 1.85 m^2. Calculate the clearance with correction for body surface.

22. Fill in the blank with the correct answer(s) from the list of answers given below.

    Tests that measure glomerular filtration rate include _____ ; tests that measure tubular secretion function include _____ .
    a. creatinine clearance
    b. PSP clearance
    c. PAH clearance
    d. inulin clearance

23. If a substance is completely filtered at the glomerulus and then is completely reabsorbed by the kidney, the clearance of that substance will be:
    a. zero
    b. 100%
    c. 50%
    d. more information is needed to evaluate

24. If a substance is completely removed by one passage through the kidney and then none is reabsorbed, the clearance of that substance will be:
    a. zero
    b. 100%
    c. 50%
    d. more information is needed to evaluate

25. The reference method for the measurement of glomerular filtration rate is:
    a. creatinine clearance
    b. PAH clearance
    c. urea clearance
    d. inulin clearance

26. Match the following. Choose one answer from each group.

    _____ 1. typical bacterial meningitis
    _____ 2. typical viral meningitis

    a. normal WBC count
    b. increased WBC count

    c. predominate cell type, neutrophil
    d. predominate cell type, mononuclear
    e. normal differential

    f. normal glucose
    g. decreased glucose

    h. normal total protein
    i. increased total protein

27. Tests that may help to differentiate bacterial meningitis from viral meningitis include all *except:*
    a. measurement of CSF lactic acid
    b. measurement of CSF isoenzymes of LD
    c. measurement of CSF glucose
    d. measurement of CSF total protein
    e. cell differential count

28. Myelin basic protein and oligoclonal banding are associated with:
    a. fungal meningitis
    b. multiple sclerosis
    c. bacterial meningitis
    d. viral meningitis

29. The crystal associated with gout is _____ and is _____; the crystal
    associated with pseudogout is _____ and is _____ .
    a. monosodium urate
    b. calcium pyrophosphate dihydrate

    c. weakly birefringent
    d. highly birefringent

30. Match the following.
    _____   1. protein-poor fluid, low specific gravity   a. transudates
    _____   2. protein-rich fluid, high specific gravity   b. exudates
    _____   3. inflammatory in origin
    _____   4. noninflammatory in origin
    _____   5. pleural effusion due to congestive
                    heart failure
    _____   6. pleural effusion due to pneumonia
    _____   7. associated with diseases of body
                    cavity
    _____   8. associated with diseases outside body
                    cavity

31. The major surface-active phospholipid of surfactant is:
    a. sphingomyelin
    b. dipalmitoyl phosphatidyl choline
    c. phosphatidyl glycerol
    d. phosphatidyl serine

32. A pregnant female late in her pregnancy developed complications. The doc-
    tor ordered a lecithin sphingomyelin ratio on amniotic fluid to determine if
    fetal lung maturity was adequate and he could therefore induce labor. The
    ratio was 2.5. These results would:
    a. indicate lung maturity with little risk of developing RDS
    b. indicate lung immaturity with risk of developing RDS

33. Match the following:
    _____   1. amniotic fluid creatinine   a. measures fetal lung
    _____   2. amniotic fluid bilirubin   maturity
    _____   3. amniotic fluid L/S ratio   b. measures fetal size
                                                      c. measures degree of
                                                         isoimmunization

ANSWERS

1. See Figure 14.1. Functions: (a) controls the concentration of most of the
   constituents of the body; (b) aids in the control of blood pH; (c) functions to

excrete end products of metabolism; (d) controls blood volume via renin–
aldosterone pathway; and (e) site of production of erythropoietin.

2. a, b
3. b, d
4. a
5. b
6. a, a, b
7. e
8. (1) b; (2) c; (3) b; (4) a; (5) a; (6) c; (7) a; (8) a
9. b
10. e
11. a
12. c
13. 107 mg/dL
14. c
15. d
16. d
17. a, b
18. d
19. b, a
20. (1) a, d; (2) b, f; (3) c, e
21. 65 mL/min
22. a, d; b, c
23. a
24. b
25. d
26. (1) b, c, g, i; (2) b, d, f, h *or* i
27. d
28. b
29. a, d; b, c
30. (1) a; (2) b; (3) b; (4) a; (5) a; (6) b; (7) b; (8) a
31. b
32. a
33. (1) b; (2) c; (3) a

CHAPTER 14

THERAPEUTIC DRUG MONITORING AND TOXICOLOGY

OBJECTIVES

Upon completion of this chapter, you will be able to:

1. Define, illustrate or explain the following terms or concepts:
   a. TDM
   b. dose-response curve

    c. $t_{1/2}$
    d. zero-order kinetics
    e. trough level
    f. bioavailability
    g. TLC
    h. Rf
    i. enzyme immunoassay
    j. FIA
2. Describe the many factors that influence the pharmacological effects of an oral medication.
3. Discuss the pharmacokinetics of drug action and the use of these concepts in therapeutic drug monitoring.
4. Discuss the various methods currently used to quantitate drugs in blood. Describe the principles of each measurement method and list the advantages and disadvantages of each method.
5. List and describe the use of various drugs currently monitored routinely in the clinical laboratory.
6. Describe the use of thin-layer chromatography screening for the identification of drugs in the urine and other biological fluids. Also describe other methods used to screen biological specimens for drugs.
7. Describe toxicology tests for volatile substances, metals and nonmetals, and poisonous gases.

GLOSSARY OF TERMS

Enzyme induction.   A stimulated increase in the activity of microsomal enzyme activity involved in the metabolism of drugs.

FIA.   Fluorescent immunoassay is a competitive binding immunoassay using a fluorescent label.

First-pass metabolism.   Following absorption, some drugs are metabolized in the intestinal mucosa and liver prior to entering the general circulation.

Loading dose.   To minimize the time required to achieve steady-state drug levels, an initial loading dose (or bolus dose) is administered.

MEC.   Minimum effective concentration of a drug.

Therapeutic range.   A range of serum levels of a particular drug associated with observed therapeutic benefit of the administered drug.

TLC.   Thin-layer chromatography is a qualitative method commonly used for the separation and identification of drugs in various body fluids.

Trough level.   After steady state, the drug level obtained on specimens collected just prior to the next dose.

V_d or volume of Distribution.   This represents the total fluid compartment of the body required for drug distribution to produce the measured serum concentration.

THERAPEUTIC DRUG MONITORING (TDM)

Therapeutic drug monitoring (TDM) refers to a rapidly expanding area of clinical chemistry concerned with the measurements of drug levels in body fluids, usually serum. Many clinical laboratories are concerned with supplying physicians with serum drug levels along with interpretive information to facilitate the physician's use of drugs in the care of their patients.

THERAPEUTIC RANGE

Many drugs produce a desired therapeutic effect over a narrow range of plasma level. When the plasma drug level falls below this therapeutic range, most patients receive no clinical benefit from the drug. On the other hand, should plasma levels exceed the upper limit of the therapeutic range, the patient often develops signs and symptoms of drug toxicity. Often the upper limit of the therapeutic range is close to toxic levels. Examples of drugs where the optimal therapeutic range has been established are listed in Table 14.1.

DRUG BIOLOGICAL EFFECT

The actual pharmacological effect of a drug is initiated by the binding of the drug to a specific receptor in the target tissue. TDM assumes the serum drug level is proportional (or directly related) to the intracellular tissue binding of the drug.

Table 14.1.   Drugs and Drug Groups with Known Therapeutic Ranges

Type	Drug	Therapeutic Range
Heart drugs	Digoxin	1–2 ng/mL
	Digitoxin	10–30 ng/mL
	Lidocaine	1.2–5.0 ng/mL
	Procainamide	4–10 ng/mL
	Quinidine	2–5 ng/mL
	Propanolol	40–100 ng/mL
Antiasthmatic	Theophylline	10–20 ng/mL
Antiepileptic	Phenytoin	10–20 ng/mL
	Phenobarbital	20–40 ng/mL
	Cabamazepine	8–12 ng/mL
	Primidone	5–12 ng/mL
	Ethosuximide	50–100 ng/mL
	Valproate	50–100 ng/mL
Antidepressant	Nortriptyline	50–140 ng/mL
	Lithium	0.9–140 mmol/L

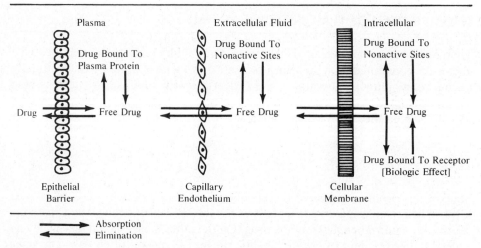

Figure 14.1. Drug absorption and distribution. (Reproduced with permission from Levine, R. R. *Drug Actions and Reactions*, 2nd ed. Boston: Little Brown, 1978.)

Hence, the serum level should correlate with the desired biological effect (see Figure 14.1).

FACTORS INFLUENCING SERUM DRUG LEVELS

Unfortunately, there are many complicating factors other than dosage that influence the serum level of a drug (see Figure 14.2). These factors include patient compliance, absorption of the drug, protein binding, drug metabolism in the liver, renal drug excretion, volume of plasma compartment, and tissue storage.

PATIENT COMPLIANCE

It has been suggested that most patients do not take their oral medications precisely as prescribed, hence they are essentially noncompliant. Often, patients either forget to take the drug at all, take either too much or too little of the drug, or take the drug, forget, and take it again in a short time interval. These latter patients often experience toxic symptoms.

DRUG ABSORPTION

When drugs are injected intravenously or intramuscularly, plasma levels are more predictable than when the drug is administered orally. As shown in Figure 14.2, oral drugs must be absorbed in the gastrointestinal (GI) tract in order to reach the vascular system. Most drugs are administered orally, hence, an understanding of the principles of drug absorption in the GI system is essential for TDM.

Drug Diffusion

Drugs diffuse across epithelial cell membranes of the GI tract by (1) passive diffusion of drugs moving from a high concentration to a lower concentration; (2) lipid binding of drug facilitating drug transport across cell membranes; and (3) an ''active-transport'' mechanism, which may even move the drug against the concentration gradient. The pH of the GI tract plays a significant role in absorption along with the gastric emptying time and intestinal motility. These three factors can be influenced greatly by other drugs and meals taken simultaneously with the administered drug.

Gastrointestinal pH

Since only ''molecular'' or nonionic drugs are absorbed, at the acid pH of the stomach, most weakly acidic drugs will be nonionic and hence readily absorbed. Mildly basic drugs would be cationic (or protonated) and as ions would not be absorbed in the stomach. The most efficient and predictable site of drug absorp-

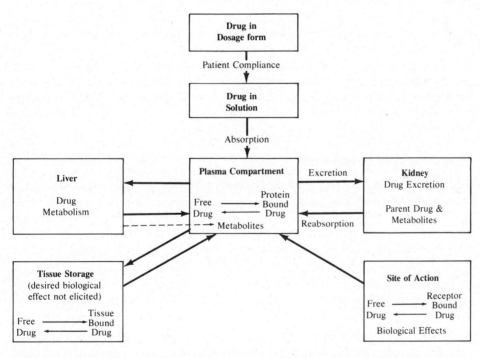

Absorption
Drug must be formulated in a manner which assures bioavailability for absorption.

Metabolism
Drug converted to a more soluble compound which may be biologically active or inactive. Metabolism can also occur in other tissues.

Excretion
Usually more water soluble drug metabolites are excreted in urine. Also drug excretion can occur via bile, feces, saliva and expired air.

Tissue Storage
Distribution of drug to sites where the desired biological effect is not elicited.

Undesirable effects may be elicited by drug interaction with a specific physiological system.

Site of Action
Free drug binds to receptor to elicit a biological effect (response). Number & type of receptors to which drug is bound determines the intensity and duration of the desired and undesired effects.

Figure 14.2.   Various factors involved in oral drug administration and desired pharmacological drug effect. (Reproduced with permission from the Syva Company, Palo Alto, CA.)

tion is in the small intestines. Slow gastric emptying time into the small intestine markedly increases the time requirements for drug absorption. In rapid intestinal motility (as in diarrhea), the drug passes absorption sites in the intestine too rapidly for absorption to take place. Obviously, lower plasma levels of the drug would be expected in either of these conditions.

PROTEIN BINDING

As indicated by Figures 14.1 and 14.2, a significant portion of the absorbed drug is bound to protein in the plasma compartment. This "bound" drug is not free to pass through membranes:

$$\text{protein} + \underset{\text{"free drug"}}{\text{drug}} \rightleftharpoons \underset{\text{"bound drug"}}{\text{protein} \cdot \text{drug}}$$

Free drug, the metabolically active form, is usually soluble in plasma and readily passes across cell membranes. The percentage of the drug actually bound to protein often varies from patient to patient. Also, other substances in the plasma (other drugs, fatty acids) can compete for carrier-protein building sites. The pH of the plasma can greatly affect drug binding as well. As the pH decreases in plasma, the binding of drugs to protein is often significantly decreased. In uremic patients, protein binding of drugs can be lowered as much as 50%, hence the dosage of the drug must be significantly reduced to avoid toxic symptoms. Toxic symptoms often occur when the level of "free drug" increases due to a shift in protein binding.

Measuring Free Drug Levels

Since the free drug is the metabolically active form of the drug, direct measurement of the free drug concentration using equilibrium dialysis is sometimes used. In this method, the free drug is allowed to diffuse across a semipermeable membrane prior to analysis. Also, the measurement of drug concentrates in "plasma ultrafiltrates" such as CSF and saliva gives an estimate of the free drug in plasma.

DRUG METABOLISM

As indicated by Figures 14.1 and 14.2, drug concentrations in the plasma compartment are reduced by drug metabolism in the liver. In general, liver metabolism of drugs results in "metabolites" of the parent drug, with subsequent formation of conjugates (sulfates and glucuronides). These drug derivatives are more soluble in plasma, and hence are more readily excreted by the kidneys. Also, metabolites of drug usually have little or no biological activity remaining.

Drug half-life is a pharmacokinetic term (to be discussed later) that relates to metabolism and excretion of drugs. Drugs that are rapidly metabolized have short half-lives while drugs that are metabolized slowly have long half-lives.

Renal function plays a vital role in the half-life of a drug, and in patients with severely depressed creatinine clearance (40 mL/min or less), the half-life of aminoglycoside antibiotics may more than double. TDM can play a vital role in determining required dosage levels in these patients.

Enzyme Induction

Drugs are metabolized in the microsomal system of hepatocytes by enzyme systems "normally" used to metabolize endogenously produced organic compounds (hormones, etc.). Another concomitantly administered drug that increases the production of hepatic enzymes involved in metabolism may, in turn, increase the rate of metabolism of the other drug present. This is an example of "drug interaction."

First- and Zero-Order Kinetics

Increasing the amount of drug administered can "induce" the production of hepatic enzymes, thus increasing the rate of metabolism in a linear fashion based on drug dose. Kinetically, this metabolic relationship is described as "first-order" (see Figure 14.3). However, when hepatic enzymes of the microsomal fraction

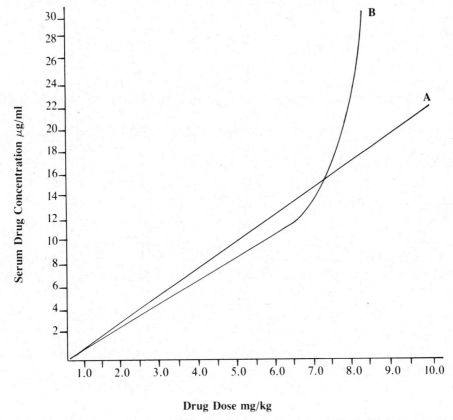

A—Dose-response curve for drug observing first-order kinetics. (linear)

B—Dose-response curve for drug observing zero-order kinetics. (non-linear or saturation)

Figure 14.3. The kinetics of drug metabolism. (Reproduced with permission from the Syva Company, Palo Alto, CA.)

become saturated with drug (at high dosage levels), the rate of drug metabolism is not influenced (or increased) by the increased plasma levels, a situation described as zero-order kinetics (see Figure 14.3). In zero-order kinetics, a very small increase in dosage may result in a markedly elevated plasma level of the drug with resulting toxicity.

Patient Metabolic Variations

Enzymic metabolic systems vary genetically from one patient to another. Fast metabolizers generally have lower than expected plasma levels of drug; slow metabolizers have much higher than expected plasma levels of the drug. Consistently, low plasma drug levels are most commonly associated with either fast metabolizers or noncompliant patients.

CHEMICAL REACTIONS OF METABOLISM

A great variety of organic reactions are involved in drug metabolism. Functional groupings on drugs include mainly amino, hydroxyl, and carboxyl groups. In metabolism, these groups may be combined with glucuronic acid and acid salts (sulfates, phosphates) to form more polar compounds. Other chemical reactions on the parent drug include oxidation, hydrolysis, hydroxylation, and dealkylation. All of these reactions tend to add polar oxygen (or hydroxyl) groupings to the parent drug making the resulting drug metabolite more polar and, hence, more soluble in plasma, where it can be readily excreted by the kidneys. As mentioned, these metabolites may be conjugated with glucuronic acid to enhance the polarity of the derivative even further. Obviously, the chemical nature of the parent drug is substantially changed by metabolism, hence, it is not surprising that metabolites often retain little or no pharmacological activity.

RENAL DRUG EXCRETION

Most drugs and drug metabolites are excreted in the urine. Plasma levels of drugs that are excreted primarily unmetabolized (e.g., the barbiturates) can become markedly elevated in patients with decreased renal function. Also, many drugs exhibit a nephrotoxic effect (e.g., the aminoglycoside antibiotics). Therefore, it is often important to evaluate renal function along with the absolute plasma level of the drug.

Drugs can also be eliminated by the bilary tract, lungs, and sweat; these routes may play a role in uremia where glomerular filtration (or tubular secretion) is impaired.

PHARMACOKINETICS

The term pharmacokinetics refers to the study of mathematical models used to predict plasma drug levels over a fixed time period and dosage regimen. Predicted plasma drug levels based on pharmacokinetics are often useful; however, because of the many variables involved, they are often inaccurate for specific clinical situations. Nevertheless, an understanding of the general principles of pharmacokinetics is useful to anyone engaged in TDM.

DOSE–RESPONSE CURVE

Oral administration of a single dose of a drug will give a "dose–response curve" as shown in Figure 14.4, where the serum drug concentration achieved is plotted against time. As shown, the serum drug level reaches a maximum (or peak) at 1.5 hours, then decreases more slowly, indicating that metabolism and excretion of a drug is generally a much slower process than drug absorption.

Half-Life $(t_{1/2})$

The half-life of a drug (which can be determined by examination of the dose–response curve) is the time required for the serum drug concentration to be reduced by one-half. In Figure 14.4, the drug concentration of 2 μg/mL decreased to 1 μg/mL in 2 hours, hence the $t_{1/2}$ for this drug is 2 hours. If the $t_{1/2}$ is long, the patient is metabolizing and excreting the drug slowly; a short $t_{1/2}$ indicates rapid drug metabolism and excretion. The negative slope of the dose–response curve to the right of the maxima reflects the rate of metabolism/excretion of the drug; the positive slope of the curve to the left of the maxima reflects the rate of absorption of the drug.

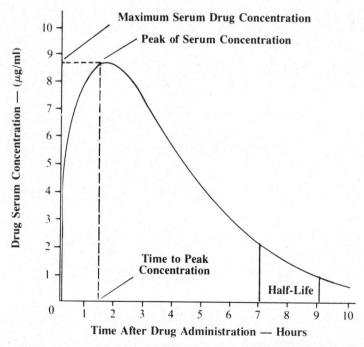

Figure 14.4. Dose-response curve for oral drugs. [Reproduced with permission from Koch-Weser, J. *New England Journal of Medicine*, **291**, 234 (1974).]

Multiple Dose–Response Curves

Repeated oral doses of a drug will result in multiple and contiguous single dose–response curves as shown in Figure 14.5. After each dose, a "peak" level is reached at 1.5 hours; just prior to the next dose, a "valley" or "trough" serum drug level is observed.

Steady State

Also, as shown in Figure 14.5, a "steady-state" serum level is reached at 20 μg/mL after five doses of the drug. Here, the amount of drug in the plasma compartment (see Figure 14.5) is relatively constant due to an established equilibrium between absorption, tissue storage, metabolism, and elimination:

$$\text{drug (absorbed)} = \text{drug (metabolized and excreted)}$$

Again, the amount of time required to achieve a steady state of concentration is usually five to six half-lives of the drug.

EXAMPLE 14.1.   Compute the time required to achieve a steady-state serum level for a drug that has a half-life of 1 day.

Answer: It takes five half-lives for a drug to reach a "steady-state" serum concentration:

$$5 \times t_{1/2} = 5 \times 1 \text{ day} = 5 \text{ days}$$

To reduce the steady-state concentration by one-half assuming first-order ki-

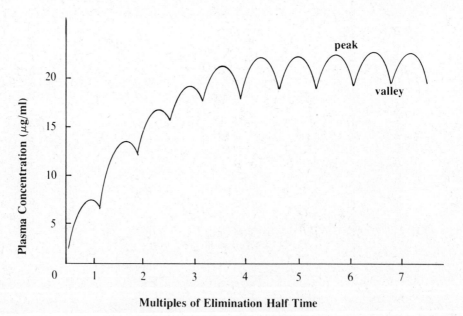

Figure 14.5.   Multiple dose-response curves for oral medications. (Reproduced with permission from the Syva Company, Palo Alto, CA.)

netics, halving the dosage level would achieve the new steady state in five half-lives. Similarly, to double the steady-state concentration (assuming first-order kinetics), doubling the dosage would double the steady-state serum concentration in five half-lives.

DOSE ADMINISTRATION

Generally, drugs are dosed based on the age and weight of the patient. In general, dosage rates vary significantly among newborns, infants, and children because of differences in factors affecting drug absorption, metabolism of drugs, protein binding of drugs, and factors affecting renal excretion of drugs. However, in the young, a major variable in TDM is that younger patients tend to metabolize drugs faster than adults. Hence, different dosages of a drug are often indicated for neonates, infants, children, adolescents, and adults. Body weight is used to account for the volume of distribution (V_d) of the drug. Larger adults obviously have a larger V_d than children who have small blood volumes and fat stores for drug distribution. For example, if an adult dosage for a drug is 4 mg/kg/day, a morning dose for a 70 kg adult would be

$$1 \text{ day} \times 70 \text{ kg} \times 4 \text{ mg/kg/day} = 280 \text{ mg}$$

A child's dosage rate may be 6 mg/kg/day to account for more rapid metabolism.

EXAMPLE 14.2.   Compute the dosage per day for a 35 kg, 11-year-old child if the recommended dosage rate is 6 mg/kg/day.

Answer:

$$\text{dose} = 1 \text{ day} \times 35 \text{ kg} \times 6 \text{ mg/kg/day}$$

$$\text{dose} = (35)(6) \text{ mg} = 210 \text{ mg}$$

Dosing Intervals

The frequency of drug administration is determined by the drug half-life. In general, the drug should be taken at time intervals no greater than one-half the half-life. At dosing intervals equal to or greater than the drug half-life, exaggerated peaks and valleys in serum drug levels are seen. Symptoms of toxicity are often seen shortly after dosage with loss of therapeutic benefits observed prior to the next dose. More frequent dosing tends to minimize peaks and troughs, however, it is often inconvenient for the patient. For example, a drug with a half-life of 8 hours may need to be taken every 4 hours.

PEAK-AND-TROUGH DRUG LEVELS

After steady-state drug levels have been achieved, blood samples for trough drug levels should be drawn just prior to the next dose (see Figure 14.6). Ideally, the

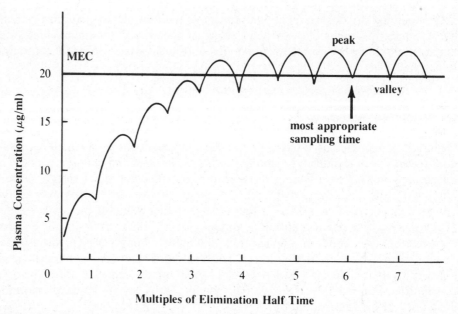

Figure 14.6. Sampling time for trough level after achieving a steady-state drug level. (Reproduced with permission from the Syva Company, Palo Alto, CA.)

drug level on this specimen should be in the low end of the therapeutic range (see Table 14.1). Trough levels toward the upper end of the therapeutic range may indicate the preceding peak levels were well into the toxic range.

MINIMUM EFFECTIVE CONCENTRATION (MEC)

This refers to the low end of the therapeutic range below which the drug has minimal therapeutic benefit for a significant number of patients (see Figure 14.6).

A serum *peak drug level* for an oral drug is usually collected at 1.5 hours after drug administration; peak drug levels ideally should approach the upper limits of the therapeutic range. Owing to the many variables involved in drug absorption and uncertainty regarding the actual time the drug was taken, "trough" blood specimens are generally preferred over "peak" specimens in most clinical situations.

BIOAVAILABILITY

The bioavailability of a drug refers to the degree of absorption of the drug and the rate of metabolism of the drug. A drug's solubility and stability in the GI tract will significantly influence the amount of drug absorbed along with so-called "first-pass" metabolism of the drug in the mucosal wall of the intestine or liver after absorption. The liver may metabolize the drug as it is delivered by the portal blood, hence the drug will never reach the general circulation. Different preparations (or formulations) of oral drugs may be absorbed at only 50% efficiency, while the same dose administered intravenously would deliver twice the drug to the circulation as the oral preparations.

JUSTIFICATION FOR THERAPEUTIC DRUG MONITORING

Although pharmacokinetic principles allow us to predict the relationship between dose and plasma level, in actual patients, the plasma level achieved is influenced by (1) disease processes [especially GI (oral drugs), hepatic, and renal]; (2) drug interactions; (3) genetic factors, especially those affecting drug metabolism; and (4) age and sex of the patient. Because of these variations in pharmacokinetics, it may be impossible to predict the plasma drug level, hence TDM may be justified. TDM is further justified (1) if the drug has a narrow, well-defined therapeutic range that is difficult to achieve without TDM; (2) the patient is not realizing the biological benefit of the drug even though dosage and compliance seem appropriate; (3) if the patient has signs and symptoms of toxicity perhaps due to changes in liver, renal, or GI function, plasma-protein binding, or drug interactions; or (4) if the patient compliance is in question. On the other hand, for some drugs, clinical evaluation along with other physical and laboratory measurements may supplant data from TDM. For example, heparin therapy may be better monitored by prothrombin time measurements than the actual measurement of serum heparin levels. Loss of pain may be a better indicator of the effectiveness of an analgesic rather than a salicylate or acetaminophen serum level. If TDM alone has the potential to change the course of therapy for the patient, then the expense of TDM is justified in most cases. Like other laboratory measurements, it is imperative that TDM data are properly interpreted prior to its application clinically.

METHODS FOR QUANTITATING DRUGS IN SERUM

UV AND CHROMATOGRAPHIC METHODS

In the period from 1950 to 1960, TDM was primarily a research tool, because analytical methods available lacked the sensitivity and specificity required. The methods primarily used were ultraviolet spectroscopy (UV methods) and gas–liquid chromatography (GLC). Volatile organic solvents were used to extract the drug from large volumes of serum, then the organic solvent was evaporated to increase the sensitivity of these UV and GLC methods. Also multiple solvent extractions (often at different pHs) were frequently required to eliminate interfering substances. These methods were laborious and required a highly skilled analyst. Turnaround times for TDM data required days and often weeks.

In the 1970s, the development and refinement of high-pressure liquid chromatography methods (HPLC) provided the needed sensitivity and specificity required for automated drug analysis; however, these methods still required expensive equipment and a highly skilled analyst.

DIRECT IMMUNOASSAY METHODS

Radioimmunoassay methods, which combine the specificity of immunological re-actions together with the sensitivity of radiochemistry, allow the analysis of drugs on microsamples (10 μL) of serum in the 10^{-12}–10^{-9} mol range. The analyses are usually performed ''directly'' on serum with no pretreatment or extraction re-quirements. However, there are several disadvantages to the use of radiolabels, hence other nonradioistopic methods for TDM have been developed.

Enzyme Immunoassays (EIA)

Homogeneous enzyme immunoassays (EMIT, Enzyme Multiplied Immunoassay Technique, Syva Co., Palo Alto, CA) are commonly used for TDM. EMIT meth-ods involve mixing the following components: (1) the patient's serum specimen containing the drug; (2) a drug labeled with an enzyme; (3) an antibody to the drug; and (4) buffer with substrate specific for the enzyme label. The drug in the pa-tient's specimen competes with the enzyme-labeled drug for binding sites on the antibody. When the enzyme-labeled drug is bound to the antibody, enzyme activ-ity is inhibited (usually glucose-6-phosphate dehydrogenase), hence, no NADH is produced. When a significant amount of the drug is present in the patient's speci-men, more of the enzyme-labeled drug is not bound by the antibody; hence, more NADH is produced from substrate. Therefore, the change in absorption at 340 nm (where NADH absorbs) is related to the concentration of the drug in the patient's specimen (see Figure 14.7).

Unlike RIA, the antibody-bound drug does not have to be physically separated from the unbound moiety; hence, this assay is called ''homogeneous.'' Also, homogeneous EIA methods are kinetic with two measurements being made in less than one minute; RIA methods reach equilibrium usually after hours of incuba-tion. Hence, homogeneous EIA methods like EMIT are rapid and can be readily automated, often with existing UV spectrophotometric equipment.

Heterogeneous EIA methods, where the bound and unbound drug must be physically separated prior to the addition of enzyme substrate, eliminate the shelf-

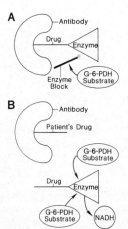

Figure 14.7. The EMIT system. *A*, Antibody–drug–enzyme complex blocks reactive site on enzyme from substrate; hence, the bound enzyme produces no NADH. *B*, Patient's drug displaces enzyme-labeled drug allowing enzymic production of NADH.

life and regulatory problems associated with RIA. However, these assays are difficult to automate and, hence, have not gained wide acceptance.

Fluorescent Immunoassay (FIA)

Substrate-labeled immunoassays (SFIA, Ames Company, Elkhart, IN) is similar to EMIT being a homogeneous method for TDM. The method involves mixing the following components: (1) the patient's specimen containing the drug; (2) a galactosyl–umbelliferone drug complex (nonfluorescent); (3) an antibody to the drug; and (4) a hydrolytic enzyme that will cleave the galactosyl–umbelliferone bond and yield a fluorescent umbelliferone–drug complex:

$$\text{glactosyl–umbelliferone–Drug} \xrightarrow{\beta\text{-galactosidase}} \text{umbelliferone–Drug}$$
$$\text{(nonfluorescent)} \qquad\qquad\qquad\qquad \text{(fluorescent)}$$

$$\text{glactosyl–umbelliferone–Drug–Antibody complex} \xrightarrow{\beta\text{-galactosidase}} \text{no reaction}$$

In the SFIA method, if the labeled drug is bound to the antibody, the β-galactosidase reaction is inhibited and, hence, little fluorescent product is produced. However, patient's drug displaces antibody-bound labeled drug, which then serves as substrate to β-galactosidase producing a fluorescent product in proportion to the amount of drug in the patient's specimen. Using a fluorometer along with a standard curve, an accurate measurement of a variety of drugs can be performed using SFIA including aminoglycoside antibiotics, antiepileptic drugs, antiasthmatic drugs, and quinidine (an antiarrhythmic drug).

Fluorescence Polarization Immunoassays (FPI)

FPI methods include the following components: (1) the patient's specimen containing the drug; (2) a fluorescein-labeled drug; (3) an antibody to the drug; and (4) an instrument for measuring fluorescence polarization. When a fluorescent molecule is excited by plane polarized light, the degree of fluorescence polarization of the emitted light is directly related to the size of the molecule since fluorescence polarization decreases with molecular rotation and large molecules rotate more slowly than small molecules. Hence, when antibody binds the fluorescein-labeled-drug conjugate forming a large molecular complex, the fluorescence polarization is high. However, when patient's drug displaces the fluorescein-labeled drug from the antibody, the smaller molecule gives rise to less fluorescence polarization. Accordingly, fluorescence polarization is inversely related to the amount of drug in the patient's specimen. FPI can be used to measure most anticonvulsant drugs, antiarrhythmic drugs, antibiotics, theophylline (an antiasthmatic drug) and cardiac glycosides (digoxin and digitoxin).

Light Scattering Immunoassays (LSI)

LSI or nephelometric methods are based on the principle that when a drug antibody combines with a drug labeled with a high-molecular-weight protein complex, an even larger molecular complex is formed capable of scattering light in proportion to the amount of labeled drug bound. When drug (in the patient's sample) is added to the above system, the patient's drug displaces the protein-labeled drug from the antibody and reduces the amount of scattered light in a proportional manner. Several automated nephelometric instruments and reagent systems are available for TDM.

All of the above homogeneous EIA and FIA methods, regardless of principle employed, have played a significant role in making TDM routinely available in most clinical chemistry laboratories.

TDM OF VARIOUS DRUGS

HEART DRUGS

Cardiac Glycosides

Digoxin (trade name, Lanoxin), a form of digitalis, is primarily used for the treatment of congestive heart failure. Digoxin increases the strength of cardiac contractions without increasing cardiac oxygen consumption. While this drug is sometimes used to manage patients with supraventricular tachyarrhythmias, toxic levels of digoxin (at serum levels greater than 2 ng/mL) can produce a variety of symptoms or conditions including heart blocks, various heart rhythm disturbances, and gastrointestinal and central nervous system disturbances. Digoxin's effect of increasing the contractility of the heart muscles is an "inotropic effect." In addition, digoxin also has a "chromotropic effect," which slows the heart rate; hence, the drug can be used to manage patients with tachycardia (rapid heart beat). However, while supraventricular tachyarrhythmias (paroxysmal atrial tachycardia, atrial flutter, and atrial fibrillation) often respond to digoxin, tachyarrhythmias of ventricular origin do not respond well to digoxin. Hence, alternate therapeutic modalities should be considered. Moreover, toxic levels of digoxin (greater than 2.0 ng/mL) often cause disturbances in cardiac rhythm. Hence, using digoxin to manage arrhythmias often requires TDM to rule out arrhythmias due to either subtherapeutic digoxin levels or toxic digoxin levels.

In addition to cardiac rhythm disturbances, toxic levels of digoxin are often associated with gastrointestinal disturbances (nausea, vomiting, anorexia) and central nervous system manifestations including fatigue, headache, and confusion.

Causes of Digoxin Toxicity
Some patients may have therapeutic digoxin levels and be clinically toxic due to low levels of certain cations including potassium, calcium, and magnesium. Chronic lung disease may also make patients "digoxin hypersensitive."

Patients with impaired renal function show a marked propensity for digoxin toxicity, since 80% of digoxin is excreted in the urine unmetabolized. Hence BUN, creatinine, and creatinine clearance measurements are helpful in digoxin TDM and patients with elevated BUN (or creatinine) should have serum digoxin levels determined more frequently. Quinidine and calcium antagonists (Vepraamil, Nifedipine) all increase digoxin levels, often to toxic levels, when given

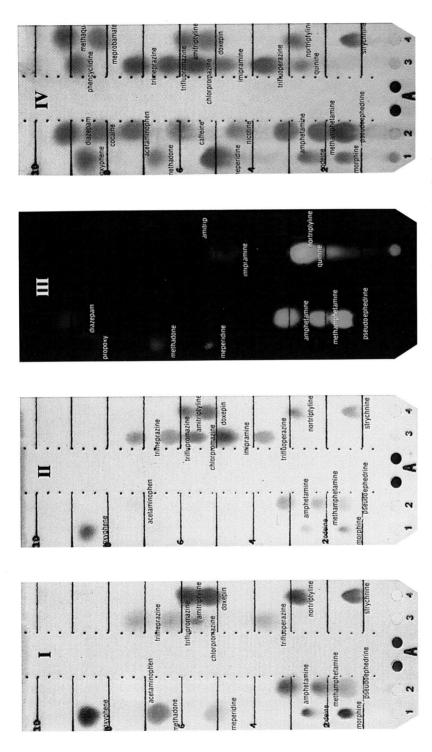

Plate 1. Color detection guide A. Thin-layer chromatography of basic and neutral urine drugs, showing representative R_f values along with drug color development and fluorescence. Some variation in the color and position of the drug spots is normal. (Reproduced with the permission of Analytical Systems, Division of Marion Laboratories, Inc., Laguna Hills, CA 92653.)

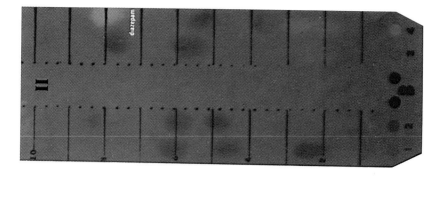

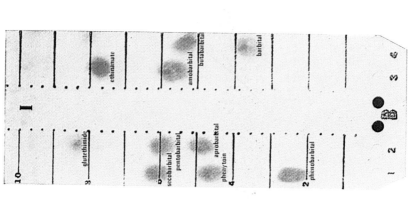

Plate 2. Color detection guide B. Thin-layer chromatography of acid drugs. Some variation in the color and position of the drug spots is normal. (Reproduced with the permission of Analytical Systems, Division of Marion Laboratories, Inc., Laguna Hills, CA 92653.)

concomitantly. Hence, multiple drug therapy for cardiac dysfunction requires more frequent digoxin TDM.

Some patients have relief from chronic atrial fibrillation with no toxic symptoms at digoxin serum levels greater than 2 ng/mL and dosage rates as high as 1.0 mg/day (or higher). These patients are called "digoxin resistant." Also, owing to abnormal liver metabolism of digoxin, a large percentage of digoxin can be converted to inactive digoxin metabolites, which crossreact with the digoxin antibodies used in many radioimmunoassays.

Sample Collection
Blood samples are collected for serum digoxin levels just prior to the next dose. Since the half-life of digoxin is 1.6 days, 8 days (or longer) may be required for steady-state kinetics. Sampling for digoxin levels may also be collected 6 hours after oral dosage; digitoxin levels can be collected 8 hours after dosage. Therapeutic levels for digoxin are 0.5–2.0 ng/mL. Digoxin is most often measured by radioimmunoassay. Automated methods for digoxin include EIA (EMIT) and fluorescence polarization immunoassay.

Quinidine

This drug is frequently used as an antiarrhythmic for tachycardia. Quinidine is usually given to relieve atrial fibrillation only after a so-called atrial–ventricular blockade with digoxin (or, more recently, beta-blockers). Properly administered, quinidine helps convert the arrhythmia to a normal sinus rhythm. Quinidine is also useful for treatment of certain ventricular arrhythmias, specifically those in tachycardia and premature ventricular contractions (PVC).

Quinidine Absorption
Oral doses of quinidine are usually given as the sulfate derivative (86.9% by weight quinidine, "fast release") and as the gluconate (62.3% quinidine, "slow release"). Both of these conjugates of quinidine have reduced bioavailability due to the "first-pass" metabolism by the liver (oral quinidine is absorbed in intestinal areas served by the portal circulation); hence, the average systemic bioavailability can be reduced by as much as 50%.

Peak serum levels of oral quinidine sulfate are drawn at 1.5 hours after dosage; peak levels for the slow absorbing gluconate moiety are collected at 3–5 hours after dosage. Generally, quinidine sulfate is administered every 6 hours with trough specimens being collected just prior to the next dose. The $t_{1/2}$ for quinidine is 6.5 hours, but varies greatly from patient to patient. Only about 20% of quinidine is excreted unmetabolized and, because the drug is a base, acidification of the urine increases clearance, while alkalinization decreases excretion of the drug.

Toxicity of Quinidine
Therapeutic levels of quinidine range from 2.3 to 5.0 μg/mL. Peak levels for quinidine should approach the upper limit; trough levels should approach the lower 2.3 limit. Gastrointestinal symptoms (nausea, vomiting, diarrhea, etc.) are common in quinidine toxicity along with cardiovascular symptoms. Hypotension has been seen in quinidine toxicity.

While quinidine can be measured by UV fluorometric methods (with double solvent extractions) and by reverse-phase HPLC, most laboratories measure quinidine using the more rapid EIA (EMIT) method. UV fluorometric methods tend to lack specificity even with multiple solvent extractions.

Lidocaine

While this drug has been used for years as a local anesthetic, more recently the drug has been given intravenously for the treatment of ventricular arrhythmias that occur (1) post myocardial infarction; (2) due to digoxin toxicity; or (3) iatrogenic arrhythmias of cardiac catherization or surgery. Because of the extremely short $t_{1/2}$ of the drug (normal, 74–140 minutes), lidocaine must be repeatedly administered or infused.

Therapeutic values of lidocaine range from 1.5 to 5 ng/mL and should be monitored every 12 hours for arrhythmia prophylaxis. Because of the rapidity of the EIA (EMIT) method, it has supplanted GLC and HPLC methods for most routine measurements. Provision for "stat" availability of lidocaine levels should also be made in hospital laboratories supporting acute coronary care.

Procainamide

Like lidocaine, procainamide is effective in the treatment of ventricular arrhythmias, exclusive of those due to digoxin toxicity. While procainamide is a potent antiarrhythmic drug, its relatively short half-life (3.5 hours) discourages its use for long-term oral administration. Also, procainamide's association with induced lupus erythematosus preclude its use except when ventricular arrythmias are refractory to other therapy (i.e., lidocaine). Liver N-acetylation of procainamide produces a pharmocologically active metabolite, N-acetylprocainamide (NAPA). Concentration of NAPA must be monitored along with procainamide for effective TDM.

Therapeutic ranges for procainamide and NAPA range from 10 to 30 ng/mL for serum; therapeutic for procainamide alone is 4–10 ng/mL. HPLC methods provide for the simultaneous measurement of procainamide and NAPA. Homogeneous EIA (EMIT) methods are available for both procainamide and NAPA, but two separate assays are required.

MONITORING ANTIEPILEPTIC DRUGS

A number of drugs have been shown to be effective in the reduction of seizure in epileptics. The drugs commonly used for seizure control include phenytoin, phenobarbital, primidone, carbamazepine, ethosuximide, and valproic acid.

Phenytoin

This drug was the first nonsedative for seizure control; however, it is only effective in a narrow serum therapeutic range of 10 to 20 ng/mL. Levels above 20 ng/mL are in the toxic range. Often the drug dosage shows little relationship to serum levels due to variations in liver metabolism.

Phenytoin is para-hydroxylated in the liver and conjugated with glucuronic acid, then excreted in urine. Several drugs are known to stimulate liver metabolism of phenytoin while others suppress metabolism. Also, phenytoin can be displaced from albumin-binding sites by other drugs (e.g., salicylates, valproic acid), thereby potentially causing toxic side effects. Most toxic symptoms of phenytoin are neurological in nature.

Phenytoin is most commonly measured by homogeneous EIA (EMIT); however, in patients with renal failure, the hydroxylated metabolite of phenytoin may accumulate and interfere by cross reacting with the EIA antibody. GLC and HPLC can effectively measure phenytoin and metabolites in a single sampling in renal failure patients.

Phenobarbital and Primidone

These drugs are both barbiturates and are effective in treating tonic mal (grand mal) seizures; they are not effective in the treatment of absence (petit mal) seizures. These barbiturates reduce mono- and polysynaptic transmission, thus reducing the chance of seizure. Therapeutic ranges for phenobarbital and primidone are 15–40 ng/mL and 5–12 ng/mL, respectively. Phenobarbital has a long half-life of 50–120 hours; primidone has a $t_{1/2}$ of 12 hours. The major side effect of phenobarbital and primidone is sedation with other toxic effects including nausea, vomiting, and dizziness.

Metabolism

Primidone is metabolized in the liver to form phenobarbital and, hence, both primidone and phenobarbital levels should be monitored. Phenobarbital is metabolized by hydroxylation with subsequent conjugation with either glucuronic acid or sulfate prior to renal excretion. Both phenobarbital and primidone have major inhibitory drug interactions due primarily to their inductive effects of liver metabolism of other drugs.

Methods

Serum levels of phenobarbital and primidone are masured routinely by homogeneous EIA (EMIT) and chromatographic methods (HPLC, GLC).

Ethosuximide

The drug is used primarily for the management of patients with absence (petit mal) seizure. Other drugs are required to cover associated clonic–tonic (grand mal) seizures in patients with absence seizures under control with ethosuximide. The therapeutic range of ethosuximide is 40–100 ng/mL. The half-life of the drug is 55 hours.

Ethosuximide, at levels greater than 100 ng/mL, may produce gastrointestinal symptoms (nausea, vomiting) along with headaches and dizziness.

Carbamazepine

This anticonvulsant drug, which has the iminostilbene structure of tricyclic antidepressants, has been shown to be effective in the management of clonic–tonic (grand mal) seizure. Carbamazepine reportedly has less severe side effects than other medications for grand mal (phenobarbital, primidone, and phenytoin); however, long-term use of carbamazepine has been rarely associated with bone marrow suppression and hepatic disease. Accordingly, it may be warranted to periodically monitor liver function tests and blood cell counts during therapy with carbamazapine.

The therapeutic range of carbamazepine is from 4 to 8 ng/mL. Carbamazepine is measured routinely by homogeneous EIA with FIA, FPIA, with nephelometric and chromatographic methods being used less commonly.

Valproic Acid

This short-chain organic carboxcylic acid can be used for treating a wide variety of seizure disorders including absent (petit mal), tonic–clonic (grand mal), and partial seizures. The drug is most effective in preventing seizure when maintained at therapeutic range of 50–100 ng/mL. Drug interactions include the following: (1) valproic acid inhibits the metabolism of phenobarbital, thereby increasing the level of the latter by as much as 50% and (2) valproic acid displaces phenytoin from protein-binding sites increasing levels of free (metabolically active) phenytoin often to toxic levels.

Liver disease markedly influences the $t_{1/2}$ of valproic acid (normally 14.9 hours) since the drug is metabolized by the liver and primarily eliminated by the liver as well.

ANTIASTHMATIC DRUG MONITORING

Theophylline

This potentially lethal drug, when properly administered and managed by TDM, is a most effective antiasthmatic. The drug, which has a methylxanthine structure similar to caffeine, can stabilize highly reactive airways to provide long-term bronchodilation in the chronic asthmatic. Also, using a loading dose of theophylline, an acute asthmatic attack can often be managed. The drug is also effective in preventing exercise-induced bronchospasm.

The therapeutic range of theophylline is narrow, being from 10 to 20 ng/mL. Toxic symptoms become more severe as theophylline concentrations increase above 20 ng/mL. Relatively mild symptoms of headache and insomnia are often seen even at therapeutic theophylline levels. At high theophylline levels (20–25 ng/mL), severe GI symptoms often develop (e.g., vomiting, diarrhea) and, at levels greater than 35 ng/mL, seizure with brain anoxia and cardiac arrhythmia often cause death. The half-life varies greatly in adults ranging from 4 to 16 hours; children (1–9 years) metabolize theophylline at a faster rate in the liver giving rise to a shorter half-life of 2–10 hours. The drug is excreted in the urine after liver metabolism. The significant variation in half-life especially with age further supports the need for frequent TDM whenever theophylline is administered.

Theophylline Preparations

Owing to the relatively short half-life, dosing of theophylline for long-term therapy involves oral drug administration every 4 to 6 hours. Slow-release oral preparations may extend the dosing interval to 8–12 hours. In addition to oral tablets, oral liquids, rectal preparations, and intravenous solutions are available.

Theophylline Testing

Regardless of the preparation, theophylline is usually monitored by collecting a blood specimen just prior to the next dose (or by a trough level). Serum theophylline levels can be measured by a variety of methods including older methods based on UV extraction and chromatographic methods (HPLC, GLC) and radioimmunoassay. Newer, more rapid, nonisotopic immuno methods require less sample, time, and skill. These advantages make theophylline available as both a routine and stat test in most clinical laboratories. The homogeneous EIA (EMIT) method is commonly used for theophylline levels with FPIA and rate nephelometric methods now available. Radioimmunoassays are also available for theophylline but are not commonly used.

ANTIBIOTIC DRUG MONITORING

Routine TDM for antibiotics has generally been limited to testing for the aminoglycosides which, include the following drugs: gentamicin, tobramycin, amikacin, and kanamycin. Generally, aminoglycosides are potent antibiotics against Gram-negative bacilli and are especially useful in the treatment of penicillin-resistant organisms.

Rational for TDM

For any antibiotic to be effective, the concentration of the antibiotic in the body fluid (or medium) supporting the infection must reach a minimum inhibitory concentration (MIC). For a blood-borne infection, the measured serum drug level must equal to or exceed the MIC for the drug to be effective. Also, aminoglycosides can be highly ototoxic and nephrotoxic, attacking primarily the proximal tubules in the latter. The combination of the narrow therapeutic range and the toxicity of aminoglycosides certainly justifies aminoglycoside TDM.

Metabolism and Excretion

Aminoglycosides are not appreciably metabolized prior to excretion in the urine. Because of the high concentration of aminoglycosides in the urine, aminoglycosides are often useful for the treatment of Gram-negative organisms detected in the urine. Also, in patients with decreased creatinine clearance, the loading dose should be decreased to reflect the resulting extended half-life of the drug.

Peak and Trough Levels

Peak levels for aminoglycosides are drawn 30 minutes after an IV dose and 1 hour after a IM dose. For gentamicin and tobramycin, peak levels should range from 4 to 10 ng/mL; amikacin and kanamycin peak levels should range from 20 to 30 ng/mL. Trough levels for aminoglycosides are collected just prior to the next dose. Generally, trough levels for gentamicin and tobramycin are considered potentially toxic if greater than 2 ng/mL, while trough levels of amikacin and kanamycin are considered toxic if greater than 8 ng/mL.

Assay Method

Older methods for measuring aminoglycosides include the microbiologic (or bio-assay), radioenzymatic, and chromatographic (HPLC, GLC) methods. Although reliable radioimmunoassay methods are also available, most laboratories use homogeneous FIA and EIA (EMIT) for aminoglycosides analysis. Automated rate nepheolmetric and FPIA methods also provide accurate and precise aminoglycoside levels.

ANTIDEPRESSANT DRUGS

Lithium

Lithium ions are used for the treatment of mania or the manic stage of manic depressive psychosis. The half-life of lithium is 24 hours; hence, 5–6 days are required to reach steady-state blood levels. Lithium is excreted in urine and, at toxic levels greater than 2 mmol/L, can cause renal damage. Therapeutic levels of lithium range from 0.5 to 1.5 mmol/L.

Lithium Analysis
Lithium levels can be accurately determined either by flame emission spectroscopy or atomic absorption spectroscopy.

Tricyclic Antidepressants

Typical tricyclic antidepressants include antitriptyline, nortriptyline, imipramine, desipramine, and protriptyline. These drugs are used in the treatment of depression through "anticholinergic" action, sedation, and inhibition of the "amine pump" in the presynaptic nerve endings. The latter drug effect is most important in the antidepressant action of tricyclics.

Variable Bioavailability
A justification of TDM of tricyclics centers on the variable bioavailability in patients that ranges from 27 to 90% due to extensive "first-pass" liver metabolism.

Therapeutic Ranges
The suggested therapeutic ranges for tricyclics, with peak time samples generally collected 2–6 hours after the last dose, are from 125 to 250 ng/mL (for nortriptyline). Half-lives vary greatly for the tricyclics and up to 10 days may be required for steady-state levels. Toxic levels of tricyclics cause a variety of mild adverse effects. Life-threatening toxic effects include cardiac arrhythmias, coma, and convulsions.

Metabolism
The tricyclics with tertiary amino side chains are metabolized to secondary amines that are also active antidepressants (e.g., demethylation of amytriptyline converts it to nortriptyline), hence the sum of the parent drug and active metabolites must be considered in TDM.

Assay Methods
Chromatographic methods (GC and GC–mass spectrometry) with derivation have
been successfully used to measure tricyclics.

TOXICOLOGY

As described in the previous sections on TDM, all drugs are potentially toxic at
high levels. Hence, a serum analysis to determine if a drug has reached toxic
levels can be described as a toxicology procedure. However, in many cases of
poisoning, the analyst is required to screen various available biological specimens
to determine both the amount and type of poison ingested. Since there exists a
vast array of potential poisons in our environment, isolation and identification of
the poison can be a formidable task. Generally, the clinical toxicologist is primarily concerned with (1) identification of the toxic substance and (2) quantitation of
the substance to facilitate proper medical management of the patient.

DIFFERENT CLASSIFICATIONS OF POISONS

There are thousands of possible poisons in our environment; these can be broken
down into categories or groups, such as: (1) drugs or nonvolatile organic substances; (2) metallic and nonmetallic elements; (3) volatile liquids and gases; and
(4) mineral acids and bases (or corrosive materials).

Toxicology of Drugs

While the clinical laboratory is concerned with the toxicology of all substances,
the major efforts of a clinical toxicologist center on the analysis of drugs in
biological fluids due to either accidental or intentional overdose.

BIOLOGICAL SAMPLES FOR TOXICOLOGY

Usually, the best possible toxicology test is a serum drug level, provided the
analyst has a suitable method of analysis and knows which drug the patient has
ingested. Otherwise, drug screening on urine, gastric washing or vomitus (if
within 4 hours of ingestion), and blood is required. Drug screening on blood by
thin-layer chromatography (TLC) is frequently not effective because drug levels
are often below detection limits. Drug screening on pills and powders can be
accomplished using TLC, GLC, or HPLC.

Detection Limits of TLC

Using 5 mL of urine for extraction, most drugs can be identified by TLC at levels
of 1 ng/mL; however, a negative result does not confirm the absence of a drug.
Also, a positive result on a TLC drug screen should be confirmed by a more
definitive method such as GLC, HPLC, RIA, and EIA.

ACIDIC, BASIC, AND NEUTRAL DRUGS AND CLINICAL CLASSIFICATIONS

Prior to a discussion of the acidic and basic properties of drugs, it is convenient to discuss various drugs in terms of their use clinically.

Antidepressant Drugs

As discussed in the TDM section, tricyclic antidepressants [amitriphline, nortriptyline (Aventyl), etc.] are used to control anxiety and depression. These drugs are basic because of an exocyclic alkyl amine group bound to the tricyclic ring system.

Analgesics

These drugs are taken to reduce pain and include narcotic drugs like morphine and meperidine (Demerol). The nonnarcotic analgesics are more popular; these include acetaminophen (Tylenol) and acetylsalicylic acid (aspirin). Narcotic drugs are basic drugs due to the presence of an alkyl amine functional group.

Anticonvulsants or Antiepileptic

These drugs are CNS depressants for seizure control. Phenobarbital and primidone, both acidic barbiturates, are used as anticonvulsants. Phenytoin (Dilantin) is a nonbarbiturate anticonvulsant which is acidic.

Stimulants

These drugs stimulate the central nervous system and are called "uppers." Cocaine, caffeine, and nicotine, all basic compounds, are examples of CNS stimulants; these drugs all have the alkaloid structure. The term alkaloid is used to name nitrogen-containing compounds having heterocyclic ring systems (i.e., more than one ring) and one or more basic nitrogen atoms serving as annular atoms. Tryptophan, a commonly found amino acid, has an alkaloid structure as does lysergic acid, a drug known to produce a schizophrenic state (see Figure 14.8). Other examples of stimulants are amphetamine (Benzedrine) and methyamphetamine; these sympathetic amines are basic compounds.

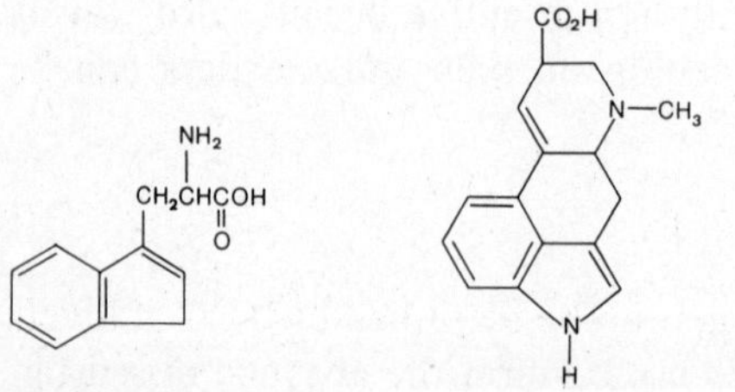

Tryptophan   Lysergic Acid

Figure 14.8. Alkaloid structure of tryptophan and lysergic acid.

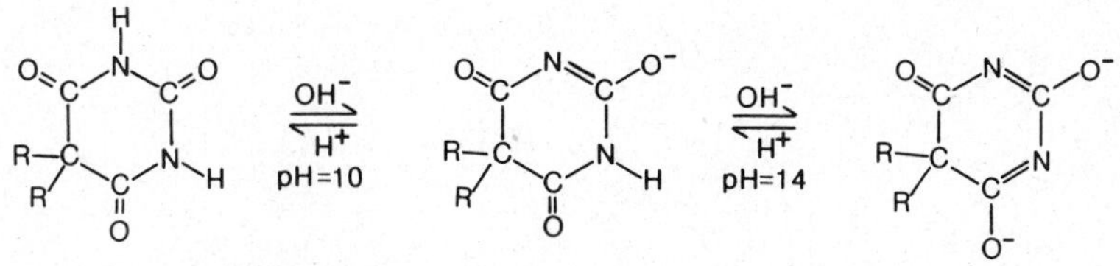

Figure 14.9. Barbituric acid serving as an acid in alkaline solution.

Hypnotics and Sedatives

These drugs tend to calm and produce sleep. Barbiturates are hypnotics along with nonbarbiturate drugs such as methaqualone (Quaalude) and glutethimide (Doriden). Methaqualone is a basic drug; glutethimide is an acidic drug.

Tranquilizers

Examples of drugs that lower anxiety include the benzodiazepines such as diazepam (Valium, neutral drug) and chlordiazepoxide (Librium). Phenothiozine such as chlorpromazine (Thorazine and trifluoropromazine) are tranquilizers; carbamates like meprobomate (Miltown, basic drug) are also tranquilizers.

Antihistamines

These drugs are antiallergenic and block the action of histamine, a potent vasodilator. Benadryl (Dramamine, basic drug) is a antihistamine.

Hallucinogens

These psychotoxic drugs produce either euphoria or hallucinations. Examples of drugs in this class include lysergic acid (LSD), mescaline, phencyclidine (PCP), cocaine, and marijuana.

SOLVENT EXTRACTION OF DRUGS BASED ON ACIDIC AND BASIC PROPERTIES

The majority of drugs are basic due to the presence of amino groups which can accept protons. Barbiturates, on the other hand, are acidic drugs (proton donors) as shown in Figure 14.9 along with lysergic acid diethyamide (LSD), phenytoin, acetominophen and acetylsalicylic acid. These drugs form anionic ("-ate") salts in alkaline medium. Basic drugs are protonated in acidic solutions forming cationic salts. Neutral drugs such as meprobamate and valium are neither acidic nor basic.

Extraction of Drugs from Urine

Urinary neutral and basic drugs can be isolated in a volatile organic solvent (methylene chloride, CH_2Cl_2) by alkalinization of the urine (to pH 9) with a NaOH

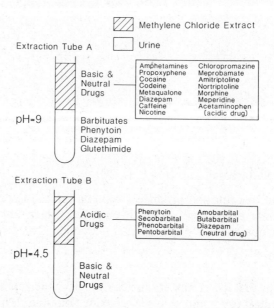

Figure 14.10.   Organic solvent extraction of various urinary drugs based on acidity and basicity.

solution. All acidic drugs will be cationic at this pH and, therefore, will be present in the aqueous phase of the extraction tube. (Note: Water is more dense than most organic-drug-extracting solvents, hence the aqueous phase is located on the bottom of the extraction tube.) Acidic drugs can be isolated in the CH_2Cl_2 solvent by acidifying the urine with hydrochloric acid to pH 4.5. At pH 4.5, basic and most neutral drugs are protonated (and positively charged) and, hence, are found in the aqueous phase (see Figure 14.10). Water is an excellent solvent for drug ions and a poor solvent for "molecular" drugs. The converse is true for the extracting organic solvent CH_2Cl_2. The extracting solvent is separated from the aqueous urine phase and evaporated to increase the concentration of the drug.

Drug Screening

The various drugs present in urine can be identified semiquantitatively by thin-layer chromatography (TLC). Once identified by TLC, the presence of a drug in the urine can be verified by gas–liquid chromatography (GLC) using essentially the same extraction procedure described above. Drugs can also be quantitated using GLC; this information may further assist in patient management.

THIN-LAYER CHROMATOGRAPHY (TLC) OF DRUGS

Drug screening by TLC involves two steps of analysis: (1) the separation step of the various drugs on a solid phase silica gel support and (2) after separation, identification of the various drugs by both the migratory position of the drug "spot" on the solid support and the color of drug "spot" after treatment with various chemicals. Usually, the TLC plate has "known drugs" for comparison so that identification of the patient's drug can be facilitated.

Solid Support

Glass plates coated with silica gel can be used. Recently, a more rapid TLC method has been introduced using a porous silica gel–glass fiber support (Toxi-Lab, St. Louis, Mo.)

Developing Solvent

After specimen extracts (containing drugs) are "spotted" onto the solid support, the latter is placed into a developing solvent, which is rapidly absorbed by the support. As the solvent moves toward the top of the support, drugs are carried with the developing solvent. Less polar drugs are more attracted to the solvent and move higher on the silica plate; more polar drugs (especially metabolites) are more attracted to the silica gel solid phase and, hence, do not migrate as high. Each drug has a reference value (R_f) based on its relative migration. The R_f value for each drug spot is calculated by dividing the distance the drug migrates by the distance the solvent migrates (see Figure 14.11). Propoxyphene has an $R_f = 0.83$ on the Toxi-Lab system; morphine has an $R_f = 0.2$. Some selected R_f values are listed in Table 14.2.

A typical developing solvent for neutral and basic drugs consists of a mixture of ethylacetate (29 mL), methyl alcohol (1 mL), water (0.5 mL), and ammonia. For acidic drugs, a typical developing solvent may contain 60 mL of chloroform ($HCCl_3$) and 40 mL of ethyl acetate.

Color Development

Color development for the various drugs is done in stages. For example, initial color development for basic and neutral drugs may be accomplished by reacting the silica gel plate with formaldehyde vapors followed by treatment with Mandelin reagent (concentrated H_2SO_4 and metavanadate). Colors and positions of drug

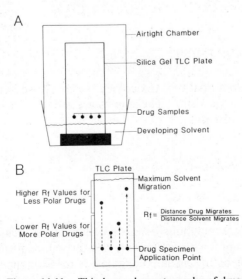

Figure 14.11. Thin-layer chromatography of drugs.

Table 14.2. Selected R_f Values for Basic, Neutral, and Acidic Drugs

Basic and Neutral Drugs	R_f	Acidic Drug	R_f
Diazepam	0.89	Secobarbital	0.71
Methaqualone	0.89	Pentobarbital	0.70
Propoxyphene	0.85	Phenytoin	0.47
Cocaine	0.83		
Caffeine	0.62		
Amitryptyline	0.62		
Amphetamine	0.40		
Codeine	0.26		

spots can then be recorded (see Plate 1, Step I). Rinsing the silica gel plate with water changes the colors of the drug spots (see Plate 1, Step II). Also fluorescent drugs can be visualized at this point under a UV lamp (366 nm) (see Plate 1, Step III). Final treatment of the plate with Dragendorff's reagent [KI, I_2, Bi $(NO_3)_3$ in acetic acid] turns most of the drug spots to brown (see Plate 1, Step IV).

Color development for acidic drugs involves a simpler procedure where the silica gel plate is treated with diphenylcarbazone, then a silver nitrate solution, and finally an acidified mercuric sulfate solution. All acidic drugs spots are lavender colored. Ultraviolet light (366 nm) of the acidic drug plate reveals fluorescent drugs (i.e., diazepam) (see Plate 2).

Experience is required in performing TLC drug screens. Because of the complexity of unexpected interfering drugs, multiple drug ingestion, and metabolite variability, interpretation of TLC drugs screens must often be verified by a more accurate method (either GLC or homogeneous EIA).

SPOT TESTS FOR SCREENING

In addition to TLC drug analysis, toxicology on comatose patients may require the use of so called "spot tests." These tests are rapid and inexpensive and can provide life-saving information.

SALICYLATE POISONING

Asprin or acetylsalicyclic acid is a commonly used analgesic and, at toxic levels, can cause severe acid–base disturbances due to hyperventilation (causing alkalosis) often followed by a metabolic acidosis. Salicylate is measured routinely by treating a serum specimen with ferric ions. The ferric ions form a violet complex with salicylate serving as the ligand. Toxic levels for salicylates are greater than 30 mg/dL.

Acetaminophen (Tylenol)

This commonly used analgesic can produce severe liver damage when toxic levels are achieved. An effective antidote is available (acetylcystine).

To quantitate acetominophen in serum, high-pressure liquid chromatography is used along with colorimetric methods. However, many colorimetric methods are subject to salicylate interference.

A modified spectrophotometric method for measuring acetaminophen involves (1) the extraction of the drug from acidified serum with ethylacetate; (2) the extracted drug is then complexed with a ferric triazine derivative [Fe(III)–2,4,6-tripyridyl-*s*-triazine]; and (3) after incubation for 25 minutes, the absorption is measured at 595 nm. This colorimetric method correlates well with HPLC methods, has a linear range to 400 ng/L, and has no salicylate interference. Prior to serum analysis of acetaminophen by either HPLC or by spectrophotometry, the urine can be screened for acetaminophen by TLC or a "spot test." One useful spot test requires treating 1 mL of urine with 1 mL of concentrated HCL and heating in a boiling water bath followed by addition of *o*-cresol solution. The formation of a blue color indicates the presence of acetominophen.

Toxic Levels
Potential hepatoxicity is considered when the serum level exceeds 200 ng/L 4 hours after ingestion or when the level exceeds 50 ng/L at 12 hours post ingestion. The acetylcystine antidote is apparently most effective if given 10–12 hours after drug ingestion.

ULTRAVIOLET SPECTROSCOPY AND DRUG SCREENING

Many drugs, owing to their aromatic rings and/or conjugated double bonds, can be readily identified and quantitated using a scanning UV spectrophotometer. In the past, many barbiturate assays involved UV spectroscopic scans at different pHs ranging from 2 to 14. Extraction of the blood sample with chloroform was required. These UV methods have been supplanted by HPLC and the newer enzyme immunoassays.

ENZYME IMMUNOASSAYS FOR DRUG SCREENING

Homogeneous enzyme immunoassays (EMIT) (Syva Co., Palo Alto, Ca.) for the determination of serum levels of various drugs have been discussed earlier. EMIT is an abbreviation for "enzyme multiplied immunoassay technique." EMIT procedures for drug assays on urine are also available. In the EMIT urine drug assays, a urine specimen containing a drug is mixed with an enzyme-labeled drug, substrate and an antibody to the drug. Antibody binding to the enzyme-labeled drug inhibits the formation of NADH from substrate. Since the patient's drug in the urine displaces enzyme-labeled drug from the inhibiting antibody, the higher the concentration of drug in the urine specimen, the more NADH produced. (The formation of NADH is monitored at 340 nm.) The assay can be conveniently

Table 14.3.   Drugs or Metabolites That are Analyzed by EMIT

EMIT-d.a.u.	EMIT-st
Amphetamine	Alcohol
Barbiturate	Amphetamine
	Barbiturate
Benzodiazepine metabolite	Benzodiazepine
Cannabinoid	Cocaine metabolite
	Cannabinoids
Cocaine metabolite	Methodone
Methodone	Opiates
	Phencylclidine
Opiate	Phenobarbital
Phencyclidine	Propoxyphene
Propoxyphene	Tricyclic Antidepressant

performed on a spectrophotometer equipped with a "flow-through" cuvette. EMIT methods can be readily automated on a variety of instruments especially centrifugal analyzers.

Table 14.3 lists the various "drugs of abuse" that can be detected using EMIT-d.a.u.

Rapid Urine Screens

A more rapid EMIT system for drug detection on urine (or serum) involves a single vial of lyophilized reagents. The reagent vial contains the typical EMIT components including enzyme-labeled drug, antibody, and substrate. A "cutoff" standard is also included with the kit. The specimen (or standards) is simply added to the vial along with water. The NADH produced by the reaction is monitored using the vial as a cuvette in a spectrophotometer designed for the method. If the absorbance change at 340 nm is greater than the cutoff standard, the EMIT test is positive for that drug. Drugs assays available by this method (i.e., EMIT-st) are listed in Table 14.3.

Serum Drug Screens

EMIT can be used to screen blood for certain drug families. EMIT-tox serum screens are available for barbiturates, benzodiazepines, and tricyclic antidepressants. The antibodies used in these methods intentionally cross react with all drugs in each of the drug groups.

RADIOIMMUNOASSAY FOR DRUGS

RIA has been used to measure a variety of drugs; however, the method has many disadvantages, such as: (1) the method is difficult to automate; (2) RIA is time consuming (especially for stat testing); and (3) RIA employs radioactive materials with associated regulations and shelf-life limitation. For these reasons, EMIT methods are overwhelmingly preferred over RIA in most laboratories. Digoxin and digitoxin are still primarily assayed in serum by RIA, however.

SCREENING FOR OTHER TOXIC AGENTS

CARBON MONOXIDE (CO)

To identify carbon monoxide poisoning, an automated CO-Oximeter can measure carboxyhemoglobin spectrophotometrically and report the latter as a percentage of total hemoglobin. A "spot test" for CO involves the use of a Conway microdiffusion cell with 1 mL of blood in the outer compartment and palladium chloride in the inner compartment. Upon addition of 10% sulfuric acid to the blood, the liberated CO reduces Pd^{2+} to silvery elemental palladium.

$$PdCl_2 + CO + H_2O \rightarrow CO_2 + 2HCl + Pd^0$$

Carboxyhemoglobin levels of 60% and higher are fatal unless treated with oxygen. Symptoms for patients with CO toxicity include coma, paralysis, and decreased respiration rate.

SCREENING FOR METALS

Heavy Metal Screens

For the comatose patient, screening for heavy metals should be performed on urine and gastric washings (if available). The Reinsch test can detect the presence of As, Hg, Sb, Bi, Te, Se, and Ag. The Reinsch test involves the copper reduction of "heavy metals" (arsenic is actually a nonmetal) ions in the solution being analyzed to their insoluble elemental form. The copper is in the form of a copper wire or foil; the heavy metals are deposited on the copper foil when reduced:

$$3Cu^0 + 2AsO_2^- + 8HCl \rightarrow 3Cu^{2+} + 2As^0 + 4H_2O + 8Cl^-$$

Elemental arsenic in the above reaction forms a visible black precipitate at concentration above 25 ng/mL. Mercury gives a "silvery" appearance to the copper foil at concentrations above 50 ng/dL.

Iron

Ferrous ions (Fe^{2+}) can be detected in gastric washings by addition of potassium ferricyanide. The formation of a blue color indicates the presence of Fe^{2+}.

Lead and Mercury Assays

Lead Poisoning

Lead is an especially toxic metal, which interferes with normal hemoglobin synthesis and produces anemia. Severe lead poisoning can cause permanent central nervous system damage.

Increased δ-aminolevulinic acid (δ-ALA), a heme precursor, accumulates in lead poisoning. Therefore, measurements of δ-ALA are used to screen for lead poisoning.

Coproporphyrin III, a porphyrin heme precursor, also accumulates in the urine in lead poisoning. Basophilic stippling of erythrocytes is also commonly seen in lead poisoning.

Direct Measurements of Lead

Lead can be measured preferably on 24-hour urine specimens or anticoagulated blood specimens. Since lead (Pb^{2+}) forms a red complex with diphenylthiocarbazone, a colorimetric method based on the latter chromogen can be used. However, lead can be more accurately measured using atomic absorption spectroscopy. The lead in the specimen must be complexed with an "organic" ligand (pyrolidine) prior to extraction by an organic solvent (usually methyl isobutylketone). The organic solvent containing the lead–ligand complex is aspirated directly into the flame for analysis. Microsampling for atomic absorption using a Delves cup or graphite furnace has simplified lead analysis by atomic absorption.

More recently, lead analysis has been accomplished by plating the lead electrolytically on a inert electrode, then "stripping" the lead from the electrode by applying a measured amount of current. The amount of current required is related to the milliequivalents of lead by Coulomb's law.

Normal levels of lead range up to 80 ng/dL for blood; normal levels in urine range to 80 ng/L.

Methemoglobin

Another derivative of hemoglobin called methemoglobin contains iron in the 3+ oxidation state rather than the ferrous (2+) oxidation state normally seen in hemoglobin. Methemoglobin levels are increased (10–30% of total hemoglobin) in hemoglobin reductase deficiency. Also, methemoglobin levels are increased after ingestion of certain drugs (e.g., acetophenetin) and, hence, methemoglobin may be rarely ordered as a toxicology test.

Mercury Poisoning

Measurement of mercury levels in urine involves the overnight wet digestion of mercury in acid and potassium permanganate followed by titration of excess with peroxide. Subsequently, the mercuric ions present in the sample are reduced to a "cold vapor" by treatment with stannous chloride ($SnCl_2$). The vapor is passed through an optical cell positioned in the hollow cathode beam of atomic absorption instrument for analysis. A strip chart recorder is required to record the atomic absorption signal.

Normal levels for mercury should be less than 20 ng/24-hour urine.

NONMETAL POISONS

Halogens

For bromide poisoning, bromide can be detected in serum by treating the specimen with gold (III) chloride ($AuCl_3$):

$$AuCl_3 + 3Br^- \rightarrow AuBr_3 + 3Cl^-$$

The $AuBr_3$ is brown in color and is measured at 440 nm.

Cyanide

Cyanide (CN^-), a pseudohalogen, is analyzed in blood specimens by liberating hydrogen cyanide gas from the specimen and trapping the cyanide in alkali (NaOH). A Conway microdiffusion apparatus can be used. The cyanide is converted first to cyanogen chloride, then to a pyridine derivative, which is treated with barbituric acid to form a colored complex. The latter complex is measured using a spectrophotometer.

VOLATILE LIQUIDS

Alcohols, short-chain aliphatic hydrocarbons and aromatic halogenated hydrocarbons, ethers, aldehydes, ketones, and other low-molecular-weight nonionic organic compounds form a large group of potentially toxic substances.

In general, for volatile organics, the best screening method for blood and urine involves the use of gas–liquid chromatography. However, a simpler "spot test" for halogenated hydrocarbons involves the treatment of a urine specimen with NaOH and pyridine; the formation of a red color suggests the presence of chlorinated hydrocarbons. For ethyl alcohol, alcohol dehydrogenase (ADH) can be used in a rapid UV method of analysis, where NADH is measured at 340 nm:

$$CH_3CH_2OH + NAD^+ \xrightarrow{ADH} CH_3CHO + NADH + H^+$$

Blood alcohol levels 100 mg/dL in serum indicates intoxication while levels higher than 300 mg/dL may be lethal.

Acetone in the urine can be identified by adding a drop of sample to an Acetest (Ames, Elkhart, Ind.) tablet. A violet color forms indicating the presence of acetone. Increased urinary ketones are more commonly due to diabetic ketoacidosis rather than ketone ingestion.

AN ENZYME TEST FOR INSECTICIDE POISONING

Cholinesterase

Cholinesterase is a vital enzyme that destroys acetylcholine by hydrolysis and thus allows additional nerve impulses to be transmitted to muscle fibers. Pseudocholinesterase is a similar enzyme found in the liver, pancreas, heart, and brain. Exposure to organic phosphorus insecticides inhibits cholinesterase activity. Hence, the test may be used to detect toxicity due to these agents.

BIBLIOGRAPHY

Baselt, R. C., Wright, J. A., and Cravey, R. H. "Therapeutic and toxic concentrations of more than 100 toxicologically significant drugs," *Clinical Chemistry* **21**(1), 44 (1975).

Dito, W. R. "Therapeutic drug monitoring," *Diagnostic Medicine,* 57 (Sept./Oct. 1979).

Dito, W. R. "Procainamide and *N*-acetyl procainamide," *Diagnostic Medicine,* 23 (Nov./Dec. 1979).

Dito, W. R. "Therapeutic drug monitoring in your laboratory," *Diagnostic Medicine,* 21 (July/Aug. 1980).

Floyd, R. A. and Read, B. *Quinidine.* Check Sample CP80-6, ASCP, 1981.

French, M. A. and Nightingale, X. *Drug Distribution into Body Compartments,* Check Sample No. CP81-5, ASCP, 1981.

Galen, R. S. "Serum digoxin," *Diagnostic Medicine,* 7 (Nov./Dec. 1980).

Henry, J. B. (ed.). *Clinical Diagnosis and Management by Laboratory Methods,* 16th ed. Philadelphia: Saunders, 1979.

Jatlow, P. *Laboratory Evaluation of Barbituate Overdose.* Check Sample No. ACC-28, ASCP, 1978.

Morrell, G. and Pribor, H. "Emergency toxicology: Rapid, rational and pragmatic approaches," *Laboratory Management,* 41 (Nov. 1979).

Morselli, P. L. and Pippinger, C. E. "Following drug disposition in children," *Clinical Chemistry,* News (Jan. 1980).

Nakamura, R. N. and Dito, W. R. "New radiosotopic immunoassays for therapeutic drug monitoring," *Laboratory Medicine* **11**(12), 807 (1980).

Ochs, H. R., Greenblatt, D. J., and Woo, E. "Clinical pharmacokinetics of quinidine," *Clinical Pharmacokinetics* **5,** 150 (1980).

Pippenger, C. E. "Drug biotransformations," *Syva Monitor,* No. 7, 1 (Oct. 1980).

Pippenger, C. E. "TDM: Observations and projections," *Syva Monitor,* No. 10, 1 (Nov. 1981).

Pippenger, C. E. "Principles of drug utilization: Absorption," *Syva Monitor,* No. 5, 1 (Nov. 1979).

Pippenger, C. E. "Basic guidelines for routine TDM," *Syva Monitor,* No. 4, 1 (May 1979).

Pippenger, C. E. "Practical pharmacokinetic applications—Part I," *Syva Monitor,* No. 3, 1 (Jan. 1979).

Pippenger, C. E. "Principles of drug utilization," *Syva Monitor,* No. 2, 1 (Nov. 1978).

Pribor, H. C. (ed.). "Plasma digoxin: Therapeutic management," *Laboratory Management,* 23 (June 1981).

Steil, C. F. "Therapeutic Drug Monitoring," Lecture Notes from Kentucky State Society for Medical Technology, Owensboro, KY, April 1981.

Sunshine, I. and Jatlow, P. I. (eds.). *Methodology for Analytical Toxicology,* Volume II. Boca Raton, FL: CRC Press, 1982.

Sunshine, I. (ed.). *Handbook Sereis in Clinical Laboratory Science: Section B—Toxicology.* West Palm Beach, FL: CRC Press, 1978.

Taylor, W. J. and Finn, A. L., (eds.). *Individualized Drug Therapy,* Volumes 1, 2, and 3. New York: Gross Towsend, Frank, 1981.

Teitz, N. (ed.). *Fundamentals of Clinical Chemistry,* 2nd ed. Philadelphia: Saunders, 1976.

"The cardiac antiarrhythmics: Quinidine and procanamide," *Clinical Chemistry,* Check Sample Program No. CC-113, ASCP, 1978.

Toxi-Lab Instruction Manual, Analytical Systems, Inc., Laguna Hills, CA.

POST-TEST

1. Factors that influence the serum drug level include:
   a. drug absorption
   b. volume of plasma compartment
   c. rate of drug metabolism
   d. patient compliance
   e. all of the above

2. Drug transport across epithelial membranes is (increased, decreased) when the drug is lipid bound.

3. The metabolically active form of a drug is generally regarded as the:
   a. metabolized form of the drug
   b. "free" drug
   c. protein-bound drug
   d. unabsorbed drug

4. In the liver, drugs are metabolized by:
   a. enzymes of lysosomes
   b. sinusoidal cells
   c. Kupffer cells
   d. microsomal enzymes

5. As the level of ingested drug increases, induction of the metabolic systems of the liver _____ in first-order kinetics; however, it _____ in zero-order kinetics.
   a. increases
   b. remains constant

6. The minimum number of half-lives of a drug required to achieve steady-state levels is:
   a. 8
   b. 2
   c. 1
   d. 5

7. After steady state is achieved, a trough drug level should be obtained at _____. For an oral drug, a peak drug level should be obtained _____.
   a. immediately following dosage
   b. just prior to next drug dose
   c. 1 hour following drug administration

    d. 1 hour before next drug dose
    e. 1.5 hours following drug administration

8. A drug has a half-life of 6 hours. How much drug would be present at the end of 24 hours if the initial drug level was 100 ng/mL?

9. The primary route for drug excretion is via:
    a. urinary tract
    b. biliary tract
    c. sweat excretion
    d. respiratory tract

10. The volume of distribution of drugs is _____ in adults as compared to children; however, adults usually metabolize drugs _____ than children.
    a. larger
    b. smaller
    c. faster
    d. slower

11. The dosing interval of a drug should be _____ the $t_{1/2}$ of the drug.
    a. greater than
    b. less than
    c. the same as

12. "First-pass" metabolism of a drug refers to:
    a. metabolism of the drug in the stomach
    b. metabolism of the drug in the mucosal wall of the intestine
    c. metabolism of the drug in the liver as the drug is delivered by portal blood
    d. b and c
    e. all of the above

13. In homogeneous enzyme immunoassay, NADH is generated when:
    a. the enzyme-labeled drug is bound
    b. the enzyme label is cleaved from the drug
    c. the enzyme-labeled drug is unbound
    d. when an enzyme cofactor is added

14. Match the following. Each question has several answers.

  _____ 1. RIA procedures, labeled complexes counted
  _____ 2. EIA (EMIT) methods using NAD coenzyme indicator system
  _____ 3. FIA procedure using a umbelliferone label
  _____ 4. FPI method

  a. homogeneous assay
  b. heterogeneous assay
  c. competitive binding type assay
  d. high fluorescence means low patient drug level
  e. high fluorescence means high patient drug level
  f. high fluorescence polarization means high patient drug level

g. high fluorescence polarization means low patient
   drug level
h. high radioactivity means
   low patient drug level
i. high radioactivity means
   high patient drug level
j. large change in absorbance means high patient
   drug level
k. large change in absorbance means low patient
   drug level

15. Match the following: Each question has one or more answers.

_____ 1. cardiac drugs	a.	nortriptyline
_____ 2. antiepileptic drugs	b.	kanamycin
_____ 3. antiasthmatic drugs	c.	amikacin
_____ 4. antimicrobials	d.	desipramine
_____ 5. antidepressants	e.	lithium
	f.	gentamicin
	g.	theophylline
	h.	lidocaine
	i.	phenobarbital
	j.	quinidine
	k.	primidone
	l.	digoxin
	m.	procainamide
	n.	tobramycin
	o.	valproic acid
	p.	phenytoin

16. Most drug metabolites are inactive. A notable exception to this statement is
    the metabolite of the cardiac drug _____ .
    a. lidocaine
    b. digoxin
    c. quinidine
    d. procainamide

17. Quinidine toxicity is frequently associated with a peak serum level above:
    a. 1 μg/mL
    b. 2 μg/mL
    c. 5 μg/mL
    d. 3 μg/mL

18. Below are statements regarding theophylline. Circle all true statements:
    a. the half-life of theophylline in adults is almost twice as long as in chil-
       dren
    b. the therapeutic range is wide
    c. theophylline acts as a bronchodilator
    d. for TDM, samples are collected at peak levels

19. Susceptibility of digoxin toxicity increases in the presence of lung disease and low levels of:
    a. K^+, Ca^{2+}
    b. K^+, Ca^{2+}, Mg^{2+}
    c. K^+, Mg^{2+}
    d. Na^+, K^+, Ca^{2+}

20. Specimen collection for toxicology purposes should include gastric, urine, and blood specimens.
    Screening tests should be done on _____ .
    Quantitative tests should be done on _____ .

21. Match the following:
    _____ 1. acid drugs will extract in water   a. acid pH
    _____ 2. neutral and basic drugs will extract   b. basic pH
                   in organic solvent
    _____ 3. basic and neutral drugs will extract in
                   water
    _____ 4. acid drugs will extract in organic
                   solvent

22. Basic drugs may be selectively extracted in an organic solvent by:
    a. acidifying urine
    b. alkalizing urine

23. TLC: Sample migrated 5.0 cm. Solvent front migrated 10 cm. R_f value is
    _____ .

24. (True or False) More polar drugs have high R_f values on the Toxi-Lab system.

25. Stimulants or "uppers" include all of the following *except*:
    a. caffeine
    b. cocaine
    c. meperidine
    d. nicotine

26. The Renisch test will detect all *except*:
    a. arsenic
    b. lead
    c. mercury
    d. silver
    e. bismuth

27. The test most likely to detect lead poisoning is:
    a. CBC
    b. heavy metals screen
    c. urine porphyrins
    d. serum iron

28. The method of choice for lead measurement is:
    a. colorimetric methods
    b. atomic absorption spectrophotometry
    c. electrolytic plating method
    d. titrametric methods

29. The method of choice for volatile liquids is:
    a. HPLC
    b. GLC
    c. spectrophotometric using enzymes as reagents
    d. colorimetric methods

30. Salicylate poisoning can cause:
    a. metabolic acidosis
    b. respiratory alkalosis
    c. hyperthermia with dehydration
    d. all of the above

31. Match the synonymous names:
    _____ 1. acetaminophen   a. LSD
    _____ 2. lysergic acid   b. aspirin
    _____ 3. phencyclidine   c. Valium
    _____ 4. acetylsalicylic acid   d. Tylenol
    _____ 5. diazepam   e. PCP

ANSWERS

1. e
2. increased
3. b
4. d
5. a, b
6. d
7. b, e
8. 6.25 ng/mL
9. a
10. a, d
11. b
12. d
13. c
14. (1) b, c, h; (2) a, c, j; (3) a, c, e; (4) a, c, g
15. (1) h, j, l, m; (2) i, k, o, p; (3) g; (4) b, c, f, n; (5) a, d, e
16. d
17. c
18. a, c
19. b
20. gastric and urine; blood
21. (1) b; (2) b; (3) a; (4) a
22. b
23. 0.50
24. false

25. c
26. b
27. c
28. b
29. b
30. d
31. (1) d; (2) a; (3) e; (4) b; (5) c

NOMOGRAM FOR THE DETERMINATION OF BODY SURFACE AREA

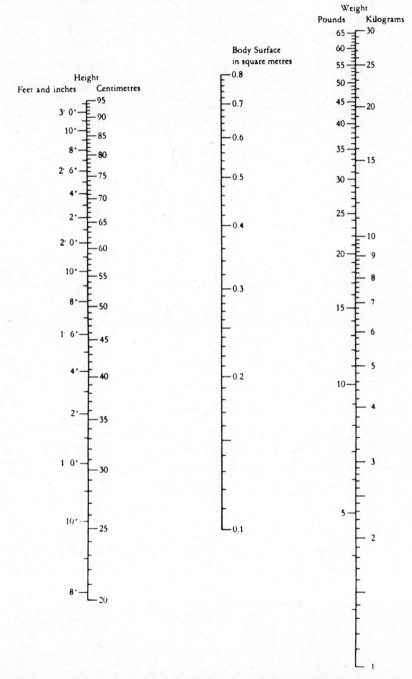

* Reprinted by permission of *The New England Journal of Medicine* **185,** 337 (1921).

REFERENCE RANGES

The following list is composed of those analytes most commonly measured in clinical chemistry laboratories. Reference ranges vary due to age, sex, diet, race, geographical location, and methodology used in analysis. Therefore, each laboratory must establish reference ranges based on population served and methods used in analysis. No reference ranges are given for enzymes because these ranges are highly method dependent.

Analyte	Specimen	Reference Range
Albumin	Serum	3.5–5.1 g/dL
Ammonia	Plasma, venous	11–35 μmol/L
Anion gap	Calculated value	8–16 mmol/L
Bicarbonate	Whole blood, arterial	20–26 mmol/L
(HCO_3^-)	Whole blood, venous	22–28 mmol/L
Bilirubin	Serum	Adult: 0.2–1.2 mg/dL (total)
		0.0–0.4 mg/dL (conjugated)
		0.0–0.7 mg/dL (unconjugated)
		Neonate: 1.0–12.0 mg/dL (total)
Calcium	Serum	8.5–10.2 mg/dL *or*
		2.12–2.55 mmol/L (total)
		1.14–1.29 mmol/L (ionized)
	Urine, 24 hour	100–250 mg/day
Carbon dioxide pressure (pCO_2)	Whole blood, arterial	32–45 mm Hg
	Whole blood, venous	39–55 mm Hg
Carbon dioxide, total ($ctCO_2$)	Whole blood, arterial	21–27 mmol/L
	Whole blood, venous	23–29 mmol/L
Chloride	Serum	98–106 mmol/L
	Sweat	0–50 mmol/L
	Urine, 24 hour	110–250 mmol/day
Cholesterol	Serum	150–250 mg/dL (highly dependent on age, sex, and methodology)
Cortisol	Serum	8:00 a.m. 7–25 μg/dL
		4:00 p.m. 4–18 μg/dL

Analyte	Specimen	Reference Range
Creatinine	Serum or plasma	0.6–1.3 mg/dL
	Urine, 24 hour	1.0–2.0 g/day
	Amniotic fluid	>2.0 mg/dL
Creatinine clearance	Serum and urine	Males: 85–125 mL/min
		Females: 75–115 mL/min
Glucose	Serum or plasma	Adult: 70–110 mg/dL
	Whole blood	60–100 mg/dL
	Serum or plasma	Neonate: 30–90 mg/dL (full term)
		20–80 mg/dL (premature)
	Whole blood	30–90 mg/dL (full term)
		20–65 mg/dL (premature)
	Cerebrospinal fluid	60–70% plasma levels
Glycosylated hemoglobin	Whole blood	5–8%
Immunoglobulin A (IgA)	Serum	140–220 mg/dL
Immunoglobulin G (IgG)	Serum	900–1500 mg/dL
Immunoglobulin M (IgM)	Serum	80–120 mg/dL
Iron	Serum	42–135 μg/dL
Iron-binding capacity (TIBC)	Serum	250–350 μg/dL
Iron, % saturation	Calculated value	30–44%
Lecithin/sphingo-myelin (L/S ratio)	Amniotic fluid	>2:1 ratio
Magnesium	Serum	1.8–2.4 mg/dL *or*
		0.8–1.3 mmol/L
Osmolality	Serum	280–300 mOsm/kg
	Urine, 24 hour	400–600 mOsm/kg
	Urine, random	250–1200 mOsm/kg
Oxygen pressure (pO_2)	Whole blood, arterial	75–100 mm Hg
	Whole blood, venous	30–50 mm Hg
Oxygen, % saturation (O_2 sat %)	Whole blood, arterial	95–98%
	Whole blood, venous	60–85%
pH	Whole blood, arterial	7.35–7.45
	Whole blood, venous	7.31–7.42
Phosphorus, inorganic	Serum	3.0–4.5 mg/dL *or*
		1.0–1.5 mmol/L
	Urine, 24 hour	<1.3 g/day
Potassium	Serum	3.5–5.0 mmol/L
	Urine, 24 hour	25–120 mmol/day
Protein, total	Serum	6.0–7.8 g/dL
	Cerebrospinal fluid	15–45 mg/dL
	Urine, 24 hour	<150 mg/day

Analyte	Specimen	Reference Range
Sodium	Serum	135–145 mmol/L
	Urine, 24 hour	40–220 mmol/day
Thyroid-stimulating hormone (TSH)	Serum	<10 μU/mL
Thyroxine (T$_4$)	Serum	Adult: 4.5–11.0 μg/dL
		Neonate: 12–22 μg/dL
T$_3$ resin uptake (T$_3$RU)	Serum	25–35%
Triglycerides	Serum	10–190 mg/dL
Urea nitrogen (BUN)	Serum or plasma	7–21 mg/dL
	Urine, 24 hour	6–20 g/day
Urea nitrogen/creatinine (BUN/Cr ratio)	Calculated value	10 : 1 to 20 : 1
Uric acid	Serum or plasma	Male: 3.7–7.9 mg/dL
		Female: 2.7–6.6 mg/dL
	Urine, 24 hour	250–750 mg/day

INDEX